Fertilizers: Science and Technology

Fertilizers: Science and Technology

Editor: Eugene Perry

www.callistoreference.com

Callisto Reference,
118-35 Queens Blvd., Suite 400,
Forest Hills, NY 11375, USA

Visit us on the World Wide Web at:
www.callistoreference.com

ISBN: 978-1-63239-907-6 (Hardback)

Trademark Notice: Registered trademark of products or corporate names are used only for explanation and identification without intent to infringe.

Cataloging-in-Publication Data

Fertilizers : science and technology / edited by Eugene Perry.
   p. cm.
Includes bibliographical references and index.
ISBN 978-1-63239-907-6
   1. Fertilizers. 2. Agricultural chemicals. 3. Soil science. I. Perry, Eugene.
S633 .F47 2018
631.8--dc23

Table of Contents

Preface

Over the recent decade, advancements and applications have progressed exponentially. This has led to the increased interest in this field and projects are being conducted to enhance knowledge. The main objective of this book is to present some of the critical challenges and provide insights into possible solutions. This book will answer the varied questions that arise in the field and also provide an increased scope for furthering studies.

Fertilizers are materials that are applied to the soil for the necessary growth of plants. They can either be natural or synthetic in nature. They are mainly used for the enhancement of soil fertility and are classified into single nutrient fertilizers, multinutrient fertilizers and micronutrients. Most of the topics introduced in this book cover new techniques and the applications of fertilizers. It includes contributions of experts and scientists which will provide innovative insights into this field. Students, researchers, experts and all associated with soil science and botany will benefit alike from this book.

I hope that this book, with its visionary approach, will be a valuable addition and will promote interest among readers. Each of the authors has provided their extraordinary competence in their specific fields by providing different perspectives as they come from diverse nations and regions. I thank them for their contributions.

Editor

Preface

Over the recent decade, advancements and applications have progressed exponentially. This has led to the increased interest in this field and projects are being conducted to enhance knowledge. The main objective of this book is to present some of the critical challenges and provide insights into possible solutions. This book will answer the varied questions that arise in the field and also provide an increased scope for furthering studies.

Fertilizers are materials that are applied to the soil for the necessary growth of plants. They can either be natural or synthetic in nature. They are mainly used for the enhancement of soil fertility and are classified into single nutrient fertilizers, multinutrient fertilizers and micronutrients. Most of the topics introduced in this book cover new techniques and the applications of fertilizers. It includes contributions of experts and scientists which will provide innovative insights into this field. Students, researchers, experts and all associated with soil science and botany will benefit alike from this book.

I hope that this book, with its visionary approach, will be a valuable addition and will promote interest among readers. Each of the authors has provided their extraordinary competence in their specific fields by providing different perspectives as they come from diverse nations and regions. I thank them for their contributions.

Editor

The *In Vitro* Mass-Produced Model Mycorrhizal Fungus, *Rhizophagus irregularis*, Significantly Increases Yields of the Globally Important Food Security Crop Cassava

Isabel Ceballos[1], Michael Ruiz[1], Cristhian Fernández[2], Ricardo Peña[2], Alia Rodríguez[1,9], Ian R. Sanders[3,*,9]

1 Soil Microbiology, Universidad Nacional de Colombia, Bogotá, Colombia, 2 Utopía, Universidad de La Salle, Yopal, Colombia, 3 Department of Ecology and Evolution, University of Lausanne, Lausanne, Switzerland

Abstract

The arbuscular mycorrhizal symbiosis is formed between arbuscular mycorrhizal fungi (AMF) and plant roots. The fungi provide the plant with inorganic phosphate (P). The symbiosis can result in increased plant growth. Although most global food crops naturally form this symbiosis, very few studies have shown that their practical application can lead to large-scale increases in food production. Application of AMF to crops in the tropics is potentially effective for improving yields. However, a main problem of using AMF on a large-scale is producing cheap inoculum in a clean sterile carrier and sufficiently concentrated to cheaply transport. Recently, mass-produced *in vitro* inoculum of the model mycorrhizal fungus *Rhizophagus irregularis* became available, potentially making its use viable in tropical agriculture. One of the most globally important food plants in the tropics is cassava. We evaluated the effect of *in vitro* mass-produced *R. irregularis* inoculum on the yield of cassava crops at two locations in Colombia. A significant effect of *R. irregularis* inoculation on yield occurred at both sites. At one site, yield increases were observed irrespective of P fertilization. At the other site, inoculation with AMF and 50% of the normally applied P gave the highest yield. Despite that AMF inoculation resulted in greater food production, economic analyses revealed that AMF inoculation did not give greater return on investment than with conventional cultivation. However, the amount of AMF inoculum used was double the recommended dose and was calculated with European, not Colombian, inoculum prices. *R. irregularis* can also be manipulated genetically *in vitro*, leading to improved plant growth. We conclude that application of *in vitro R. irregularis* is currently a way of increasing cassava yields, that there is a strong potential for it to be economically profitable and that there is enormous potential to improve this efficiency further in the future.

Editor: Matthias Rillig, Freie Universität Berlin, Germany

Funding: The authors thank the Swiss National Science Foundation (http://www.snf.ch) for funding this study with a Joint Research Project awarded to AR and IRS (project number: IZ70Z0-131311/1). The funders had no role in study design, data collection and analysis, decision to publish, or preparation of the manuscript.

Competing Interests: The authors have declared that no competing interests exist.

* E-mail: ian.sanders@unil.ch

9 These authors contributed equally to this work.

¶ These authors are joint senior authors on this work.

Introduction

The global human population is expected to reach over 9 billion by 2050, according to United Nations estimates [1]. Feeding 9 billion people represents a major challenge that requires global yield increases of approximately 70% to 100% [2]. However, most of these yield increases will be necessary in globally important food crops in developing countries in tropical and sub-tropical regions, where population growth is high. Developing both existing and new technologies for more efficient production of globally important food crops is key to achieving these goals [2,3]. One of the most potentially useful technologies is to apply the fungi involved in the arbuscular mycorrhizal symbiosis. The symbiosis occurs between plant roots and arbuscular mycorrhizal fungi (AMF). It is potentially useful because most plants, including all the major globally important food crops, naturally form this symbiosis and the symbiosis can increase plant biomass because the fungi help plants obtain phosphate from the soil [4]. Phosphate is an

essential nutrient for plant growth. While stocks of phosphate fertilizer are rapidly being depleted there is a simultaneous increase in demand for phosphate to help feed the growing population [5,6]. Thus, the mycorrhizal symbiosis is potentially applicable at a global scale for increasing food production and more efficiently using phosphate reserves.

The potential benefits of inoculating crops with AMF have been known for decades. Yet there are very few published examples clearly demonstrating that large-scale inoculation of globally important crops in a real agricultural situation results in significant increases in food production that are either economically viable or practically achievable. There are three main reasons for this. Firstly, AMF are present in almost all soils. Crops naturally become colonized by native AMF. An overwhelming majority of published studies demonstrating the benefits of AMF inoculation on plant growth and phosphate acquisition have been conducted in sterile soil and compare the growth of mycorrhizal plants with non-mycorrhizal plants. However, effective AMF inoculation in agriculture requires inoculating plants with AMF that results in yield increases over those attained in the uninoculated crop that that is naturally colonized by the native AMF community. Secondly, most field trials with AMF have been conducted in temperate cropping systems. In many temperate agricultural regions, soil nutrient availability is relatively high compared with tropical soils. In most temperate agricultural soils inorganic phosphate fertilization is an effective way to increase yields. In contrast, most tropical regions are comprised of highly acidic Oxisols and Ultisols that typically have low total and available phosphate contents and a very high phosphate retention capacity [7]. In such soils farmers have to add large amounts of phosphate fertilizer to achieve significant yield increases. It is in these soils where the application of AMF could potentially increase food production and reduce application of phosphate fertilizer. Thirdly, AMF inoculum can only be grown with plants. Culturing AMF has traditionally been labour-intensive, requiring large-scale production of plants, from which the AMF inoculum can be harvested. Often this AMF inoculum is in the form of soil containing AMF propagules. It is difficult to ensure consistent inoculum quality, impossible to ensure that the soil is free from other potentially harmful microorganisms and the weight and volume of substrate for inoculum can make transport costs prohibitively high. However, these problems have recently been partially resolved by the development of a biotechnological mass production system for AMF. The fungi are grown *in vitro* on *Agrobacterium rhizogenes*-transformed carrot roots in a sterile artificial medium. Efficient production systems and the ability to concentrate very large numbers of AMF propagules into a small volume of sterile medium make the product easy to transport, free of unwanted microorganisms and potentially economically viable for large-scale application to important food crops.

One obvious target crop for the application of biotechnologically produced AMF is cassava (*Manihot esculenta* Crantz). Cassava is a traditional crop in latin America that has become globally important, annually feeding almost a billion people in 105 countries and representing almost a third of their daily caloric contribution [8]. The Food and Agriculture Organization (FAO) of the UN promotes cassava cropping in developing countries. It is considered vital for food security and as an alternative energy source [9].

Cassava is highly dependent on AMF for its growth and nutrition. Large growth increases due to mycorrhizal inoculation have been observed in cassava. Plants inoculated with AMF were 10–20 times larger than non-mycorrhizal plants growing in sterile soil [10] [11]. While cassava plants are never non-mycorrhizal in the field, these studies demonstrate that cassava is dependent on AMF for its growth. In a series of pioneering experiments by Sieverding and co-workers (summarized in [12] impressive effects of AMF inoculation were observed in the field in Colombia, where yields of cassava roots could be increased by up to approximately 5 tons ha^{-1}.

Numerous studies show that different AMF species have highly variable effects on plant growth [13,14] [15,16]. Indeed, yields of cassava in the field were highly variable following inoculation with different AMF species, ranging from no effect, compared to uninoculated plants colonized by the native AMF community, up to an approximate 20% yield increase [12]. This shows the need to consider the AMF identity for inoculation of cassava. Thus, it is not a foregone conclusion that a given biotechnologically-produced AMF species will significantly enhance cassava yields.

The most efficient large-scale biotechnological production of an AMF that can be highly concentrated in a sterile carrier has been achieved for one AMF species *Rhizophagus irregularis*, making it a strong candidate for use in cassava cropping. An additional strong incentive to use this species to increase cassava production is that *R. irregularis* has become the model AMF species for researchers as it can easily be grown in an *in vitro* culture system in the laboratory that is very similar to that used in large-scale commercial *in vitro* AMF cultivation. Consequently, more information is known about the genome and transcriptome of *R. irregularis* than any other AMF species [17,18]. The availability of only one *in vitro*-cultivated AMF species may appear very limited. However, populations of this fungus exhibit high natural genetic variation [19,20]. Recent studies showed that crossing genetically different *R. irregularis* isolates *in vitro* gave rise to genetically novel lines [21]. Allowing segregation of genetic material in subsequent generations by so-called single spore culturing also gave rise to a large number number of genetically novel *R. irregularis* lines [22]. Both crossed and segregated *R. irregularis* lines were shown to have differential effects on the growth of rice, in the greenhouse, with up to a five-fold increase in rice growth with some lines [23,24]. The ability to easily generate a large number of genetically novel *R. irregularis* lines *in vitro* that can have differential effects on plant growth, and then rapidly produce them in commercial quantities, would make this fungus an ideal candidate for a future molecular-assisted AMF breeding and improvement program, if *in vitro*-produced *R. irregularis* has the capacity to actually increase cassava yield in the field over uninoculated cassava. Therefore, it is highly pertinent to know whether *in vitro*-produced inoculum of this particular model AMF species can significantly increase cassava yields in commercial cassava cropping.

We, therefore, tested whether *in vitro*-produced *R. irregularis* increases cassava yield over uninoculated cassava in large-scale commercial cassava cropping in Colombia. We performed the experiment at two different locations with different soils, with no phosphate fertilization and with two different levels of phosphate fertilization. The highest application of phosphate represented the amount normally used locally by farmers in commercial cassava cropping in Colombia. A full economic analysis of the results allowed us to test not only whether *in vitro* produced *R. irregularis* can be used to significantly increase food production but also at what point such technology could become economically viable to produce food more cheaply. Thus, we were able to address the potential of using this technology in the short term to help reduce both hunger and poverty; the two main goals of the FAO of the United Nations. Additionally we were able to assess whether this fungus is a potentially good candidate for a molecular assisted AMF improvement program in an attempt to increase cassava yields over that which is currently possible and reduce food production costs even more in the future.

Materials and Methods

Field Sites

Two field experiments were established to test the effects of *in vitro*-produced AMF inoculum on the growth of cassava. The first was at the Utopía campus of La Salle University (at Yopal, Casanare) in the Los Llanos region of Colombia (72° 17′ 48′′ W, 5° 19′ 31′′ N). The second was on a commercial farm located in Santana, Boyacá in a mountainous region of Colombia (73° 29′ 59′′ W, 6° 03′ 27′′ N). Physical and chemical soil properties at both field sites are shown in Table S1.

The climate in Yopal is tropical with average temperatures of 18°C (night) to 28°C (day), with an average air humidity of 75% and total annual precipitation of 2335 mm with 172 rain days.

The field in Yopal had not been cultivated for 14 years prior to this experiment.

In Santana the temperature ranges between 15°C (night) and 23°C. The average air humidity is 78% and the total annual precipitation is 1901 mm with 221 rainy days. The plot in Santana had been regularly cultivated with sugar cane before cassava crops were established five years ago.

Plant and Fungal Material

Cassava (*Manihot esculenta* Crantz) varieties used in Yopal and Santana were MCOL2737 and COL2215, respectively. These are the most frequently used varieties by local farmers for cassava as a food crop.

The AMF used for these experiments was *Rhizophagus irregularis* produced in an artificial *in-vitro* AMF culture system with *Agrobacterium rhizogenes*-transformed carrot roots. The commercial product used is known as Glomygel® Hortalizas (http://www. mycovitro.com). Most isolates were previously ascribed to the species *Glomus intraradices* but were then found phylogenetically to be a separate species which was subsequently named *G. irregulare* [25]. Ribosomal DNA sequences obtained for this fungus allowed us to verify that the fungus used in Glomygel® fitted to the *G. irregulare* group described by Stockinger et al. (2009). *Glomus irregulare* has, however, recently been renamed as *Rhizophagus irregularis* [26].

Design and Establishment of the Field Experiment

The experiment in Yopal was a two-factor design arranged into blocks using a split-plot layout, with mycorrhizal treatment as the main plot and phosphorus treatment as the sub-plot. There were four blocks, each containing three replicate minor plots of 6 treatment combinations. The blocks were arranged perpendicular to the slope of the field. There were two mycorrhizal treatments, inoculated with AMF (+AMF) or non-inoculated (−AMF). There were 3 phosphate treatments; no phosphate fertilization (0 P), 50% phosphate fertilization (50% P) and 100% phosphate fertilization (100% P). Thus, there were 6 treatment combinations. The treatment 100% P represented the normal amount to P fertilizer applied to cassava crops in the region. There were 12 replicates of each of the six treatments with 25 plants within each minor plot. Two rows of plants were planted around each minor plot to reduce edge effects. The experimental design was the same in Santana except that there were five replicates of each of the six treatments, with 20 plants inside each minor plot.

Cassava was planted as stem cuttings (stakes). These were taken from the middle third of parental plants and had two or three vegetative buds and a thickness of between 1.5 and 2.0 cm. Ends of the stakes were cut diagonally to improve plant emergency. In Yopal (May 2011), 20 cm deep holes were made at an angle of 25°–30° and cassava stakes were completely buried with vegetative buds facing upward. In Santana (April 2011), cassava stakes were completely buried horizontally, 5 cm below ground. Total plant density was 7600 ha^{-1}. For both experiments, there was no artificial irrigation and conventional crop management for the region was applied depending on pests, diseases and weed incidence.

Fertilizers were applied in Yopal and Santana at 45 and 80 days after planting, respectively. The amount of fertilizer applied was determined by the initial soil nutrient content, nutritional requirements of the cassava variety and fertilizer efficiency. Plants in the 100% P treatments in Yopal received 104 Kg ha^{-1} urea, 201 Kg ha^{-1} di-ammonium phosphate, 54 Kg ha^{-1} potassium chloride (KCl), 42 Kg ha^{-1} of Kieserite (a fertilizer comprising 3% soluble potassium, 24% magnesium and 19% sulphur) and 40 Kg

ha^{-1} of Vicor (a granular fertilizer comprising 15% nitrogen, 5% calcium, 3% magnesium, 2% sulphur, 0.02% boron, 0.02% copper, 0.02% manganese, 0.005% molybdenum, 2.5% zinc). In Santana, plants in the 100% P treatment received 380 Kg ha^{-1} urea, 330 Kg ha^{-1} di-ammonium phosphate, 240 Kg ha^{-1} potassium chloride (KCl) and 100 Kg ha^{-1} of Kieserite (a fertilizer comprising 3% soluble potassium, 24% magnesium and 19% sulphur).

Inoculation of cassava plants in the inoculated treatment (+AMF) was carried out at planting. Plants were inoculated with approximately 12500 AMF propagules per plant, which corresponded to double recommended dose. The product was diluted in water (1:200) in a 150 l container. Cassava stakes were completely submerged in the diluted inoculum before planting. Stakes were then planted and the excess inoculum in was distributed in an equal amount in the hole where each plant in the +AMF treatment was placed. Non-inoculated plants received the same amount of water.

Plant and Fungal Growth Measurements

In the experiment in Yopal, plant growth variables were measured every 45 days. One plant was collected from each minor plot of each treatment. The number of roots per plant, root diameter and root fresh weight per plant were measured. Total leaf area per cassava plant (m^2/plant) was determined by measuring the leaf area of all leaves on a plant using a Li-Cor leaf area meter Li-Cor, Nebraska, USA). Total plant dry mass and the separate dry mass of leaves, stems, petioles and tuberous roots was measured for each plant. Roots were cut into pieces to facilitate drying. Plant material was dried at 70°C until constant weight (approximately 49 hours).

In the experiment in Yopal, total AMF colonization in roots was measured in fine roots with a thickness of ≤2 mm. Fungal structures were visualised with Shaeffer black ink acording to [27]. The percentage of root colonization was determined by the grid line-intersect method [28], on roots of one plant per minor plot. Values of colonization in the roots of the three plants from minor plots per block were pooled for further statistical analysis.

At the final harvest in Yopal, all the same measurements were made as at each previous time point. In addition, the fresh weight (yield) of cassava roots was measured. Final yield was calculated as cassava root fresh weight per hectare. In Santana, the only measurements made were at the final harvest, 14 months after planting, and these were root fresh weight (yield) and the colonization of the roots by AMF.

Statistical Analyses

All data were analysed using the JMP® statistical discovery software (Statistical Analysis Systems Institute, version 10). To test for significant differences in the cassava growth variables or colonization of the roots by AMF, analysis of variance (ANOVA) was performed using a split-split plot model in JMP®. Prior to ANOVA, data were first checked for equal variances and normal distribution. When significant differences were observed in ANOVA, a Tukey honest significant difference (HSD) test was performed to determine which treatments were different from each other at $P \leq 0.05$.

Economic Analysis

The economic analysis included a full assessment of all production costs in the different treatments, profits, cash flow and profitability (return on investment - ROI%). The cost-benefit and ROI %, of the crop under the six treatments (+AMF, 0 P; +AMF, 50% P; +AMF, 100% P; −AMF, 0 P; −AMF, 50% P;

−AMF, 100% P), were calculated at both locations. Agricultural interest rates were used for calculations involving monetary value, corresponding to an effective rate of 11% annually. This represents the most conservative interest rate for such analyses.

An economic simulation was performed using different inoculum prices to calculate the economic feasibility of using the product and also to find the maximum inoculum price at which the farmer would create more profit than by using the traditional crop management. Additionally, the same simulation was made with varying phosphate fertilizer prices, to calculate the economic feasibility of using this product when considering potential future increases in the price of P fertilizer. The inoculum price used in these calculations is the European price of the product as it is not currently available on the Colombian market.

Ethics Statement

According to Colombian law and with advice from the Colombian Environmental Licensing Authority, it was found that for this project, permission was not required for use of Glomygel® in scientific research, because the inoculum was used in agricultural research activities without involving fauna or flora specimens. CITIES permission signed by Colombia for export and import of inoculum was not required because the species in the study did not belong to any of the lists of endangered species. Both field experiments were conducted on private land and permission was given by the owner of the land at each location. Furthermore, no additional special permission was required by Colombian law as the organisms in this study also occur in Colombia. None of the species used in this study are endangered or protected.

Results

There was a significant effect of inoculation with *in vitro*-produced AMF on cassava yields at both sites. Additionally, the yields were higher than the locally expected yields at both sites.

Cassava Growth and Yield in Yopal

In Yopal, growth of the plants was measured at 45 day intervals throughout the experiment. As expected, the growth of cassava increased over time for each of the variables and for overall plant dry mass. However, there was no significant overall AMF effect on the whole dataset over all time periods (data not shown). However, at the final harvest, both root fresh weight and root dry weight were significantly affected by AMF inoculation. Inoculation with AMF significantly increased root dry biomass at the final harvest in Yopal with an 18.3% increase in root dry mass and an increase of over 2 tons more cassava roots per hectare (ANOVA F ratio = $F_{(1,3)}$ 22.84, $P \leq 0.017$; Fig. 1a). At the final harvest in Yopal, increasing P fertilization levels also significantly altered root dry weight, irrespective of AMF inoculation (data not shown). However, there was no AMF inoculation x P treatment interaction meaning that AMF inoculation significantly influenced cassava dry weight at all levels of P fertilization.

Cassava root fresh weight (yield) in Yopal was also significantly affected by AMF inoculation with a highly significant effect (ANOVA F ratio = $F_{(1,3)}$ 16104.27, $P \leq 0.001$). Overall, inoculated plants were 20.4% heavier than roots of non-inoculated plants. However, this was not the same at each of the 3 P fertilization levels as indicated by a significant AMF inoculation x P fertilization interaction (ANOVA F ratio = $F_{(2,60)}$ 3.16, $P \leq 0.049$; Fig. 1b). The combination of inoculation with AMF and 100% P resulted in the significantly highest yield. The significantly lowest yields were in the treatment combinations without AMF inoculation and with 0 P and 100% P. Surprisingly, the treatment

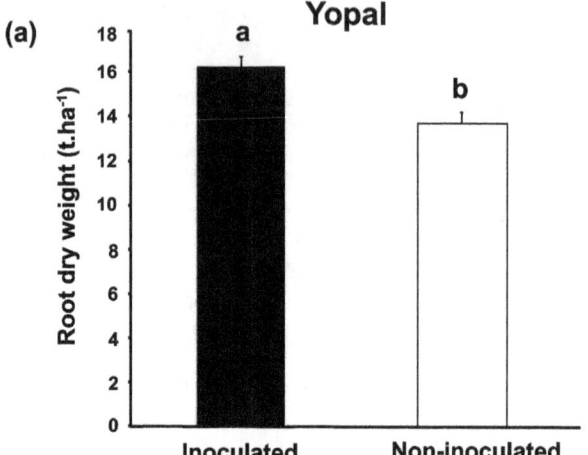

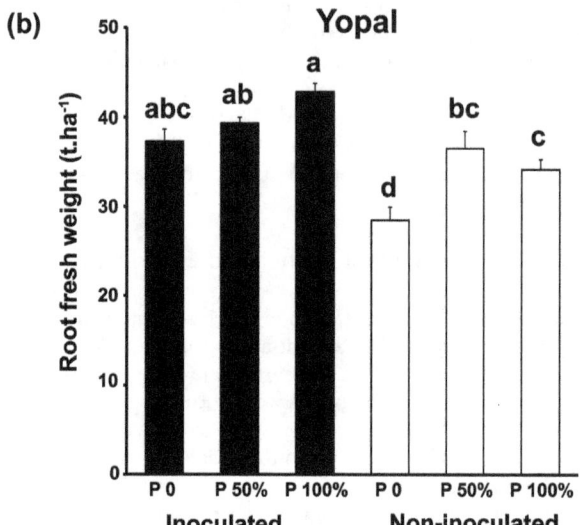

Figure 1. Effects of inoculation with AMF on cassava root growth. (a) Effects of inoculation with AMF on cassava root dry weight (t.ha⁻¹) in Yopal. (b) Effects of inoculation with AMF and P fertilization on cassava root fresh weight or yield (t.ha⁻¹) in Yopal. Black shaded bars represent the weight of inoculated cassava and white bars represent the weight of non-inoculated cassava. Error bars represent +1 S.E. Different letters above bars represent significant differences at $P \leq 0.05$.

combination with AMF inoculation and 0 P resulted in yields as high, or higher, than non-inoculated plants with either 50% P or 100% P (Fig. 1b).

Only root weight was affected by inoculation with AMF. None of the aboveground cassava growth variables were significantly affected by inoculation with AMF at the final harvest (data not shown).

Cassava Yield in Santana

In Santana, cassava yield (root fresh weight) was only measured at the final harvest. Inoculation with AMF had a significant effect on cassava yield but this was not the same at the different P fertilization levels, as indicated by an AMF inoculation x P fertilization interaction (ANOVA F ratio = $F_{(2,16)}$ 4.12, $P < 0.036$; Fig. 2). The highest yields were obtained with cassava inoculated with AMF and at 50% P fertilization. The highest yield in non-inoculated plants was obtained at 100% P. Meanwhile, the lowest

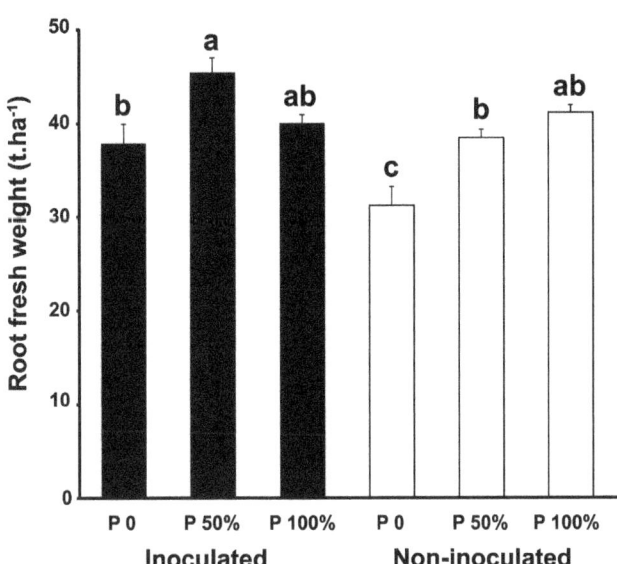

Figure 2. Effects of inoculation with AMF and P fertilization on cassava root fresh weight or yield (t.ha^{-1}) in Santana. Black shaded bars represent the weight of AMF-inoculated cassava and white bars represent the weight of non-inoculated cassava. Error bars represent +1 S.E. Different letters above bars represent significant differences at $P\leq0.05$.

yields were obtained with non-inoculated plants at 0 P fertilization. Plants that were inoculated with AMF but received no P fertilizer achieved a yield that was not significantly different that of non-inoculated plants that received either 50% or 100% P fertilization.

AMF Colonization in Yopal and Santana

Cassava naturally becomes colonized by mycorrhizal fungi. Therefore, in both AMF inoculated and non-inoculated treatments, we expected plants to be colonized by AMF. In Yopal, AMF colonization of cassava roots increased over the duration of the experiment, reaching a peak of colonization by the end of the experiment (Fig. 3). There was, however, a decline in AMF colonization in all the treatments at 225 days after planting that coincided with a particularly dry period. There was no significant difference in colonization in cassava plants that were inoculated with AMF or non-inoculated for most of the experiment. At the final harvest AMF colonization differed in inoculated and non-inoculated plants but this was not the same effect at each level of P fertilization (ANOVA for the AMF inoculation x P fertilization interaction was F ratio = $F_{(2,12)}$ 3.9, $P<0.049$). At the final harvest, AMF colonization was higher in non-inoculated plants at 50% and 100% P fertilization than non-inoculated plants. The opposite effect on colonization occurred at 0 P fertilization (data not shown).

In Santana, AMF inoculation and P fertilization also had a combined effect on the colonization of cassava roots by AMF at harvest (ANOVA for the AMF inoculation x P fertilization interaction was F ratio = $F_{(2,16)}$ 18.16, $P\leq0.001$; Fig. 4). The highest AMF colonization was observed in inoculated plants with no added P fertilizer. The lowest colonization values were obtained with non-inoculated and plants that received no phosphate fertilizer (Fig. 4).

The methodology used to measure colonization in this experiment does not allow us to differentiate between colonization by the local the AMF community and the inoculated AMF.

Economic Analyses

All production costs were used to calculate the return on investment (ROI) for production of cassava in one year in each of the six different treatments. The return on investment, represented as a percentage, is shown in Table 1. With the amount of inoculum used (which was double the recommended dose), and using a European inoculum price for the analysis, the highest ROI was observed in uninoculated treatments. The highest ROI in Yopal was in the treatment with no AMF inoculation and 50% P fertilizer for both cassava root fresh weight and dry weight (Table 1). In Santana, the highest ROI was achieved in the treatment with no AMF inoculation and 100% P fertilizer (Table 1). This was the case even though AMF inoculation significantly increased cassava yield at both sites. Interestingly, at both sites inoculation with AMF, in the absence of any P fertilizer, resulted in a higher ROI than when cassava was not inoculated (Table 1). The traditional practice is for the farmer to apply 100% P fertilizer but no AMF inoculum even though this is not necessarily the most profitable practice. Thus, in Yopal, inoculation with AMF and 100% P fertilization gave a higher return with root fresh weight than the traditional practice.

Because we used an unrealistic inoculum price and double the recommended dose, we made a simulation to calculate and project the following:

1. At what price the inoculum should be sold to make inoculation profitable for the farmers at the dose used in this experiment;
2. Project how profitable AMF inoculation will be in the event of future increasing P fertilizer prices;
3. Project how profitable AMF inoculation could be for farmers if the recommended dose would give similar yields.

To do this we used a baseline ROI of the most profitable treatment which was no inoculation with 100% P fertilizer in Santana and no inoculation with 50% P fertilizer in Yopal (Fig. 5). We simulated how the ROI would be affected at both sites when inoculating with AMF of varying price. We also calculated how profitable inoculation with the different priced inoculum would be in the event of rising P fertilizer prices (Fig. 5).

In Yopal, and with the sale of dry cassava roots, inoculation with AMF would become profitable if the same amount of inoculum required for one hectare of cassava were sold below 1 million Colombian pesos (COP) (Fig. 5a). Small price increases P fertilizer would not greatly affect the profitability of inoculating with AMF. This is also true for the sale of fresh cassava roots (Fig. 5b). In Santana, a small inoculum price reduction, to below COP 1200000, makes the use of AMF inoculum economically viable and any small prices increases in P fertilizer would have a large effect on the profitability of inoculating plants with AMF (Fig. 5c).

Because we used double the amount of AMF inoculum than that recommended by the inoculum producer, we also calculated the return on investment if the yield remained the same with only half the dose of inoculum that was used in this experiment (shown by the vertical dashed line in Fig. 5 representing half the inoculum cost). In all cases, this would make inoculation with AMF highly profitable, even at European inoculum prices and at current P fertilizer prices.

It is important to note that all the economic analyses were made on yields from one inoculation with AMF and for one crop of

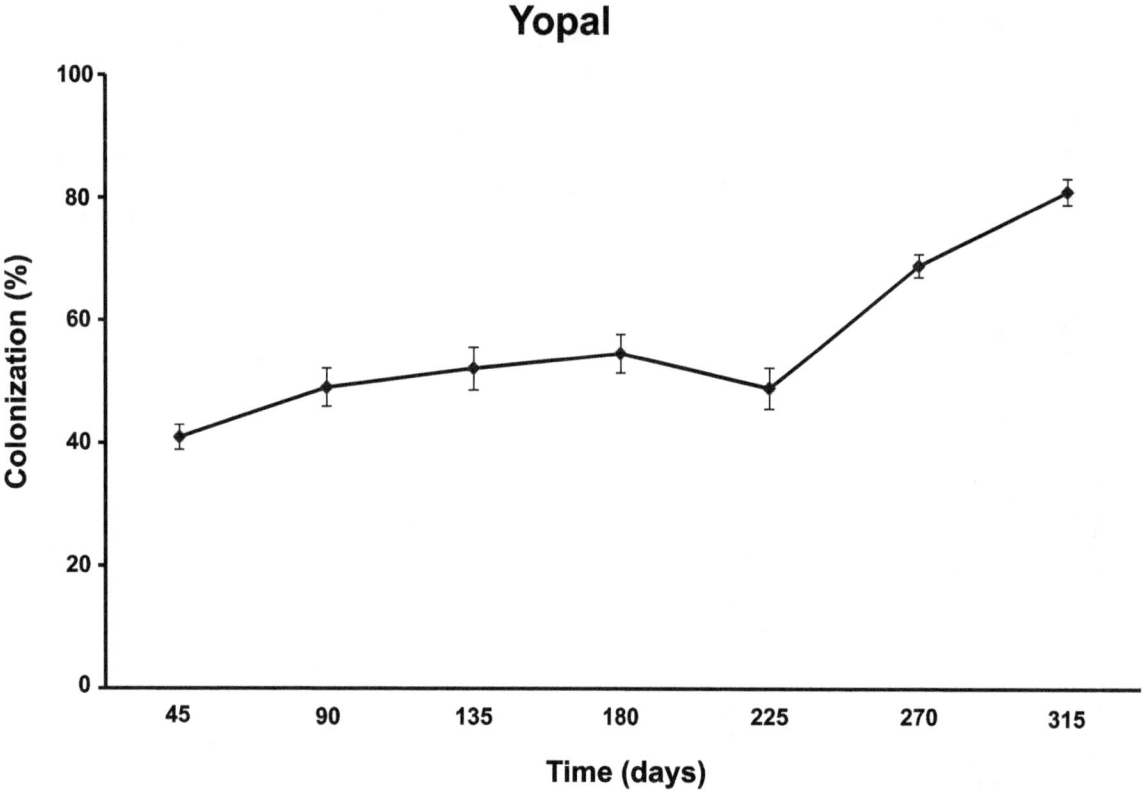

Figure 3. Colonization (represented as a % of root length) by AMF in the roots of cassava for the duration of the experiment in Yopal. The axis Time represents the number of days after planting. Error bars represent ±1 S.E.

cassava. Further yield increases in the next cassava crop due to inoculation in the previous year are not accounted for in these economic analyses.

Discussion

The results of this study show that inoculating the globally important food security crop cassava with AMF significantly improves cassava yields. The differences between this study and previous studies are that: 1. The fungus has been produced biotechnologically in an *in vitro* system that affords many advantages. 2. The fungal species was one that can very easily be grown and genetically manipulated for improvement *in vitro* and has become the model AMF for molecular and genomic research. 3. A detailed economic analysis allows us to set realistic targets for how to make the application of *in vitro* produced AMF economically viable for cassava production. The goal of the FAO is to increase food security and reduce hunger and poverty. Below we address how our results could help to realistically address these three goals in the future and potentially in a large number of developing countries.

Applying *in vitro*-produced AMF Leads to Higher Food Production

Mycorrhizal effects on food production and food security. Applying *in vitro*-produced AMF to cassava was effective in increasing cassava yields in both regions in Colombia, even though the soils, climate and cassava variety were different. The effects were not exactly the same in the two regions.

In Yopal, the most effective treatment was to add the same amount of P fertilizer as a farmer would normally use (P 100%

treatment) and inoculate with AMF. However, an important result is that inoculating with AMF gave significantly greater dry mass yield than uninoculated cassava at all P fertilization levels (Figure 1a). Furthermore, a higher cassava yield was achieved with AMF, without P fertilization, in comparison with plants with the full P dose. This is an important result as it indicates that the farmer can achieve significant increases in food production by using *in vitro* produced *R. irregularis* even if P fertilizer becomes more scarce or fluctuates greatly in price. Thus, the use of AMF in a cropping system, like cassava, represents an increment in the efficiency of phosphate fertilizer use, since yields achieved in inoculated plants exceeded yields obtained with maximum P fertilizer level. Thus, in these soils applying AMF can significantly increase food production and may increase food security by ensuring stable yields in the event of rising or fluctuating P fertilizer prices/availability.

In Santana, the significantly highest cassava yields were obtained by applying *in vitro* produced AMF inoculum in combination with 50% of the P fertilizer that is normally used by farmers in the region. This is a particularly interesting result as it shows that in these soils, it is possible to reduce P fertilizer applications and simultaneously increase food production. This is also particularly important as it means that more food can be produced more securely as the farmer would rely less on P fertilizer price fluctuations and availability.

There are some notable features of the data, particularly in Yopal, where measurements of plant growth and AMF colonization were made at 45-day intervals throughout the growing season. Inoculation with AMF in Yopal only had a significant positive effect on cassava root production; the part that is harvested for food. Inoculation did not have any effect on above-ground cassava

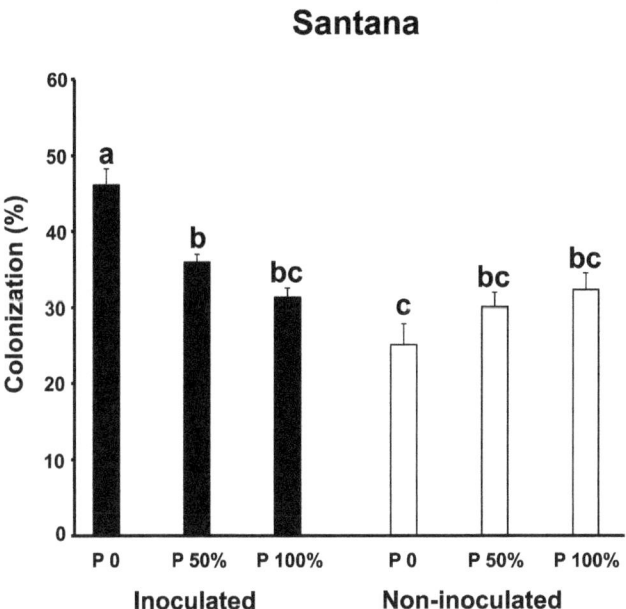

Santana

Figure 4. Colonization (represented as a % of root length) by AMF in the roots of inoculated and non-inoculated cassava and with different levels of P fertilization at the final harvest in Santana. Error bars represent +1 S.E. Different letters above bars represent significant differences at $P \leq 0.05$.

Table 1. Comparison of return on investment (ROI %) among the six treatments in Yopal and Santana.

Treatment		ROI (%)		
		Root Fresh Weight (yield)		Root Dry Weight
% P	AMF inoculation	Santana	Yopal	Yopal
0	Without	210.4	82.8	2.5
50	Without	265.5	126.8	15.1
100	Without	276.6	113.2	14.7
0	With	220.9	95.9	1.9
50	With	270.7	103.1	10.5
100	With	218.0	117.0	11.0

growth. Additionally, the positive effect of AMF inoculation was only observed at the end of the growing season when the roots were filling out with starch. Both of these results are consistent with previous findings from some field trials in Colombia on AMF and cassava [12]. In Yopal, inoculating cassava with AMF did not have any effect on overall AMF colonization levels in the roots of cassava during most of the experiment. In Santana, AMF colonization levels were overall higher when cassava was inoculated with AMF. However, how AMF colonization responded to P fertilization in the inoculated and uninoculated cassava in Santana followed different patterns. There are several reasons that could explain these differences. *Rhizophagus irregularis* may be more efficient at colonizing one cassava variety than the other, or the colonization levels could represent a difference in ability of *R. irregularis* to co-exist with the native AMF community at the two sites. From our data, we cannot distinguish between these different scenarios.

It is not possible to directly compare the effects of AMF inoculation at the two sites. The soils and cassava varieties differed at the two sites. Additionally, farmers typically use very different amounts of fertilizer in the two regions. To make the experiments agriculturally realistic the 100% P treatment represented the amount of P fertilizer typically applied by farmers in the region. Consequently, 100% P in Yopal is not an equivalent amount of P to 100% P in Santana.

Mycorrhizal effects in a realistic agricultural management system. Increased plant growth following AMF inoculation has been observed in many plant species and in numerous scientific publications [4]. However, most of these demonstrations have not been performed in an agriculturally realistic situation with unsterilized soil that already contains AMF. Furthermore, there are very few examples of where AMF inoculation has produced significant yield increases in a realistic agricultural management system over that which can be achieved using conventional fertilizer. In this respect, cassava is one of the

only crops that has been shown to benefit from AMF inoculation and that feeds a very significant number of the global population (approx. 1 billion people in 105 countries). However, many previous applications of AMF were not very applicable on a wide-scale to many different farmers as they required inoculating plants with a large amount of inoculum contained in soil with a farmer managed on-site inoculum production system. For example, in the pioneering experiments documented by Sieverding [12], 500 g of soil containing AMF had to be added to the roots of each cassava plant meaning that on-site fungus production and local specialist knowledge was necessary. We argue that such systems are susceptible to poor quality inoculum production and are unlikely to be sustained as they require a greatly different management practice to the farmers traditional management system. Also, the amount of inoculum required for a farming unit (i.e. per hectare) would result in an expensive practice in view of the amount of material needed to be transported/stored and applied in real farming conditions. In contrast, adding inoculum of an assured quality that is easy to transport and which is very easy to apply at planting is more likely to be adopted by farmers. The *in vitro*-produced fungus used in these experiments was shipped in a concentration where 0.25 ml volume of non-diluted liquid product was needed per plant. Furthermore, the cost of labour for applying AMF at planting was negligible.

The Importance of using the AMF Species *R. irregularis*

One major importance of this study is that we used the fungus *R. irregularis*. This species has become the model AMF studied by molecular biologists because, unlike many other AMF species, it can be efficiently produced *in vitro*. Also, the genome of this fungus has now been sequenced (F. Martin, personal communication). Thus, laboratory-based studies on this species, have rendered much of the knowledge we have about AM fungi. More importantly, studies have shown that natural genetic variation is very high in this fungus. Such naturally occurring genetic variation can be used to give rise to genetically novel strains of the fungus by crossing the fungi *in vitro* [29]. Previously, researchers had assumed that this was not possible. Inoculation of rice with genetically novel varieties of *R. irregularis* resulted in up to five-fold differences in rice growth in the greenhouse [23]. Even though non-bred strains of the fungi actually reduced rice growth several of the genetically novel lines greatly increased rice biomass. While we show that this fungus can be used now to increase cassava yields, the results of genetic studies on this species highlight the enormous potential for using such a breeding program to greatly improve the effect of this

(a)
Return on investment (ROI %)
25
23
21
19
17
15
13
11
9
Dry roots (Yopal)
Price of recommended dose
Price of dose used
ROI P 50% -AMF
300000
500000
700000
900000
1100000
1300000
1500000
Inoculum price (Colombian Pesos.ha-1)
(b)
Return on investment (ROI %)
150
145
140
135
130
125
120
115
Fresh roots (Yopal)
Price of recommended dose
Price of dose used
ROI P 50% -AMF
300000
500000
700000
900000
1100000
1300000
1500000
Inoculum price (Colombian Pesos.ha-1)
(c)
Return on investment (ROI %)
320
310
300
290
280
270
260
250
240
230
220
Fresh roots (Santana)
Price of recommended dose
Price of dose used
ROI P 100% -AMF
300000
500000
700000
900000
1100000
1300000
1500000
Inoculum price (Colombian Pesos.ha-1)

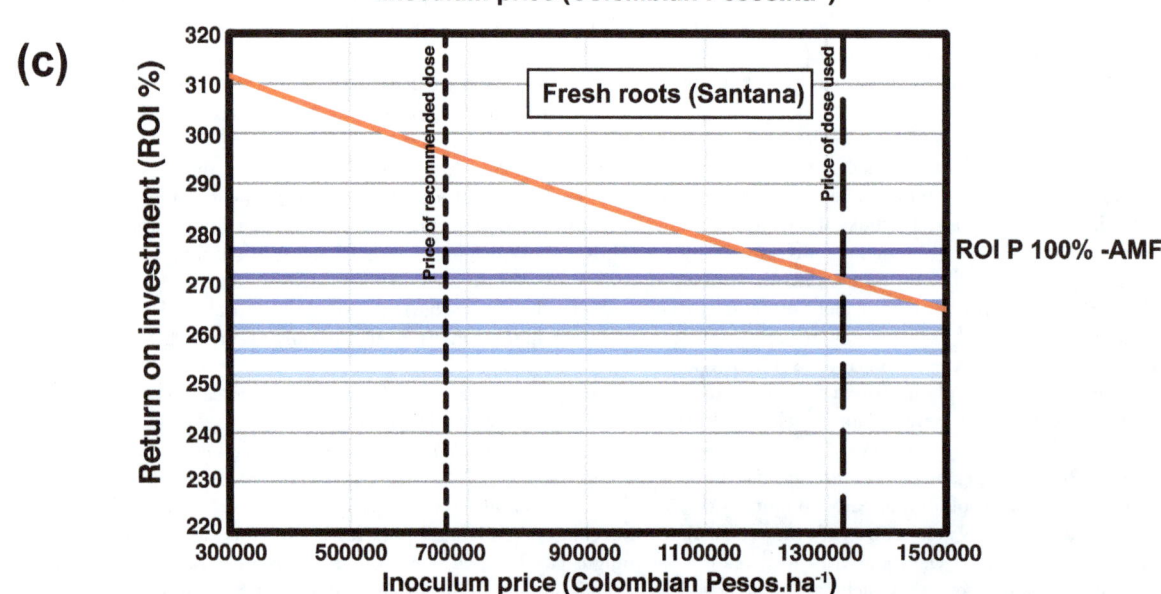

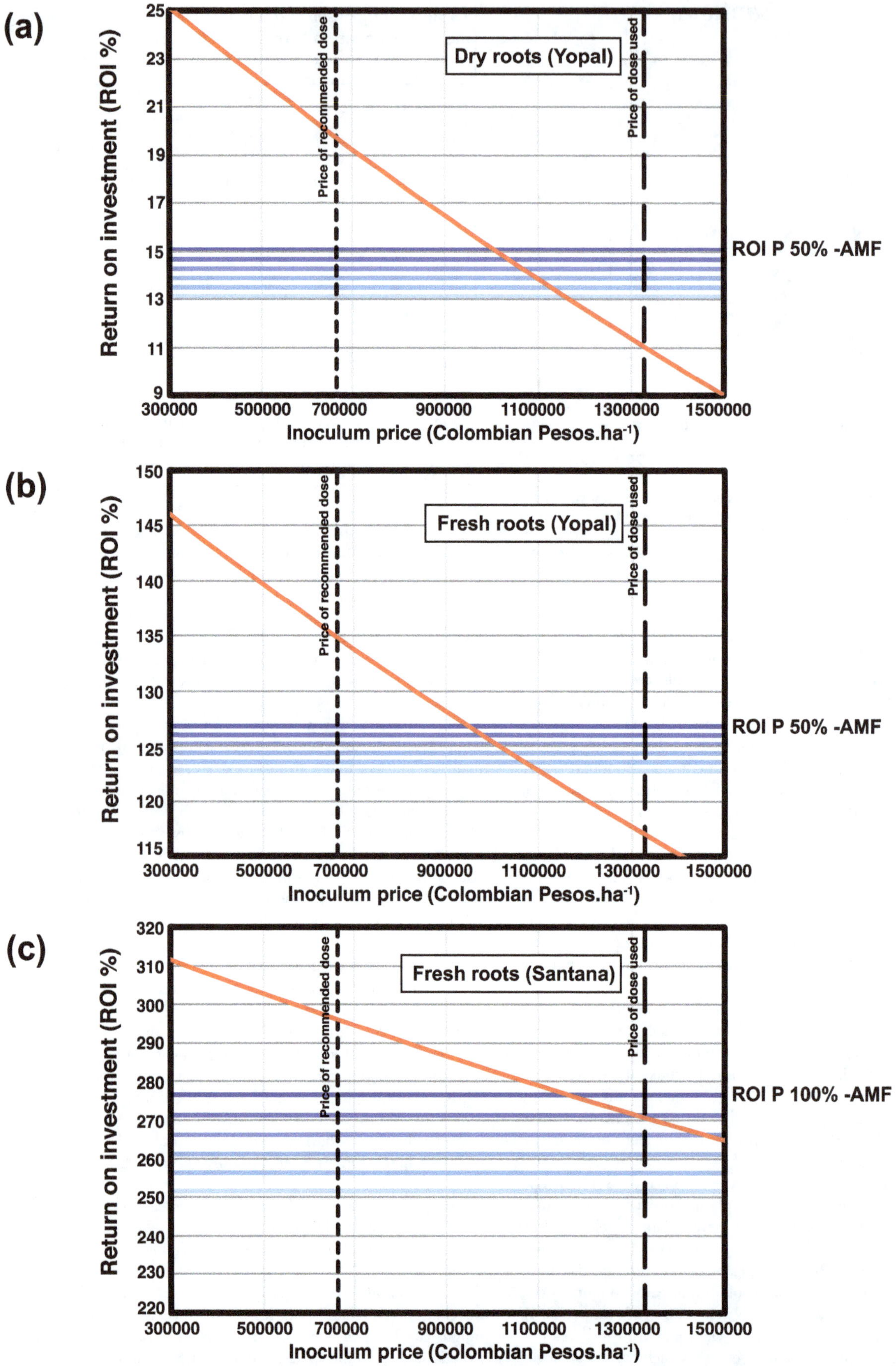

Figure 5. Simulation of return on investment (ROI %) with varying prices of *in vitro-* **produced AMF inoculum and increasing P fertilizer prices.** (a) Simulation based on sale of dry cassava root produced in Yopal. (b) Simulation based on sale of fresh cassava root produced in Yopal. (c) Simulation based on sale of fresh cassava root produced in Santana. The diagonal red line represents the ROI for the most profitable treatment with AMF inoculation, at that site, when simulating varying inoculum price. The top horizontal blue line represents the ROI for the most profitable non-inoculated treatment at current P fertilizer prices, even though this may not necessarily be what farmers normally practice. Blue horizontal lines below represent a simulation of ROI for this treatment with a scenario of increasing P fertilizer prices (+20%, +40%, +60%, +80% and +100% price increase). Vertical dashed lines represent the price for the amount of inoculum used in these experiments (right) and the price for half of the amount of inoculum which represents the recommended dose (left).

fungus on cassava yields in the future and be able to expand these results to other parts of the world.

Environmental Impact of Adding a Non-local AM Fungus

One point which has received almost no attention in the commercial application of AMF is the potential dangers of adding any AMF inoculum into a region from which it was not isolated [30]. In this study, we did not address the potential impact of such a practice. However, we consider two important tests are made before such inocula are used commercially. First, an assessment of whether the addition of the AMF inoculum alters the structure and diversity of the native AMF community. Second, a population genetics-based assessment of whether adding the commercial inoculum introduces a significant number of new alleles into the local *R. irregularis* population. To our knowledge, this type of environmental impact assessment for AMF application has never been undertaken.

Economics of using *in vitro* Produced AMF in Cassava Cropping

The economic analyses were performed for cassava production in two regions of Colombia. Although the experiments documented here were not set up to test the economic viability of applying this inoculum to cassava, the economic analyses in this study allow us to formulate clear targets for achieving this in the future. While it was not economically viable to use *in vitro* produced AMF with the amounts of inoculum we used, the economic analyses indicate that such inoculum application could easily be economically viable in the near future, and effectively alleviate strong dependency on P fertilizers, in the cassava cropping system. We highlight realistic predictions from our analysis that indicate how to make such applications economically viable:

Inoculum amount. In these experiments we applied double the amount of inoculum to that recommended by the manufacturer. This means that each plant received 12500 propagules of the fungus. This is a very large amount given that in the greenhouse we inoculate plants with 300 spores of *R. irregularis* to obtain consistent colonization in the roots. Previous experiments in the field with cassava have demonstrated that inoculating cassava with about 1500–2500 propagules gives significant yield increases that cannot be increased more by adding larger numbers of effective propagules [12]. Therefore, it is very likely that the amount of inoculum used in these experiments could have been greatly reduced. Using 2500 propagules per plant would have reduced the inoculum price to a fifth of that which we used. This would make the use of *in vitro* AMF extremely profitable for farmers in both regions of Colombia (see Fig. 5).

Effects over multiple seasons. These experiments were conducted for only one field season. Therefore, the economic analysis only takes into account the effects on profitability from inoculating once and harvesting once. Previous detailed experiments over 2 cropping periods in Colombia have shown that one inoculation in the first year consistently gives rise to significant crop yields, in the first and second cropping [12]. In fact, in those experiments only one inoculation in the first year resulted in cassava yields that were higher in the second year than the first year. Thus, it is possible that our economic analyses greatly underestimate the profitability of inoculating cassava with *in vitro* produced AMF. However, it must be recognized that in this study, we have not measured the effects in a second cropping and therefore, it is also possible that there could be no carry over effect or that it could be negative. This needs to be tested in further experiments.

Inoculum price. The inoculum used was bought at European market price as the product is not distributed in Colombia. Therefore, the economic analyses use European inoculum prices. It is highly likely that inoculum prices on the market in a developing country could be greatly reduced in several ways: 1. The inoculum could be produced under licence locally, thereby, taking advantage of local salary levels that would be much lower than those in Europe; 2. Currently the inoculum price in Europe is based on relatively low sales for a specialised market. If this technology were adopted by farmers, the market for very widely-grown food crops such as cassava in the tropics would be much larger than on specialized crops in Europe, thus allowing a significant price decrease of the product because of much larger effective product sales. 3. Inoculum producers could lower their prices for sales of the product in developing countries.

Fluctuating P fertilizer prices and availability. Phosphate fertilizer prices and availability are expected to change over the next decades [5,6]. The economic analyses also indicate that relatively small changes in P fertilizer prices would make the use of this inoculum economically viable, even at the high price and inoculum quantity used in these experiments and even if there was no inoculation effect in a second year.

Conclusions

We conclude that application of the *in vitro*-produced AMF *R. irregularis* in cassava cropping can help in achieving the FAOs main goals of reducing hunger and poverty by significantly increasing cassava yields in a way is very likely to be easily made economically viable and profitable for farmers in developing countries. We also predict that its application could lead to economically more secure food production because cassava yield can be increased at very different levels of P fertilization making the farmer much less dependent on P fertilizer availability or price. Because the inoculation was observed in different soils, and with different cassava varieties, our study shows the potential to use this technology to increase cassava production in other parts of the world, especially in Africa where many cassava-cropped soils are very similar to those in Yopal. Finally, while we show that *in vitro*-produced *R. irregularis* can immediately be used to significantly increase cassava yields, the fact that the effects of this fungus can be increased by *in vitro* crossing makes it a very strong candidate for an improvement program to further increase cassava yields in the future over that which can be achieved with these fungi at present.

Acknowledgments

We thank the students at the Utopía campus for their help in the planting, maintenance and measurement of plants, and in particular Leonela Yised Nañes Cortés, Maria Idaly Ibañez, Victor Fernando Camacho, Camilo Alberto Gutiérrez Rodríguez, Leidy Andrea Barrera Franco, José Irenarco Tapias Fernández, Diego Alexander Arias Zorro, Erika Tatiana Naranjo Barragán, Alenis Guillermina Chaparro Pulido.

References

1. FAO (2013) FAOSTAT. Rome, Italy: FAO publications. Available: http://faostat3.fao.org/home/index.html. Accessed 2013 June 17.

2. Godfray HCJ, Beddington JR, Crute IR, Haddad L, Lawrence D, et al. (2010) Food security: The challenge of feeding 9 billion people. Science 327: 812–818.

3. Foresight: The future of farming (2011) Final project report. London, UK: The Government Office for Science. 211 p.

4. Smith SE, Read DJ (2008) The mycorrhizal symbiosis. San Diego, USA: Academic Press.

5. Gilbert N (2009) The disappearing nutrient. Nature 461: 716–718.

6. Gross G (2010) Fears over phosphorus supplies. Curr Biol 20: 386–387.

7. Friesen DK, Rao IM, Thomas RJ, Oberson A, Sanz JI (1997) Phosphorus acquisition and cycling in crop and pasture systems in low fertility tropical soils. Plant Soil 196: 289–294.

8. FAO (2010) Available: http://www.fao.org/ag/AGP/AGPC/gcds/index_en.html.Accessed 2013 June 17.

9. FAO (2005) A review of cassava in Africa with country case studies on Nigeria, Ghana, the United Republic of Tanzania, Uganda and Benin: Proceedings of the validation forum on the global cassava development strategy. Rome, Italy: Food and Agriculture Organization of the United Nations.

10. Howeler RH, Sieverding E (1983) Potentials and limitations of mycorrhizal inoculation illustrated by experiments with field-grown cassava. Plant Soil 75: 245–261.

11. Sieverding E, Howeler RH (1985) Influence of species of VA mycorrhizal fungi on cassava yield response to phosphorus fertilization. Plant Soil 88: 213–221.

12. Sieverding E (1991) Vesicular-arbuscular mycorrhiza management in tropical agrosystems. Eschborn, Germany: Deutche Gesellschaft für Technische Zusammenarbeit (GTZ). 371 p.

13. van der Heijden MGA, Klironomos JN, Ursic M, Moutoglis P, Streitwolf-Engel R, et al. (1998) Mycorrhizal fungal diversity determines plant biodiversity, ecosystem variability and productivity. Nature 396: 69–72.

14. van der Heijden MGA, Boller T, Wiemken A, Sanders IR (1998) Different arbuscular mycorrhizal fungal species are potential determinants of plant community structure. Ecology 79: 2082–2091.

15. Bever JD, Schultz PA, Pringle A, Morton JB (2001) Arbuscular mycorrhizal fungi: More diverse than meets the eye, and the ecological tale of why. Bioscience 51: 923–931.

16. Vogelsang KM, Reynolds HL, Bever JD (2006) Mycorrhizal fungal identity and richness determine the diversity and productivity of a tallgrass prairie system. New Phytol 172: 554–562.

17. Martin F, Gianinazzi-Pearson, Hijri M, Lammers P, Requena N, et al. (2008) The long hard road to a completed *Glomus intraradices* genome. New Phytol 180: 747–750.

18. Tisserant E, Kohler A, Dozolme-Seddas P, Balestrini R, Benabdellah K, et al. (2012) The transcriptome of the arbuscular mycorrhizal fungus *Glomus intraradices* (DAOM 197198) reveals functional tradeoffs in an obligate symbiont. New Phytol 193: 755–769.

19. Croll D, Wille L, Gamper HA, Mathimaran N, Lammers PJ, et al. (2008) Genetic diversity and host plant preferences revealed by simple sequence repeat and mitochondrial markers in a population of the arbuscular mycorrhizal fungus *Glomus intraradices*. New Phytol 178: 672–687.

20. Börstler B, Raab PA, Thiéry O, Morton JB, Redecker D (2008) Genetic diversity of the arbuscular mycorrhizal fungus *Glomus intraradices* as determined by mitochondrial large subunit rRNA gene sequences is considerably higher than previously expected. New Phytol: 452–465.

21. Croll D, Giovannetti M, Koch AM, Sbrana C, Ehinger M, et al. (2009) Nonself vegetative fusion and genetic exchange in the arbuscular mycorrhizal fungus *Glomus intraradices*. New Phytol 181: 924–937.

22. Ehinger M, Croll D, Koch MA, Sanders IR (2012) Significant genetic and phenotypic changes arising from clonal growth of a single spore of an arbuscular mycorrhizal fungus over multiple generations. New Phytol 196: 853–861.

23. Angelard C, Colard A, Niculita-Hirzel H, Croll D, Sanders IR (2010) Segregation in a mycorrhizal fungus alters rice growth and symbiosis-specific gene transcription. Current Biology 20: 1216–1221.

24. Colard A, Angelard C, Sanders IR (2011) Genetic exchange in an arbuscular mycorrhizal fungus results in increased rice growth and altered mycorrhiza-specific gene transcription. Appl Environ Microbiol 77: 5004–5007.

25. Stockinger H, Walker C, Schussler A (2009) '*Glomus intraradices* DAOM197198', a model fungus in arbuscular mycorrhizal research, is not *Glomus intraradices*. New Phytol 183: 1176–1187.

26. Krüger M, Krüger C, Walker C, Stockinger H, Schüssler A (2012) Phylogenetic reference data for systematics and phylotaxonomy of arbuscular mycorrhizal fungi from phylum to species level. New Phytol 193: 970–984.

27. Vierheilig H, Coughlan AP, Wyss U, Piche Y (1998) Ink and vinegar, a simple staining technique for arbuscular- mycorrhizal fungi. Appl Environ Microbiol 64: 5004–5007.

28. Giovannetti M, Mosse B (1980) Evaluation of techniques for measuring vesicular arbuscular mycorrhizal infection in roots. New Phytol 84: 489–500.

29. Sanders IR, Croll D (2010) Arbuscular mycorrhiza: The challenge to understand the genetics of the fungal partner. Ann Rev Gen 44: 271–292.

30. Schwartz MW, Hoeksema JD, Gehring CA, Klironomos JN, Johnson NC, et al. (2006) The promise and the potential consequences of the global transport of mycorrhizal fungal inoculum. Ecol Lett 9: 501–515.

Author Contributions

Conceived and designed the experiments: IRS AR IC MR CF RP. Performed the experiments: IC MR. Analyzed the data: IC MR. Contributed reagents/materials/analysis tools: IRS AR MR RP. Wrote the paper: IRS AR IC.

2

Tillage, Mulch and N Fertilizer Affect Emissions of CO_2 under the Rain Fed Condition

Sikander Khan Tanveer, Xiaoxia Wen, Xing Li Lu, Junli Zhang, Yuncheng Liao*

College of Agronomy, Northwest A&F University Yangling, Shaanxi, P.R. China

Abstract

A two year (2010–2012) study was conducted to assess the effects of different agronomic management practices on the emissions of CO_2 from a field of non-irrigated wheat planted on China's Loess Plateau. Management practices included four tillage methods i.e. T_1: (chisel plow tillage), T_2: (zero-tillage), T_3: (rotary tillage) and T_4: (mold board plow tillage), 2 mulch levels i.e., M_0 (no corn residue mulch) and M_1 (application of corn residue mulch) and 5 levels of N fertilizer (0, 80, 160, 240, 320 kg N/ha). A factorial experiment having a strip split-split arrangement, with tillage methods in the main plots, mulch levels in the sub plots and N-fertilizer levels in the sub-sub plots with three replicates, was used for this study. The CO_2 data were recorded three times per week using a portable GXH-3010E1 gas analyzer. The highest CO_2 emissions were recorded following rotary tillage, compared to the lowest emissions from the zero tillage planting method. The lowest emissions were recorded at the 160 kg N/ha, fertilizer level. Higher CO_2 emissions were recorded during the cropping year 2010–11 relative to the year 2011–12. During cropping year 2010–11, applications of corn residue mulch significantly increased CO_2 emissions in comparison to the non-mulched treatments, and during the year 2011–12, equal emissions were recorded for both types of mulch treatments. Higher CO_2 emissions were recorded immediately after the tillage operations. Different environmental factors, i.e., rain, air temperatures, soil temperatures and soil moistures, had significant effects on the CO_2 emissions. We conclude that conservation tillage practices, i.e., zero tillage, the use of corn residue mulch and optimum N fertilizer use, can reduce CO_2 emissions, give better yields and provide environmentally friendly options.

Editor: Ben Bond-Lamberty, DOE Pacific Northwest National Laboratory, United States of America

Funding: The study was supported by the National Natural Science foundation of China (31071375 and 31171506). The funders had no role in study design, data collection and analysis, decision to publish, or preparation of the manuscript.

Competing Interests: The authors have declared that no competing interests exist.

* E-mail: yunchengliao@163.com or yunchengliao@nwsuaf.edu.cn

Introduction

Studies regarding soil CO_2 emissions have attracted significant attention because the concentration of CO_2 in the atmosphere is increasing very rapidly as a consequence of fossil fuel combustion and deforestation. The past two centuries of human activities have reportedly contributed as much as approximately half of the increase in CO_2 emissions [1], [2]. Global terrestrial ecosystems absorbed carbon at the rate of 1–4 Pg yr^{-1}, during 1980s and 1990s, which made up approximately 10–60% of the fossil fuel emissions [3], [4]. Currently, significant attention is given to CO_2 emissions from soils because this source significantly affects the global carbon cycle and the function of the terrestrial ecosystem [5]. Fluxes of greenhouse gases (CO_2, N_2O and CH_4) between the atmosphere and agricultural soils considerably influence the stock of anthropogenic greenhouse gases [6]. Agriculture is an important source of emissions for these different gases, and its contribution to climate change is approximately 20% on an annual basis [7]. It has been reported that soils have already contributed approximately 50 Pg of anthropogenic CO_2 to the atmosphere in the past, through cultivation processes [8].

Tillage is an integral part of agriculture which not only significantly affects crop production but is also considered one of the leading factors in soil degradation. This technique is a fundamental operation that has affected both the soil and the environment and is considered one of the most important sources of CO_2 emissions into the atmosphere [9] because humans have tilled the soil for crop production for thousands of years [10] and approximately 23–44% of total CO_2, is emitted into the atmosphere through soil preparation-related operations [11]. Approximately 30–50% of soil C has already been lost through the adaptations of intensive tillage practices [12], and major C losses from, soils in the form of CO_2 occur immediately after the tillage operations [13].

Agricultural management practices affect different soil processes (i.e., soil temperature, soil moisture and soil pH), and other ongoing soil decomposition processes, which ultimately result in the conversions of plant-derived C to soil organic matter and CO_2 [14]. Applications of inorganic as well as organic fertilizers [15] and different degrees of soil moisture and temperature strongly affect the fluxes of soil CO_2 [16], [17] & [18]. Similarly, the application of N fertilizer also affects soil CO_2 emissions [19]. Instead of burning crops residues, farmers, applications of inorganic fertilizers and use of green manures as well as organic manures can be of great use in maintaining soil fertility [20]. These practices can provide essential nutrients to crops and reductions in the burning of crops can reduce CO_2 emissions into the atmosphere [21].

Agricultural tillage practices can be helpful in the sequestering of atmospheric CO_2 [22], [23] and [24]. Conservation tillage has the potential to increase soil C and N [25] and other types of

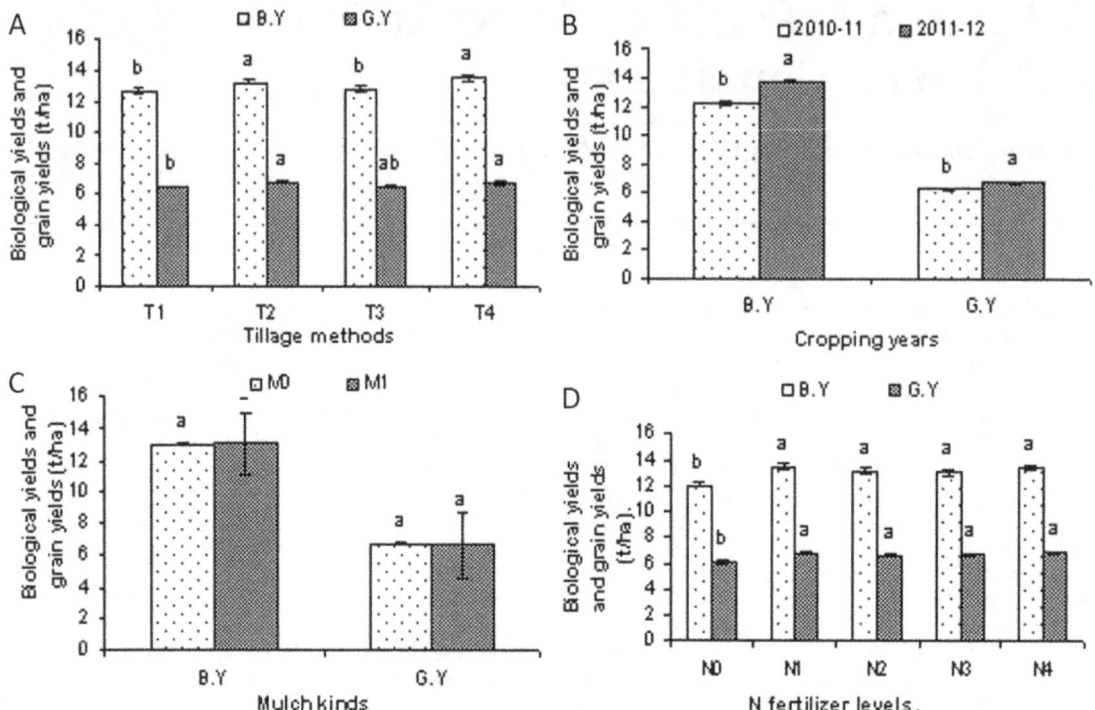

Figure 1. Wheat crop biological and grain yields as affected by different tillage methods, different mulch kinds and different N fertilizer levels during different cropping years (2010–12). (A), Wheat crop biological and grain yields as affected by different tillage methods i.e. T_1, chisel plow tillage., T_2, zero tillage., T_3, rotary tillage and T_4, mold board plow tillage., (B), Mean wheat crop biological and grain yields as affected by different tillage methods, mulch kinds and different N fertilizer levels during two cropping years (2010–12). (C), Wheat crop biological and grain yields as affected by different mulch kinds i.e. M_0, No mulch and M_1, corn residue mulch., (D), Wheat crop biological and grain yields as affected by different N fertilizer levels during two cropping years (2010–12) including, N_0, 0 kg N/ha., N_1, 80 kg N/ha, N_2, 160 kg N/ha., N_3, 240 kg N/ha and N_4, 320 kg N/ha.

conservation practices can be helpful in reducing the loss of soil organic carbon from the soils [26], [27]. Similarly, the retention of crop residue, nitrogen fertilization and no-tillage are generally supposed to enhance the soil organic carbon (SOC) stocks in the soil [28] because these farms, management practices not only increase crops biomass, but are also considered very important for the microbial decomposition of crop residues [29]. As far as N fertilization is concerned, some scientists have reported that increased N fertilization can depress CO_2 emissions [30], [31] however, others [32] have reported that N fertilization has no effect on SOC, while some other scientists [33] have reported that higher N fertilization improves the SOC of the soil.

The Loess Plateau has an area of approximately 63, 5000 km². It covers many provinces in China, and is home to millions of people. It is one of the most highly eroded areas of the world, and traditional agriculture, i.e., intensive tillage, is considered one of the leading man-made factors responsible for this erosion. However many crops residues are produced in this region. A small portion of these residues is used for forage or fuel consumption and the remaining residues are generally burned. Mold board plowing followed by harrowing is commonly used for the tillage operations in this region [34]. Few studies showing CO_2 emissions to the adaptation of different agronomic management practices have been previously reported from this region of China. In this area intensive tillage methods i.e. rotary tillage and mold board plow tillage methods are commonly used for land preparation. Commonly higher levels of N fertilizers are applied and crop residues are removed from the fields at the time of soil preparation.

The main aim of this two year study was to identify the effects of different tillage methods i.e. chisel plow and zero tillage in comparison with intensive tillage practices i.e. rotary tillage and mold board plow tillage methods, different N fertilizer levels and the application of corn residue mulch on CO_2 emissions. The results from this study can be of great help in improving the management of soils not only in this area of China but also in other regions of the world.

Results

Wheat crop yields

Significant variations in wheat crops biomass and grain yields were recorded during both study years (2010–12). When compared to the other tillage methods, there were better yields overall with the zero tillage planting method. No grain yields differences were recorded under different mulch treatments and similarly low yields were recorded for N_0 nitrogen fertilizer levels. Statistically equal grain yields were recorded for all of the other higher tested levels of N fertilizer (Fig. 1).

Soil CO_2 flux

All tested treatments had significant effects on the CO_2 emissions (Figs. 2, 3). The details are given below.

Tillage method effects on CO_2 emissions

Two years of combined data show that the rotary tillage and mold board plow tillage methods had their highest and statistically equal CO_2 emissions during the first week of planting the wheat

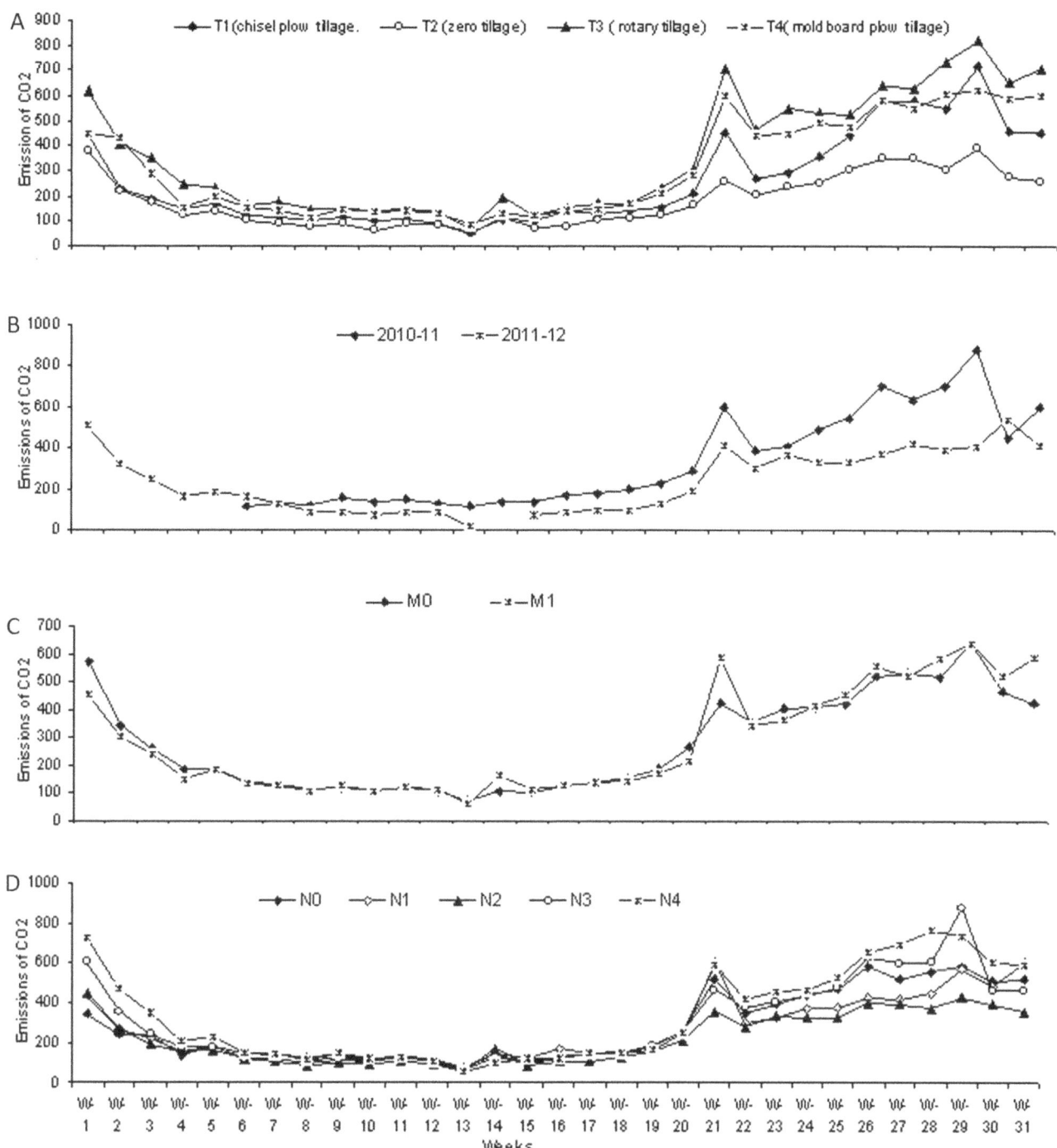

Figure 2. CO$_2$ emission trends as affected by different tillage methods, mulch kinds and N fertilizer levels during two cropping years (2010–12). (A). Emissions trends of CO$_2$ from different tillage methods i.e. T$_1$, chisel plow tillage, T$_2$, zero tillage, T$_3$, rotary tillage and T$_4$, mold board plow tillage., (B).CO$_2$ emission trends as affected by different kinds of tillage methods, different types of mulch and different N fertilizer levels, during two cropping years (2010–12). (C). CO$_2$ emission trends due to different mulch kinds i.e. M$_0$, no mulch and M$_1$, corn residue mulch., (D). CO$_2$ emissions trends due to different N fertilizer levels during two cropping years (2010–12) including, N$_0$, 0 kg N/ha., N$_1$, 80 kg N/ha, N$_2$, 160 kg N/ha, N$_3$, 240 kg N/ha and N$_4$, 320 kg N/ha.

crop, compared to the chisel plow tillage and zero tillage planting methods. When compared to all the other tillage methods, the lowest CO$_2$ emissions were recorded for the zero tillage planting method (Fig. 2; Fig. 3). Emission trends recorded during the whole wheat crop seasons show that although there were variations in the CO$_2$ emissions in response to different tillage methods, the overall

highest emissions were recorded for the rotary tillage planting method, followed by mold board plow tillage. Although higher CO$_2$ emissions were recorded for the chisel plow tillage method, the lowest CO$_2$ emissions were generally recorded for zero tillage planting method (Fig. 2; Fig. 3).

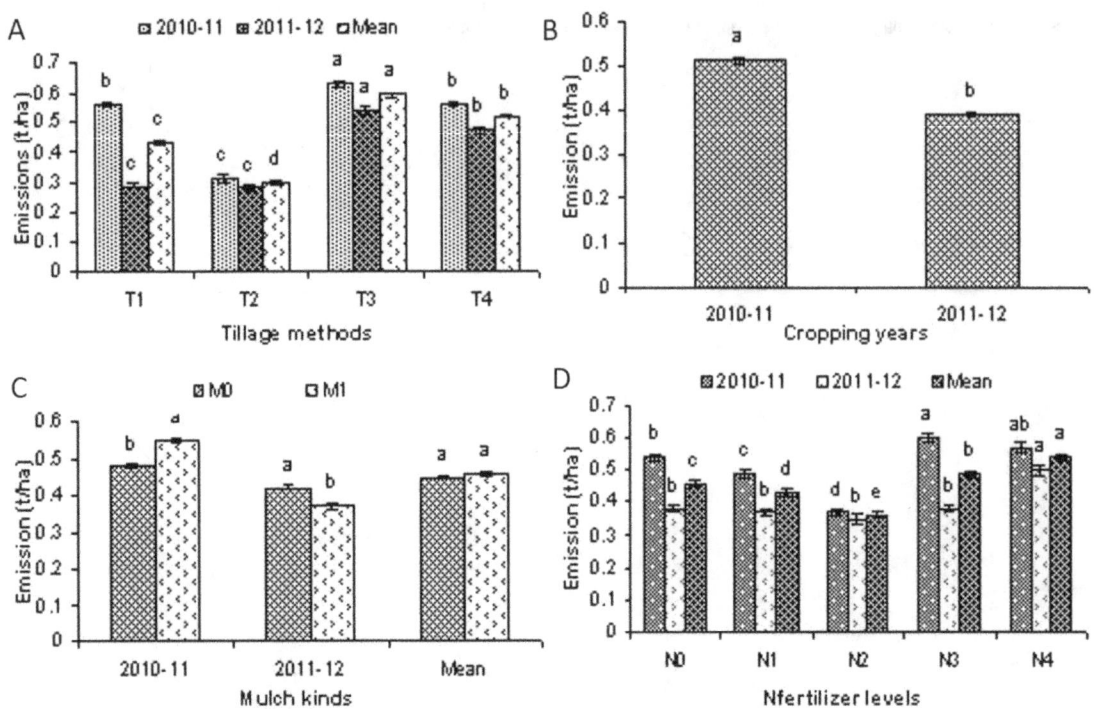

Figure 3. Total/mean emissions of CO_2 as affected by different tillage methods, mulch kinds, and N fertilizer levels during two cropping years (2010–12). (A).Emissions of CO_2 from different tillage methods i.e. T_1, chisel plow tillage., T_2, zero tillage.,T_3, rotary tillage and T_4., mold board plow tillage., (B). Total emissions of CO_2 as affected by different kinds of tillage methods, different kinds of mulch and different levels of N fertilizer, during two cropping years (2010–12)., (C), Emissions of CO_2 from the different mulch kinds i.e. M_0, no mulch and M_1, corn residue mulch., (D).Total/mean emissions of CO_2 as affected by different levels of N fertilizer, during two cropping years (2010–12) including, N_0, 0 kg N/ha., N_1, 80 kg N/ha., N_2,160 kg N/ha., N_3, 240 kg N/ha and N_4, 320 kg N/ha.

Effects of cropping years on CO_2 emissions

With the exception of one week (i.e., week 23), the highest weekly emissions of CO_2 were recorded during cropping year 2010–11 in comparison to the weekly emissions of CO_2 during cropping year 2011–12 (Figs. 2, 3).

Mulch effects on CO_2 emissions

Weekly CO_2 emissions during both wheat crops growing seasons (2010–11 and 2011–12) and the mean of the two years (2010–12) of data (Figs. 2, 3) show that there were more CO_2 emissions recorded from the corn residue-mulched treatments during cropping year 2010–11 than from the non-residue mulched treatments (Figs. 2, 3). However, fewer CO_2 emissions were recorded from the corn residue mulched treatments during cropping year 2011–12 and in comparison to the non residue-mulched treatments, over all less emissions of CO_2 were recorded from the corn residue mulched treatments (Figs. 2, 3). However the two year mean data show mixed types of emissions were recorded on a weekly basis following the applications of corn residue mulch or no mulch (Fig. 2).

N fertilizer level effects on CO_2 emissions

CO_2 emissions data recorded during both wheat cropping seasons, i.e., 2010–11 and 2011–12 and the two year (2010–12)) mean data show that there were significant differences in the weekly CO_2 emissions in response to different N fertilizer levels (Fig. 2; Fig. 3). This finding also indicates that the lowest CO_2 emissions were recorded for the N_0, N fertilizer level at the start of the wheat crop growing seasons compared to all the other higher

N fertilizer levels (Fig. 2). During the winter months the CO_2 emissions decreased under all treatments but when the temperatures rose again, higher CO_2 emissions were recorded (Fig. 2). However, when compared to all the other N fertilizer level treatments, the lowest overall CO_2 emissions were recorded for 160 kg N/ha (Fig. 2). CO_2 emission fluxes increased with the increase in crop growth and temperatures, so during the last weeks of wheat crop growth equal CO_2 emissions were recorded for all of the N fertilizer treatments (Fig. 2).

Cumulative CO_2 emissions

Two years (2010–12) of CO_2 emissions data show that on a mean cumulative basis, except in the case of corn residue mulch treatments, significant differences in the emissions of CO_2 were recorded for all of the different tillage methods and different N fertilizer level treatments (Fig. 3). Statistically significant variations in the total and mean CO_2 emissions were recorded for all of the tillage methods, and the emissions trend for the different tillage methods was $T_3>T_4>T_1>T_2$ (Fig. 3). Different N fertilizer levels had significant effects on the total and mean CO_2 emissions. The emissions trend for the different nitrogen fertilizer levels was $N_4>N_3>N_0>N_1>N_2$ (Fig. 3). On a cumulative basis, more CO_2 emissions were recorded during the cropping year 2010–11 than during cropping year 2011–12 (Fig. 3), and on the whole approximately 30% more CO_2 emissions were recorded during cropping year 2010–11 than during 2011–12 (Fig. 3).

CO_2 emissions varied for all of the tillage methods and N fertilizer levels. For the chisel plow tillage treatment, using 160 kg N/ha reduced the emissions of CO_2 (data not shown) and the CO_2

emissions trend in case of the chisel plow tillage method and different N fertilizer levels was $N_4 > N_3 > N_0 > N_1 > N_2$ (data not shown) (Table 1). For the Zero tillage planting method and N fertilizer level interactions, the CO_2 emission varied but the emissions trend was $N_0 > N_3 > N_2 > N_1 > N_4$ (data not shown) (Table 1). For the rotary tillage planting method and N fertilizer level interactions, the CO_2 emissions trend was $N_4 > N_3 > N_1 > N_0 > N_2$ (data not shown) (Table 1). Similarly for the mold board plow tillage method and N fertilizer level interactions, the CO_2 emissions trend was $N_0 > N_4 > N_1 > N_3 > N_2$ (data not shown) (Table 1).

With the exception of the mold board plow tillage method, corn residue mulch applications increased the CO_2 emissions in all of the other three tillage methods. Lower CO_2 emissions were recorded for all the tillage methods during cropping year 2011–12 in comparison to cropping year 2010–11 and throughout cropping year 2010–11, corn residue mulch applications increased the CO_2 emissions to 16.5% compared to the non residue mulched treatments. Similarly during cropping year, 2011–12, 12.6% fewer CO_2 emissions were recorded for the corn residue mulched treatments in comparison to the non-residue mulched treatments (data not shown) (Table 1).

For the N_0 (0 kg N/ha), N_1 (80 kg N/ha), N_2 (160 kg N/ha) and N_4 (320 kg N/ha), treatments, the use of corn residue mulch increased the CO_2 emissions by approximately 6.2%, 5.5%, 0.6% and 35.6%, respectively, and for N_4 (240 kg N/ha), the application of corn residue mulch reduced the CO_2 emissions by approximately 35.6% (data not shown) (Table 1).

Soil temperatures versus CO_2 emissions

Temperature changes for the top 5 cm of soil from the different treatments during the two years (2010–12) of study are given in Fig. 4. Although with the passage of time, there were variations in soil temperatures for the different tillage methods, however intermediary types of temperature changes were recorded following zero tillage planting in comparison to the other three tillage methods (Fig. 4). Similarly, although statistically comparable temperatures were recorded for both types of mulch treatments, slightly increased temperatures were recorded in the corn residue mulched treatments relative to the non-residue mulched treatments (Fig. 4). However, no soil temperature differences were recorded for the different N fertilizer level treatments (Fig. 4). Generally speaking, higher soil temperatures were recorded during the cropping year 2010–11 than during cropping year 2011–12 (Fig. 4). CO_2 emission trends showed that CO_2 emissions increased with an increase in soil temperatures and vice versa (Figs. 2, 4).

Soil moisture versus CO_2 emissions

When compared to all of the other tillage methods, the highest soil water contents were recorded in response to the rotary tillage planting method (Fig. 5). Similarly, higher water contents were recorded during the cropping year 2011–12 for the different crop growth stages relative to cropping year 2010–11 (Fig. 5). Corn residue mulch increased the soil moisture contents of the crop revival stage and on the booting stage compared to the non-residue mulched treatments (Fig. 5). Lower soil moisture contents were recorded for the 80 kg N/ha (N_1) and 160 kg N/ha (N_2), treatments compared to all of the other N fertilizer level treatments (Fig. 5). Two years of mean data show that the CO_2 emissions were lower in those tillage methods or N fertilizer levels treatments that had lower water contents (Fig. 2; Fig. 4).

Table 1. ANOVA (Mean Square Values) of biological yields, grain yields, soil organic carbon, and cumulative emissions of CO_2 during two cropping years (2010–12).

Source	D.F	B.Y	G.Y	SOC	CO_2
Tillage methods	3	6593271.9***	122260.74**	32.12145866***	9605737865***
Planting years	1	13496603.4***	29504910.64***	97.90534734***	8211670776***
Mulch kinds	1	132977.5NS	20786.14NS	3.74523078***	93546720 NS
N Fertilizer levels	4	1470235.5***	4549170.27***	3.78932599***	21321766841***
Tillage methods X Mulch kinds	3	6780289.1***	1690508.29***	26.55366456***	1455896530***
Tillage methods X N Fertilizer levels	12	1443071.8 NS	302485.89NS	5.42559025***	1418530119***
Tillage methods X planting years	3	3443481.8NS	1276931.92*	12.85179871***	1455478300***
Planting years X Mulch kinds	1	66692.7NS	418751.64NS	0.67995212 NS	2547564904***
Planting years X N Fertilizer levels	4	10758689.1***	3983648.62***	3.45638215***	720140735***
Mulch kinds X N Fertilizer levels	4	975872.1NS	461889.16NS	3.23681715***	3518560905***
Tillage methods X Planting years X Mulch kinds	3	1162751.4NS	190004.99NS	12.61344035***	406313591***
Tillage methods X Planting years X N fertilizer levels	12	12108232.1NS	369168.67NS	4.50848299***	959919429***
Tillage methods X Mulch kinds X N fertilizer levels	12	1785389.2NS	420989.67NS	6.20244131***	2881830882***
Tillage methods X Mulch kinds X Planting years X N fertilizer levels	16	2029198.1 NS	782603.91***	4.05394235***	1326809061***

*Significant at 0.05 probability levels.
**Significant at 0.01 probability levels.
***Significant at 0.001 probability levels.
B.Y, Biological yields., G.Y, Grain yields., SOC, Soil organic carbon.
X, indicates interactions between different factors i.e. Tillage's X Mulches indicates interactions between different tillage methods and mulch kinds.

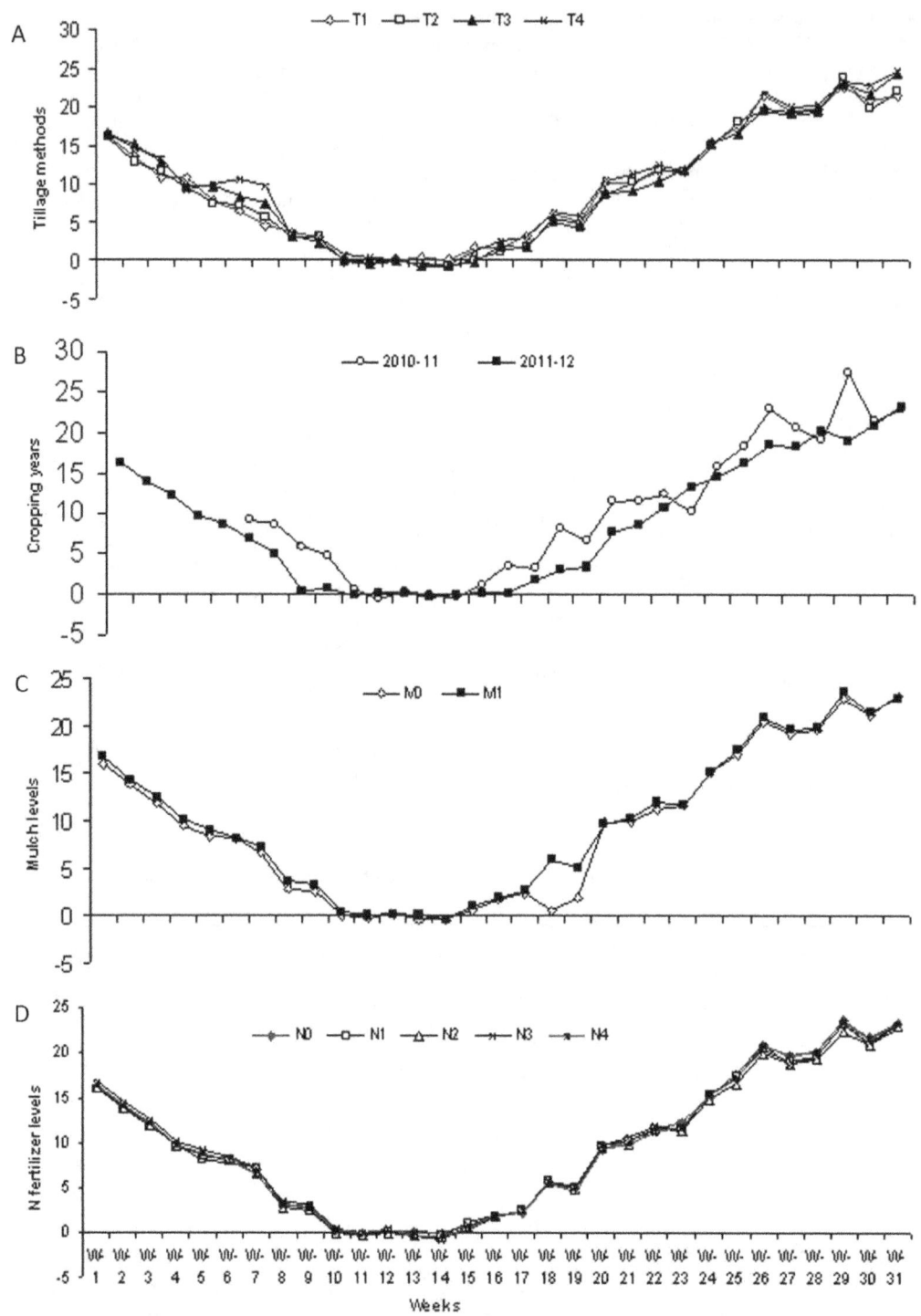

Figure 4. Changes in soil temperatures (0–5 cm depth) due to different tillage methods, corn residue mulch and N fertilizer levels during two cropping years (2010–12). (A). Changes in soil temperatures (0–5 cm depth) due to different tillage methods i.e. T_1, chisel plow tillage, T_2, zero tillage, T_3, rotary tillage and T_4, mold board plow tillage method., (B) Changes in soil temperatures (0–5 cm depth), as affected by different tillage methods, different mulch kinds and different N fertilizer levels during two cropping years (2010–12)., (C), Changes in soil temperatures (0–5 cm depth) during two cropping years (2010–12) due to different mulch kinds i.e. M_0, no mulch and M_1, corn residue mulch., (D). Changes in soil temperatures (0–5 cm depth) due to different N fertilizer levels during two cropping years (2010–12) including, N_0, 0 kg N/ha., N_1, 80 kg N/ha., N_2, 160 kg N/ha., N_3, 240 kg N/ha and N_4, 320 kg N/ha.

Soil organic carbon versus CO_2 emissions

Two years of mean data show that the tillage methods had significant effects on the SOC from 0–10 cm soil depths and that compared to all of the tillage methods, the highest SOC was recorded following chisel Plow tillage (Fig. 6). Higher SOC contents were recorded for all of treatments during cropping

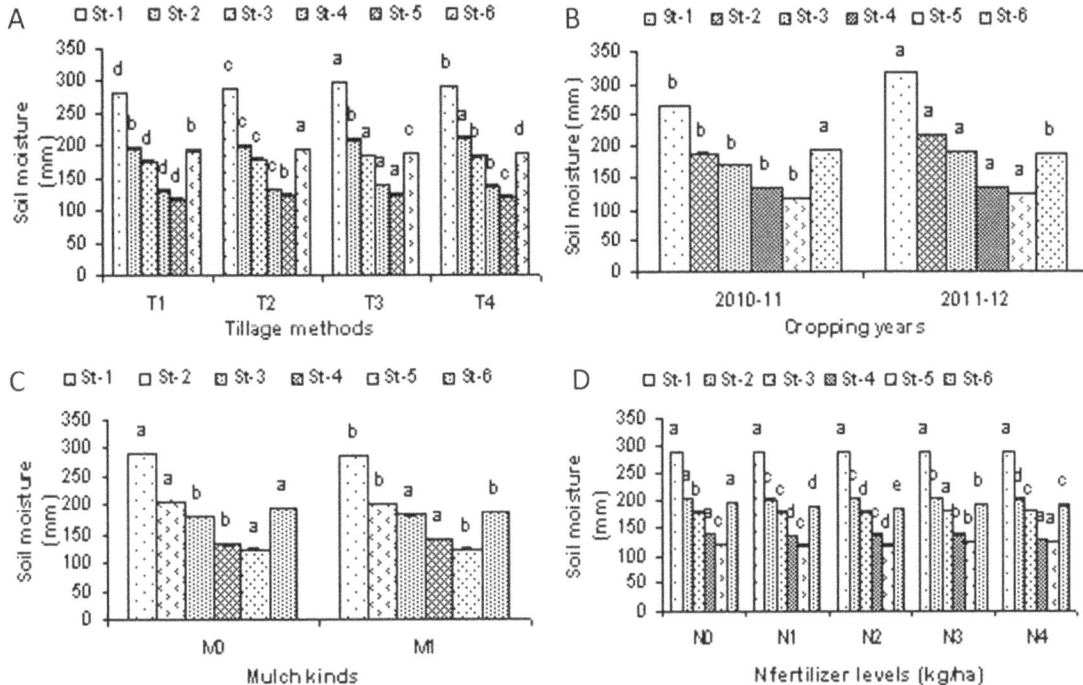

Figure 5. Total soil moisture contents (0–100 cm soil depth) as affected by different tillage methods, mulch kinds, and N fertilizer levels during two cropping years (2010–12). (A). Total soil moisture contents (0–100 cm soil depth) due to different tillage methods i.e. T_1, chisel plow tillage., T_2, zero tillage., T_3, rotary tillage and T_4, mold board plow tillage method., (B).Total soil moisture contents (0–100 cm soil depth) as affected by different tillage methods, different mulch kinds and different N fertilizer levels, during two cropping years (2010–12).(C), Total soil moisture contents (0–100 cm soil depth) as affected by different mulch kinds during the two cropping years (2010–12) i.e. M_0, no mulch and M_1, corn residue mulch., (D). Total soil moisture contents (0–100 cm soil depth) as affected by different N fertilizer levels, during two cropping years (2010–12) including, N_0, 0 kg N/ha., N_1, 80 kg N/ha, N_2, 160 kg N/ha., N_3, 240 kg N/ha and N_4, 320 kg N/ha., (*) Stage-1, (Crop sowing stage), Stage-2, (Crop revival stage), Stage-3, (Stem elongation stage), Stage-4, (Booting stage), Stage-5, (Grain formation stage) and Stage-6 (Crop harvesting stage).

year 2011–12 in comparison to cropping year 2010–11 (Fig. 6). The use of corn residue mulch increased the SOC compared to the non-mulched treatments (Fig. 6), and higher over all SOC contents were recorded in the cases of N_1 (80 kg N/ha) and N_2 (160 kg N/ha) in comparison to all of the other N fertilizer levels treatments (Fig. 6). The data show that CO_2 emissions were lower in those treatments that had higher SOC contents (Fig. 3; Fig. 6).

Effects of seasonal variations on CO_2 emissions

Seasonal temperature variations had significant effects on the CO_2 emissions. Because there were normal temperatures when the wheat crops were sown, higher CO_2 emissions were recorded, but reduced emissions were recorded with the decline in temperature during the winter seasons. With increased of crop growth and ascending of seasonal temperatures higher CO_2 emissions were recorded during both years (Figs. 2, 3, 4, 7, 8,).

Discussion

Soil CO_2 fluxes

Variations in seasonal temperatures had significant effects on soil temperatures, which ultimately affected the CO_2 emissions (Figs. 2, 3, 4, 7, 8). Higher CO_2 fluxes were recorded immediately after the tillage operations, which continued for a few days, and these emissions decreased with the passage of time (Fig. 2). Our results are in agreement with other findings [35]. Other investigators have reported changes in CO_2 emissions with seasonal variations. According to these investigators, seasonal variations in CO_2 emissions are controlled not only by the soil

temperatures and soil moistures but also by the tillage practices. Changes in CO_2 emissions from seasonal variations have been reported for almost all ecosystems. These emissions mainly depend both on the type of climate and the ecosystem [36].

Similar findings regarding variations in CO_2 fluxes with changes in soil temperatures and crop growth stages have been reported for rice crops [37]. In our study, the air and soil temperatures had significant effects on CO_2 emissions. Thus with the decrease in soil temperatures during the winter months, the CO_2 emissions decreased, and the CO_2 emissions again increased with the rise in soil temperatures during the summer months (Figs. 2, 3, 4, 7, 8).

Generally, the seasonal CO_2 emissions variations found in our experiment were similar to other findings [38]. Other investigators reported that these emission variations might be related to variations in autotrophic and in heterotrophic respiration because both are involved in soil CO_2 emissions. In addition a large amount of CO_2 is released from plant roots, during the continuation of the plant-energy system. Microbial and root respiration can also significantly contribute to CO_2 emissions.

Tillage method effects on CO_2 emissions

Higher CO_2 emissions were recorded immediately after the tillage operations, which continued for a few days (Fig. 2). Our results are in agreement with findings from other researchers, who reported that CO_2 emissions following tillage increased up to 2–15 times [39], [40], [41], [42] and [43]. According to these findings [39], instead of microbial activity, the basic reason for higher CO_2 emissions immediately after the tillage was actually the release of entrapped CO_2 from the soil pores as a result of physical

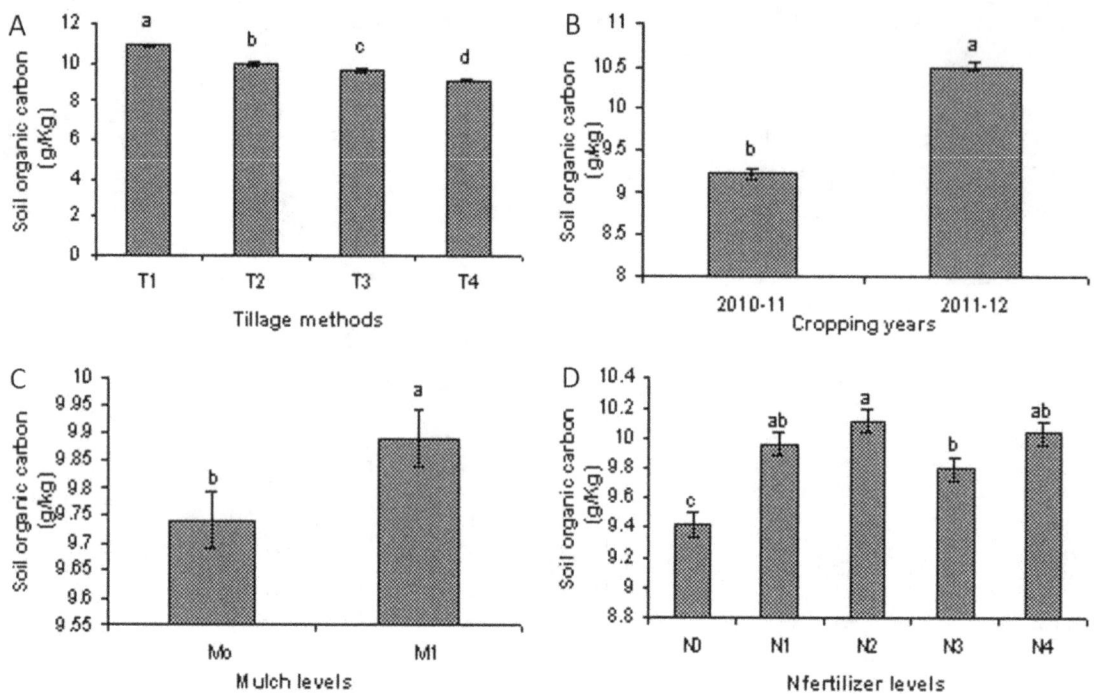

Figure 6. Soil organic carbon (SOC) contents (g/kg) as affected by different tillage methods, different mulch kinds and different N fertilizer levels during different cropping years (2010–12). (A), Soil organic carbon (SOC) contents as affected by different tillage methods i.e. T_1, chisel plow tillage., T_2, zero tillage., T_3, rotary tillage and T_4, mold board plow tillage method., (B), Mean soil organic carbon (SOC) contents as affected by different tillage methods, mulch kinds and different N fertilizer levels during two cropping years (2010–12)., (C), Soil organic carbon (SOC) contents as affected by different mulch kinds i.e. M_0, no corn residue mulch and M_1, corn residue mulch., (D), Soil organic carbon (SOC) contents as affected by different N fertilizer levels during two cropping years (2010–12) including, N_0, 0 kg N/ha., N_1, 80 kg N/ha., N_2, 160 kg N/ha., N_3, 240 kg N/ha and N_4, 320 kg N/ha.

operations. The other reasons for these higher emissions might be that (1) tillage operations break soil aggregates and expose their organic matter to microbial attack [44], [45]; (2) tillage operations encourage the mineralization of soil organic matter by incorporating crops residues into the soil [46]; and (3) tillage operations enhance soil aeration [43]. In our study, the tillage methods had significant effects on the CO_2 emissions and, overall, the rotary tillage and mold board plow tillage methods led to higher CO_2 emissions compared to the chisel plow and zero tillage planting methods (Fig. 2 and Fig. 3). Similar findings under different tillage systems have been previously reported [13]. These researchers reported significantly more CO_2 emission fluxes from the fields tilled by a mold board plow relative to the fields prepared by the chisel plow methods. According to these researchers, the basic reason for the greater emission fluxes from the mold board plow tillage method compared to the chisel plow tillage method was the depth and extent to which the soil was disturbed by using the different tillage implements. In our experiment, higher soil temperatures in the top 5 cm depth (Fig. 4) and generally higher moisture contents (Fig. 5) in the rotary tillage and mold board plow methods might be responsible for the higher emissions, in addition to the soil preparation depths. The tillage depths also resulted in a reduction of the SOC in the top 0–10 cm soil layers for the rotary tillage and mold board plow methods, compared to the chisel plow tillage and zero tillage planting methods (Fig. 6).

Corn residue mulch effects on CO_2 emissions

Weekly CO_2 emissions data (Figs. 2, 4) show that during cropping year 2010–11, the application of corn residue mulch caused an overall increase in CO_2 emissions compared to the non-

residue mulched treatments, but during the cropping year 2011–12, fewer CO_2 emissions were recorded in response to the application of corn residue mulch relative to the non-residue mulched treatments (Fig. 2 and Fig. 4). Although more CO_2 emissions fluxes were recorded in the corn residue mulched treatments in comparison to the non-residue mulched treatments, these might be due to the more microbial activities in the corn residue mulched treatments, which might have increased the SOC with the passage of time (Fig. 6). However, the two year mean data show that statistically non significant differences were recorded for CO_2 emissions following the use of the corn residue mulched or non mulched treatments. These findings might be explained by noting that both years CO_2 collecting chambers were fixed before the application of corn residue mulch. As a result there were fewer corn residues within the CO_2 collecting chambers from the corn residue mulched treatments. This smaller amount might be the reason why no CO_2 emission differences were recorded following the applications of different corn residue mulch treatments. The other reason might be that during the cropping year 2010–11, the application of corn residue mulch increased the CO_2 emissions but this mulch also increased the SOC contents of the corn residue mulched treatments, possibly resulting in the lower emission of CO_2 from the corn residue mulched treatments during the year 2011–12 (Fig. 3). A modeling study reported [47] that instead of applying higher rates of fertilizers, the use of crop residues or manure amendments would mitigate GHG emissions more efficiently. Similarly, it has been reported [48] that applications of straw increased the SOC sequestration in the soil which ultimately influenced the temporal patterns of CO_2 emissions from the soil.

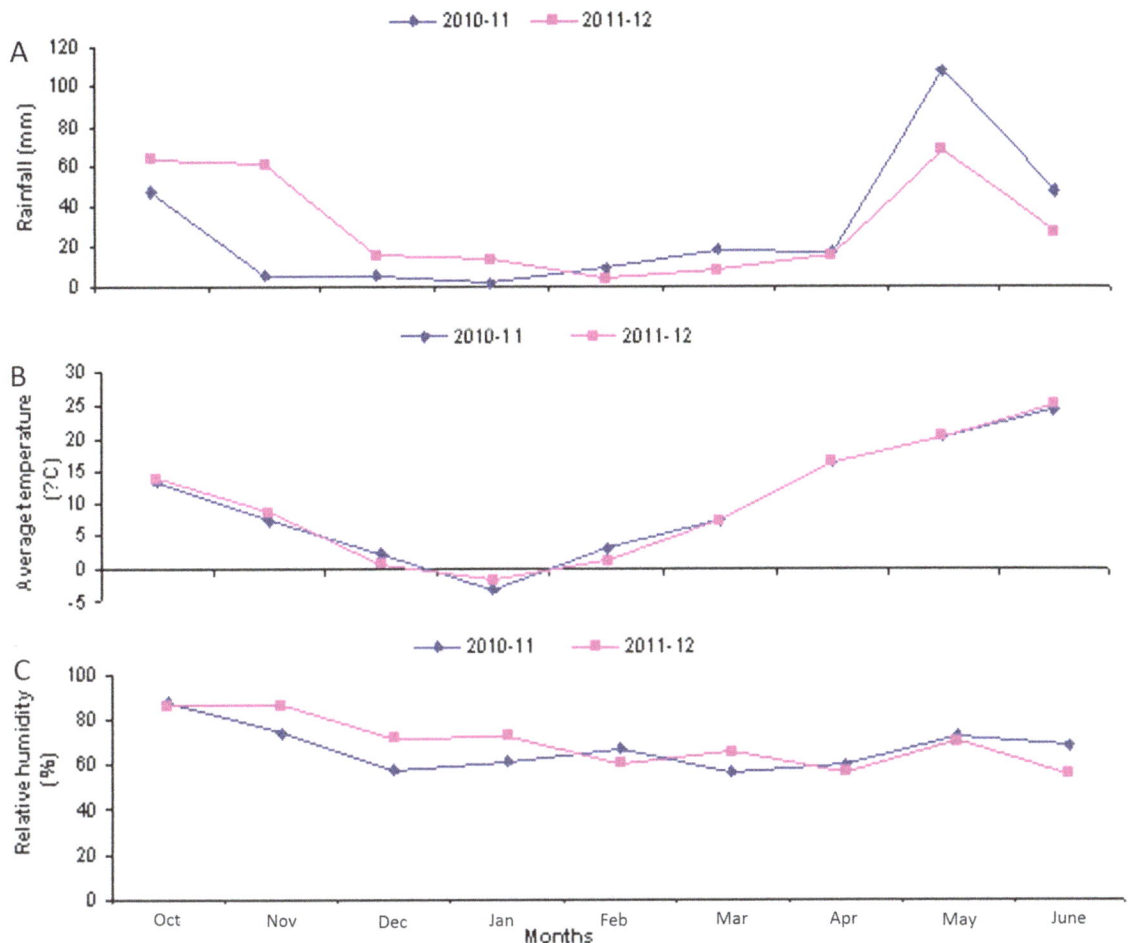

Figure 7. Monthly rainfalls, average temperatures and average relative humidity's of the study area during two wheat crop growing seasons (2010–12). (A).Rainfalls of the study area during two wheat crop growing seasons (2010–12), (B) Average temperatures of the study area during two wheat crop growing seasons (2010–12) (C) Average relative humidities of the study area during two wheat crop growing seasons (2010–12). (*). R.F-1, Rainfalls during cropping year 2010–11., R.F-2, rainfalls during cropping year 2011–12., Av.temp-1, Monthly average temperatures during the cropping year 2010–11., Av.temp-2, Monthly average temperatures during the cropping year 2011–12., R.H (%)-1, Average monthly relative humidities during the cropping year 2010–11., R.H (%)-2, Average monthly relative humidities during the cropping year 2011–12.

N fertilizer level effects on CO_2 emissions

Very few studies regarding CO_2 emissions in relation to the different tillage methods, corn residue mulch and N fertilizer levels have previously been reported in this region of China. It is expected that the application of inorganic N fertilizers along with organic materials will affect the mineralization of soil organic matter and crop productions, which will ultimately affect CO_2 emissions [8].

However variations in CO_2 emissions following fertilizer applications have been reported for different areas of China. Some scientists [15] have reported that the fertilizer applications suppresses CO_2 emissions and others [49] have reported that fertilizer application enhances CO_2 emissions. Moreover, some other scientists [19] have reported that fertilizer applications have no effects on CO_2 emissions.

Our study shows that the use of 80 kg and 160 kg N/ha, suppressed CO_2 emissions when compared with the 0 N fertilizer level, but that further increases in N fertilizer application rates enhanced CO_2 emissions. Higher emissions from the N_0, nitrogen fertilizer treatments might be explained by noting that plants under unfertilized N treatments are considered to respond to a relative shortage of N by increasing the plant's carbon allocation to

its structures and functions, which are responsible for N acquisition [50]. In our study, the use of different levels of N fertilizer relative to the no nitrogen fertilizer level significantly increased crop yields (Fig. 1).This result shows that the higher CO_2 emission fluxes in response to higher levels of N fertilizer might be explained by the increased use of C for microbial growth [51] and it might also be explained by the less efficient use of carbon by the microbial biomass, which resulted in a greater proportion of carbon loss in the form of CO_2 fluxes [52].

In our case, the use of corn residue mulch in combination with different levels of N fertilizer increased CO_2 emissions following the chisel plow, zero tillage and rotary tillage planting methods (Table 1). Our results regarding CO_2 emissions following the combination of N fertilizer and organic amendments are in agreement with the results of other scientists [53], [38] and [54]. These investigators reported a higher CO_2 emission flux from the treatments that utilized both fertilizers and organic manures. This finding shows that up to a certain extent the use of higher N fertilizer levels suppresses CO_2 emissions but in our case, higher levels of N fertilizer (i.e., 240 and 320 kg N/ha) enhanced CO_2 emissions. These results are contrary to the findings of other

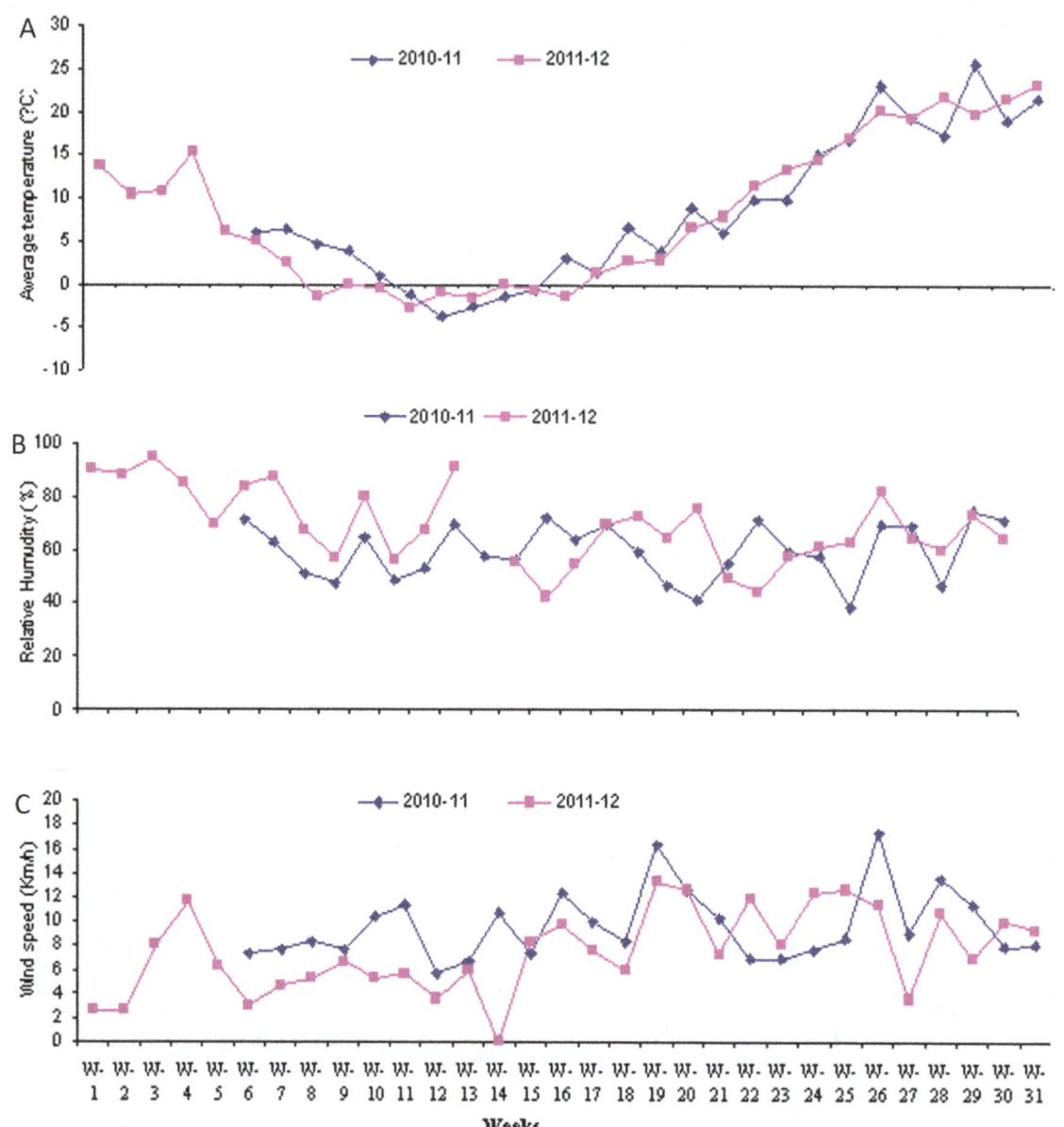

Figure 8. Average weekly temperatures, relative humidity's and wind speeds during the two wheat crop growing seasons (2010–12). (A) Average weekly temperatures of the study area during the two wheat crop growing seasons (2010–12), (B).Average weekly relative humidity's of the study area during the two wheat crop growing seasons (2010–12), (C) Average weekly wind speeds of the study area during the two wheat crop growing seasons (2010–12).

scientists [30] and [31]. These investigators reported no clear reasons for the CO_2 emission reductions.

Cropping year effects on CO_2 emissions

Two years of data on a weekly as well as on a cumulative basis show that there were more CO_2 emissions from all of the treatments during cropping year 2010–11 compared to the CO_2 emissions during cropping year 2011–12 (Figs. 2, 3). The main reasons for these lower emissions during cropping year 2011–12 were, on the whole, lower air temperatures, higher relative humidities and higher over all rainfall concentrations during the early periods of wheat crop growth relative to cropping year 2010–11 (Fig. 7). These factors all ultimately reduced the soil temperatures, which resulted in reduced CO_2 emissions during

the cropping year 2011–12 compared to cropping year 2010–11 (Figs. 4, 7, 8). Another reason for the lower emissions might be an increase in the SOC from the corn residue mulched treatments, which might have ultimately helped to reduce the CO_2 emissions during cropping year 2011–12 in comparison to cropping year 2010–11 (Figs. 2, 3, 6).

Effects of soil temperatures, soil moisture and soil organic carbon on CO_2 emissions

Respiration of ecosystem mainly depends on both the heterotrophic (microbe) and autotrophic (plant) activities and both of these factors are controlled by the prevailing environmental conditions (basically temperature and water availability), availability of carbohydrates and substrates and others [55], [56] and

Table 2. Q10 values of the different treatments during the different cropping years (2010–12).

Treatments	chisel plow		zero tillage		rotary tillage		mold board plow tillage	
	2010–11	2011–12	2010–11	2011–12	2010–11	2011–12	2010–11	2011–12
N_0	$Y=69.220e^{0.0516x}$	$Y=47.81\,e^{0.0898x}$	$Y=73.371e^{0.0736x}$	$Y=74.836e^{0.807x}$	$Y=140.49e^{0.0427x}$	$Y=67.628\,e^{0.0749x}$	$Y=287.91e^{0.0707x}$	$Y=78.941e^{0.1001x}$
	$R^2=0.7076$	$R^2=0.8064$	$R^2=0.0823$	$R^2=0.7213$	$R^2=0.7472$	$R^2=0.7372$	$R^2=0.6076$	$R^2=0.7439$
	Q10=1.68	Q10=2.43	Q10=2.09	Q10=1.90	Q10=2.54	Q10=2.55	Q10=2.72	Q10=2.72
N_1	$Y=79.76e^{0.0744x}$	$Y=73.979e^{0.0816x}$	$Y=57.341e^{0.0845x}$	$Y=62.738e^{0.0996x}$	$Y=142.39e^{0.0691x}$	$Y=93.744e^{0.092x}$	$Y=247.36e^{0.0385x}$	$Y=129.37e^{0.0818x}$
	$R^2=0.6672$	$R^2=0.7158$	$R^2=0.0869$	$R^2=0.752$	$R^2=0.8187$	$R^2=0.8074$	$R^2=0.5688$	$R^2=0.736$
	Q10=2.10	Q10=2.26	Q10=2.33	Q10=2.71	Q10=1.63	Q10=2.11	Q10=2.03	Q10=2.72
N_2	$Y=80.442e^{0.0827x}$	$Y=0.787e^{0.1012x}$	$Y=114.12e^{0.0509x}$	$Y=57.382e^{0.0891x}$	$Y=90.813e^{0.0616x}$	$Y=122.76e^{0.0669x}$	$Y=108.97e^{0.0634x}$	$Y=79.756e^{0.0873x}$
	$R^2=0.5944$	$R^2=0.8201$	$R^2=0.528$	$R^2=0.7701$	$R^2=0.790$	$R^2=0.6534$	$R^2=0.8134$	$R^2=0.7563$
	Q10=2.29	Q10=2.75	Q10=1.66	Q10=2.70	Q10=2.00	Q10=2.51	Q10=1.47	Q10=2.27
N_3	$Y=125.25e^{0.1085x}$	$Y=70.222e^{0.0931x}$	$Y=82.409e^{0.0478x}$	$Y=55.391e^{0.1006x}$	$Y=176.74e^{0.097x}$	$Y=149.95e^{0.0859x}$	$Y=103.92e^{0.0614x}$	$Y=69.581e^{0.0843x}$
	$R^2=0.7977$	$R^2=0.7054$	$R^2=0.7439$	$R^2=0.7937$	$R^2=0.6429$	$R^2=0.7876$	$R^2=0.852$	$R^2=0.6457$
	Q10=2.96	Q10=2.47	Q10=2.33	Q10=2.02	Q10=1.85	Q10=1.95	Q10=1.89	Q10=2.32
N_4	$Y=55.139e^{0.0653x}$	$Y=61.6e^{0.0904x}$	$Y=43.884e^{0.0903x}$	$Y=60.312e^{0.8051x}$	$Y=90.419e^{0.058x}$	$Y=221.26e^{0.0624x}$	$Y=116.85e^{0.0613x}$	$Y=96.28e^{0.1086x}$
	$R^2=0.6443$	$R^2=0.7961$	$R^2=0.7996$	$R^2=0.9015$	$R^2=0.6951$	$R^2=0.6654$	$R^2=0.7312$	$R^2=0.7969$
	Q10=1.02	Q10=2.18	Q10=1.61	Q10=2.16	Q10=2.64	Q10=2.32	Q10=1.85	Q10=2.96

N_0, 0 kg N/ha., N_1, 80 kg N/ha., N_2, 160 kg N/ha., N_3, 240 kg N/ha and N_4, 320 kg N/ha.

Table 3. Q10 values of the different treatments during the different cropping years (2010–12).

Treatments	chisel plow		zero tillage		rotary tillage		mold board plow tillage	
	2010-11	2011-12	2010-11	2011-12	2010-11	2011-12	2010-11	2011-12
N_0+M	$Y = 43.137e^{0.1316x}$	$Y = 87.49e^{0.0778x}$	$Y = 163.72e^{0.0363x}$	$Y = 61.295e^{0.09x}$	$Y = 62.946e^{0.1111x}$	$Y = 99.78e^{0.0961x}$	$Y = 183.66e^{0.0518x}$	$Y = 55.625e^{0.0947x}$
	$R^2 = 0.8849$	$R^2 = 0.6632$	$R^2 = 0.4067$	$R^2 = 0.8372$	$R^2 = 0.7113$	$R^2 = 0.8096$	$R^2 = 0.753$	$R^2 = 0.8216$
	Q10 = 3.72	Q10 = 2.69	Q10 = 2.47	Q10 = 2.10	Q10 = 1.79	Q10 = 2.96	Q10 = 1.85	Q10 = 2.58
N_1+M	$Y = 254.21e^{0.0484x}$	$Y = 0.0988e^{0.0516x}$	$Y = 84.882e^{0.0413x}$	$Y = 49.272e^{0.09965x}$	$Y = 175.25e^{0.0557x}$	$Y = 76.273e^{0.0858x}$	$Y = 112.84e^{0.0703x}$	$Y = 51.454e^{0.0957x}$
	$R^2 = 0.7689$	$R^2 = 0.8367$	$R^2 = 0.7109$	$R^2 = 0.7881$	$R^2 = 0.5183$	$R^2 = 0.8149$	$R^2 = 0.695$	$R^2 = 0.7376$
	Q10 = 1.62	Q10 = 2.52	Q10 = 1.44	Q10 = 2.22	Q10 = 3.04	Q10 = 2.58	Q10 = 1.68	Q10 = 2.60
N_2+M	$Y = 72.653^{0.0809x}$	$Y = 45.736e^{0.0926x}$	$Y = 80.339e^{0.0722x}$	$Y = 61.262e^{0.0907x}$	$Y = 163.02e^{0.0519x}$	$Y = 104.06e^{0.083x}$	$Y = 93.668e^{0.0666x}$	$Y = 46.012e^{0.1166x}$
	$R^2 = 0.8237$	$R^2 = 0.7309$	$R^2 = 0.8477$	$R^2 = 0.7992$	$R^2 = 0.4896$	$R^2 = 0.7173$	$R^2 = 0.6503$	$R^2 = 0.816$
	Q10 = 2.25	Q10 = 2.87	Q10 = 1.52	Q10 = 2.18	Q10 = 1.95	Q10 = 2.36	Q10 = 2.02	Q10 = 3.17
N_3+M	$Y = 1.015e^{0.0898x}$	$Y = 46.994e^{0.1054x}$	$Y = 164.85e^{0.0502x}$	$Y = 30.192e^{0.1197x}$	$Y = 163.9e^{0.405x}$	$Y = 40.179e^{0.09983x}$	$Y = 75.02e^{0.0742x}$	$Y = 72.535e^{0.1075x}$
	$R^2 = 0.8604$	$R^2 = 0.8214$	$R^2 = 0.4050$	$R^2 = 0.8784$	$R^2 = 0.6464$	$R^2 = 0.84$	$R^2 = 0.7363$	$R^2 = 0.8221$
	Q10 = 2.45	Q10 = 1.11	Q10 = 2.06	Q10 = 2.18	Q10 = 1.68	Q10 = 2.29	Q10 = 1.95	Q10 = 2.93
N_4+M	$Y = 8.69e^{0.10179x}$	$Y = 69.932e^{0.0859x}$	$Y = 51.822e^{0.0641x}$	$Y = 36.672e^{0.1073x}$	$Y = 287.97e^{0.0937x}$	$Y = 121e^{0.1050x}$	$Y = 127.88e^{0.0702x}$	$Y = 73.455e^{0.1177x}$
	$R^2 = 0.8342$	$R^2 = 0.7264$	$R^2 = 0.7247$	$R^2 = 0.8364$	$R^2 = 0.6465$	$R^2 = 0.854$	$R^2 = 0.7656$	$R^2 = 0.8821$
	Q10 = 2.76	Q10 = 2.36	Q10 = 1.65	Q10 = 2.74	Q10 = 1.50	Q10 = 2.67	Q10 = 2.10	Q10 = 3.24

N_0+M, 0 kg N/ha + corn residue mulch., N_1+M, 80 kg N/ha + corn residue mulch., N_2+M, 160 kg N/ha + corn residue mulch., N_3+M, 240 kg N/ha + corn residue mulch., N_4+M, 320 kg N/ha + corn residue mulch.

[57]. Many studies have shown that seasonal variations in CO_2 emissions were mainly caused by the soil temperature, soil moisture or the combination of both these factors [58], [36].

Our study also indicates that soil temperature was an important driving force for the increased CO_2 emissions, which is also supported by the Q10 values given in Table 2 and Table 3. It has also been reported that the CO_2 evolution rate significantly increases with an increase in temperature and moisture [37]. Our results are also in agreement with many other field studies, which have shown strong relationships between soil temperatures and CO_2 flux rates [59], [60], [17] and [18]. Additionally, our results agree with the findings of [61], who reported a stronger polynomial for temperature and moisture interaction ($r^2 = 0.89$) than for temperature alone ($r^2 = 0.47$). Many previous studies have reported that changes in crop management practices, i.e., the appropriate use of tillage operations, proper fertilization, crop residue applications and crop rotations can be helpful for managing soil organic matter, e.g., [62].

Our study also shows that the results at the end of cropping year 2010–11 revealed more SOC in the top 0–10 cm of soil than that at the end of cropping year 2011–12. This result shows that the adaptation of different management practices, i.e., the application of crop residues, increased the SOC contents, especially following chisel plow tillage and zero tillage planting compared with rotary tillage and mold board plow tillage. SOC sequestration is a long term processes and various results have been previously reported, i.e. [63] and [64], have reported that conservation tillage is a recommended management practice for agricultural ecosystems that can enhance the pool of soil organic carbon (SOC), in the soil. Following an analysis of global data, NT reportedly sequestered carbon at an average rate of 0.57 Mg C ha^{-1} compared with the mold board plow [65]. It has also been reported that increases in the SOC pools can be credited to either reductions in the CO_2 efflux from the soil or to increases in the C inputs [65]. When comparing the soil surface across different tillage systems, conservation tillage systems retain more crop residues, which ultimately result in the formation of more SOC [66] and [67]. In addition to this finding, the decomposition process of surface applied plants residues as a part of conservation tillage is slow compared to conventional tillage systems because of lower contact with the soil.

In our study a negative but highly significant correlation coefficient (r) value i.e. $-0.19403**$ was recorded between the SOC and cumulative CO_2 emission. This finding might help in the reduction of CO_2 emissions from the chisel plow tillage and zero tillage planting methods compared to the rotary tillage and mold board plow methods (Figs. 2 and 3). In our experiment, increases in the SOC contents from different tillage methods, especially chisel plow tillage and zero tillage, might be related to the lower disturbance of the soil, retention of more crop residues on the soil surface and reductions in the efflux of CO_2 from the soil. Similar results have been reported by [65], [66] and [67].

Conclusions

Intensive tillage and higher N fertilization are not only detrimental to the soil but are also destructive for the entire environment. Adaptations of appropriate tillage methods, crop residue applications and proper fertilization are beneficial for the soil as well as for the environment. These practices are also economically beneficial for resource-poor farmers. The findings from our study clearly indicate that the tillage methods significantly affected CO_2 emissions and the zero tillage planting method emitted the lowest CO_2 compared with the other three tillage

methods. No significant differences in CO_2 emissions were recorded for the applications of corn residue mulch, but the applications of corn mulch significantly improved the soil organic carbon (SOC) contents of the soil for all of the tillage systems. In addition, corn residue mulch application reduced the weed infestation by up to 40% (data not shown). Therefore, the application of corn residue mulch can be helpful for reducing the use of herbicides, which will also be helpful in establishing a healthy environment. Applications of different N fertilizer levels also significantly affected CO_2 emissions and, overall, the lowest emissions were recorded for the 160 kg N/ha treatment. Higher CO_2 emissions were recorded immediately after the tillage operations. This study also indicated that both soil temperatures and moistures strongly affected CO_2 emissions and that compared with the other tillage methods, zero tillage planting gave better grain yields. The lowest N fertilizer use gave equal yields, for the two year mean, as did the application of higher N fertilizer levels. These results clearly indicate that proper changes in farm management practices i.e., the adoption of zero tillage, crops residue application and optimum use of N fertilizers can reduce CO_2 emissions from soils. Therefore this type of long term study can be further helpful in reducing the emissions of CO_2 from soils, which will be helpful in reducing the use of inorganic fertilizers. These practices will be helpful in reducing production costs and will be beneficial for the entire environment.

Materials and Methods

Experimental site

A two-year (2010–12)), field study was conducted at the experimental area of Northwest A&F University, Shaanxi Province, northwestern China (latitude of 34°20′ N, longitude of 108° 04′E and elevation of 466.7 m above sea level) on the Eum-Orthrosols (Chinese soil Taxonomy) soil, with a mean bulk density of approximately 1.29 g cm^3. The soil in the top 40 cm had an SOC of approximately 14.26 g/kg, total nitrogen 0.74 g/kg and the pH was approximately 7.85. This area is under the corn-wheat rotation system. During both years, the wheat crop was planted after harvesting the corn crop. Both fertilizers, i.e., phosphorous in the form of calcium phosphate (Ca$_2$ (PO$_4$)$_3$ with 16% P and nitrogen in the form of urea with N≥46%, were applied to the corn crop at the rate of 750 kg/ha and 375 kg/ha, respectively. The total rainfall during the wheat crop growing season (October-June) was 231.6 mm and 242.7 mm during, cropping years 2010–11 and 2011–12 respectively (Fig. 7).

Experimental design and treatments

A factorial experiment having a strip split-split arrangement, with tillage methods in the main plots, mulch levels in the sub plots and N-fertilizer levels in the sub-sub plots with three replicates, was used for this study. Different tillage methods, i.e., chisel plow tillage (T$_1$), zero tillage (T$_2$), rotary tillage (T$_3$) and mold board plow tillage (T$_4$) methods, were kept in the main plots, different mulch kinds, i.e., M$_0$ (no residue mulch) and M$_1$ (corn residue mulch), in the sub plots, while different nitrogen fertilizer rates, i.e., 0, 80, 160, 240 and 320 kg N/ha, were kept in the sub-sub plots. The area was uniform in terms of fertility. The total experimental plot area (3300 m^2) was equally divided into four main tillage treatments. The area of each tillage treatment (33 m×25 m) was further sub-divided into sub–plots for mulch treatments, and finally the sub plots were further divided into sub-sub plots, and each sub-sub individual plot had an area of 3 m×25 m. Treatments were randomized within each sub-plot.

Chisel plow tillage (T_1), was performed using a chisel plow. Following fertilizer applications, a chisel plow with a shank spacing of approximately 40 cm apart and 30–35 cm deep was used once. Later on, fertilizers were mixed by using the rotavator for up to a 5 cm depth. For zero tillage (T_2) fertilizers were mixed by using the rotavator up to a 5 cm depth due to the lack of availability of a proper zero tillage drill. For rotary tillage (T_3), the seed bed was prepared by using the rotavator up to a depth of 15–20 cm, and in the case of the mold board. plow tillage method (T_4), the soil was plowed up to a 20–30 cm depth by using the mold board plow, followed by the rotavator for the final seed bed preparation.

Urea fertilizer with N≥46%, was used as the source of the nitrogen, and phosphorous (P) fertilizer in the form of calcium phosphate ($Ca_2 (PO_4)_3$ with 16% P was equally applied to all of the treatments at a rate of 750 kg/ha at the time of soil preparation. The treatments arrangements were kept the same during both years (2010–12) of study. Previously harvested air-dried corn crops residues were used as the source of corn residue mulch during both years. When the wheat crop was at the 3–4 leaf stage, mulch was applied at a rate of 750 g/m^2. The field was flat with a uniform topography. This area is rain fed, and a wheat-corn rotation is the main cropping system. No irrigation was applied to either crop. No changes were made to the areas with different tillage treatments, and a corn crop was planted after the wheat crop harvest by using the same tillage methods. The wheat crop was harvested using a combine harvester, and after harvesting the wheat crop, the corn crop was planted using a corn planter.

Winter wheat (C.V Shaan mai −139) was planted on October 17, 2010 and October 18, 2011 by using wheat drills. The line to line distance was maintained at approximately 16 cm apart. The seed had a 13% moisture contents and a 85% germination rate during both years. During cropping season 2010–11, a seed rate of about @ 190–200 kg/ha was used, while during cropping season 2011–12, a seed rate of approximately 205–210 kg/ha was employed. Experimental treatments were separated from each other by making boundaries between the treatments. Both years an herbicide application i.e. carfentrazone-ethyl ($C_{15}H_{14}C_{12}F_3N_3O_3$), was used to control the weeds. At physiological maturity, which occurred on June 8, 2011 and on June 10, 2012, three samples were randomly selected from each treatment and manually harvested by using the 1 m^2 quadrants to calculate the grain yields. Finally the wheat crop was harvested using a combine harvester.

Measurements

Meteorological factors. Meteorological data for the study area are given in Fig. 7 and Fig. 8, which indicate that during cropping year 2011–12, more rains were concentrated during the early period of crop growth relative to 2010–11, during which more rains were concentrated in the later crop growth stages. Higher average seasonal temperatures were recorded during the year 2010–11 than in 2011–12. However, more relative humidity was recorded during 2011–12 than in 2010–11. Weekly meteorological data showed higher, average temperatures during the year 2010–11 than during 2011–12. On the other hand, higher weekly relative humidities were recorded during the cropping year 2011–12 than during 2010–11. Similarly, higher weekly basis more wind speeds were recorded during the cropping year 2010–11 than during 2011–12.

Monitoring CO_2 emissions. Because of the high number of treatments, both years of CO_2 emissions data were recorded three times per week and one day of CO_2 emissions data was used as one replicate for statistical analysis. CO_2 emissions data were recorded using a GXH-3010E1 portable gas analyzer. This gas

analyzer is made by the Beijing Huayun Yiqi Company, and the CO_2 emission was recorded by using the method described by Gao et al [68]. During both years (2010–12) CO_2 emissions data from all of treatments was recorded 6840 times, which included 3120 times during the cropping year 2010–11 and 3720 times during cropping year 2011–12.

One round PVC chamber (21 cm in diameter and 13.5 cm in height), having total area of approximately 0.0047 m^3 was permanently fixed in the center of each treatment plot. The chamber was completely fixed in the soil up to a 4.5 cm depth. The plants growing with in the chambers were removed. As a consequence of some technical problems with the Gas analyzer, during wheat growing season 2010–11, the CO_2 emissions data were recorded after the wheat had been planted for one month, and during cropping season, 2011–12, CO_2 emissions were recorded starting during the first week of wheat planting. However, due to severe snow fall, data were not recorded on the 14th week during this year. Taking data from wheat planting until harvest for 2010–11, the CO_2 emissions data were recorded for 26 weeks, and during cropping year 2011–12, these data were recorded for 31 weeks. Every week, the data were collected 3 times depending upon the environmental condition, i.e., if the field was too wet from rainfall, then the data recording was stopped. Each time the data collection started at 900 a.m., and each sequence of CO_2 flux measurements took at least 4 hours. Due to the large number of treatments the data were randomly collected from different treatments. The main purpose of this randomization was to minimize the effects of different days as well as changing soil temperatures on the emissions of CO_2. The GXH-3010E1 gas analyzer was attached to the data collector chambers with intake and an outtake tubes. These tubes were made up of soft plastic pipes and each one was approximately one meter in length. At the time of data recording, first the CO_2 data, i.e., X_1 were recorded without closing/covering the chamber, and then the chamber top was tightly closed with a cover that had a fixed small fan. The gas within the chamber was mixed for three minutes with this fan. After this CO_2 emission, data (X_2) were recorded using the gas analyzer.

Chambers tops were closed for only three minutes at the time of data recording. To avoid any chemical change in the soil, these chambers were kept open for the whole remaining time. Along with CO_2 data, soil temperatures data were also recorded from each treatment from depths of 5 cm. For this purpose, thermometers were permanently fixed in each plot each year for the whole crop growth period. The CO_2 emission rate was calculated by using equation (1) as described by Gao et al. [68].

$$F = k(X_2\text{-}X_1)H/\Delta t \qquad (1)$$

Where F is the CO_2 emission in mg/(m^2.h); K is a constant with a value of 1.80 (25°C) and X_1 and X_2 are the CO_2 emissions rates from the chambers before and after covering of the chambers. H is the height of the chambers in meters and Δt is the time in hours (h). The cumulative emission of CO_2 was calculated using the following relationship, as described by Wilson H.M and Alkaisi W.W (2008) [69],

$$CO_2\text{-}C\left(kgha^{-1}\right)$$
$$= \sum_{i=first}^{n=last} X_i + X_{i+1} * N + X_{i+2} * N + \ldots + X_{i+n} * N \qquad (2)$$

where (i) is the first week of the growing season when the first CO2 emission rate was taken, (n) is the last week of the growing season when the last CO2 emission rate was taken, X is the CO2 rate (Kg ha−1 day−1), and N is the number of days between two consecutive CO2 emission rates measurements. Finally, these CO2 emission rates for the whole wheat crop growing period (taken between 9.00 A.M until approximately 1.00 P.M) were converted into tons/ha.

Soil moisture measurements. Both years soil moisture contents were measured during different crop growth stages, i.e., before planting the wheat crops, at the 5 leaf stage (Zadoks stage 1.0–1.9), stem elongation stage (Zadoks stage 3.0–3.9), booting stage (Zadoks stage 4.0–4.9), grain formation stage (Zadoks stage 7.0–7.9) and harvesting stage. For this purpose, soil samples were collected from each treatment from 0–100 cm soil depths with three replicates, with increments of 0–10 cm, 10–20 cm, 20–30 cm. 30–40 cm, 40–60 cm, 60–80 cm and 80–100 cm. Soil samples were collected in aluminum boxes using a hand auger, and fresh soil samples were immediately transported to the laboratory. After recording the fresh weights, these samples were dried in an oven at 105°C for at least 48 hours, and the soil water contents were then measured using equation (3) given below.

$$\text{Soil water content}(\%) = \frac{\text{fresh weight of soil} - \text{dry weight of soil}}{\text{dry weight of soil}} \times 100 \quad (3)$$

Soil water contents (mm) were calculated as gravimetric moisture contents using the equation (soil water ×B.D× thickness of soil layer). Soil bulk densities from the different tillage treatments and from the different soil depths were measured using the core method. Total soil water contents in the 0–100 cm soil depths of the different treatments were measured on the basis of the bulk densities of these different soil layers.

Soil organic carbon (SOC) measurements. Both year soil SOC samples were collected from the top 0–10 cm soil depth after the wheat harvest. Soil samples were collected from two randomly chosen points from each plot using a hand augur. The samples from each treatment were then mixed together to make a composite sample of each treatment. These samples were then air-dried at room temperature, crushed gently and passed through a 2 mm sieve for further chemical analysis. Soil organic carbon was determined by the oxidation method with $K_2 Cr_2 O_7$ -H_2SO_4. For chemical analysis, 0.5 grams of soil was digested with 5 mL of 1 M $K_2 Cr_2 O_7$ and 5 mL of concentrated H_2SO_4 and was heated at 175°C for 5 minutes followed by titration of the digests with $FeSO_4$ [70].

Statistics

Annual data collected for the CO_2 emission rates and for other related parameters over the whole 2-year period were subjected to an analysis of variance (ANOVA) by using the factorial experiment with the strip-split-split arrangement having the tillage methods in the main plots, mulch in the sub plots and N-fertilizer levels in the sub-sub plots. The SAS analytical software package GLM (8.01) was used for the analyses. Mean values and standard errors (SE) were calculated for each treatment, and an ANOVA was used to assess the treatment effects on the measured variables. Means were declared statistically significant at a 0.05 probability level, or $P \leq (0.05)$, using the DUNCAN test (DNMRT).

Author Contributions

Conceived and designed the experiments: SKT YL. Performed the experiments: SKT XLL. Analyzed the data: SKT JLZ. Wrote the paper: SKT XXW YL.

References

1. Post WM, Peng TH, Emanuel WR, King AW, Dale VH, et al (1990) The global carbon cycle. Am Sci 78: 310–326.
2. Houghton RA, Skole DL (1990) Carbon, In: Turner BL, Clark WC, Kates RW, Richards JF, Mathews JT, Meyer WB, editors. The earth as Transformed by Human Action. Cambridge University Press. 393–408.
3. Houghton RA (2007) Balancing the global carbon budget. Ann Rev Earth Planet Sci 35: 313–47.
4. Denman KL, Brasseur G, Chidthaisong A, Ciais P, Cox PM, et al. (2007) In Climate Change 2007: The Physical Science Basis. Contribution of Working Group 1 to the Fourth Assessment Report of the International Panel on Climate Change. Solomon, S., et al, editors. Cambridge Univ. Press. 499–587.
5. Valentini R, Matteucci G, Dolman AJ, Schulze ED, Rebmann C, et al. (2000) Respiration as the main determinant of carbon balance in European forests. Nature 404: 862–864.
6. IPCC (2007) Climate Change 2007: The Physical Science Basis. Cambridge University Press. 996.
7. Cole CV, Duxbury J, Freney J, Heinemeyer O, Minami K, et al. (1997) Global estimates of potential mitigation of greenhouse gas emissions by agriculture. Nut. Cycl. Agroecosyst. 49: 221–228.
8. Paustian K, Collins HP, Paul EA (1997) Management controls on soil carbon. In: Paul EA, Paustian K, Elliot ET, Cole CV, editors. Soil Organic Matter in Temperate Agro ecosystems (Long-term Experiments in North America). CRC Press, Boca Raton, FL. 15–49.
9. Lal R (1997) Long–term tillage and maize monoculture effects on a tropical alfisol in western Nigeria.1. Crop yield and soil physical properties. Soil and Tillage Res 42: 145–160.
10. Ahmadi H, Mollazade K (2009) Effect of plowing depth and soil moisture content on reduced secondary tillage. Agricultural Engineering International: The CIGR E Journal 11: 1–9.
11. Koga N, Tsuruta H, Tsuji H, Nakanoa H (2003) Fuel consumption – derived CO2 emission under conventional and reduced tillage cropping systems in northern Japan, Agriculture Ecosystems and Environment 99: 213–219.
12. Schlesinger WH (1986) Changes in soil carbon storage and associated properties with disturbance and recovery. In: Trabalka JR, Reichle DE, editors. The changing Carbon Cycle- A global analysis. Springer Verlag, New York. 194–220.
13. Reicosky DC, Lindstrom MJ (1993) Fall tillage method: Effect on short – term carbon dioxide flux from soil. Agron J 85(6): 1237–1243.
14. Franzluebbers AJ, Hons FM, Zubberer DA (1995) Tillage induced seasonal changes in soil physical properties affecting soil CO2 evolution under intensive cropping. Soil Tillage Res 34: 41–60.
15. Ding W, Cai Y, Cai Z, Zheng X (2006) Diel pattern of soil respiration in N-amended soil under maize cultivation. Atmos Environ 40: 3294–3305.
16. Ren X, Wang Q, Tong C, Wu J, Wang K, et al. (2007) Estimation of soil respiration in a paddy ecosystem in the subtropical region of China. Chin Sci Bull 52 (19): 2722–2730.
17. Iqbal J, Ronggui H, Lijun D, Lan L, Shan L, et al. (2008) Differences in soil CO2 flux between different land use types in mid-subtropical China. Soil Bio Biochem 40: 2324–2333.
18. Liu H, Zhao P, Lu P, Wang YS, Lin YB, et al. (2008) Greenhouse gas fluxes from soils of different land-use types in a hilly area of south China. Agric Ecosyst Environ 124(1–2): 125–135.
19. Lee DK, Doolittle JJ, Owens VN (2007) Soil carbon dioxide fluxes in established switch grass land managed for biomass production. Soil Biol Biochem 39: 178–186.
20. Ladd JN, Amato M, Zhou LK, Schultz JE (1994) Differential effects of rotation, plant residue and nitrogen fertilizer on microbial biomass and organic matter in an Australian alfisol. Soil Biol Biochem 26: 821–831.
21. Edmeades DC (2003) The long-term effects of manures and fertilizers on soil productivity: a review. Nutr Cycl Agroecosys 66: 165–180.
22. Kern JS, Johnson MG (1993) Conservation tillage impacts on natural soil and atmospheric carbon levels. Soil Sci SOC Am J 57: 200–210.
23. Reeves DW (1997) The role of soil organic matter in maintaining soil quality in continuous cropping system. Soil Tillage Res 43: 131–167.
24. Smith P, Powlson DS, Glendining MJ, Smith JU (1998) Preliminary estimates of potential for carbon mitigation in European soils through no-till farming. Global Chang Biol 4: 679–685.
25. Schlesinger WH (1999) Carbon sequestration in soils. Science 248: 2095.
26. Johnson MG, Levine ER, Kern JS (1995) Soil organic matter, distribution, genesis, and management to reduce greenhouse gas emissions. Water Air Soil Pollut 82: 593–615.
27. Lal R, Follett RF, Kimble J, Cole CV (1999) Managing US crop land to sequester carbon in soil. J. Soil Water Conserv 54: 374–381.

28. Paustian K, Andren O, Janzen H, Lal R, Smith P, et al. (1997) Agricultural soils as a C sink to offset CO_2 emissions. Soil Use Manage 13: 230–244.

29. Green CJ, Blackmer AM, Horton R (1995) Nitrogen effects on conservation of carbon during corn residue decomposition in soil. Soil Sci Soc Am J 59: 453–459.

30. Kowallenko CG, Ivarson KG, Cameron DR (1978) Effect of moisture content, temperature and nitrogen fertilization on carbon dioxide evolution from field soils. Soil Biol Biochem J 10: 417–423.

31. Fogg K (1988) The effect of added nitrogen on the rate of organic matter decomposition. Biol Rev 63: 433–472.

32. Halvorson AD, Wienhold BJ, Black AL (2002) Tillage, nitrogen, and cropping system effects on soil carbon sequestration. Soil Sci Soc Am J 63: 906–912.

33. Liang BC, Mackenzie AF (1992) Changes in the soil organic carbon and nitrogen after six years of corn production. Soil Sci 153: 307–313.

34. Wang XB, Cai DX, Hoogmoed WB, Oenema O, Perdok UD (2007) Developments in conservation tillage in rainfed regions of North China. Soil & Tillage Res 93: 239–250.

35. Zhang H, Wang X, Feng Z, Pang J, Lu F, et al. (2011) Soil temperature and moisture sensitivities of soil CO_2 efflux before and after tillage in a wheat field of Loess Plateau, China. J Environ Sci 23: 79–86.

36. Lou YZ, Zhou X (2006) Soil respiration and the environment. Academic, San Diego. 110–111.

37. Iqbal J, Hu R, Lin S, Hatano R, Feng M, et al. (2009) CO_2 emission in a subtropical red paddy soil (Ultisol) as affected by straw and N- fertilizer applications: A case study in Southern China. Agriculture Ecosyst and Environ 131: 292–302.

38. Zou J, Huang Y, Zheng X, Wang Y, Chen Y (2004) Static opaque chamber – based technique for determination of net exchange of CO_2 between terrestrial ecosystem and atmosphere. Chin Sci Bull 49 (4): 381–388.

39. Reiccosky DC, Dugas WA, Torbert HA (1997) Tillage- induced soil carbon dioxide loss from different cropping system. Soil Tillage Res 41: 105–118.

40. Calderon FJ, Jackson LE, Scow KM, Rolston DE (2001) Short term dynamics of nitrogen, microbial activity, and phospholipid fatty acids after tillage. Soil Sci Soc Am J 65: 118–126.

41. Reiccosky DC (2002) Long term effect of moldboard plowing on tillage-induced CO_2 loss. In: Kimble JM. Lal R, Follett RF editors. Agricultural practices and policies for carbon sequestration in soil. CRC Press, Boca, 87–96.

42. La Scala N Jr, Lopes A, Panosso AR, Camara FT, Periera GT (2005) Soil CO_2 efflux following rotary tillage of a tropical soil. Soil Tillage Res 84: 222–225.

43. Alvaro-Fuentes J, Cantero-Martinez C, Lopez MV, Arrue JL (2007) Soil carbon dioxide fluxes following tillage in semiarid Mediterranean agro-ecosystems. Soil Tillage Res 96: 331–341.

44. Beare MH, Cabrera ML, Hendrix PF, Coleman DC (1994) Aggregate-protected and unprotected organic matter pools in conventional and no-tillage soils. Soil Sci Soc Am J 58: 787–795.

45. Kladivko EJ (2001) Tillage systems and soil ecology. Soil Tillage Res 61: 61–76.

46. Zhang HX, Zhou X, Lu F, Pang J, Feng Z, et al (2011) Seasonal dynamics of soil CO_2 efflux in a conventional tilled wheat field of the Loess Plateau, China. Ecol Res. 26: 735–743. DOI 10. 1007/s11284-011-0832-5.

47. Li H, Qiu J, Wang L, Tang H, Li C, et al. (2010) Modelling impacts of alternative farming management practices on greenhouse emissions from a winter wheat – maize rotation system in China. Agriculture, Ecosyst and Environ 135: 24–33.

48. Jacinthe PA, Lal R, Kimble JM (2002) Carbon budget and seasonal carbon dioxide emission from a Central Ohio Luvisol as influenced by wheat residue amendment. Soil & Tillage Res 67: 147–157.

49. Xiao Y, Xie G, Lu G, Ding X, Lu Y (2005) The value of gas exchange as a service by rice paddies in suburban Shanghai, PR China. Agric Ecosyst Environ 109: 273–283.

50. Chapin FS III (1991) Effects of multiple environmental stress upon nutrient availability In: Mooney HA, Winner WA, Pell EJ, editors. Integrated Responses of Plants to Environmental Stress. Academic New York. 67–86.

51. Fisk MC, Fahey TJ (2001) Microbial biomass and nitrogen cycling responses to fertilization and litter removal in young northern hardwood forests. Biogeochemistry 53: 201–223.

52. Anderson TH (1994) Physical analysis of microbial communities in soil: applications and limitations. In: Ritz K, Dighton J, Giller KE, editors. Beyond the Biomass: Compositional and Functional Analysis of Soil Microbial Communities. John Wiley and Sons. New York. 67–76.

53. Galantini J, Rosell R (2006) Long-term fertilization effects on soil organic matter quality and dynamics under different production systems in semiarid Pampean soils. Soil Tillage Res 34: 41–60.

54. Ding W, Meng L, Yin Y, Cai Z, Zheng X (2007) CO_2 emission in an intensively cultivated loam as affected by long-term application of organic manure and nitrogen fertilizers. Soil Biol Biochem 39: 669–679.

55. Raich JW, Schlesinger WH (1992) Modeling soil organic matter in organic amended and N-fertilized, long-term plots. Soil Sci Soc Am J 56: 476–488.

56. Davidson EA, Belk E, Boone RD (1998) Soil water content and temperature as independent or confounded factors controlling soil respiration in a temperate mixed hardwood forest. Glob Chang Biol 4: 217–227.

57. Reichstein M, Tenhunen JD, Roupsard O, Ourcival JM, Rambal S, et al. (2002) Ecosystem respiration in two Mediterranean evergreen Holm oak forests: drought effects and decomposition dynamics. Funct Ecol 16: 27–39.

58. Buchmann N (2000) Biotic and abiotic factors controlling soil respiration rates in *Picea abies* stands. Soil Biol Biochem 32: 1625–1635.

59. Saiz G, Black K, Reidy B, Lopez S, Farrell EP (2007) Assessment of soil CO_2 efflux and its components using a process-based model in a young temperate forest site. Geoderma 139: 79–89.

60. Bauer J, Herbst M, Huisman JA, Weihermuller L, Vereecken H (2008) Sensitivity of simulated soil heterotrophic respiration to temperature and moisture reduction functions. Geoderma 145: 17–27.

61. Bowden RD, Newkirk KM, Rullo GM (1998) Carbon dioxide and methane fluxes by a forest soil under laboratory-controlled moisture and temperature conditions. Soil Biol. Biochem 30: 1591–1597.

62. Haynes RJ (2000) Labile organic matter as an indicator of organic matter quality in arable and pastoral soils in New Zealand. Soil Biol Biochem 32: 211–219.

63. Lal R, Kimble JM (1997) Conservation tillage for carbon sequestration. Nutr. Cycling Agroecosyst 49: 243–253.

64. Lal R, Kimble JM, Follet RF, Cole CV (1998) The potential for US crop land to Sequester Carbon and Mitigate the Greenhouse Effect. Chelsea: Ann Arbor Science.

65. West TO, Post WM (2002) Soil organic carbon sequestration rates by tillage and crop rotation: a global data analysis. Soil Sci. Soc. Am J 66: 1930–1946.

66. Drury CF, Tan CS, Welacky TW, Oloya TO, Hamill AS, et al. (1999) Red clover and tillage influence on soil temperature, water content and corn emergence. Agron J 91: 101–108.

67. Hutchinson JI, Campbell CA, Desjardins RL (2007) Some perspectives on carbon sequestration in agriculture. Agric Meteorol 142: 288–302.

68. Gao C, Sun X, Cao J, Luan Y, Hao H, et al. (2008) A method and apparatus of measurement of carbon dioxide flux from soil surface from situ [J]. Journal of Beijing Forestry University 30 (2): 102–105. (In Chinese with English abstract).

69. Wilson HM, Al-kaisi MM (2008) Crop rotation and nitrogen fertilization effect on soil CO_2 emissions in Central Lova. Applied Soil Ecology 39: 264–270.

70. Bao SD (2008) Soil Agricultural Chemistry Analysis [M]. Beijing: China Agricultural press (In Chinese).

3

Simulated Nitrogen Deposition Reduces CH$_4$ Uptake and Increases N$_2$O Emission from a Subtropical Plantation Forest Soil in Southern China

Yongsheng Wang[1,2], Shulan Cheng[2], Huajun Fang[1]*, Guirui Yu[1], Minjie Xu[2], Xusheng Dang[1,2], Linsen Li[1], Lei Wang[1,2]

1 Key Laboratory of Ecosystem Network Observation and Modeling, Institute of Geographical Sciences and Natural Resources Research, Chinese Academy of Sciences, Beijing, China, **2** University of Chinese Academy of Sciences, Beijing, China

Abstract

To date, few studies are conducted to quantify the effects of reduced ammonium (NH$_4^+$) and oxidized nitrate (NO$_3^-$) on soil CH$_4$ uptake and N$_2$O emission in the subtropical forests. In this study, NH$_4$Cl and NaNO$_3$ fertilizers were applied at three rates: 0, 40 and 120 kg N ha^{-1} yr^{-1}. Soil CH$_4$ and N$_2$O fluxes were determined twice a week using the static chamber technique and gas chromatography. Soil temperature and moisture were simultaneously measured. Soil dissolved N concentration in 0–20 cm depth was measured weekly to examine the regulation to soil CH$_4$ and N$_2$O fluxes. Our results showed that one year of N addition did not affect soil temperature, soil moisture, soil total dissolved N (TDN) and NH$_4^+$-N concentrations, but high levels of applied NH$_4$Cl and NaNO$_3$ fertilizers significantly increased soil NO$_3^-$-N concentration by 124% and 157%, respectively. Nitrogen addition tended to inhibit soil CH$_4$ uptake, but significantly promoted soil N$_2$O emission by 403% to 762%. Furthermore, NH$_4^+$-N fertilizer application had a stronger inhibition to soil CH$_4$ uptake and a stronger promotion to soil N$_2$O emission than NO$_3^-$-N application. Also, both soil CH$_4$ and N$_2$O fluxes were driven by soil temperature and moisture, but soil inorganic N availability was a key integrator of soil CH$_4$ uptake and N$_2$O emission. These results suggest that the subtropical plantation soil sensitively responses to atmospheric N deposition, and inorganic N rather than organic N is the regulator to soil CH$_4$ uptake and N$_2$O emission.

Editor: Dafeng Hui, Tennessee State University, United States of America

Funding: This project was supported by National Natural Science Foundation of China (No. 31290222, 31130009, 31290221, 31070435, and 41071166), Development Plan Program of State Key Basic Research (No. 2012CB41710, 32010CB833502, and 2010CB833501), Bingwei's Funds for Young Talents from Institute of Geographical Sciences and Natural Resources Research, CAS, (No. 2011RC202) and CAS Strategic Priority Program (No. XDA 05050600). The funders had no role in study design, data collection and analysis, decision to publish, or preparation of the manuscript.

Competing Interests: The authors have declared that no competing interests exist.

* E-mail: fanghj@igsnrr.ac.cn

Introduction

Humid tropical biome stores approximately 10% of global soil carbon (C) [1] and plays a vital role in the budget of ecosystem C and nitrogen (N) fluxes. The amount of nitrous oxide (N$_2$O) emission from the subtropical and tropical forest soils is estimated at 0.9–3.6 T g yr^{-1}, accounting for 14% to 23% of the global N$_2$O budget [2]. Simultaneously, well-aerated soils in the subtropical and tropical forests potentially function as a significant sink of atmospheric methane (CH$_4$) during the dry season [3–6]. The uptake of CH$_4$ from the subtropical and tropical forest soils is estimated to be 6.2 T g yr^{-1}, accounting for 28% of the global CH$_4$ sink [7]. Although the importance of the subtropical and tropical forest soils as atmospheric CH$_4$ sink and N$_2$O source is well understood, few observations can be available in this region [8–10]. Moreover, low-frequency measurement of gas fluxes in the few studies is unable to accurately estimate the annual amount of soil CH$_4$ uptake and N$_2$O emission, which leads to a high uncertainty in the budget of global soil CH$_4$ and N$_2$O fluxes.

Chronic N deposition input into terrestrial ecosystems alters plant physiology and soil microbial community, thereby changes the soil biogenic CH$_4$ and N$_2$O fluxes [11–13]. Based on a meta-analysis of N addition experimental data in globe, Liu and Greaver [14] concluded that N addition reduced CH$_4$ uptake by 38% and increased N$_2$O emission by 216%. In general, chronic N deposition will increase NH$_4^+$ and NO$_3^-$ availability in the forest ecosystems, thereby affects CH$_4$ uptake from the forest soils through changing the activity and composition of methanotrophic community [15–17]. Soil NH$_4^+$ accumulation can decrease, increase or have no effects on soil CH$_4$ uptake in the forest ecosystems, depending on forest types [5], duration of N application [18], and N fertilizer types and doses [19]. Three potential mechanisms have been proposed to clarify the inhibition of NH$_4^+$ accumulation to soil CH$_4$ uptake: (1) the competition of soil NH$_3$ to use CH$_4$ monooxygense with soil CH$_4$ [20], (2) the toxic inhibition of hydroxylamine and nitrite produced during soil NH$_4^+$ oxidation [21], and (3) the indirect effects of applied N and associated salt ions through osmotic stress [22]. On the contrary, elevated soil NH$_4^+$ availability can increase soil CH$_4$ uptake, which is related to the increase in the quantity of soil ammonia-oxidizing microorganisms [23]. Soil NO$_3^-$ accumulation can also decrease or increase soil CH$_4$ uptake [18,24]. Osmotic stress caused by NO$_3^-$-N fertilizer-associated salts [22] and anaerobic

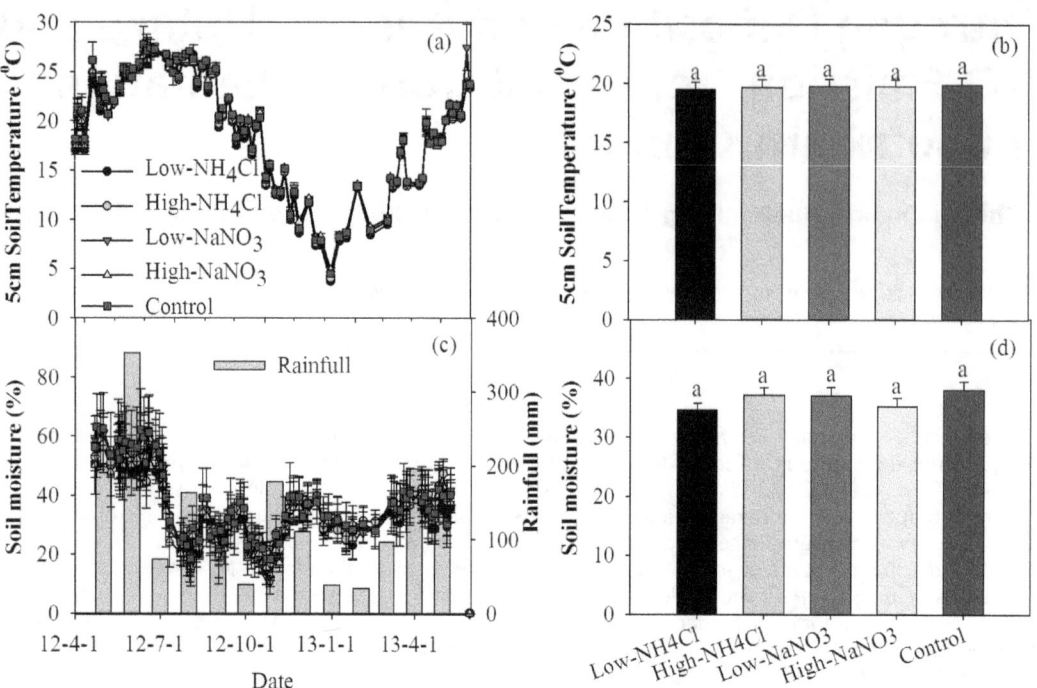

Figure 1. Temporal variations of 5 cm soil temperature and 10 cm soil moisture and their responses to N addition. Different letters above the columns mean significant differences between experimental treatments.

'microsites' produced by NO_3^- reduction [25] are toxic to CH_4-oxidizing bacteria. The mechanism responsible for the promotion of NO_3^- addition to soil CH_4 uptake is still unclear and need a number of experimental evidences to support [26]. A positive relationship between the amount of N addition and N_2O fluxes from the subtropical forest soils is mainly attributed to the promotion of soil nitrification or/and denitrification rates caused by increased N availability [27,28]. Some studies reported that denitrification was the main source of N_2O emission from the subtropical forest soils [29–31], whereas other studies claimed that nitrification dominated soil N_2O fluxes [32]. To date, single N fertilizer (i.e., NH_4NO_3) is widely used to simulate the effects of N deposition in all N manipulative experiments of subtropical forests in China [8,33,34]. The above studies have not evaluated the relative contributions of the deposited N ions (NH_4^+ and NO_3^-) to soil CH_4 uptake and N_2O emission. Moreover, most of soil CH_4 and N_2O fluxes are measured by low-frequency sampling over the short term, which is difficult to accurately assess the budget of soil CH_4 and N_2O fluxes and leads to great uncertainty.

In China, the plantations cover an area of 6.2×10^7 ha, accounting for 31.8% of China's forest area and ranking first in the world [35]. Approximately 63% of plantations concentrate in the subtropical region of southern China [36]. Meanwhile, the southern China is the most economically developed regions with high population density, and plantations, cities and farmlands are interspersed. Because a number of reactive N originated from fossil fuel combustion and fertilizer use is emitted to atmosphere, the forests in this region are receiving a high level of anthropogenic N deposition, mostly as ammonium [37]. Atmospheric N deposition rate via precipitation in southern China has been reported and ranges from 30 to 73 kg N ha^{-1} yr^{-1} [8]. So far, few studies are conducted to examine the effects of N deposition on CH_4 uptake and N_2O emission from the plantation of this region [33,34].

Humid subtropical forest soils are generally characterized by high N availability and high N turnover [38]. Therefore, we hypothesize that increased NH_4^+ and NO_3^- availability via experimental N deposition will inhibit soil CH_4 uptake and promote N_2O emissions from the subtropical plantation. Furthermore, NH_4^+-N fertilizer addition will decrease CH_4 uptake and increase N_2O emission due to soil NH_4^+-N accumulation. In contrast, the effects of NO_3^--N fertilizer addition on soil CH_4 uptake and N_2O emission depend on the concentration of soil NO_3^--N as well as associated salt ions. Our objectives were (1) to quantify the effects of NH_4^+-N and NO_3^--N fertilizer application on soil CH_4 and N_2O fluxes and soil variables in the subtropical plantation; (2) to examine the relationships between soil CH_4 and N_2O fluxes and the relevant soil properties.

Materials and Methods

Site description

This study was conducted in a subtropical slash pine plantation at the Qianyanzhou Ecological Station (QYZ, 26°44′39″N, 115°03′33″E) in southern China. The station belongs to the Institute of Geographic Sciences and Natural Resources Research, Chinese Academy of Sciences. All necessary permits were obtained for this field study. The field study did not involve endangered or protected species. According to local climate records from 1989 to 2008, mean temperature of QYZ site varies between 17 and 19°C. Mean annual precipitation ranges from 945 to 2145 mm, of which 24%, 41%, 23% and 12% occurs in four quarters in turn. The rainfall scarcity and high temperature in late summer often result in seasonal drought [39]. The exotic slash pine plantation was established in 1985. Mean tree height, diameter at breast height, stand basal area, and leaf area index were 12.0 m, 15.8 cm, 35 m^2 ha^{-1}, and 4.5 m^2 m^{-2}, respectively [40]. The main understory and midstory species are *Woodwardia japonica* (L.f.) Sm., *Dicranopteris dichotoma* (Thunb) Bernh, *Loropetalum*

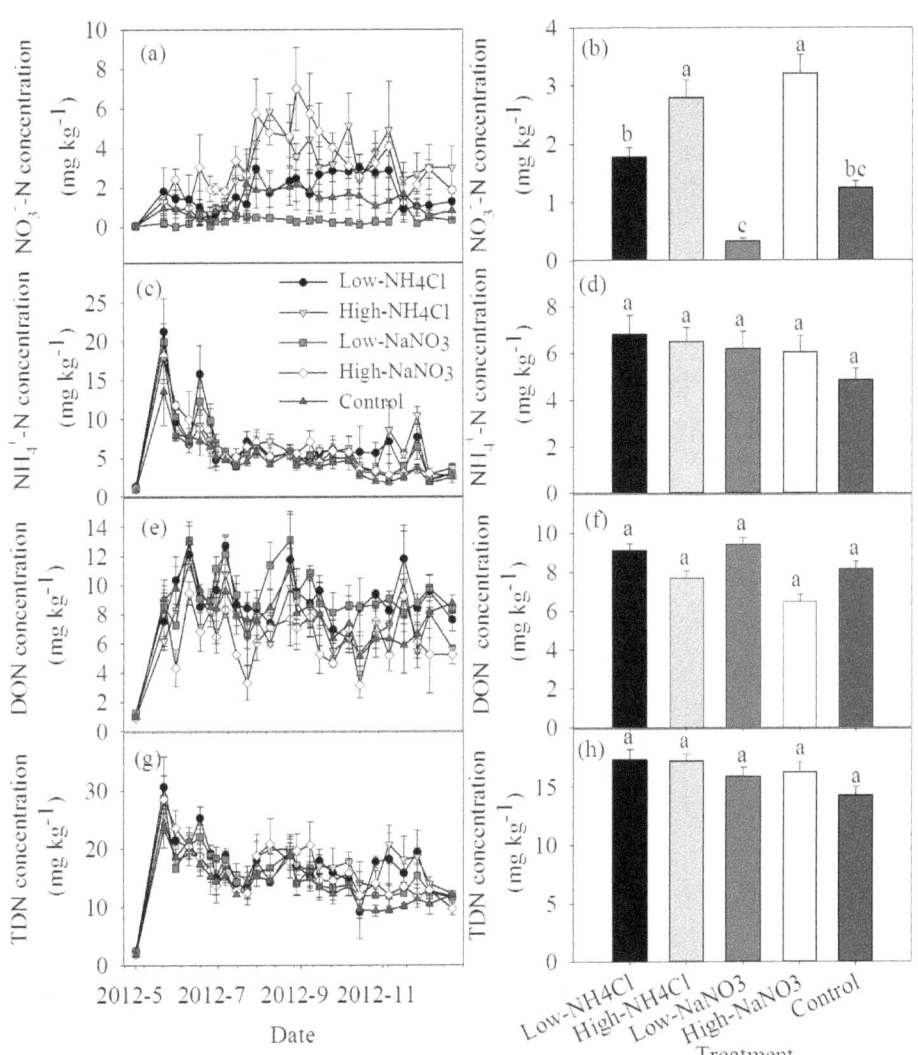

Figure 2. Temporal variations of soil NO$_3^-$-N, NH$_4^+$-N, DON, and TDN concentrations and their responses to N addition. Different letters above the columns mean significant differences between experimental treatments.

chinense (R.Br.) Oliv, and *Quercus fabric* Hance. The red soil is weathered from red sand rock, and soil texture is divided into 2.0-0.05 mm (17%), 0.05-0.002 mm (68%), and <0.002 mm (15%) [39].

Experimental design

The N addition experiment is a randomized block design. In May 1, 2012, two N fertilizers (NH$_4$Cl and NaNO$_3$) were used to simulate the effects of deposited NH$_4^+$ and NO$_3^-$ on ecosystem processes and functions. According to the level of atmospheric N deposition at the QYZ site, two levels referred to as low N (40 kg N ha^{-1} yr^{-1}) and high N (120 kg N ha^{-1} yr^{-1}) were used to simulate a future increase in the atmospheric N deposition by 1-, and 3-fold. A control treatment was designed at each block to calculate the net effect of N addition. Each N treatment was replicated three times, and a total of 15 plots were included. Each plot with 20 m×20 m was divided into four subplots with 5 m×5 m, and the plots were separated by 10 m wide buffer strips. Three subplots were used to measure soil CH$_4$ and N$_2$O fluxes, and the other one was used to investigate aboveground biomass and diversity. N fertilizer solutions were sprayed on the plots once a month in 12 equal applications over the entire year, and the control plots received equivalent deionized water only.

Measurement of soil CH$_4$ and N$_2$O fluxes

Flux measurements of soil CH$_4$ and N$_2$O fluxes were performed by using a static opaque chamber and gas chromatography method [41]. The static chambers were made of stainless steel and consisted of two parts: a square base frame (length×width×height = 50 cm×50 cm×10 cm) and a removable top (length×width×height = 50 cm×50 cm×15 cm). The installed equipments on the static chambers were detailed by Fang et al. [42]. The frames were inserted directly into the soil to a depth of 10 cm and remained intact during the entire observation period. To assess the spatial heterogeneity of soil C and N fluxes, a pre-experiment was conducted to examine the difference of CH$_4$ and N$_2$O fluxes among the three subplots of each plot before N addition. No significant difference of CH$_4$ and N$_2$O fluxes among the three subplots was found during the observation, suggesting a negligible effect of soil heterogeneity. Considering the practical reasons such as high labor intensity, we collected gas samples through changing the subplots within a month. The soil CH$_4$ and

Table 1. Results of repeated-measures ANOVAs on the effects of experimental treatment, month and their interaction on soil dissolved N concentrations.

Source of variation	Soil NO_3^--N		Soil NH_4^+-N		Soil DON		Soil TDN	
	F	p	F	p	F	p	F	p
Month	11.49	0.016	38.30	<0.001	16.71	<0.001	37.97	0.002
Treatment	4.40	0.026	1.62	0.24	1.81	0.20	1.51	0.27
Month×Treatment	1.31	0.24	1.41	0.13	1.33	0.17	0.75	0.77

N_2O fluxes were measured twice a week and conducted between 9:00 and 11:00 am (China Standard Time, CST). Five gas samples were taken using 100 ml plastic syringes at intervals of 0, 10, 20, 30, and 40 min after closing the chambers. CH_4 and N_2O concentrations of gas samples were analyzed within 24 h with a gas chromatography (Agilent 7890A, USA) equipped with an electron capture detector (ECD) for N_2O analysis and a flame ionization detector (FID) for CH_4 analysis. The high purity N_2 and H_2 were used as carrier gas and fuel gas, respectively. The ECD and FID were heated to 350°C and 200°C, respectively, and the column oven was kept at 55°C. The soil fluxes were calculated based on their rate of concentration change within the chamber, which was estimated as the slope of linear or nonlinear regression between concentration and time [41]. All the coefficients of determination (r^2) of the regression were greater than 0.90 in our study.

Measurements of soil temperature and moisture

Simultaneously, soil temperature at 5 cm (T_s) and soil moisture at 10 cm below soil surface (SM) were monitored at each chamber. Soil temperature was measured using a portable temperature probes (JM624 digital thermometer, Living–Jinming Ltd., China). Volumetric soil moisture ($m^3\ m^{-3}$) was measured using a moisture probe meter (TDR100, Spectrum, USA).

Soil sampling and mineral N analysis

During the measurement of soil CH_4 and N_2O fluxes, soil samples were collected weekly nearby the static chambers from a depth of 0–20 cm using an auger (2.5 cm in diameter). Five soils were collected and were pooled to one composite sample for each soil layer at each plot. Soils were immediately passed through a 2 mm sieve to remove roots, gravel and stones. Soil samples were extracted in 1.0 M KCl solution (soil: water = 1:10) and shaken for 1 h. The soil suspension was subsequently filtered through Whatman No. 40 filter papers for NH_4^+-N, NO_3^--N, and total dissolved nitrogen (TDN) determination on a continuous-flow autoanalyzer (Seal AA3, Germany). Dissolved organic nitrogen (DON) concentration was calculated as the difference between TDN and total inorganic nitrogen (NH_4^+-N and NO_3^--N).

Statistical analyses

Repeated measures analysis of variance (AVOVA) with Duncan test was applied to examine the differences of soil temperature, soil moisture, soil dissolved N, and soil CH_4 and N_2O fluxes between control and N addition plots. Experimental treatments were set as factors of between-subjects and measurement date was selected as a variable of within-subjects. Linear and nonlinear regression analyses were used to examine the relationships between soil CH_4 and N_2O fluxes and the measured soil variables in monthly scale. All statistical analyses were conducted using the SPSS software package (version 16.0), and statistical significant differences were set with P values<0.05 unless otherwise stated. All figures were drawn using the Sigmaplot software package (version 10.0).

Results

Soil temperature, moisture and precipitation

During the whole observation period, soil temperature at 5 cm depth fluctuated greatly, which correlated with the weather condition. Soil temperature varied as a single-peak and single-sink curve, i.e. temperature was the highest in early July, gradually reached the lowest value in early January, and then increased (Fig. 1a). There was no significant difference in surface temperature among various treatments (Fig. 1b).

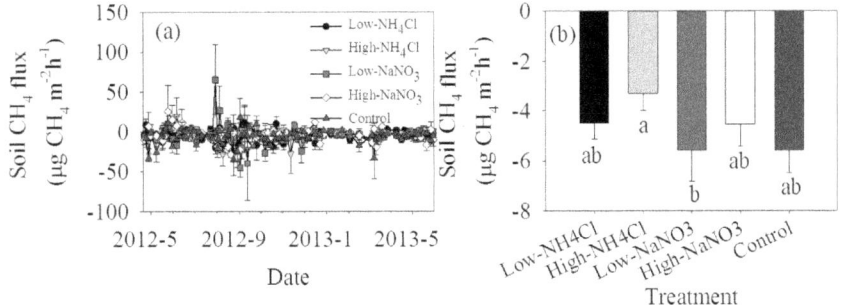

Figure 3. Temporal variations of soil CH$_4$ fluxes and their responses to N addition. Different letters below the columns mean significant differences between experimental treatments.

Soil moisture at 10 cm depth behaved as significant seasonal variation, with the maximum occurred in May and June and the minimum occurred in August and October (Fig. 1c). The seasonality of soil moisture was well consistent with that of precipitation (Fig. 1c). Similar to surface temperature, no significant difference in soil moisture was found among various treatments (Fig. 1d).

Soil dissolved N concentrations

Soil NO$_3^-$-N concentration showed significant seasonal variation, with the minimum and maximum occurring in May and August (Fig. 2a, Table 1, $P = 0.016$). In the control, the concentration of soil NO$_3^-$-N ranged from 0.06 to 2.19 mg kg^{-1}, with an average of 1.25 mg kg^{-1} (Fig. 2a–b). N addition tended to alter soil NO$_3^-$-N concentration, and the difference was significant among five experimental treatments (Table 1, $P = 0.026$). Compared with the control, high level of NaNO$_3$ addition tended to increase soil NO$_3^-$-N concentration, while an opposite pattern was found in the low level of NaNO$_3$ addition treatment (Fig. 2b). Furthermore, the promotion of high level of NH$_4$Cl addition to soil NO$_3^-$-N concentration seemed to be stronger than that of low level of NH$_4$Cl addition (Fig. 2b).

Soil NH$_4^+$-N concentration peaked in the middle of May, and then continued to decrease (Fig. 2c). The seasonal variation of soil NH$_4^+$-N concentration was significant (Table 1, $P<0.001$). In the control, soil NH$_4^+$-N concentration ranged from 1.91 to 10.80 mg kg^{-1}, with an average of 4.84 mg kg^{-1} (Fig. 2d). Overall, although N addition treatments tended to increase soil NH$_4^+$-N concentration, the difference between N addition treatments and control was not significant (Fig. 2d, Table 1, $P = 0.244$).

Soil DON concentration exhibited a significant seasonal variation (Fig. 2e, Table 1, $P<0.001$), and its seasonality was the same as that of soil NO$_3^-$-N concentration (Fig. 2a and Fig. 2e). In

the control, soil DON concentration ranged from 5.30 to 14.11 mg kg^{-1}, with an average of 8.18 mg kg^{-1} (Fig. 2e–f). Low level of N addition tended to increase the concentration of soil DON, while high level of N addition tended to reduce the concentration of soil DON (Fig. 2f). However, N addition did not change soil DON concentration at the level of 0.05 (Table 1, $P = 0.203$).

The seasonal variation of soil TDN concentration was consistent with that of soil NH$_4^+$-N concentration, dramatically decreasing from May to December (Fig. 2c and Fig. 2g). The seasonal variation of soil TDN concentration was significant (Table 1, $P = 0.002$). N addition tended to increase soil TDN concentration; moreover, the promotion of NH$_4$Cl application to soil TDN concentration was slightly higher than that of NaNO$_3$ addition (Fig. 2h). However, the difference of soil TDN concentration among the five experimental treatments was not significant (Table 1, $P = 0.273$).

Soil CH$_4$ and N$_2$O fluxes

Soil CH$_4$ fluxes showed a significant seasonal pattern (Table 2, $P = 0.008$). We observed both soil CH$_4$ uptake and emission in the control plots, ranging from -34.9 to 17.9 μg CH$_4$ m^{-2} h^{-1}, with an average of -5.56 μg CH$_4$ m^{-2} h^{-1} (Fig. 3a–b). A weak interaction between measurement date and treatment was found (Table 2, $P = 0.079$). Significant differences in CH$_4$ fluxes between the control and N addition treatments were only found in July and September (Fig. 3a). For the same level of N addition, NH$_4$Cl fertilizer exhibited a higher inhibition to soil CH$_4$ uptake than NaNO$_3$ fertilizer. However, there was no significant difference in soil CH$_4$ fluxes between the control and N addition treatments (Fig. 3b).

Soil N$_2$O fluxes also showed a significant seasonality with the minimum occurring from early October to March next year (Fig. 4 a, Table 2, $P<0.001$). In the control, Soil N$_2$O flux ranged from $-$

Table 2. Results of repeated-measures ANOVAs on effects of experimental treatment, month and their interaction on soil CH$_4$ and N$_2$O fluxes.

Source of variation	CH$_4$ flux		N$_2$O flux	
	F	p	F	p
Month	2.34	0.008	23.83	<0.001
Treatment	0.66	0.064	2.47	0.011
Month×Treatment	1.37	0.079	2.35	<0.001

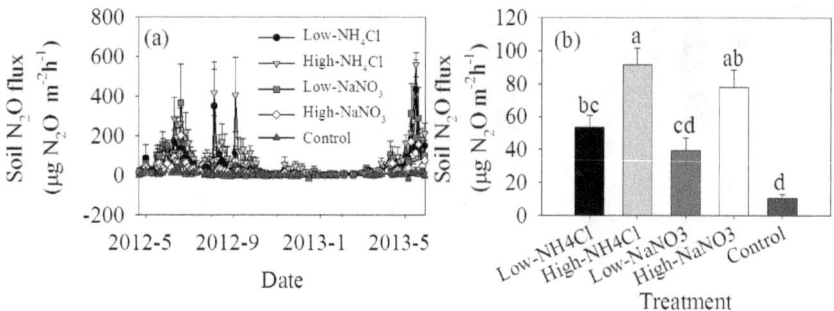

Figure 4. Temporal variations of soil N₂O fluxes and their responses to N addition. Different letters above the columns mean significant differences between experimental treatments.

15.26 to 559.30 µg N_2O m^{-2} h^{-1}, with an average of 10.60 µg N_2O m^{-2} h^{-1} (Fig. 4b). Nitrogen addition produced obvious peaks of soil N_2O emission, which was detected within one week after N addition (Fig. 4a). Soil N_2O fluxes positively responded to N addition, and the promotion increased with the levels of N addition (Fig. 4b). In addition, there was a significant interaction between month and N treatment in the entire observation period (Table 2, $P<0.001$). For the same level of N addition, NH_4Cl fertilizer had a higher promotion to soil N_2O emission than $NaNO_3$ fertilizer (Fig. 4b).

Relationships between soil CH₄ and N₂O fluxes and soil properties

Both soil CH_4 and N_2O fluxes were positively correlated with soil temperature at 5 cm depth and soil moisture at 10 cm depth (Fig. 5, Table 3). The relationships between soil CH_4 fluxes and soil temperature and between soil CH_4 fluxes and soil moisture could be well fitted with quadratic and linear equations, respectively (Fig. 5a–b, Table 3). Similarly, soil N_2O fluxes were linearly correlated with soil temperature and soil moisture (Fig. 5c–d, Table 3).

Soil CH_4 fluxes were positively correlated with soil NO_3^--N concentrations, whereas no significant correlations between soil CH_4 fluxes and other dissolved N species were found (Fig. 6a–d,

Table 3). Soil N_2O fluxes were linearly correlated with soil NO_3^--N and TDN concentrations (Fig. 6e, Fig. 6h, Table 3), and the relationship between soil N_2O fluxes and soil NH_4^+-N concentrations was well fitted with a logarithm equation (Fig. 6f, Table 3).

Discussion

Effects of N addition on soil CH₄ fluxes

The subtropical plantation soils can act as a sink of atmospheric CH_4. The mean annual soil CH_4 uptake in the control (0.49 kg CH_4 ha^{-1} yr^{-1}) was lower than those of subtropical plantation in Pingxiang (3.84 kg CH_4 ha^{-1} yr^{-1}) and Dinghushan of southern China (1.34 kg CH_4 ha^{-1} yr^{-1}) [5,6] as well as that of subtropical rainforest in Australia (3.13 kg CH_4 ha^{-1} yr^{-1}) [3]. Except low level of $NaNO_3$ treatment, the other three treatments decreased, on average, the rates of soil CH_4 uptake by 18.38% to 41.04% relative to control (Fig. 3). The decrease in soil CH_4 uptake caused by N addition in our site was higher than those of plantations in Dinghushan [5] and Heshan stations of southern China [8], despite the levels of N addition are similar in the three forest sites (120 vs. 150 kg ha^{-1} yr^{-1}). This result indicated that the response of soil CH_4 uptake to N addition was more sensitive in the northern subtropical plantations than in the southern subtropical

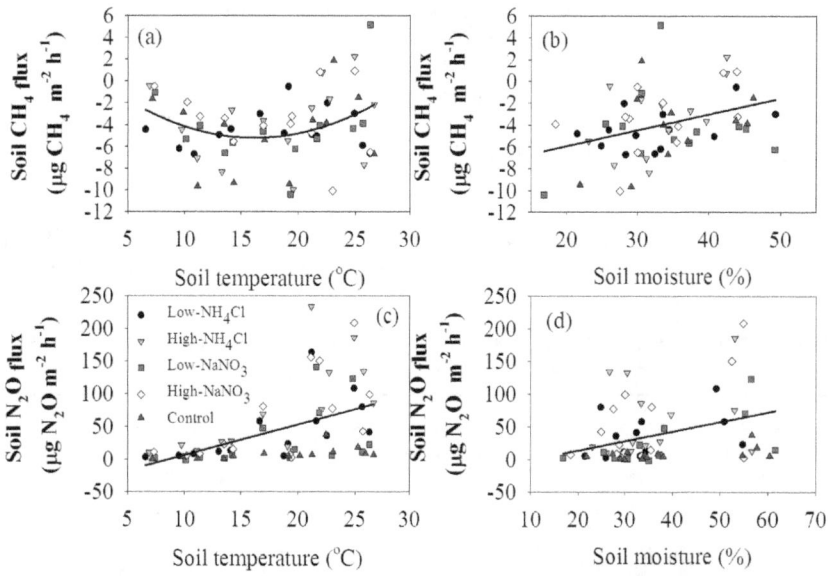

Figure 5. Relationships between soil CH₄ and N₂O fluxes, 5 cm soil temperature, and 10 cm soil moisture (n = 70).

Table 3. Regression models between soil CH_4 and N_2O fluxes and soil properties.

Flux	Soil variables[a]	Regression equation	R^2	P value
CH_4	T_s	$y = 0.028\ T_s^2 - 0.90\ T_s + 2.02$	0.094	0.044
	M_s	$y = 0.15\ M_s - 8.84$	0.123	0.006
	NO_3^-	$y = -1.28\ NO_3^- - 1.77$	0.21	0.004
	Combined	$y = 0.17\ M_s - 10.66$	0.21	0.003
N_2O	T_s	$y = 4.67\ T_s - 40.11$	0.25	<0.0001
	M_s	$y = 1.99\ M_s - 29.76$	0.10	0.001
	NO_3^-	$y = 12.29\ NO_3^- + 10.58$	0.22	0.005
	NH_4^+	$y = 81.52\ \ln(NH_4^+) - 85.34$	0.37	<0.0001
	TDN	$y = 9.26\ TDN - 93.50$	0.17	0.008
	Combined	$y = 0.01\ NH_4^+ + 0.013\ NO_3^- - 0.041$	0.50	<0.0001

[a]: T_s is soil temperature at 5 cm depth, M_s is soil moisture at 10 cm depth, NH_4^+, NO_3^-, and TDN are the concentrations of soil NH_4^+, NO_3^-, and TDN at 20 cm depth.

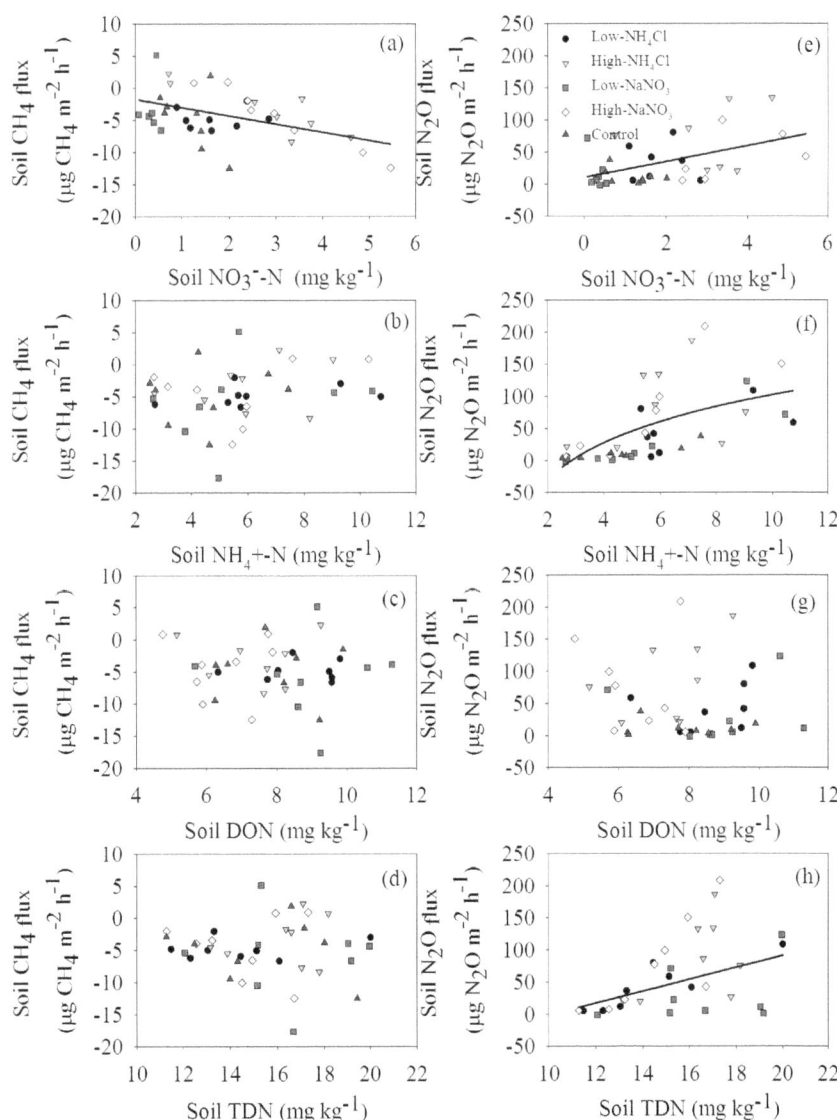

Figure 6. Relationships between soil CH_4 and N_2O fluxes and soil dissolved N concentrations (n = 40).

plantations. This could be attributed to the lower soil N availability, lower atmospheric deposition rate, and the shorter duration of N application in QYZ, compared with the southern subtropical plantations [5,8]. Furthermore, the subsurface mineral soils generally have higher capacity of oxidizing CH_4 than surface litter layer [22,24]. In our site, exogenous N input would directly affect the soil methanotrophic community as well as the amount of CH_4 oxidation due to the lacking of litter layer.

Generally, atmospheric N deposition increases NH_4^+ accumulation and thereby inhibits CH_4 uptake in the well-drained forest soils [8,12,43], despite contrasting results such as promotion and no effect have also been documented [44,45]. In this study, we found that various levels and forms of N addition did not significantly change soil CH_4 uptake over one year (Fig. 3). This could be related to the following third aspects. First, the short-term N fertilizers application did not significantly lead to soil NH_4^+-N accumulation (Fig. 2b), and no significant relationship between soil CH_4 fluxes and soil NH_4^+-N concentrations was found (Fig. 6b). Whalen and Reeburgh [46] also concluded that N inputs did not influence CH_4 uptake until they significantly increased soil NH_4^+ availability in the boreal forest soils. Although an inhibitory trend of soil CH_4 uptake under the NH_4^+-N addition treatments was found, the competition and toxic inhibition of accumulated NH_4^+ did not occur over the short term. Second, N addition enhances the availability of NH_4^+ to soil nitrifiers, which will accordingly decrease the extent to which CH_4 consumers are exposed to NH_4^+ [20]. A slight accumulation in soil NO_3^--N concentration under NH_4^+-N fertilizer treatments indirectly supported our deduction (Fig. 3a). Third, we also found that soil NO_3^--N accumulation could significantly promote soil CH_4 uptake (Fig. 6a), which had been documented in the subtropical plantations of southern China [6]. Especially, the low level of $NaNO_3$ treatment tented to reduce soil NO_3^--N concentration, and thereby it slightly stimulated soil CH_4 uptake (Fig. 2b, Fig. 3b). Moreover, stronger relationships were found between soil CH_4 fluxes and soil NO_3^--N concentrations than between soil CH_4 fluxes and other soil dissolved N concentrations (Fig. 6), suggesting that soil NO_3^- played a more important role in soil CH_4 uptake than other soil dissolved N species in the subtropical plantation.

Soil CH_4 flux is controlled by methanogens operating at anaerobic conditions and methanotrophs taking oxygen as a terminal electron acceptor [47]. Activities and population sizes of these microbial communities depend on a series of soil factors, including soil temperature, moisture, pH, substrate availability, and aeration of soil profile [19,48,49]. Soil CH_4 uptake is dominated by an optimal soil temperature [50]. In our study, the optimal soil temperature was about 15°C (Table 3), and the capacity of soil methanotrophs to oxidize CH_4 would decline when soil temperature was lower or higher than the threshold [51]. Also, soil moisture controls the mass flow of air and diffusion of atmospheric CH_4 into the soil by altering the water filled pore space (WFPS) of soils [52]. We also found that soil CH_4 fluxes under the N addition and control treatments were significantly related to soil moisture (Table 3). Based on the result of stepwise regression analysis, we found that the variation in soil CH_4 uptake was less affected by soil moisture (Table 3). Because N addition did not change soil moisture (Fig. 1), we reasonably deduced that the variation of CH_4 uptake elicited by N treatments was mainly attributed to the change in soil N availability.

Effects of N addition on soil N_2O fluxes

Our result showed that the subtropical slash pine plantation in QYZ exhibited a source of atmospheric N_2O under natural conditions. The average soil N_2O flux in the control (0.93 kg N_2O

ha^{-1} yr^{-1}) was comparable with that of Heshengqiao station in Hubei province (0.71 kg N_2O ha^{-1} yr^{-1}) [53], but lower than that of Dinghushan station in South China (2.11 kg N_2O ha^{-1} yr^{-1}) [33]. In our study, NH_4Cl and $NaNO_3$ addition at rates of 40 and 120 kg N ha^{-1} yr^{-1} increased soil N_2O emission by 403% to 762%. On the contrary, in the pine, mixed and evergreen broadleaved forests of Dinghushan station, NH_4NO_3 addition at rates of 50, 100 and 150 kg N ha^{-1} yr^{-1} only increased soil N_2O fluxes by 38% to 58% [33]. These results indicated that the subtropical plantation had high turnover rates of soil N and sensitively responded to increased N deposition. The potential reasons include that the optimal hydrothermal conditions [54], low soil pH [55], and high clay content [39], which favor both soil nitrification and denitrification as well as soil N_2O emission.

Except soil DON concentration, soil N_2O fluxes were significantly correlated with concentrations of soil NH_4^+, NO_3^-, and TDN (Fig. 6), suggesting soil N_2O flux was dominated by both soil nitrification and denitrification processes. Furthermore, the promotion of $NaNO_3$ addition to N_2O emission was slightly lower than that of NH_4Cl addition (Fig. 4). Two potential mechanisms can be responsible for this phenomenon: (1) the high rates of NO_3^- immobilization and nitrification [38], and the low denitrification potential are found in the same type of subtropical plantations [56]; and (2) temperature regulates soil N_2O flux through influencing soil N_2O-producing microorganisms, such as nitrifers and denitrfiers [57]. Soil moisture effects on soil N_2O fluxes are a result of the limitation of O_2 diffusion into the soil and the expansion of soil anaerobic microbial community [58]. The relatively high temperature in wet season was benefit for soil nitrifers and denitrfiers activities, which partly explained the seasonal variation of soil N_2O fluxes with maximum occurring in between May and June (Fig. 4a). Because N addition did not change soil temperature and soil moisture (Fig. 1), the changes in soil N_2O emission under N addition treatments were unlikely to be caused by the changes in soil temperature and soil moisture. Therefore, soil NH_4^+-N and NO_3^--N concentrations were the dominant factors controlling soil N_2O emission in our study, and could explain 49.9% of the temporal variability of soil N_2O fluxes (Table 3).

Conclusions

This study emphasizes the contrasting effects of oxidized NO_3^- and reduced NH_4^+ inputs on the fluxes of CH_4 uptake and N_2O emission from a subtropical plantation soil based on high frequency observations. We found that N addition tended to inhibit soil CH_4 uptake, and dramatically promoted soil N_2O emission. Compared with NO_3^--N fertilizer application, NH_4^+-N fertilizer application had a stronger inhibition to soil CH_4 uptake and a stronger promotion to soil N_2O emission. Also, both soil CH_4 and N_2O fluxes were driven by soil moisture and temperature, but soil inorganic N availability was a key integrator of soil CH_4 uptake and N_2O emission. Overall, short-term N addition has already changed soil CH_4 and N_2O fluxes, which indicated that the subtropical plantation soil was sensitive to N deposition input. In the future, the long-term observation of soil fluxes and the measurement of key microbial functional groups are necessary to clarify the mechanisms responsible for the coupling between soil CH4 and N2O fluxes.

Author Contributions

Conceived and designed the experiments: YW HF SC GY. Performed the experiments: YW MX XD LL LW. Analyzed the data: YW HF SC. Wrote the paper: YW HF SC.

References

1. Post WM, Emanuel WR, Zinke PJ, Stangenberger AG (1982) Soil carbon pools and world life zones. Nature 298: 156–159.
2. IPCC (2001) Climate change 2001: the scientific basis. Contribution of working group I of intergovernmental panel on climate change. Cambridge University Press, Cambridge.
3. Rowlings DW, Grace PR, Kiese R, Weier KL (2012) Environmental factors controlling temporal and spatial variability in the soil-atmosphere exchange of CO₂, CH₄ and N₂O from an Australian subtropical rainforest. Global Change Biology 18: 726–738.
4. Tang XL, Liu SG, Zhou GY, Zhang DQ, Zhou CY (2006) Soil-atmospheric exchange of CO₂, CH₄, and N₂O in three subtropical forest ecosystems in southern China. Global Change Biology 12: 546–560.
5. Zhang W, Mo JM, Zhou GY, Gundersen P, Fang YT, et al. (2008) Methane uptake responses to nitrogen deposition in three tropical forests in southern China. Journal of Geophysical Research-Atmospheres 113, D11116. doi:10.1029/2007JD009195.
6. Wang H, Liu SR, Wang JX, Shi ZM, Lu LH, et al. (2013) Effects of tree species mixture on soil organic carbon stocks and greenhouse gas fluxes in subtropical plantations in China. Forest Ecology and Management 300: 4–13.
7. Dutaur L, Verchot LV (2007) A global inventory of the soil CH₄ sink. Global Biogeochemical Cycles 21: GB4013. doi: 10.1029/2006GB002734.
8. Zhang W, Zhu XM, Liu L, Fu SL, Chen H, et al. (2012) Large difference of inhibitive effect of nitrogen deposition on soil methane oxidation between plantations with N-fixing tree species and non-N-fixing tree species. Journal of Geophysical Research-Biogeosciences 117. doi:10.1029/2012JG002094.
9. Martinson GO, Corre MD, Veldkamp E (2013) Responses of nitrous oxide fluxes and soil nitrogen cycling to nutrient additions in montane forests along an elevation gradient in southern Ecuador. Biogeochemistry 112: 625–636.
10. Steudler PA, Garcia-Montiel DC, Piccolo MC, Neill C, Melillo JM, et al. (2002) Trace gas responses of tropical forest and pasture soils to N and P fertilization. Global Biogeochemical Cycles 16.
11. Butterbach-Bahl K, Gasche R, Huber C, Kreutzer K, Papen H (1998) Impact of N-input by wet deposition on N-trace gas fluxes and CH₄-oxidation in spruce forest ecosystems of the temperate zone in Europe. Atmospheric Environment 32: 559–564.
12. Kim YS, Imori M, Watanabe M, Hatano R, Yi MJ, et al. (2012) Simulated nitrogen inputs influence methane and nitrous oxide fluxes from a young larch plantation in northern Japan. Atmospheric Environment 46: 36–44.
13. Jassal RS, Black TA, Roy R, Ethier G (2011) Effect of nitrogen fertilization on soil CH₄ and N₂O fluxes, and soil and bole respiration. Geoderma 162: 182–186.
14. Liu LL, Greaver TL (2009) A review of nitrogen enrichment effects on three biogenic GHGs: the CO₂ sink may be largely offset by stimulated N₂O and CH₄ emission. Ecology Letters 12: 1103–1117.
15. Castro MS, Peterjohn WT, Melillo JM, Steudler PA, Gholz HL, et al. (1994) Effects of Nitrogen-Fertilization on the Fluxes of N₂O, CH₄, and CO₂ from Soils in a Florida Slash Pine Plantation. Canadian Journal of Forest Research-Revue Canadienne De Recherche Forestiere 24: 9–13.
16. Jang I, Lee S, Zoh KD, Kang H (2011) Methane concentrations and methanotrophic community structure influence the response of soil methane oxidation to nitrogen content in a temperate forest. Soil Biology & Biochemistry 43: 620–627.
17. Mohanty SR, Bodelier PLE, Floris V, Conrad R (2006) Differential effects of nitrogenous fertilizers on methane-consuming microbes in rice field and forest soils. Applied and Environmental Microbiology 72: 1346–1354.
18. Aronson EL, Helliker BR (2010) Methane flux in non-wetland soils in response to nitrogen addition: a meta-analysis. Ecology 91: 3242–3251.
19. Reay DS, Nedwell DB (2004) Methane oxidation in temperate soils: effects of inorganic N. Soil Biology & Biochemistry 36: 2059–2065.
20. Chan ASK, Steudler PA (2006) Carbon monoxide uptake kinetics in unamended and long-term-amended temperate forest soils. Fems Microbiology Ecology 57: 343–354.
21. Nyerges G, Stein LY (2009) Ammonia cometabolism and product inhibition vary considerably among species of methanotrophic bacteria. Fems Microbiology Letters 297: 131–136.
22. Bodelier PLE, Laanbroek HJ (2004) Nitrogen as a regulatory factor of methane oxidation in soils and sediments. Fems Microbiology Ecology 47: 265–277.
23. King GM, Schnell S (1994) Effect of increasing atmospheric methane concentration on ammonium inhibition of soil methane consumption. Nature 370: 282–284.
24. Fang H, Cheng S, Yu G, Cooch J, Wang Y, et al. (2014) Low-level nitrogen deposition significantly inhibits methane uptake from an alpine meadow soil on the Qinghai–Tibetan Plateau. Geoderma 213: 444–452.
25. Xu XK, Inubushi K (2004) Effects of N sources and methane concentrations on methane uptake potential of a typical coniferous forest and its adjacent orchard soil. Biology and Fertility of Soils 40: 215–221.
26. Bodelier PLE (2011) Interactions between nitrogenous fertilizers and methane cycling in wetland and upland soils. Current Opinion in Environmental Sustainability 3: 379–388.
27. Matson P, Lohse KA, Hall SJ (2002) The globalization of nitrogen deposition: Consequences for terrestrial ecosystems. Ambio 31: 113–119.
28. Venterea RT, Groffman PM, Verchot LV, Magill AH, Aber JD, et al. (2003) Nitrogen oxide gas emissions from temperate forest soils receiving long-term nitrogen inputs. Global Change Biology 9: 346–357.
29. Chen GC, Tam NFY, Ye Y (2012) Spatial and seasonal variations of atmospheric N₂O and CO₂ fluxes from a subtropical mangrove swamp and their relationships with soil characteristics. Soil Biology & Biochemistry 48: 175–181.
30. Zhang JB, Cai ZC, Zhu TB (2011) N₂O production pathways in the subtropical acid forest soils in China. Environmental Research 111: 643–649.
31. Zhu J, Mulder J, Solheimslid SO, Dorsch P (2013) Functional traits of denitrification in a subtropical forest catchment in China with high atmogenic N deposition. Soil Biology & Biochemistry 57: 577–586.
32. Wang LF, Cai ZC (2008) Nitrous oxide production at different soil moisture contents in an arable soil in China. Soil Science and Plant Nutrition 54: 786–793.
33. Zhang W, Mo JM, Yu GR, Fang YT, Li DJ, et al. (2008) Emissions of nitrous oxide from three tropical forests in Southern China in response to simulated nitrogen deposition. Plant and Soil 306: 221–236.
34. Hu ZH, Zhang H, Cheng ST, Li T, Sheng SH (2011) Effects of simulated nitrogen deposition on N₂O and CH₄ fluxes of soils in forest belt. China Environmental Science 31: 892–897.
35. Department of Forest Resources Management S (2010) The 7th National forest inventory and status of forest resources. Forest Resource and Management 1: 3–10.
36. SFA (2007) China's Forestry 1999–2005. China Forestry Publishing House, Beijing.
37. Liu XJ, Duan L, Mo JM, Du EZ, Shen JL, et al. (2011) Nitrogen deposition and its ecological impact in China: An overview. Environmental Pollution 159: 2251–2264.
38. Zhang JB, Cai.Z.C,Zhu TB, Yang WY, Christoph M (2013) Mechanisms for retention of inorganic N in acid forest soils in southern China. Nature 3: 1–8.
39. Wen XF, Wang HM, Wang JL, Yu GR, Sun XM (2010) Ecosystem carbon exchanges of a subtropical evergreen coniferous plantation subjected to seasonal drought, 2003–2007. Biogeosciences 7: 357–369.
40. Wang YD, Wang ZL, Wang HM, Guo CC, Bao WK (2012) Rainfall pulse primarily drives litterfall respiration and its contribution to soil respiration in a young exotic pine plantation in subtropical China. Canadian Journal of Forest Research-Revue Canadienne De Recherche Forestiere 42: 657–666.
41. Wang YS, Wang YH (2003) Quick measurement of CH₄, CO₂ and N₂O emissions from a short-plant ecosystem. Advances in Atmospheric Sciences 20: 842–844.
42. Fang HJ, Cheng SL, Yu GR, Zheng JJ, Zhang PL, et al. (2012) Responses of CO₂ efflux from an alpine meadow soil on the Qinghai Tibetan Plateau to multi-form and low-level N addition. Plant and Soil 351: 177–190.
43. Le Mer J, Roger P (2001) Production, oxidation, emission and consumption of methane by soils: A review. European Journal of Soil Biology 37: 25–50.
44. Basiliko N, Khan A, Prescott CE, Roy R, Grayston SJ (2009) Soil greenhouse gas and nutrient dynamics in fertilized western Canadian plantation forests. Canadian Journal of Forest Research-Revue Canadienne De Recherche Forestiere 39: 1220–1235.
45. Borken W, Beese F, Brumme R, Lamersdorf N (2002) Long-term reduction in nitrogen and proton inputs did not affect atmospheric methane uptake and nitrous oxide emission from a German spruce forest soil. Soil Biology & Biochemistry 34: 1815–1819.
46. Whalen SC, Reeburgh WS (2000) Methane oxidation, production, and emission at contrasting sites in a boreal bog. Geomicrobiology Journal 17: 237–251.
47. Topp E, Pattey E (1997) Soils as sources and sinks for atmospheric methane. Canadian Journal of Soil Science 77: 167–178.
48. Werner C, Kiese R, Butterbach-Bahl K (2007) Soil-atmosphere exchange of N₂O, CH₄, and CO₂ controlling environmental factors for tropical rain forest sites in western Kenya. Journal of Geophysical Research-Atmospheres 112.
49. Merino A, Perez-Batallon P, Macias F (2004) Responses of soil organic matter and greenhouse gas fluxes to soil management and land use changes in a humid temperate region of southern Europe. Soil Biology & Biochemistry 36: 917–925.
50. Fang HJ, Yu GR, Cheng SL, Zhu TH, Wang YS, et al. (2010) Effects of multiple environmental factors on CO₂ emission and CH₄ uptake from old-growth forest soils. Biogeosciences 7: 395–407.
51. Steinkamp R, Butterbach-Bahl K, Papen H (2001) Methane oxidation by soils of an N limited and N fertilized spruce forest in the Black Forest, Germany. Soil Biology & Biochemistry 33: 145–153.
52. Lin XW, Wang SP, Ma XZ, Xu GP, Luo CY, et al. (2009) Fluxes of CO₂, CH₄, and N₂O in an alpine meadow affected by yak excreta on the Qinghai-Tibetan plateau during summer grazing periods. Soil Biology & Biochemistry 41: 718–725.
53. Lin S, Iqbal J, Hu RG, Ruan LL, Wu JS, et al. (2012) Differences in nitrous oxide fluxes from red soil under different land uses in mid-subtropical China. Agriculture Ecosystems & Environment 146: 168–178.
54. Xu YB, Xu ZH, Cai ZC, Reverchon F (2013) Review of denitrification in tropical and subtropical soils of terrestrial ecosystems. Journal of Soils and Sediments 13: 699–710.

55. Xu YB, Cai ZC (2007) Denitrification characteristics of subtropical soils in China affected by soil parent material and land use. European Journal of Soil Science 58: 1293–1303.

56. Zhang JB, Cai ZC, Cheng Y, Zhu TB (2009) Denitrification and total nitrogen gas production from forest soils of Eastern China. Soil Biology & Biochemistry 41: 2551–2557.

57. Bijoor NS, Czimczik CI, Pataki DE, Billings SA (2008) Effects of temperature and fertilization on nitrogen cycling and community composition of an urban lawn. Global Change Biology 14: 2119–2131.

58. Luo GJ, Kiese R, Wolf B, Butterbach-Bahl K (2013) Effects of soil temperature and moisture on methane uptake and nitrous oxide emissions across three different ecosystem types. Biogeosciences 10: 3205–3219.

4

Predicting Greenhouse Gas Emissions and Soil Carbon from Changing Pasture to an Energy Crop

Benjamin D. Duval[1,2¤a], Kristina J. Anderson-Teixeira[1¤b], Sarah C. Davis[1¤c], Cindy Keogh[3], Stephen P. Long[1,2,4], William J. Parton[3], Evan H. DeLucia[1,2,4]*

1 Energy Biosciences Institute, University of Illinois at Urbana-Champaign, Urbana, Illinois, United States of America, 2 Global Change Solutions, Urbana, Illinois, United States of America, 3 Natural Resource Ecology Laboratory, Fort Collins, Colorado, United States of America, 4 Department of Plant Biology, University of Illinois at Urbana-Champaign, Urbana, Illinois, United States of America

Abstract

Bioenergy related land use change would likely alter biogeochemical cycles and global greenhouse gas budgets. Energy cane (*Saccharum officinarum* L.) is a sugarcane variety and an emerging biofuel feedstock for cellulosic bio-ethanol production. It has potential for high yields and can be grown on marginal land, which minimizes competition with grain and vegetable production. The DayCent biogeochemical model was parameterized to infer potential yields of energy cane and how changing land from grazed pasture to energy cane would affect greenhouse gas (CO_2, CH_4 and N_2O) fluxes and soil C pools. The model was used to simulate energy cane production on two soil types in central Florida, nutrient poor Spodosols and organic Histosols. Energy cane was productive on both soil types (yielding 46–76 Mg dry mass·ha^{-1}). Yields were maintained through three annual cropping cycles on Histosols but declined with each harvest on Spodosols. Overall, converting pasture to energy cane created a sink for GHGs on Spodosols and reduced the size of the GHG source on Histosols. This change was driven on both soil types by eliminating CH_4 emissions from cattle and by the large increase in C uptake by greater biomass production in energy cane relative to pasture. However, the change from pasture to energy cane caused Histosols to lose 4493 g CO_2 eq·m^{-2} over 15 years of energy cane production. Cultivation of energy cane on former pasture on Spodosol soils in the southeast US has the potential for high biomass yield and the mitigation of GHG emissions.

Editor: Chenyu Du, University of Nottingham, United Kingdom

Funding: This research was funded by the Energy Bioscience Institute. The funders had no role in study design, data collection and analysis, decision to publish, or preparation of the manuscript.

Competing Interests: Please note that Duval, Long and DeLucia are affiliated with a company, Global Change Solutions LLC. This affiliation does not represent a "competing interest".

* E-mail: delucia@illinois.edu

¤a Current address: Dairy Forage Research Center, United States Department of Agriculture, Agricultural Research Service, Madison, Wisconsin, United States of America
¤b Current address: Conservation Ecology Center, Smithsonian Conservation Biology Institute, National Zoological Park, Front Royal, Virginia, United States of America
¤c Current address: Voinovich School for Leadership and Public Affairs, Ohio University, Athens, Ohio, United States of America

Introduction

Land use has a pervasive influence on atmospheric greenhouse gas (GHG) concentrations and thereby on climate [1,2,3]. Carbon emissions from land use change, often to make way for agriculture, have contributed substantially to anthropogenic increases in the atmospheric CO_2 concentration [2]. For example, C emissions from tropical deforestation have been estimated at 10.6 ± 1.8 Pg CO_2 per year between 1990 and 2007, equal to ~40% of global fossil fuel emissions [3]. Likewise, it is estimated that 40–52 Pg CO_2 have been released by plowing high-C native prairie soils [4]. Agricultural practices are important to global GHG budgets, with agroecosystems contributing ~14% of global anthropogenic GHG emissions [1]. Agricultural practices can also reduce GHG emissions and enhance soil carbon, and have the potential to mitigate climate change [4,5].

Land use and land management changes associated with the emerging bioenergy industry are likely to have substantial impacts on global GHG budgets [6,7,8]. A change from fossil fuels to an energy economy more reliant on plant-derived biofuels has the potential to reduce GHG emissions [9]. The prospect of lowering emissions is one factor leading to the United States' mandate to produce 136 billion liters of renewable fuel by 2022 [10]. However, meeting this mandate will require substantial land area [11,12], which implies potentially major changes to regional biogeochemical cycling [12,13].

Corn grain (*Zea mays*) is the dominant crop used for ethanol production in the US [14]. However, the ability of corn ethanol to reduce GHG emissions is questionable [15,16], and corn production exacerbates nitrogen pollution and other environmental problems [17–19]. Of particular concern is the possibility that diversion of corn for ethanol production will increase global grain prices and trigger agricultural expansion and deforestation elsewhere in the world [7]. The emerging commercial technology to convert ligno-cellulose to ethanol could redress the reliance on corn grain as an ethanol feedstock [20]. This could be particularly beneficial if cellulosic biofuel crops are grown on land that is not important for food production, while having lower GHG emissions than traditional row-crop agriculture [21,22]. Therefore, considerable research has focused on understanding the soil C and greenhouse gas consequences of replacing traditional agriculture used for bioenergy with perennial grasses like switchgrass (*Panicum virgatum* L.), Miscanthus (*Miscanthus x giganteus* J. M. Greef & Deuter ex Hodk. & Renvoize), or restored prairie cropping systems in the Midwestern United States [17,23–27].

The Southeastern United States holds particular potential for cultivation of second-generation biofuel crops [28,29]. In comparison with the corn-soy and wheat belts of the Midwestern US, this region's longer growing season, high precipitation and relatively lower land costs make it attractive for biofuel crop production. However, far less is known about the biogeochemical consequences of land-use change to biofuel crop production in this region.

Energy cane, a promising crop for ligno-cellulosic fuel production, is a variety of sugarcane (*Saccharum officinarum*) that is higher yielding, more cold tolerant and has lower sucrose content than commercially produced sugarcane [28]. Because of its lower sugar concentration, it has not been widely cultivated, but has been of interest commercially as a genetic stock for improving cold tolerance in higher sucrose sugarcane strains [28]. With the development of ligno-cellulosic ethanol conversion technologies, sucrose concentration is less important for ethanol production, and energy cane could become an important biofuel feedstock as yields are high, ranging from 25–74 Mg·ha^{-1}·yr^{-1} dry mass (Table 1).

Florida is the largest sugarcane producing state in the US and is therefore a likely location for large-scale energy cane production [30]. Currently, 466,000 hectares of land in Florida are used for low-intensity grazing, and converting some portion of this land could provide an option for growing energy cane [31,32]. However, it is unknown if converting pasture to cultivated land will affect GHG exchange with the atmosphere and soil carbon storage. More frequent soil disturbance and the presence of larger quantities of litter from growing energy cane could increase CO_2 efflux to the atmosphere [33,34], while removing cattle from the landscape will displace methane (CH_4) efflux [35]. If fields are fertilized, nitrous oxide (N_2O) emissions may increase because of greater substrate availability for denitrifying microbes [36], and

indeed, high rates of N_2O efflux have been measured from sugarcane grown on highly fertilized soils in Australia [37]. However, considering the entire suite of greenhouse gasses, there may be an overall reduction in GHG flux due to the offset provided by greater atmospheric carbon uptake into the crop.

The region of Florida where energy cane is likely to be grown has two distinct soil types. The most common soils are Spodosols, which are low nutrient and low organic matter sands requiring significant fertilizer to maintain agricultural productivity [38]. Substantial sugarcane production in Florida also occurs on Histosols, which are high organic matter "mucks" that are not typically fertilized, as production on these soils can be maintained by N mineralization from organic matter [39]. The cultivation of Histosols began by draining swamplands, where organic matter had accumulated under anaerobic conditions. Drainage accelerates decomposition and further cultivation of these organic soils is associated with rapid oxidation of organic matter, resulting in significant soil C loss and emissions of CO_2 and N_2O to the atmosphere [9,40,41].

Theoretical [13,25] and empirical research [42,43] indicate that the conversion of land in the rain-fed Midwest currently used to produce corn for ethanol to perennial biofuel feedstocks such as switchgrass or Miscanthus (a close relative of sugarcane) would greatly reduce or reverse the emission of GHG to the atmosphere and rebuild depleted carbon stocks in the soil. Prior studies with Miscanthus in Europe have measured substantial decreases in nitrogen use, and large increases in soil biomass and organic matter relative to other agricultural land uses [44,45]. There have been no experimental studies that address how changing a landscape to cultivate energy cane will impact GHG emissions and soil C stocks. This is addressed here by using the process-based biogeochemical model DayCent to run *in silico* experiments to ask how land use change from pasture to energy cane production changes ecosystem GHG flux and soil C storage. We test the hypotheses that converting pastures to energy cane will lead to reductions in GHG flux to the atmosphere and increase soil C stocks, and that soil type is an important modulator of that change.

Methods

Plant and Soil Analyses

To parameterize the DayCent model, plants and soils were collected on private land in Highlands County, Florida (27° 21′ 49″ N, 81° 14′ 56″ W) in May 2011. Paired 4-m^2 plots ($n = 3$) were randomly located in energy cane fields that had been recently (<2 months) converted from pasture and in adjacent non-cultivated pasture on both Spodosols (hyperthermic Arenic Alaquods) and Histosols (hyperthermic Histic Glossaqualfs). We harvested all aboveground biomass from each plot. Soil samples were taken from the pastures in areas not yet under energy cane cultivation. Three soil cores to a depth of 1 m were extracted from each plot with a 1.75-cm diameter wet sampling tube (JMC product # PN010, Newton, IA). Soil cores were separated by depth (0–30 cm, >30 cm). Plant material and soils were oven dried at 65°C (plant material) and 105°C (soils) until they reached constant mass. Dried soils were coarse ground with a mill (model F-4, Quaker City, Phoenixville, PA), and then fine ground with a coffee grinder (Sunbeam Products Inc., Boca Raton, FL). Total C and N content may have been slightly underestimated from the dried Histosols due to volatilization, but the values we measured (Table 1) fall well within the range reported by NRCS Web Soil Survey [46]. Plant material was ground to pass a 425-μm mesh (Wiley mill, Thomas Scientific, Swedesboro, NJ, USA). Plant and soil subsamples

Table 1. Input parameters (mean and one standard error of the mean; SEM) for carbon and nitrogen concentration of energy cane and soils collected from the Highlands Ethanol farm, Highlands County, Florida.

		%C		%N	
		Mean	SEM	Mean	SEM
Energy cane	Live leaves	43.68	0.22	1.80	0.18
	Dead leaves	39.77	0.22	0.52	0.03
	Stalks	41.18	0.33	0.87	0.10
Soils	Soil Depth				
Histosols	0–60 cm	7.77	2.48	0.50	0.20
	60–100 cm	7.77	2.48	0.50	0.20
Spodosols	0–30 cm	0.77	0.17	0.04	<0.01
	30–60 cm	0.36	0.03	0.02	0.01
	60–100 cm	0.36	0.03	0.02	0.01

When site-specific data were not available, plant information was used from reference [65], and soil data were collected from the NRCS Web Soil Survey (http://websoilsurvey.nrcs.usda.gov/).

Table 2. Site information for studies used in DAYCENT model validation.

Site	Lit. Yield	Model Yield	Max. Temp.	Min. Temp.	Precipitation	Latitude	Longitude	Reference
Auburn, AL	26.1	25.4	24.2	9.8	1160	32.67	−85.44	Woodard and Prine, 1993
Belle Glade, FL	25.0	28.3	27.8	16.4	1378	26.68	−80.67	Korndorfer, 2009
EREC, FL	51.3	43.5	29.1	17.7	1181	26.65	−80.63	Gilbert et al., 2006
Gainesville, FL	35.6	27.3	27.0	13.7	1123	29.68	−82.27	Woodard and Prine, 1993
Hendry, FL	39.2	55.9	28.4	18.3	1362	27.78	−82.15	USDA, 2011
Hillsboro, FL	60.7	62.5	28.5	18.3	1547	27.90	−82.49	Gilbert et al., 2006
Houma, LA (1st ratoon)	36.6	38.2	25.2	14.8	500	29.57	−90.65	Legendre and Burner, 1995
Houma, LA (2nd ratoon)	34.9	37.6	25.2	14.8	500	29.57	−90.65	Legendre and Burner, 1995
Hundley, FL	73.5	62.0	28.5	18.1	1457	26.30	−80.16	Gilbert et al., 2006
Jay, FL (plant cane)	35.8	33.6	26.9	16.4	1321	28.65	−80.82	Woodard and Prine, 1993
Jay, FL (1st ratoon)	27.8	32.8	26.9	16.4	1321	28.65	−80.82	Woodard and Prine, 1993
Lakeview, FL	71.3	62.5	28.5	17.4	1275	26.30	−80.15	Gilbert et al., 2006
Hidalgo, TX	34.6	42.5	28.7	18.1	576	26.17	−97.93	Weidenfeld, 1995
Ona, FL (1st ratoon)	40.5	38.1	28.6	16.0	1160	27.48	−81.92	Woodard and Prine, 1993
Ona, FL (2nd ratoon)	30.2	31.6	28.6	16.0	1160	27.48	−81.92	Woodard and Prine, 1993
Pahokee, FL	60.5	65.5	28.4	17.5	1269	26.82	−80.66	Glaz and Ulloa, 1993
Palm Beach, FL	32.3	35.4	29.0	16.6	851	26.67	−80.15	USDA, 2011
Quincy, FL (1st ratoon)	26.3	25.1	25.8	12.9	1445	30.59	−84.58	Woodard and Prine, 1993
Quincy, FL (2nd ratoon)	27.8	21.7	25.8	12.9	1445	30.59	−84.58	Woodard and Prine, 1993
Shorter, AL	26.4	25.4	24.9	10.9	1119	32.40	−85.94	Sladden et al., 1991
Sundance, FL	42.1	43.5	28.6	17.5	1303	26.60	−80.87	Gilbert et al., 2006

Yield values from the literature and modeled yields for energy cane and sugarcane represent total aboveground biomass expressed as Mg ha^{-1} on a dry mass basis. Climate variables include mean annual maximum and minimum temperature (°C) and mean annual precipitation (mm).

within each plot were combined, and C and N concentrations were measured for depth-stratified soil samples (Table 1) and total above ground biomass with a flash combustion chromatographic separation elemental analyzer (Costech 4010 CHNSO Analyzer, Costech Analytical Technologies Inc. Valencia, CA). The instrument was calibrated with acetanilide obtained from Costech Analytical Technologies, Inc. Other physical soil attributes, including texture, bulk density and water holding capacity were obtained from the NRCS Web Soil Survey [46] for Highlands County, Florida.

The DayCent Model

The DayCent model [47,48] was developed to simulate ecosystem dynamics for agricultural, forest, grassland and savanna ecosystems [49–51]. The model is a daily time step version of the Century model [52,53], using the same soil carbon and nutrient cycling submodels to simulate soil organic matter dynamics (C and N) and nitrogen mineralization. DayCent uses more mechanistic submodels than Century to simulate daily plant production, plant nutrient uptake, trace gas fluxes (N_2O, CH_4), NO_3 leaching, and soil water and temperature [48,54–58].

The DayCent soil organic matter model is widely used to simulate the impacts of management practices on soil carbon dynamics and nutrient cycling. Specifically, the soil organic matter submodel has been used to simulate the impacts of soil tillage practices; no-tillage, minimum tillage and conventional tillage [59,60], crop rotations [59], and biofuel crops; woody biomass, switchgrass (*Panicum virgatum*), Miscanthus (*Miscanthus X giganteus*), and sugarcane [13,61] on soil carbon dynamics for agricultural systems. These studies test model performance against observed

data and demonstrate general success in simulating changes in soil carbon levels associated with management practices.

The soil trace gas submodel has been extensively tested using observed soil CH_4 and N_2O data sets from agricultural and natural ecosystems, and once parameterized with plant production data, provides accurate predictions of trace gas fluxes. Specifically, DayCent has successfully simulated the observed impacts of N fertilizer additions and cropping systems [50,58,59] on soil N_2O and CH_4 fluxes. The model results and observed data sets demonstrate that increasing N fertilizer levels increases soil N_2O fluxes and that soil N_2O fluxes are much lower for perennial crops as compared to annual crops.

The DayCent model has been used extensively to simulate grassland and crop yields [50,58,59,62], and to evaluate the environmental impacts of growing crops. Adler et al. [63] used the DayCent model to simulate net greenhouse gas fluxes (soil C status and soil CH_4 and N_2O fluxes) associated with the use of corn, soybeans, alfalfa, hybrid popular, reed canary grass and switchgrass for biofuel energy production in Pennsylvania. Davis et al. [25] used the DayCent model to simulate the environmental impacts of growing switchgrass and Miscanthus in Illinois and compared simulated plant production for switchgrass and Miscanthus with observed yield data. The authors also compared the net soil greenhouse gas fluxes (soil C changes and soil CH_4 and N_2O fluxes) associated with growing switchgrass and Miscanthus and growing corn and soybeans. Davis et al. [13] recently used the DayCent model to simulate the environmental impact of replacing the corn currently grown for ethanol production in the Corn Belt with perennial grasses (Miscanthus and switchgrass) for second-generation biofuel production. The authors found that the

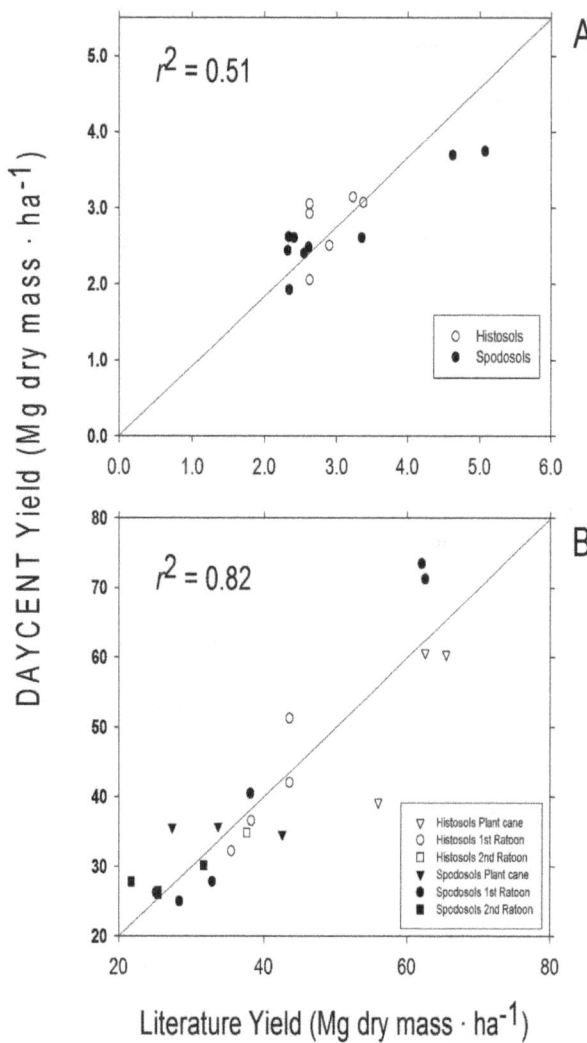

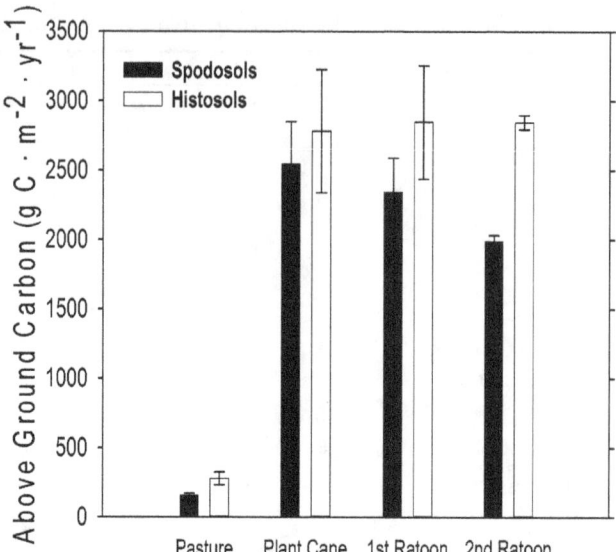

Figure 2. Modeled above ground production of grazed pasture and energy cane in Highlands County, Florida. Values are mean above ground carbon (g C·m^{-2}·yr^{-1}, ± SD) for 15 years in pasture, and for 5 × 3-year ratoon cycles in energy cane (each bar represents the average of 5 values, one for each year for each stage in the planting cycle).

Figure 1. Regression analysis used in DayCent model validation. Model output for dry mass yield was compared to literature values for A) pasture yield values from the USDA-NASS database, B) sugarcane and energy cane dry mass yield. Data points are compared to a 1:1 line.

DayCent model successfully predicted corn, Miscanthus, and switchgrass biomass production for U.S. sites with multiple N fertilizer levels. They also showed that the DayCent model successfully simulated observed annual soil N$_2$O fluxes from corn and switchgrass grown with multiple N fertilizer levels and showed that soil N$_2$O fluxes are much lower for fertilized switchgrass than for corn.

Furthermore, the basis for the DayCent model, Century, has been used to simulate sugarcane production in Brazil [61,64] and Australia [65]. These authors show that the Century/DayCent soil organic matter sub-model can correctly simulate the impacts of fertilizer, and organic matter additions on soil carbon levels and surface litter decay.

Model Parameterization

Energy cane is a variety of sugarcane, and thus parameterizing DayCent for this crop required only minor changes to the previously published input data used for sugarcane [61,65]. Energy cane differs from sugarcane in that it has increased cold

tolerance, decreased sucrose content, and higher cellulose content. We adjusted parameters based on direct measurement of energy cane tissue traits described above. The principal changes from sugarcane to energy cane were reducing the minimum C:N ratio of leaves from 28.6 to 22.1 and changing the C:N of stems from 160 to 30.5. Because of this change in C:N, the parameter for C allocation to stems in DayCent also was modified (from 60% to 40%), to reflect the lower C content of stems relative to N for energy cane versus the previously modeled sugarcane parameters [65]. Bahiagrass (*Paspalum notatum* Flueggé) pasture was simulated using the existing DayCent model parameters for warm season grasses [66,67].

Histosols are challenging to model as organic matter and C content typically are uniform throughout the soil profile [68], but they also are known to subside because of oxidation of the highly labile organic C pools characteristic of these soils [69]. This subsidence was calculated from the modeled rate of organic matter loss and bulk density. DayCent simulates soil C flux to a depth of 30 cm [47], so as soil was lost with subsidence new soil and organic matter became part of this upper 30 cm column from below. This assumes that loss only occurred in the upper 30 cm, which is reasonable since this is the disturbed and aerated part of the soil. The C and N added from low in the soil profile was calculated from the rate of subsidence and the measured elemental contents and bulk density of the soil that was below 30 cm, when sampled, which is at time zero in our model. However, model output calculates GHG and soil C to soil depths to 30 cm.

Model Validation

Literature values of aboveground production (dry mass) for grazed pasture, sugarcane and energy cane (Table 2) were used to validate DayCent. Validation focused on aboveground biomass production because this variable has been measured widely across a range of sites. While there were insufficient data on trace gas flux or changes in soil C in sugarcane or energy cane for validation of

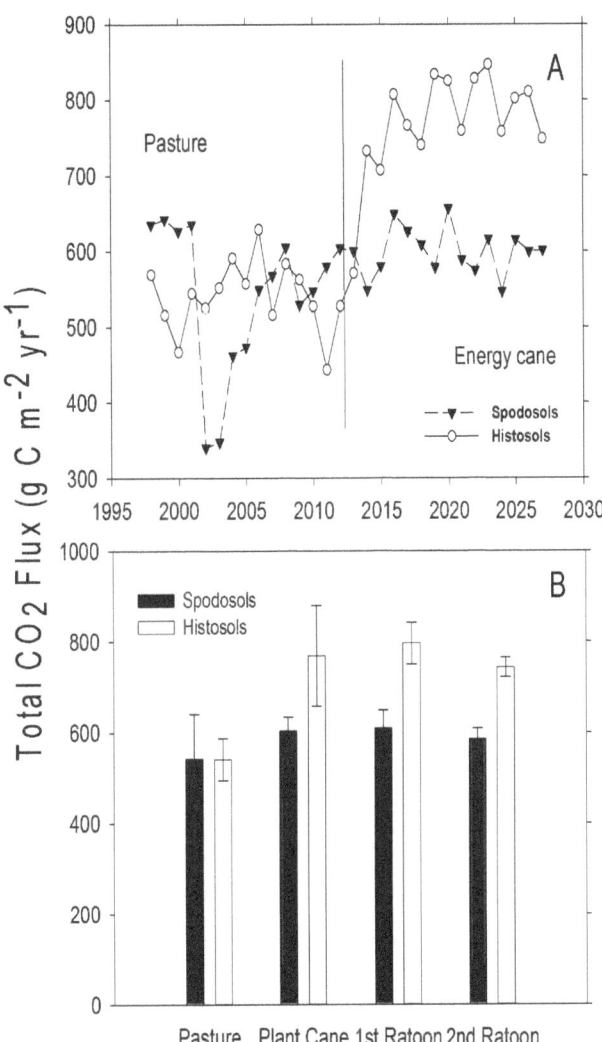

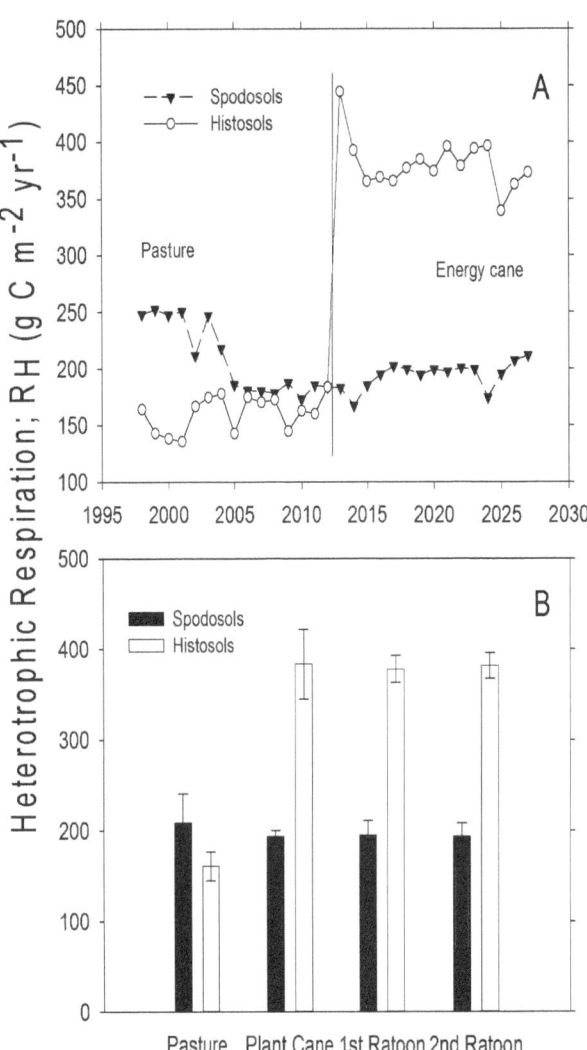

Figure 3. Modeled total soil CO₂ flux from pasture and land converted to energy cane in Highlands County, Florida. A) Total annual soil CO_2 flux (expressed as g C·m^{-2}). Dashed line represents year of land use conversion from pasture to energy cane. B) Mean total soil CO_2 flux (g C·m^{-2}·yr^{-1}, ± SD) for 15 years in pasture, and for 5, 3-year ratoon cycles in energy cane (each bar represents the average of 5 values, one for each year for each stage in the planting cycle).

Figure 4. Modeled heterotrophic respiration (R_H) from pasture and land converted to energy cane in Highlands County, Florida. A) Total annual heterotrophic respiration (g C·m^{-2}). Dashed line represents year of land use conversion from pasture to energy cane. B) Mean heterotrophic respiration (g C·m^{-2}·yr^{-1}, ± SD) for 15 years in pasture, and for 5, 3-year ratoon cycles in energy cane (each bar represents the average of 5 values, one for each year for each stage in the planting cycle).

these variables, validation based on productivity for other crops reliably predicts trace gas flux [59,60,63,70–72].

We compiled a literature database of 17 sites that had reliable data on sugarcane and energy cane yield. There were also pasture productivity data for 15 of those sites [73]. In some instances we were able to contact researchers directly to access unpublished data (Table 2). The geographic range of sites represents the breadth of sugarcane production in the continental United States, and the potential range of energy cane production on currently grazed pastures. For all sites, daily weather data inputs (minimum and maximum temperature, daily precipitation) from 1980 to 2002 were obtained from the DayMet database [74]. The model was run using the DayCent growing degree-day subroutine to determine plant emergence, senescence and death, based on plant phenological characters and daily weather data. Soil data for the validation sites were obtained from the NRCS Web Soil Survey [75]. Using the same schedule of management events used for the

in silico experiments (described below), DayCent was run with site-specific soil and weather data for each sites. The fit of modeled to measured above ground dry mass production (Mg dry matter ha^{-1}) of our simulations of grazed pasture and energy cane were separately tested via linear regression, using the linear model function in R [76].

Initial Simulation Conditions

A "spin-up" period in DayCent based on historical land use and vegetation type was used to set initial soil conditions. The dominant, historic vegetation type for this area of south-central Florida was savanna, with a mixture of grasses and several species of scrub-oak, or sawgrass for the swamp areas [77]. A mix of perennial C_3 grasses species and symbiotic N_2 fixing plants, were used as initial conditions for the savanna simulation (initial vegetation type "savanna" in DayCent). A period of 2000 years was simulated to obtain an initial soil C and N conditions prior to

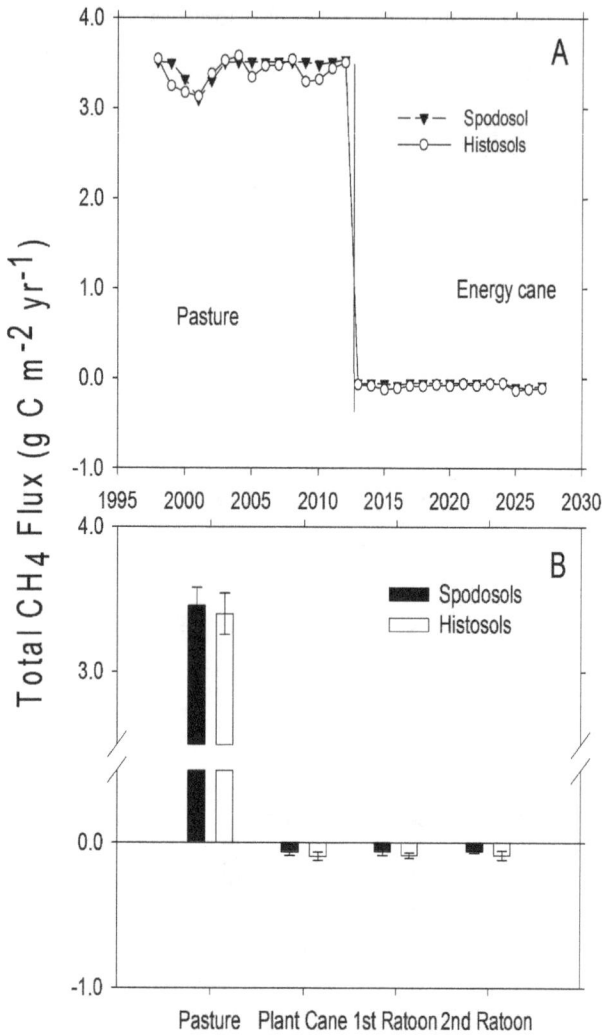

Figure 5. Modeled CH₄ flux from pasture and land converted to energy cane in Highlands County, Florida. A) Total annual CH₄ flux (g C·m⁻²). The solid vertical line represents year of land use conversion from pasture to energy cane, positive values indicate CH₄ efflux and negative values indicate CH₄ uptake. B) Mean CH₄ flux (g C·m⁻²·yr⁻¹, ± SD) for 15 years in pasture, and for 5, 3-year ratoon cycles in energy cane (each bar represents the average of 5 values, one for each year for each stage in the planting cycle).

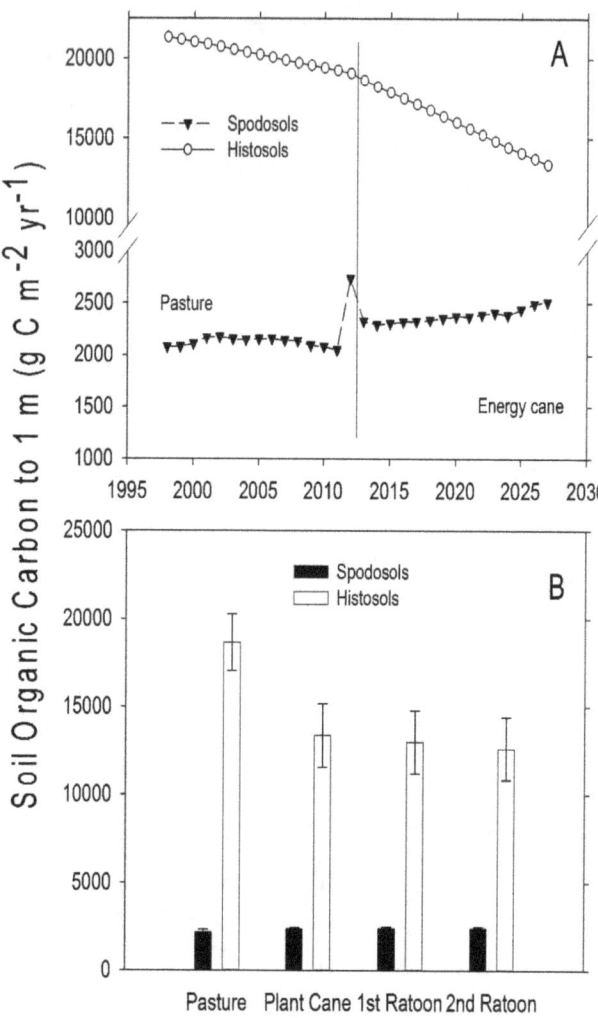

Figure 6. Changes in total soil organic C from pasture and land converted to energy cane in Highlands County, Florida. A) Total annual SOC flux (g C·m⁻²). The solid vertical line represents year of land use conversion from pasture to energy cane. B) Mean SOC flux (g C·m⁻²·yr⁻¹, ± SEM) for 15 years in pasture, and for 5, 3-year ratoon cycles in energy cane (each bar represents the average of 5 values, one for each year for each stage in the planting cycle).

our *in silico* experiments. The model was run for spin ups and all subsequent experiments using the growing degree-day sub-routine.

In silico Experiments

Model simulations were then run to determine the GHG soil-atmosphere exchange and change in soil C predicted for conversion of pasture to energy cane on the two dominant soil types, Spodosols and Histosols. We used daily weather data inputs (minimum and maximum temperature, daily precipitation) from 1951 to 2002, which was the longest time period available for Highlands County, Florida obtained from the DayMet database [74]. This weather file is used by DayCent to create a mean and standard deviation of weather parameters, thus the more weather data available for a given site, the more accurately the variability of a site will be captured by the model.

To initiate the experimental simulations, in 1998 we converted the savanna by removing all above ground biomass and plowing to a depth of 30 cm. A landscape conversion to a grazed Bahiagrass (*Paspalum notatum* Flueggé) ecosystem was then simulated. Bahiagrass is a common forage grass for this part of Florida that would be considered "improved pasture", although usually not fertilized or irrigated [78]. We simulated grazing in our modeling experiment by annually removing 10% of live shoot and 1.0% of standing dead shoots. Prior to planting energy cane, another plow event to 30 cm was initiated to remove the pasture vegetation and simulate the physical land use change.

The simulated cycle of energy cane planting and harvest was based on the sugarcane literature [65,79,80] and discussions with University of Florida and USDA sugarcane agronomists [81,82]. In the simulations, energy cane was planted in January of the first year (2013), followed by a two-year ratoon (crop regenerated from remaining biomass) from which 80% of the above ground biomass was harvested in December. At the end of the second ratoon, the crop was removed and the land plowed before planting a new

Table 3. Modeled ecosystem carbon, nitrogen and greenhouse gas fluxes after converting pasture to energy cane on nutrient poor Spodosols and organic matter rich Histosols.

	Spodosols			Histosols		
	Pasture	Energy cane	Δ	Pasture	Energy cane	Δ
SOC (g C·m^{-2})	2736	2513	−224	16087	10373	−5715
Nitrogen Mineralization (g N·m^{-2})	134	203	69	216	293	77
Heterotrophic Respiration (g C·m^{-2})	3130	2913	−218	2413	5715	3302
Total Soil CO$_2$ Efflux (g C·m^{-2})	8148	8993	845	8111	11540	3429
CH$_4$ (g CO$_2$eq·m^{-2})	2980	−33	−3013	2958	−46	−3004
N$_2$O (g CO$_2$eq·m^{-2})	214	649	435	6713	1742	−4970
Total System C Flux (g CO$_2$eq·m^{-2})	−1159	−2812	−1653	−1367	924	2291
Total Greenhouse Gas Flux (g CO$_2$eq·m^{-2})	2035	−2196	−4231	8304	2620	−5684

Greenhouse gas and N mineralization values are the sum of values from pasture 15 years prior to conversion to energy cane and the sum values for 15 years following the conversion to energy cane. Positive values indicate a flux to the atmosphere and negative values indicate uptake from the atmosphere by the ecosystem. Soil organic matter values are the differences between the last year of energy cane production and the last year of pasture. Total GHG values are the sums of CH$_4$, N$_2$O and total system C flux (calculated in DayCent as the difference between all C uptake and storage versus efflux from respiration) expressed as CO$_2$e. Differences (Δ) represent the values for energy cane minus pasture.

plant crop. This cycle of ratooning and planting was repeated in the simulation for fifteen years following conversion from pasture; i.e. five cycles of three years each. This three-year planting cycle is typical for sugarcane production in Florida [83,84].

Irrigation events were scheduled every month throughout the dry season, and every two months during the rainy season to maintain soil water at field capacity. Fertilizer $(NH_4^+ - NO_3^-)$ was applied in mid February and mid June of each year of the simulation, at a rate of 102 kg N·ha^{-1} per fertilization event for Spodosols. No fertilizer was added to the organic rich Histosols. This fertilization schedule was based on studies that suggest that a split fertilization regime at this rate maximizes sugarcane yield, and that fertilizing above this level does not increase yield but increases N$_2$O efflux [38,65,85]. The input files used to drive DayCent (e.g. schedule files, plant input parameters, and soil input files) are available online [86].

Calculations and Statistical Analyses

We summed daily GHG and soil C fluxes from DayCent to calculate yearly fluxes and report those in g C or N·m^{-2} yr^{-1}, with the exception of total GHG values which are reported as CO$_2$ equivalents [87] and factored by warming potential (CO$_2$ = 1, CH$_4$ = 23, N$_2$O = 296; ref. 85). Total ecosystem C flux was calculated as the annual change in total ecosystem C storage between the beginning and end of a year and represents the net ecosystem carbon balance expressed in CO$_2$eq [88,89].

Because the model experiments were performed using the same site with the same weather data, but controlled for soil type, the simulations had the structure of a paired design where each year was a replicate [90]. We therefore used paired t-tests to determine differences between soil types within a plant type (n = 15) and between plant types within a soil type (n = 15). The variation reported with mean annual values represents inter-annual variation in the predicted variables. Heteroscedasticity was examined with the Fligner-Killeen test, and output data distributions, which did not meet variance assumptions, were compared with the Wilcoxon rank-sum test. The routines t.test (paired = TRUE) and wilcox.test were performed using R [76,90].

Because of the large number of pair-wise comparisons of our model results, the False Discovery Rate (FDR) test was used to account for multiple comparisons. The FDR test is less conser-

vative than a P-value adjustment such as the Bonferroni correction, and determines the probability of a Type I error. We calculated a FDR of 0.024 for our matrix of tests, and therefore justified the use of multiple paired t-tests without P-value adjustment [91].

Results

Predicted harvested yields for both pasture and energy cane in our validation sites agreed well with measured values from the literature (Pasture: $r^2 = 0.52$, Energy cane: $r^2 = 0.82$, Figure 1A & 1B), indicating that our modeled predictions provided a good representation of the productivity that drives the biogeochemical dynamics of DayCent.

For our modeled site, DayCent estimated a large increase in aboveground plant biomass production after conversion of pasture to energy cane (Figure 2); annual aboveground biomass production increased by a factor of 14 on Spodosols and by a factor of 10 on Histosols, relative to pasture. Energy cane production ranged from 1911–3153 g C m^{-2} yr^{-1} (46–76 Mg dry biomass·ha^{-1}). Predicted energy cane production remained high through the three harvests on Histosols, but declined through the modeled ratoon cycle on Spodosols (Figure 2).

There was considerable temporal variation in predicted soil CO$_2$ efflux from pasture in the 15 years simulated prior to the conversion to energy cane (Figure 3a). This variation was particularly evident for pasture on Spodosols and was driven primarily by variation in precipitation. Total soil CO$_2$ efflux was similar for pasture on both soil types, but significantly increased when averaged over 15 years after conversion to energy cane on the Histosols (Figure 3a; $t = 10.65$, d.f. = 14, $P<0.001$). Land use conversion did not increase CO$_2$ efflux on Spodosols ($t = 0.58$, d.f. = 14, $P = 0.57$). Following conversion to energy cane CO$_2$ efflux from Histosols was significantly higher than energy cane on Spodosols (Figure 3b; $t = 9.56$, d.f. = 14, $P<0.001$).

The conversion of land from pasture to energy cane had no significant effect on the predicted heterotrophic component of soil respiration (R$_H$) on Spodosols (Figure 4a), but caused a large increase in R$_H$ from the Histosols (Figure 4a; $t = 31.86$, d.f. = 14, $P<0.001$) and resulted in higher R$_H$ on Histosols than Spodosols following the conversion to energy cane (Figure 4b; $t = 23.68$, d.f.

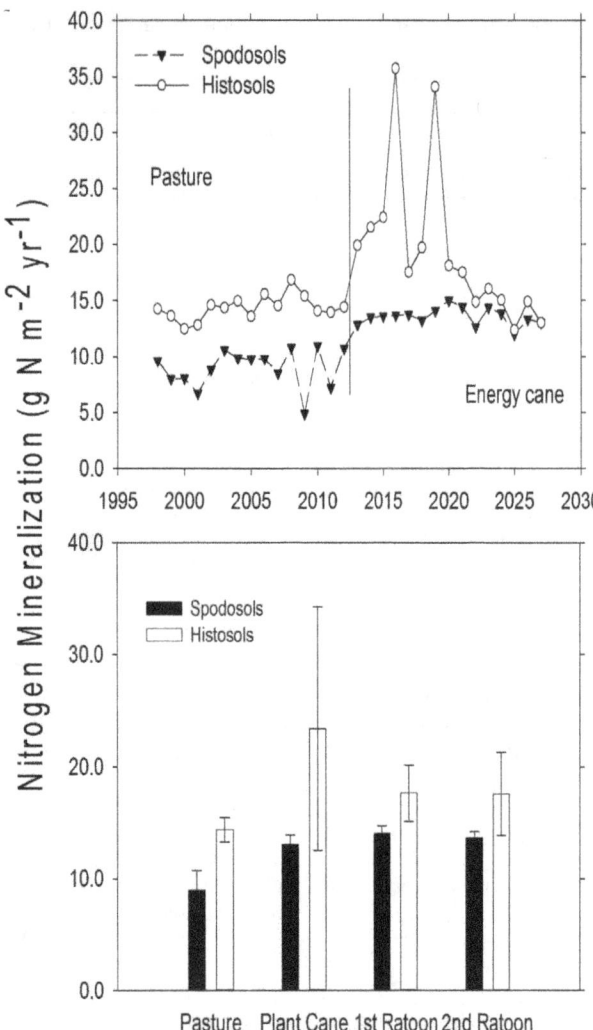

Figure 7. Modeled nitrogen mineralization rates from pasture and land converted to energy cane in Highlands County, Florida. A) Total annual N mineralization rate (g N·m^{-2}). The solid vertical line represents year of land use conversion from pasture to energy cane. B) Mean N mineralization rate (g N·m^{-2}·yr^{-1}, ± SD) for 15 years in pasture, and for 5, 3-year ratoon cycles in energy cane (each bar represents the average of 5 values, one for each year for each stage in the planting cycle).

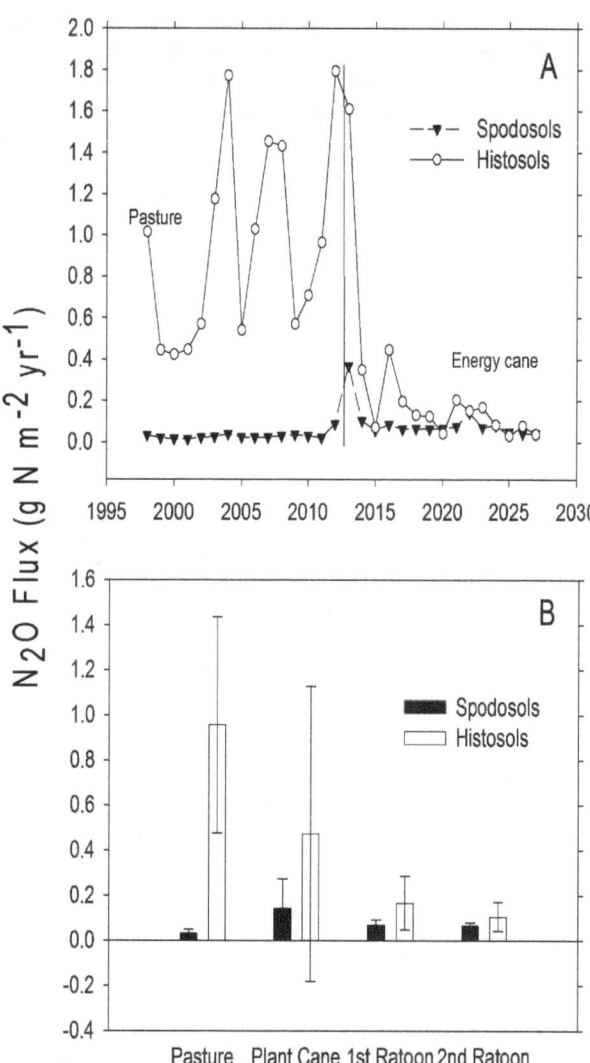

Figure 8. Modeled N$_2$O flux from pasture and land converted to energy cane in Highlands County, Florida. A) Total annual N$_2$O flux (g N·m^{-2}). The solid vertical line represents year of land use conversion from pasture to energy cane, positive values indicate N$_2$O efflux and negative values indicate N$_2$O uptake. B) Mean N$_2$O flux (g N·m^{-2}·yr^{-1}, ± SD) for 15 years in pasture, and for 5, 3-year ratoon cycles in energy cane (each bar represents the average of 5 values, one for each year for each stage in the planting cycle).

= 14, P<0.001). Prior to the conversion to energy cane, modeled (R$_H$) was slightly higher in pasture on Spodosols than on Histosols (Figure 4; t = 31.86, d.f. = 14, P<0.001).

On both soil types, the removal of cattle associated with the conversion of pasture to energy cane caused a substantial change in predicted CH$_4$ flux (t = 185, d.f. = 14, P<0.001 on Spodosols; t = 167, d.f. = 14, P<0.001 on Histosols; Figure 5a). Without cattle, pastures were a small CH$_4$ sink (0.16–0.60 g C·m^{-2}·yr^{-1} uptake in Spodosols, 15 year sum = 112 g CO$_2$eq·m^{-2}, 0.12–0.57 gC·m^{-2}·yr^{-1} uptake for Histosols, 15 year sum = 135 g CO$_2$eq·m^{-2}). Introducing cattle at stocking rates and grazing intensity typical for this region (1 head cattle·ha^{-1}: ref. 31), caused pasture on both soil types to be a substantial source of CH$_4$ to the atmosphere (Figure 5).

Changes in vegetation and management practices altered soil organic carbon (SOC), and these changes were particularly evident on the Histosols (Table 2; Figure 6). Histosols had a larger pool of active C (weekly to monthly turnover) than Spodosols under both pasture and energy cane (pasture, t = 19.25, d.f. = 14, P<0.001; energy cane, t = 14.21, d.f. = 14, P<0.001). Comparing the remaining total SOC pools between the end of pasture and the last year of the energy cane simulation, Histosols lost a large amount of soil organic C; 5714 g C·m^{-2} to 1 m depth (Figure 6; t = 296, d.f. = 14, P<0.001), compared to the SOC loss from Spodosols of 224 g C·m^{-2} to 1 m (Table 3).

Nitrogen mineralization increased after pasture was converted to energy cane on both the fertilized Spodosols (t = 9.02, d.f. = 14, P<0.001) and the non-fertilized Histosols (t = 2.72, d.f. = 14, P = 0.02). After conversion to energy cane, Histosols had higher rates of N mineralization than Spodosols (Figure 7; t = 3.43, d.f. = 14, P = 0.004), and this increase in available N likely accounted for the continued high yields on Histosols.

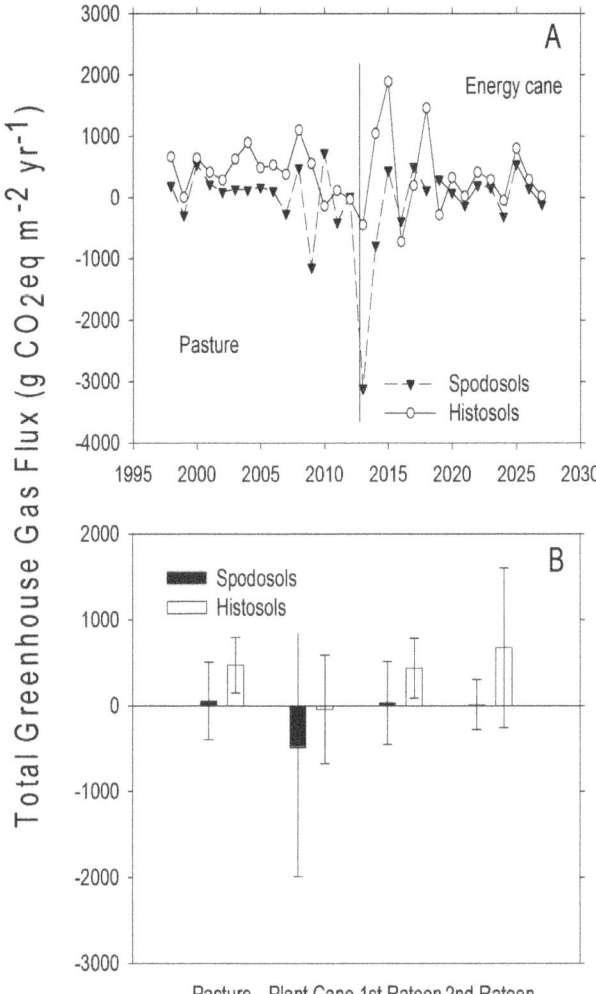

Figure 9. Changes in total greenhouse gas (GHG) from pasture and land converted to energy cane in Highlands County, Florida. Positive values indicate GHG efflux and negative values indicate GHG uptake. A) Total annual GHG flux, reported as CO_2 equivalents converted to account for differences in warming potential (g $CO_2 \cdot m^{-2}$). The solid vertical line represents year of land use conversion from pasture to energy cane. B) Mean greenhouse gas flux in CO_2 equivalents converted to account for differences in warming potential (g $CO_2 e \cdot m^{-2} \cdot yr^{-1}$, ± SD) for 15 years in pasture, and for 5, 3-year ratoon cycles in energy cane (each bar represents the average of 5 values, one for each year for each stage in the planting cycle).

Prior to conversion to energy cane, N_2O efflux was higher in pastures on Histosols compared to Spodosols (Figure 8; Wilcoxon rank sum, W = 225, $P<0.001$). After conversion to energy cane, Histosols remained greater sources of N_2O than Spodosols (t = 12.15, d.f. = 14, $P<0.001$). Conversion of pasture to energy cane decreased N_2O efflux on Histosols (Figure 8a; t = 4.30, d.f. = 14, $P<0.001$), but increased N_2O efflux on Spodosols (Figure 8b; t = 2.87, d.f. = 14, P = 0.01). It is likely that N_2O emission from Histosols decreased following conversion because the increase in productivity resulted in a higher uptake of nitrate that would otherwise be available for denitrification.

Total GHG exchange (global warming potential) was calculated by converting the fluxes of CH_4 and N_2O to CO_2 equivalents based on their warming potential relative to CO_2 [87] and

summing these with total system C flux (Table 2). Variation in weather caused substantial inter-annual variation in total GHG flux, with both pasture and energy cane varying between net GHG sinks and sources (Figure 9); no significant differences in annual GHG flux were resolved on either soil type (Spodosols: t = 1.15, d.f. = 14, P = 0.27; Histosols: t = 0.13, d.f. = 14, P = 0.90). When the cumulative GHG emission were calculated for the fifteen years prior to conversion, pasture was a net source to GHGs to the atmosphere on both soil types, and pasture was a stronger source on Histosols (8304 gCO_2eq$\cdot m^{-2}$) than on Spodosols (2035 gCO_2eq$\cdot m^2$; Table 3). Conversion of pasture to energy cane caused the Spodosols to transition from a source to a sink for GHGs and reduced the flux of GHGs to the atmosphere on Histosols. On both soil types, the reduction GHG emission to the atmosphere was associated with a large decrease in CH_4 emissions caused by the elimination of cattle grazing. On the Histosols, the reduction in N_2O emissions to the atmosphere also contributed to reduced emission of GHGs. This analysis of GHG emissions and their corresponding global warming potentials did not account for the displacement of fossil fuel emissions by the biofuel product.

Discussion

Parameterization of the DayCent model for energy cane, an emerging bioenergy crop, successfully simulated biomass production across the southeast United States (Figure 1). Our simulations suggested high yields for energy cane on former pastureland in a subtropical climate when Spodosols are highly fertilized (200 kg N$\cdot ha^{-1} \cdot yr^{-1}$), and when microbial activity in Histosols leads to high rates of N mineralization (rates were 44% higher on Histosols). When integrated over 15 years (Table 3), conversion of pasture to energy cane on Spodosols converted a net source of GHG (due to cattle CH_4 emissions) to a sink driven by the removal of cattle and the increase in C uptake by energy cane. While Histosols were a net GHG source under both pasture and energy cane, the source was reduced by the land use conversion (Table 3). The GHG improvement resulting from this conversion from pasture to energy cane would be even greater if fossil fuel displacement by cellulosic ethanol had been included.

The range of our simulated energy cane yields was 46–76 Mg ha^{-1} dry mass per year on fertilized Spodosols and unfertilized Histosols. Using published values for the conversion efficiency for the production of cellulusoic ethanol [22], a hypothetical energy cane farm of 10,000 ha could therefore produce between 142–236 million liters of ethanol [92]. In comparison, equal areas of land devoted to corn grain and Miscanthus in the Midwest would yield between 25 and 73% this amount of ethanol, respectively, assuming the maximum yields reported by other authors [11,22].

Typically, sugarcane yield declines with ratooning, the repeated harvests of aboveground material generated by vegetative growth [93]. The model reproduced the yield decline for energy cane on Spodosols but not on Histosols, but the model in its current configuration probably failed to capture the mechanisms that would normally cause a decline in yield. Various factors ranging from increases in nematode populations and ratoon stunting disease, to mechanical compaction of the soil have been implicated in ratoon decline [94,95], and these were not accounted for in the model. Although sugarcane in Florida typically is grown for three years and three annual harvests before re-planting, if it were grown for more years between re-planting, we would expect a continuing yield decline on the Spodosols. In contrast, continued mineralization of organic matter on Histosols may sustain high yields beyond the 3-year period simulated in the model. On both soils the GHG benefits would be improved with longer ratoon cycles because of less soil disturbance due to decreased frequency of soil disturbance for replanting.

Organic matter (OM) content of soils is important for sustaining high yields of sugarcane, in part because OM mineralization provides the labile N necessary to sustain plant growth. Spodosols had much less OM than Histosols (Appendix I; Figure 6; [65,96]). The high CO_2 efflux rates from Histosols (Figures 3–4) and the patterns of SOC loss following land use change (Figure 6) correspond to higher rates of OM mineralization. The associated higher rates of N mineralization (Table 3; Figure 7) on Histosols provided additional N to energy cane and improved crop yield (Figure 2). Although energy cane on Spodosols was fertilized to offset the low N content of these soils, rates of nitrification (the process by which NH_4^+ is converted into the highly mobile NO_3^- anion) were higher on these soils. The fertilizer applied to energy cane crops on Spodosols in the simulations was NH_4^+ - NO_3^-, a labile substrate for nitrification [66]. Spodosols had consistently higher nitrification rates than Histosols, and therefore higher NO_3^- content because of fertilization, and it is possible that some fraction of fertilizer was lost before plant uptake [97]. We hypothesize that a combination of NO_3^- leaching from fertilizer before plant uptake, lower initial N content, and lower mineralization rates may have created a stronger N limitation to yield on Spodosols but not on Histosols.

Before land use change, pasture on both soil types was a net source of GHGs to the atmosphere (Table 3). This is consistent with both direct measurements [98] and modeling efforts [99] that have found grazed pastures to be net sources of GHGs, but this is also a function of grass species present and animal stocking density [100]. The model estimated that pastures were sinks for CO_2, with total C uptake of 1159 g CO_2 m^{-2} and 1367 g CO_2 m^{-2} over 15 years on Spodosols and Histosols, respectively (Table 3). In the absence of cattle, both soil types were CH_4 sinks (112 and 135 g CO_2eq, respectively), but including reasonable estimates of CH_4 efflux from cattle (Figure 5) and N_2O efflux from soils (Figure 8) resulted in net GHG emission to the atmosphere on both pasture soils (Table 3). Following conversion to energy cane, the production of N_2O on Spodosols increased (Figure 8) within the range of N_2O flux rates previously reported for Australian sugarcane fertilized at similar rates to this study [37]. The increase in N_2O was offset by uptake of CO_2 and the change from a source to a sink for CH_4 (Figure 5), with the net effect that Spodosols became a net GHG sink (Table 3). Indeed, over 15 years energy cane on Spodosol was a GHG sink of >40 Mg CO_2eq per hectare (Table 3). On Histosols, eliminating grazing following the conversion of pasture to energy cane caused a similar decrease in CH_4 efflux to the atmosphere (Figure 5) and this land use change also reduced N_2O emissions (Figure 8; Table 3). However, following land conversion this system switched from a net CO_2 sink to a source, and this change in total system C prevented energy cane on Histosols from becoming a net sink for GHGs. The driver for GHG production on Histosols was higher R_H, and significant losses of soil organic matter [69] that resulted in total C efflux from these soils (Table 3).

The model successfully simulated energy cane biomass production across a range of sites across the southern United States (Figure 1). Previous studies have shown that DayCent reliably predicts soil biogeochemistry and GHG exchange when parameterized for net primary production [51], suggesting that the estimates of GHG flux and soil C dynamics were reasonable. Eddy-flux measurements of GHG exchange that are now being initiated at this site will provide an independent test of the predictions of GHG effects of conversion made here.

Indirect land use change (ILUC) – the stimulation of deforestation or increased agriculture in other parts of the world driven by diversion of current agricultural land to bioenergy production – potentially poses an environmental risk of bioenergy production [7,101]. Growing energy cane on land converted from low stocking density pasture would be unlikely to trigger significant increases in food price or ILUC in the way that large-scale shifts from corn or soy production in the Midwestern United States would motivate greater production of those crops elsewhere [13]. Indeed, the recommended stocking density for Bahiagrass pasture in this region is ~1 animal·ha^{-1} [31], and cattle and calf operations in Florida account for less than 6% of the state's annual agricultural revenue [102]. The loss in meat production could be redressed with minimal increases in current stocking rates, and would be unlikely to trigger the type of large-scale landscape changes that may occur through the diversion of midwestern agricultural land [7]. However, displacing cattle for energy cane production may potentially increase methane emissions elsewhere, which would negate the local benefit of reduced methane flux to the atmosphere.

The environmental impacts of changing land use from pasture to energy cane were highly dependent on the soil type. Whereas the cultivation of Histosols results in high CO_2 efflux and the reduction of soil carbon (Figures 3, 4, and 6), the model predicted that energy cane crops on Spodosols would act as a net C and GHG sink (Figure 6, Table 3). From both a biofuel and biogeochemical perspective, these results suggest that energy cane grown on nutrient poor soils, as opposed to organic soils, has the potential to be a high-yielding bio-ethanol feedstock that creates a GHG sink in the Southeastern United States.

Acknowledgments

We thank Michael Masters for lab analysis of plant and soil C and N. Lykes Brothers Inc. graciously provided access to energy cane plantations and pastures, which greatly helped in parameterizing our model. We also thank Dr. Robert Gilbert, Dr. Barry Glaz, and Mr. Pedro Korndorfer for sharing their knowledge on sugar and energy cane agronomy and data that aided our modeling effort.

Author Contributions

Conceived and designed the experiments: BDD KJAT WJP EHD. Performed the experiments: BDD SCD CK WJP. Analyzed the data: BDD SCD KJAT WJP SPL EHD. Contributed reagents/materials/analysis tools: CK WJP. Wrote the paper: BDD KJAT WJP SPL EHD.

References

1. IPCC (2007) Climate Change 2007: The Physical Science Basis. Contribution of Working Group I to the Fourth Assessment Report of the Intergovernmental Panel on Climate Change. Solomon S, Qin D, Manning M, Chen Z, Marquis M et al. Eds. Cambridge, UK, Cambridge, UK.
2. Le Quéré C, Raupach MR, Canadell JG, Marland G, Bopp L, et al. (2009) Trends in the sources and sinks of carbon dioxide. Nat Geosci 2: 831–836.
3. Pan Y, Birdsey RA, Fang J, Houghton R, Kauppi PE, et al. (2011) A large and persistent carbon sink in the world's forests. Science 333: 988–993.
4. Lal R (2004) Soil carbon sequestration impacts on global climate change and food security. Science 304: 1623–1627. doi: 10.1126/science.1097396.
5. Tilman D (1998) The greening of the green revolution. Nature 396: 211–212.
6. Fargione JE, Hill JD, Tilman D, Polasky S, Hawthorne P (2008) Land clearing and the biofuel carbon debt. Science 319: 1235–1238.
7. Searchinger T, Heimlich R, Houghton RA, Dong F, Elobeid A, et al. (2008) Use of U.S. croplands for biofuels increases greenhouse gases through emissions from land-use change. Science 319: 1238–1240.
8. Melillo JM, Reilly JM, Kicklighter DW, Gurgel AC, Cronin TW (2009) Indirect emissions from biofuels: How important? Science 326: 1397–1399.
9. Börjesson P (2009) Good or bad bioethanol from a greenhouse gas perspective – What determines this? Applied Energy 86: 589–594.

10. United States Congress (2007) The Energy Independence and Security Act of 2007 (H.R. 6). Available: http://frwebgate.access.gpo.gov/cgibin/getdoc. cgi?dbname = 110_cong_bills&docid = f: h6enr.txt.pdf. Accessed 2010 Dec 19.

11. Heaton EA, Dohleman FG, Long SP (2008) Meeting US biofuel goals with less land: the potential of Miscanthus. Glob Change Biol 14: 2000–2014.

12. Fargione JE, Plevin RJ, Hill JD (2010) The ecological impact of biofuels. Ann Rev Ecol Evol Syst 41: 351–377.

13. Davis SC, Parton WJ, Del Grosso SJ, Keough C, Marx E, et al. (2012) Impacts of second-generation biofuel agriculture on greenhouse gas emissions in the corn-growing regions of the US. Front Ecol Environ 10: 69–74. doi:10.1890/110003.

14. Dien BS, Bothast RJ, Nichols NN, Cotta MA (2002) The U.S. corn ethanol industry: an overview of current technology and future prospects. Int Sugar J 103: 204–211.

15. Davis SC, Anderson-Teixeira KJ, DeLucia EH (2009) Life-cycle analysis and the ecology of biofuels. Trends Plant Sci 14: 140–146.

16. O'Hare M, Plevin RJ, Martin JI, Jones AD, Kendall A, et al. (2009) Proper accounting for time increases crop-based biofuels' greenhouse gas deficit versus petroleum. Environ Res Lett 4: doi:10.1088/1748–9326/4/2/024001.

17. Donner SD, Kucharik CJ (2008) Corn-based ethanol production compromises goal of reducing nitrogen export by the Mississippi River. Proc Natl Acad Sci USA 105: 4513–4518.

18. Hill J, Polasky S, Nelson E, Tilman D, Huo H (2009) Climate change and health costs of air emissions from biofuels and gasoline. Proc Nat Acad Sci USA 106: 2077–2082.

19. Smeets EMW, Bouwman LF, Stehfest E, van Vuuren DP, Posthuma A (2009) Contribution of N_2O to the greenhouse gas balance of first-generation biofuels. Glob Change Biol 15: 1–23.

20. Solomon BD, Barnes JR, Halvorsen KE (2007) Grain and cellulosic ethanol: history, economics, and energy policy. Biomass Bioenergy 31: 416–425.

21. Tilman D, Socolow R, Foley JA, Hill J, Larson E (2009) Beneficial biofuels-the food, energy and environment trilemma. Science 325: 270–271.

22. Somerville C, Youngs H, Taylor C, Davis SC, Long SP (2010) Feedstocks for lignocellulosic biofuels. Science 329: 790–792.

23. Tilman D, Hill J, Lehman C (2006) Carbon-negative biofuels from low-input high-diversity grassland biomass. Science 314: 1598–1600.

24. Anderson-Teixeira KJ, Davis SC, Masters MD, DeLucia EH (2009) Changes in soil organic carbon under biofuel crops. Glob Change Biol Bioenergy 1: 75–96.

25. Davis SC, Parton WJ, Dohleman FG, Smith CM, Del Grosso S, et al. (2010) Comparative biogeochemical cycles of bioenergy crops reveal nitrogen fixation and low greenhouse gas emissions in a Miscanthus x giganteus agro-ecosystem. Ecosystems 13: 144–156.

26. Robertson GP, Hamilton SK, Del Grosso SJ, Parton WP (2011) The biogeochemistry of bioenergy landscapes: carbon, nitrogen, and water considerations. Ecol App 21: 1055–1067. doi: 10.1890/09–0456.1.

27. Zeri M, Anderson-Teixeira KJ, Masters MD, Hickman G, DeLucia EH, et al. (2011) Carbon exchange by establishing biofuel crops in central Illinois. Agric Ecosyst Environ 144: 319–329.

28. Sladden SE, Bransby DI, Aiken GE, Prine GM (1991) Biomass yield and composition, and winter survival of tall grasses in Alabama. Biomass Bioenergy 1: 123–127.

29. Mark T, Darby P, Salassi M (2009) Energy cane usage for cellulosic ethanol: estimation of feedstock costs. Southern Agricultural Economics Association Annual Meeting, Atlanta, Georgia, January 31-February 3, 2009.

30. Baucum LE, Rice RW (2009) An overview of Florida sugarcane. University of Florida IFAS Extension document SS-AGR-232.

31. Hersom M (2005) Pasture stocking density and the relationship to animal performance. Animal Science Department, Florida Cooperative Extension Service, Institute of Food and Agricultural Sciences, University of Florida, Document number AN155.

32. Steiner J (2012) personal communication.

33. Bowden RD, Nadelhoffer KJ, Boone RD, Melillo JM, Garrison JB (1993) Contributions of aboveground litter, belowground litter, and root respiration to soil respiration in a temperate mixed hardwood forest. Can J For Res 23: 1402–1407.

34. Paustian K, Six J, Elliott ET, Hunt HW (2000) Management options for reducing CO_2 emissions from agricultural soils. Biogeochemistry 48: 147–163.

35. DeRamus HA, Clement TC, Giampola DD, Dickison PC (2003) Methane emissions of beef cattle on forages: efficiency of grazing management systems. J Environ Qual 32: 269–277.

36. Mosier A, Kroeze C, Nevison C, Oenema O, Seitzinger S, et al. (1998) Closing the global N_2O budget: nitrous oxide emissions through the agricultural nitrogen cycle. Nutr Cycl Agroecosys 52: 225–248.

37. Thorburn PJ, Biggs JS, Collins K, Probert ME (2010) Nitrous oxide emissions from Australian sugarcane production systems – are they greater than from other cropping systems? Agric Ecosyst Environ 136: 343–350.

38. Rice RW, Gilbert RA, Lentini RS (2002) Nutritional requirements for Florida sugarcane. Florida Cooperative Extension Service. UF/IFAS, Document SS-ARG-228. University of Florida Institute of Food and Agricultural Science.

39. Morgan KT, McCray JM, Rice RW, Gilbert RA, Baucum LE (2009) Review of current sugarcane fertilizer recommendations: a report from the UF/IFAS sugarcane fertilizer standards task force. Document SL 295, Soil and Water

40. Morris DR, Gilbert RA, Reicosky DC, Gesch RW (2004) Oxidation potentials of soil organic matter in Histosols under different tillage methods. Soil Sci Soc Am J 68: 817–826.

41. Stehfest E, Bouwman LF (2006) N_2O and NO emission from agricultural fields and soils under natural vegetation: summarizing available measurement data and modeling of global annual emissions. Nutr Cycl Agroecosys 74: 207–228.

42. Anderson-Teixeira KJ, Masters MD, Black CK, Zeri M, Hussain MZ, et al. (2012) Altered belowground carbon cycling following land use change to perennial bioenergy crops. Ecosystems, in press.

43. Smith CM, David MB, Mitchell CA, Masters MD, Anderson-Teixeira KJ, et al. (2013) Reduced nitrogen losses following conversion of row crop agriculture to perennial biofuel crops. J Env Qual 42: 219–228, doi: 10.2134/jeq2012.0210.

44. Beale CV, Long SP (1997) Seasonal dynamics of nutrient accumulation and partitioning in the perennial C-4-grasses Miscanthus x giganteus and Spartina cynosuroides. Biomass Bioenergy 12: 419–428.

45. Hansen EM, Christensen BT, Jensen LS, Kristensen K (2004) Carbon sequestration in soil beneath long-term Miscanthus plantations as determined by ^{13}C abundance. Biomass Bioenergy 26: 97–105.

46. NRCS Web Soil Survey website. Available: http://websoilsurvey.nrcs.usda. gov/app/HomePage.htm. Accessed 2010 Oct 7

47. Parton WJ. Hartman MD, Ojima DS, Schimel DS (1998) DAYCENT and its land surface submodel: description and testing. Glob Planet Change 19: 35–48.

48. Parton WJ, Holland EA, Del Grosso SJ, Hartmann MD, Martin RE, et al. (2001) Generalized model for NO_x and N_2O emissions from soils. J Geophys Res-Atmos 106: 17403–17420.

49. DayCent: Daily Century Model website. Available: http://www.nrel.colostate. edu/projects/daycent/. Accessed 2010 Jun 4.

50. Del Grosso SJ, Halvorson AD, Parton WJ (2008a) Testing DayCent model simulations of corn yields and nitrous oxide emissions in irrigated tillage systems in Colorado. J Environ Qual 37: 1383–1389, doi:10.2134/jeq2007.0292.

51. Parton WJ, Hanson PJ, Swanston C, Torn M, Trumbore SE, et al. (2010) ForCent model development and testing using the Enriched Background Isotope Study experiment. J Geophys Res 115: G04001.

52. Parton WJ, Schimel DS, Cole CV, Ojima DS (1987) Analysis of factors controlling soil organic levels of grasslands in the Great Plains. Soil Sci Soc Am J 51: 1173–1179.

53. Parton WJ, Ojima DS, Cole CV, Schimel DS (1994) A general model for soil organic matter dynamics: sensitivity to litter chemistry, texture and management, R.B. Bryant,R.W. Arnoldm, Editors, Quantitative Modeling of Soil Forming Processes, Soil Science Society of America, Madison, WI. Pp. 147–167.

54. Eitzinger J, Parton WJ, Hartman M (2000) Improvement and validation of a daily soil temperature submodel for freezing/thawing periods. Soil Sci 165: 525–534.

55. David MB, Del Grosso SJ, Hu X, McIsaac GF, Parton WJ, et al. (2009) Modeling denitrification in a tile-drained, corn and soybean agroecosystem of Illinois, USA. Biogeochemistry 93: 7–30.

56. Del Grosso SJ, Parton WJ, Mosier AR, Ojima DS, Hartmann MD (2000a) Interaction of soil carbon sequestration and N2O flux with different land use practices. In: van Ham J, Baede APM, Meyer LA, Ybema R (eds.), Non-CO_2 Greenhouse Gases: Scientific Understanding, Control and Implementation. Kluwer Academic Publishers, The Netherlands. 303–311.

57. Del Grosso SJ, Parton WJ, Mosier AR, Ojima DS, Kulmala AE, et al. (2000b) General model for N_2O and N_2 gas emissions from soils due to denitrification. Global Biogeochem Cycles 14: 1045–1060.

58. Del Grosso SJ, Mosier AR, Parton WJ, Ojima DS (2005) DayCent model analysis of past and contemporary soil N_2O and net greenhouse gas flux for major crops in the USA. Soil Tillage Res 83: 9–24.

59. Del Grosso SJ, Ojima DS, Parton WJ, Mosier AR, Peterson GA, et al. (2002) Simulated effects of dryland cropping intensification on soil organic matter and GHG exchanges using the DAYCENT ecosystem model. Environ Pollution 116, S75–S83.

60. Del Grosso SJ, Parton WJ, Ojima DS, Keough CA, Riley TH, et al. (2008b) DAYCENT simulated effects of land use and climate on county level N loss vectors in the USA. Pages 571–595 in: R.F. Follett, and J.L. Hatfield (eds.) Nitrogen in the Environment: Sources, Problems, and Management, 2nd ed. Elsevier Science Publishers, The Netherlands.

61. Galdos MV, Cerri CC, Cerri CEP, Paustian K, Van Antwerpen R (2010) Simulation of sugarcane residue decomposition and aboveground growth, Plant Soil 326: 243–259. DOI 10.1007/s11104–009–004–3.

62. Hartmann MD, Merchant EK, Parton WJ, Gutmann MP, Lutz SM, et al. (2011) Impact of historical land use changes in the U.S. Great Plains, 1883 to 2003. Ecol App 21: 1105–1119.

63. Adler PR, Del Grosso SJ, Parton WJ (2007) Life-Cycle assessment of net greenhouse-gas flux for bioenergy cropping systems. Ecol Appl 17: 675–691.

64. Galdos MV, Cerri CC, Cerri CEP, Paustian K, Van Antwerpen R (2009) Simulation of soil carbon dynamics under sugarcane with the CENTURY Model, Soil Sci Soc Am J 73: 802–811.

65. Vallis I, Parton WJ, Keating BA, Wood AW (1996) Simulation of the effects of trash and N fertilizer management on soil organic matter levels and yields of sugarcane. Soil Tillage Res 38: 115–132.

66. Pepper DA, Del Grosso SJ, McMurtrie RE, Parton WJ (2005) Simulated carbon sink response of shortgrass steppe, tallgrass prairie and forest ecosystems to rising [CO_2], temperature and nitrogen input, Global Biogeochem Cycles 19: GB1004 doi:10.1029/2004GB002226.

67. Kelly RH, Parton WJ, Hartman MD, Stretch LK, Ojima DS, et al. (2000), Intra-annual and interannual variability of ecosystem processes in shortgrass steppe, J Geophys Res 105(D15) 20093–20100 doi:10.1029/2000JD900259.

68. Brady NC, Weil RR (2002) The nature and properties of soils. Prentice Hall, Upper Saddle River, New Jersey. 960 p.

69. Morris DR, Gilbert RA (2005) Inventory, crop use, and soil subsidence of Histosols in Florida. J Food Agr Environ 3: 190–193.

70. Newman Y, Vendramini J, Blount A (2010) Bahiagrass (*Paspalum notatum*): overview and management. University of Florida IFAS Extension. Publication #SS-AGR-332. http://edis.ifas.ufl.edu/ag342. Accessed 2011 Aug 7.

71. Valentine DW, Holland EA, Schimel DS (1994) Ecosystem and physiological controls over methane production in northern wetlands. J Geophys Res 99: 1563–1571.

72. Del Grosso SJ, Parton WJ, Mosier AR, Walsh MK, Ojima DS, et al. (2006) DayCent national scale simulations of N_2O emissions from cropped soils in the USA. J Environ Qual 35: 1451–1460.

73. US Department of Agriculture, National Agricultural Statistics Service website. Available: http://www.nass.usda.gov/Quick_Stats/. Accessed 2010 Feb 25.

74. DAYMET United States Data Center-A source for daily surface weather data and climatological summaries website. Available: www.daymet.org. Accessed 2010 Jun 1.

75. US Department of Agriculture Natural Resources Conservation Service, Web Soil Survey website. Available: http://websoilsurvey.sc.egov.usda.gov/app/HomePage.htm. Accessed 2010 Dec 12.

76. R Development Core Team (2007) R: A language and environment for statistical computing. R Foundation for Statistical Computing, Vienna, Austria.

77. Barbour MG, Billings WD (1988) North American terrestrial vegetation. Press Syndicate of the University of Cambridge. Melbourne, Australia.

78. Pitman WD, Portier KM, Chambliss CG, Kretschmer AE (1992) Performance of yearling steers grazing bahia grass pastures with summer annual legumes or nitrogen fertilizer in subtropical Florida. Trop Grasslands 26: 206–211.

79. Glaz B, Ulloa MF (1993) Sugarcane yields from plant and ratoon sources of seed cane. J Am Soc Sugar Cane Tech 13: 7–13.

80. Wiedenfeld RP, Encisco J (2008) Sugarcane responses to irrigation and nitrogen in semiarid south Texas. Agron J 100: 665–671.

81. Gilbert RA, Shine JM, Miller JD, Rice RW, Rainbolt CR (2006) The effect of genotype, environment and time of harvest on sugarcane yields in Florida, USA. Field Crop Res 95: 156–170.

82. Glaz B (2012) Personal communication.

83. Glaz B, Morris DR (2010) Sugarcane Responses to water-table depth and periodic flood. Agron J 102: 372–380.

84. US Environmental Protection Agency (2011) Florida Sugarcane Metadata. Environmental Protection Agency, Washington DC. Available: http://www.epa.gov/oppefed1/models/water/met_fl_sugarcane.htm. Accessed 2011 August 1.

85. Muchovej RM, Newman PR (2004) Nitrogen fertilization of sugarcane on a sandy soil: II soil and groundwater analysis. J Am Soc Sugar Cane Tech 24: 225–240.

86. University of Illinois, DeLucia Laboratory Public Data Archive website. Available: http://www.life.illinois.edu/delucia/Public%20Data%20Archive/. Accessed 2013 Jun 4.

87. Department of Energy and Climate Change (DECC) and the Department for Environment, Food and Rural Affairs website. Available: https://www.gov.uk/government/publications/2012-guidelines-to-defra-decc-s-ghg-conversion-factors-for-company-reporting-methodology-paper-for-emission-factors. Accessed: 2013 Jul 23.

88. Forster P, Ramaswamy V, Artaxo P, Berntsen T, Betts R, et al. (2007) Climate Change 2007: The Physical Science Basis. Contribution of Working Group I to the Fourth Assessment Report of the Intergovernmental Panel on Climate Change. Cambridge University Press, Cambridge, United Kingdom and New York, NY, USA.

89. Chapin FS, Woodwell G, Randerson J, Rastetter E, Lovett G, et al. (2006) Reconciling carbon-cycle concepts, terminology, and methods. Ecosystems 9: 1041–1050.

90. Crawley MJ (2007) The R Book. John Wiley and Sons, West Sussex, England. 942 p.

91. Storey JD (2003) The positive false discovery rate: a Bayesian interpretation and the q-value. Ann Stat 31: 2013–2035.

92. Graham-Rowe D (2011) Agriculture: beyond food versus fuel. Nature 474: S6–S8. doi: 10.1038/474S06a.

93. Ball-Coelho B, Sampaio EVSB, Tiessen H, Stewart JWB (1992) Root dynamics in plant and ratoon crops of sugar cane. Plant Soil 142: 297–305.

94. Hoy JW, Grisham MP, Damann KE (1999) Spread and increase of ratoon stunting disease of sugarcane and comparison of disease detection methods. Plant Disease 83: 1170–1175.

95. Stirling GR, Blair BL, Pattemore JA, Garside AL, Bell MJ (2001) Changes in nematode populations on sugarcane following fallow, fumigation and crop rotation, and implications for the role of nematodes in yield decline. Australas Plant Pathol 30: 323–335.

96. Yadav RL, Prasad SR (1992) Conserving the organic matter content of the soil to sustain sugarcane yield. Exp Ag 28: 57–62.

97. Chapin FS, Matson PA, Mooney HA (2002) Principals of Terrestrial Ecosystem Ecology. Springer Science, New York, New York, USA.

98. Rowlings D, Grace P, Kiese R, Scheer C (2010) Quantifying N_2O and CO_2 emissions from a subtropical pasture. 19th World Congress of Soil Science, Soil Solutions for a changing World. 1–6 August 2010, Brisbane, Australia.

99. Howden SM, White DH, Mckeon GM, Scanlan JC, Carter JO (1994) Methods for exploring management options to reduce greenhouse gas emissions from tropical grazing systems. Clim Change 27: 49–70.

100. Liebig MA, Gross JR, Kronberg SL, Phillips RL (2010) Grazing management contributions to net global warming potential: A long-term evaluation in the Northern Great Plains. J Environ Qual 39: 799–809.

101. Plevin RJ, O'Hare M, Jones AD, Torn MS, Gibbs HK (2010) Greenhouse gas emissions from biofuels: Indirect land use change are uncertain but may be much greater than previously estimated. Env Sci Tech 44: 8015–8021.

102. Florida Department of Agriculture and Consumer Services website. Available: http://www.fl-ag.com/agfacts.htm. Accessed 2011 Oct 1.

5

Phthalic Acid Esters in Soils from Vegetable Greenhouses in Shandong Peninsula, East China

Chao Chai[1], Hongzhen Cheng[1], Wei Ge[2], Dong Ma[1], Yanxi Shi[1]*

1 College of Resources and Environment, Qingdao Agricultural University, Qingdao, China, 2 College of Life Sciences, Qingdao Agricultural University, Qingdao, China

Abstract

Soils at depths of 0 cm to 10 cm, 10 cm to 20 cm, and 20 cm to 40 cm from 37 vegetable greenhouses in Shandong Peninsula, East China, were collected, and 16 phthalic acid esters (PAEs) were detected using gas chromatography-mass spectrometry (GC-MS). All 16 PAEs could be detected in soils from vegetable greenhouses. The total of 16 PAEs (Σ_{16}PAEs) ranged from 1.939 mg/kg to 35.442 mg/kg, with an average of 6.748 mg/kg. Among four areas, including Qingdao, Weihai, Weifang, and Yantai, the average and maximum concentrations of Σ_{16}PAEs in soils at depths of 0 cm to 10 cm appeared in Weifang, which has a long history of vegetable production and is famous for extensive greenhouse cultivation. Despite the different concentrations of Σ_{16}PAEs, the PAE compositions were comparable. Among the 16 PAEs, di(2-ethylhexyl) phthalate (DEHP), di-n-octyl phthalate (DnOP), di-n-butyl phthalate (DnBP), and diisobutyl phthalate (DiBP) were the most abundant. Compared with the results on agricultural soils in China, soils that are being used or were used for vegetable greenhouses had higher PAE concentrations. Among PAEs, dimethyl phthalate (DMP), diethyl phthalate (DEP) and DnBP exceeded soil allowable concentrations (in US) in more than 90% of the samples, and DnOP in more than 20%. Shandong Peninsula has the highest PAE contents, which suggests that this area is severely contaminated by PAEs.

Editor: Raffaella Balestrini, Institute for Plant Protection (IPP), CNR, Italy

Funding: This work was supported by the "Science and Technology Plan Projects of Qingdao (No. 12-1-3-64-nsh), the Two Districts" Foundation of Shandong Province, China (No. 2011-Yellow-19) and the Talent Foundation of Qingdao Agricultural University (No. 630642). The funders had no role in study design, data collection and analysis, decision to publish, or preparation of the manuscript.

Competing Interests: The authors have declared that no competing interests exist.

* E-mail: yanxiyy@126.com

Introduction

Phthalic acid esters (PAEs) are used extensively as plasticizers of plastic products, such as polyvinyl chloride, and as nonplasticizers in consumer products, including medical devices, building materials, paints, pesticides, fertilizes, food packaging, and so on [1]. The large-scale production and application of 6.0 million tons/yr [2] of PAEs have made these materials ubiquitous environment pollutants [3–8]. Some PAEs have endocrine disruptive effects [9], and six PAEs are categorized as priority environmental pollutants by the United States Environmental Protection Agency [10].

Greenhouse cultivation has expanded dramatically in China since the 1980s, reaching up to 3.5 million ha by 2011 [11]. Greenhouse cultivation is mainly for vegetable production in China, and plastic greenhouses account for more than 99% of greenhouse cultivation relative to glass greenhouses [12–13]. Several studies detected PAEs in soils of vegetable greenhouses in Nanjing and Hangzhou [14–15], as well as in other agricultural soils, such as vegetable soils in Guangzhou and paddy soils in Leizhou Peninsula in China [16–17]. The buildup of PAEs in agricultural soils may contaminate agricultural products, and further raise the human health risk [18].

Shandong Peninsula is the largest Peninsula in China with rapid urbanization and high population density of 550 people/km². The Peninsula includes the cities of Qingdao, Yantai, Weifang, and Weihai. Shandong Peninsula has a long history of vegetable greenhouse cultivation and is a main vegetable-producing region,

with its greenhouse coverage accounting for approximately 50% of that of China. The vegetable greenhouses in this peninsula are close to the highly populated urban areas, and plastic film is widely used. Plastic film of 30000 tons/yr is estimated to be used only in one county, i.e., Shouguang in Weifang of Shandong Peninsula [15]. PAEs account for 10 wt% to 60 wt% of plastic products [9,19], thus giving rise to concerns about the potential risk of PAEs in recent years. However, few studies focused on the characteristics of PAEs in soils of vegetable greenhouses in Shandong Peninsula.

This study provides information on the concentrations, compositions, and distributions of 16 PAEs in soils from vegetable greenhouses in Shandong Peninsula and discusses possible sources, influence factors, and potential environment risk.

Materials and Methods

Chemicals and materials

Mixed standard solutions of 16 PAEs containing dimethyl phthalate (DMP), diethyl phthalate (DEP), diisobutyl phthalate (DiBP), di-n-butyl phthalate (DnBP), dimethylglycol phthalate (DMGP), di(4-methyl-2-pentyl) phthalate (DMPP), di(2-ethylhexyl) phthalate (DEHP), di(2-ethoxyethyl) phthalate (DEEP), dipentyl phthalate (DPP), di-n-hexyl phthalate (DHXP), butylbenzyl phthalate (BBP), di(2-n-butoxyethyl) phthalate (DBEP), dicyclohexyl phthalate (DCHP), di-n-octyl phthalate (DnOP), diphenyl phthalate (DPhP), and di-n-nonyl phthalate (DNP) were supplied by O2SI, Inc. (USA). The concentration of each PAE in this mixture solution was 1000 mg/L. Glassware was steeped with

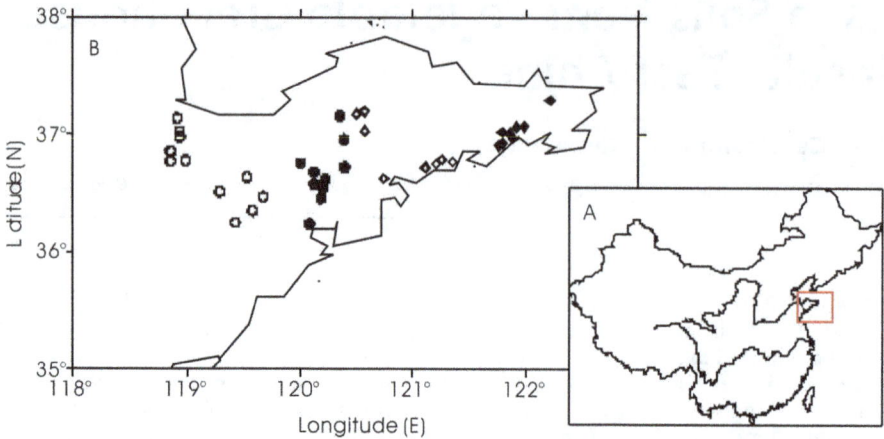

Figure 1. Schematic map showing the geographical location of (A) Shandong Peninsula and (B) the vegetable soil sampling sites in 4 regions in the Shandong Peninsula (solid round: Qingdao; solid diamond: Weihai; circle: Weifang; diamond: Yantai).

K_2CrO_7/H_2SO_4 solution for 12 h, washed with redistilled water, and then baked at 300°C for 4 h. Acetone, petroleum ether, and diethyl ether were of analytical grade and re-distilled before use to avoid PAEs contamination. Hexane was of HPLC grade and purchased from Anpel Company Inc. Florisil (60 mesh to 80 mesh) was activated at 650°C, and anhydrous sodium sulfate was baked at 420°C for 4 h.

Sampling

No specific permissions were required for sampling locations/ activities. The field studies did not involve endangered or protected species. A total of 111 soil samples were collected from 37 vegetable greenhouses in Qingdao (number of samples: 30), Weihai (number of samples: 24), Weifang (number of samples: 33), and Yantai (number of samples: 24) in Shandong Peninsula in from 28 to 30 May 2012. The sampling locations are shown in Fig. 1.

Each sampling site consisted of five sub-samples (0.2 kg each) in the middle and four corners at depths of 0 cm to 10 cm, 10 cm to 20 cm, and 20 cm to 40 cm. The five sub-samples were mixed

immediately after sampling, and then the soils were collected using aluminum foil envelopes through a pre-cleaned stainless steel auger and transported to laboratory in an ice box. Soils were stored in glass bottles at −20°C until analysis after being freeze-dried, ground, and homogenized with a stainless steel sieve (60 mesh). PAE contamination was avoided during sampling and further processing.

Soil physical and chemical analyses

Soil pH was measured using a pH meter with a soil/water ratio of 1:2.5. Soil cation exchange capacity (CEC) was analyzed using the Ba^{2+} compulsive exchange method [20]. Particle-size fraction was determined using the pipette method, and the soil texture was classified according to the Soil Survey Division Staff [21]. Total organic carbon (TOC) was determined using the wet oxidation method with chromate [22] and total nitrogen (TN) using micro-Kjeldahl digestion method [23].

Table 1. The main characteristics of the soils from vegetable greenhouses in Shandong Peninsula.

Area	Soil depth (cm)	pH	TOC (g/kg)	TN (g/kg)	C/N	CEC (mol/kg)	Sand (%)	Silt (%)	Clay (%)
Qingdao	0~10	6.62±0.64	31.7±9.8	1.3±0.4	26.44±0.90	0.14±0.05	54.1±3.8	27.6±6.6	17.8±3.7
	10~20	6.52±0.56	29.6±13.3	1.1±0.3	26.02±0.36	0.15±0.06	53.2±5.4	25.9±4.6	15.5±4.6
	20~40	6.64±0.42	25.1±7.1	0.8±0.3	33.49±9.12	0.11±0.05	46.7±4.6	24.6±7.8	15.9±4.2
Weihai	0~10	6.31±0.56	30.0±3.8	1.4±0.6	23.89±6.93	0.07±0.03	50.7±3.4	22.1±0.7	17.1±2.7
	10~20	6.10±0.43	27.7±4.1	1.0±0.1	27.33±2.72	0.09±0.01	55.3±6.2	29.6±9.9	15.6±1.4
	20~40	5.90±0.55	19.3±7.2	0.7±0.2	27.32±9.13	0.08±0.04	52.6±4.5	33.2±1.6	16.0±2.5
Weifang	0~10	6.88±0.46	32.2±6.0	1.7±3.1	40.49±5.39	0.13±0.09	61.3±0.2	37.8±7.7	19.9±6.0
	10~20	6.96±0.36	27.9±12.1	0.9±0.9	41.81±8.63	0.13±0.06	57.0±9.8	33.6±2.6	17.9±5.6
	20~40	6.99±0.49	23.4±4.3	1.2±2.3	42.55±5.78	0.14±0.08	59.2±2.1	28.1±5.9	21.1±7.4
Yantai	0~10	6.46±0.66	24.0±6.4	1.3±0.4	18.95±5.92	0.10±0.02	57.8±3.4	38.7±9.5	18.9±4.6
	10~20	6.56±0.96	23.7±6.3	1.1±0.4	22.99±5.90	0.10±0.03	57.4±4.2	38.6±4.4	18.0±3.0
	20~40	7.10±0.99	20.5±5.0	0.7±0.1	29.22±9.46	0.11±0.03	55.1±6.1	33.6±6.6	16.3±1.7

Table 2. The detection rate and concentration of PAEs in all soil samples from vegetable greenhouses in Shandong Peninsula (n = 111).

PAEs	Detection rate (%)	Mean (mg/kg)	SD (mg/kg)	Minimum (mg/kg)	Maximum (mg/kg)
DMP	99.1	0.364	0.276	ND	1.245
DEP	100	0.108	0.169	0.002	1.051
DiBP	96.4	1.118	1.928	ND	11.434
DnBP	100	1.471	2.715	0.016	15.722
DMGP	23.4	0.015	0.031	ND	0.170
DMPP	48.6	0.246	0.405	ND	1.971
DEHP	100	1.465	1.207	0.073	5.323
DEEP	23.4	0.041	0.243	ND	2.556
DPP	58.6	0.088	0.098	ND	0.516
DHXP	64.9	0.084	0.157	ND	1.448
BBP	86.5	0.194	0.557	ND	5.691
DBEP	18.9	0.015	0.038	ND	0.267
DCHP	44.1	0.035	0.048	ND	0.204
DnOP	97.3	1.239	1.796	ND	14.397
DPhP	82.0	0.240	0.290	ND	2.371
DNP	19.8	0.026	0.060	ND	0.251
Σ_{16}PAEs		6.748	5.716	1.939	35.442

ND: not detected. The data labeled as "ND" were treated as zero in further statistical treatment.

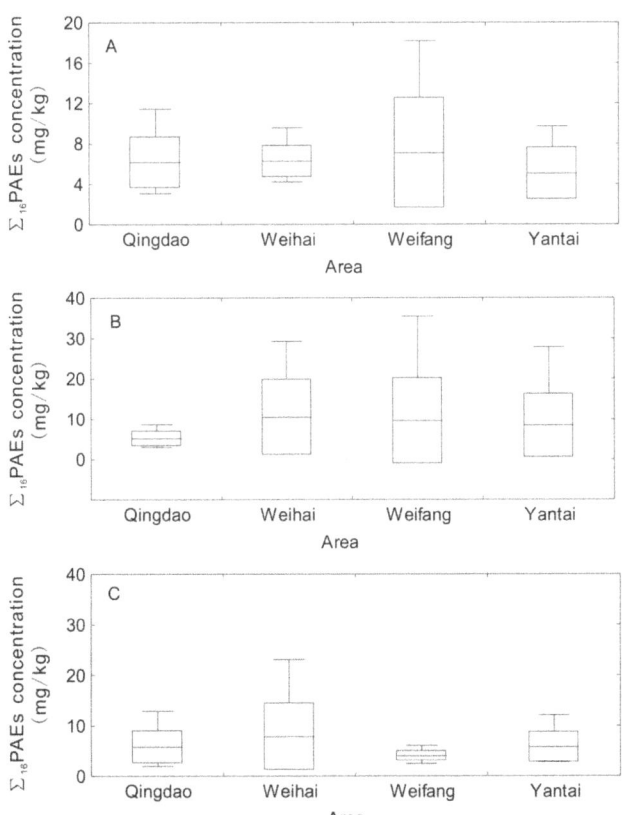

Figure 2. The concentrations of Σ_{16}PAEs in (A) soils of 0–10 cm, (B) soils of 10–20 cm and (C) soils of 20–40 cm from vegetable greenhouses in Shandong Peninsula.

Sample extraction of PAEs and instrumental analysis

PAEs extraction was conducted according to Wang's methods [24]. 5.0 g soil was spiked with surrogate standard (benzyl benzoate) and extracted through ultrasonication for 15 min thrice with 90 mL of acetone/petroleum ether (1:3, v:v). The extracts were combined, filtered, and concentrated to approximately 1 mL. The extracts were cleaned with anhydrous sodium sulfate (3 g), florisil (6 g), and anhydrous sodium sulfate (3 g) on a glass column (1 cm i.d.). The column was washed with 10 mL of petroleum ether/diethyl ether (10:0.4, v:v), and then PAEs were eluted with 90 mL of petroleum ether/diethyl ether (10:3, v:v). The extracts were reduced to 1.0 mL in hexane, and internal standard (diisophenyl phthalate) was added before instrumental analysis.

Instrumental analysis was performed on an Agilent 6890 GC-5973 MSD gas chromatography-mass spectrometry system (GC-MS) in electron impact and selective ion monitoring modes according to Zeng et al. [25]. The GC column used was DB-5MS capillary column (30 m×0.25 mm i.d. ×0.25 mm film thickness, J&W Scientific). The column temperature program was 80°C (1 min), to 180°C (10°C/min, 1 min), to 300°C (2°C/min, 10 min). The carrier gas was helium with flow rate of 0.8 mL/min. Then, 1 μL of the extracts was injected into GC-MS in splitless injection mode, and the injector temperature was 250°C. The GC-MS transfer line was 280°C, and the post run temperature was 285°C for 2 min.

Quality control and quality assurance

Quality assurance was performed by analyzing a procedural and solvent blank, a spiked blank every 10 samples, surrogate standards for each sample, and sample duplicate. DiBP, DnBP, and DEHP were subtracted from those in the soil samples because of the small amount in procedural blanks. The surrogate recoveries were 84.1%±8.5%, and no surrogate corrections were

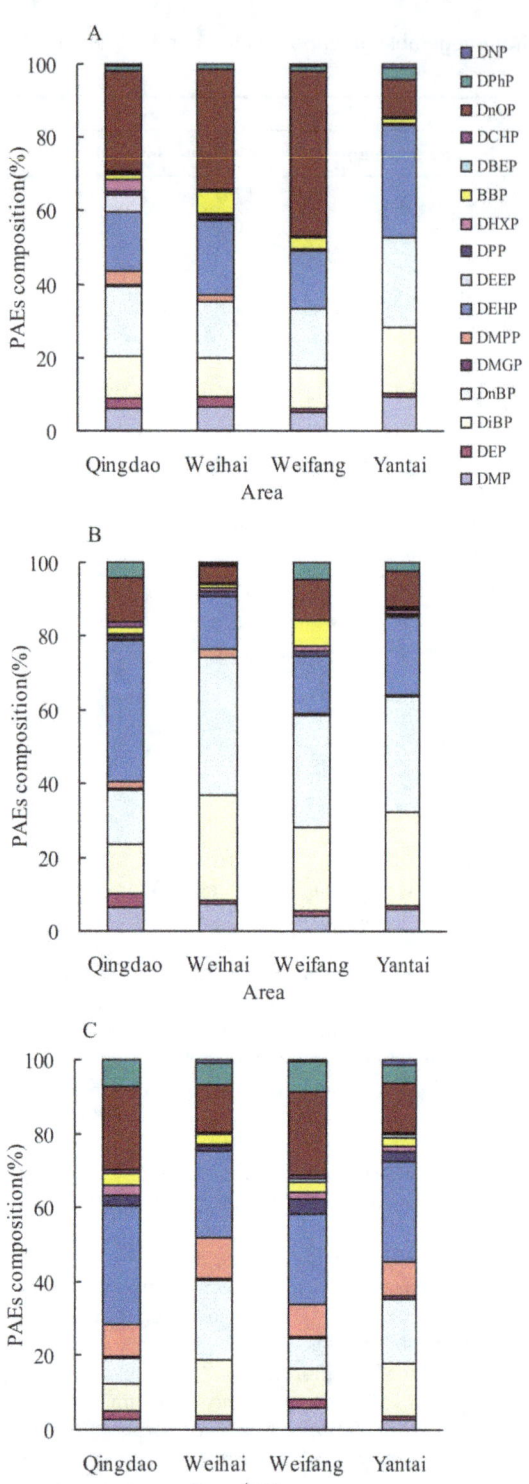

Figure 3. The compositions of PAEs in (A) soils of 0–10 cm, (B) soils of 10–20 cm and (C) soils of 20–40 cm from vegetable greenhouses in Shandong Peninsula.

made to the final PAE concentrations reported. The calibration curves were used with six concentration levels of a standard mixture for PAEs quantitation. The standard mixture was analyzed every 10 samples to determine instrument stability and to confirm the calibration curve. The instrumental detection limits

ranged from 1~9 pg, calculated by signal to noise ratio of 10. The method detection limits for PAEs were determined as mean field blanks plus three times the standard deviation of the field blanks [25], ranging from 0.002 mg/kg for DEP to 0.024 mg/kg for DEHP.

Results

Properties of the soils from vegetable greenhouses

The major characteristics of soils from vegetable greenhouses in Shandong Peninsula are presented in Table 1. The pH (H_2O) of soils was neutral in three sample areas, but was less than 6.5 in Weihai, which was moderately acidic. The TOC ranged from 19.3 g/kg to 32.2 g/kg and presented a decreasing trend with soil depth. The C/N ratio was approximately 20 to 30, except in Weifang, which had a value of more than 40 on average. This value indicated that the organic matter content was more C-rich. The C/N ratio presented an increasing trend with soil depth. The CEC followed a similar pattern as pH, with lower values in soil samples from Weihai. Most of the soils were sandy loam, and some were loam.

PAE concentrations in soils from vegetable greenhouses in Shandong Peninsula

All 16 PAEs were detected in soils from vegetable greenhouses (Table 2). Among them, three PAEs, namely, DEP, DnBP, and DEHP, were detected in all samples. The detection rates of another three PAEs (DMP, DiBP, and DnOP) were more than 90%. By contrast, the detection rates of DBEP and DNP were lower than 20%. The mean concentrations of DiBP, DnBP, DEHP, and DnOP were more than 1 mg/kg, higher than other PAEs. On the whole, the mean was almost systematically inferior to standard deviation, suggesting a very high heterogeneity between soils. The total of 16 PAEs (Σ_{16}PAEs) in Shandong Peninsula ranged from 1.939 mg/kg to 35.442 mg/kg, with an average of 6.748 mg/kg.

The concentrations of Σ_{16}PAEs in soils from vegetable greenhouses in different areas of Shandong Peninsula are presented in Fig. 2. The variability of Σ_{16}PAEs was high in Weifang in the two upper layers, but low in Qingdao (10 cm to 20 cm) and Weifang (20 cm to 40 cm). The maximum value of Σ_{16}PAEs in soils at 0 cm to 10 cm and 10 cm to 20 cm both appeared in Weifang, with values of 18.179 and 35.442 mg/kg, respectively.

PAE composition in soils of vegetable greenhouses from Shandong Peninsula

Despite the different concentration of Σ_{16}PAEs, the PAE compositions in soils from vegetable greenhouses in Shandong Peninsula were comparable (Fig. 3). DnOP had the highest proportion (27.1% to 45%) in soils at 0 cm to 10 cm in Qindao, Weihai, and Weifang, whereas DEHP had the highest proportion (30.4%) in Yantai. The proportion of DnBP and DiBP ranged from 10.6% to 24.5%, suggesting significant proportion. In soils at 10 cm to 20 cm, DEHP had the largest proportion of 38.4% in Qingdao, but DnBP had the largest and DiBP had the second largest proportion in the other three areas. In soils at 20 cm to 40 cm, DEHP was a dominant congener, ranging from 23.4% to 31.8% in four areas. DnOP in Qingdao and Weifang had the second highest proportion, whereas DnBP in Weihai and Yantai had the second highest proportion. Therefore, DEHP, DnOP, DnBP, and DiBP had the largest proportion in soils in all areas. In addition, DMP accounted for about 5~10% in the upper two

Table 3. Comparisons of PAEs contents in agricultural soils in China (mg/kg).

Area	DMP	DEP	DnBP	DEHP	BBP	DnOP	Σ_6PAEs	Σ_{16}PAEs	Type of soils	References
Shandong Peninsula	0.36 (ND~1.24)	0.11 (ND~1.21)	1.47 (0.02~15.72)	1.47 (0.07~5.32)	0.19 (ND~5.69)	1.24 (ND~14.40)	4.48 (1.18~23.35)	6.73 (1.94~35.44)	Soils of vegetable greenhouses	In this study
Nanjing	0.006 (ND~0.016)	0.005 (ND~0.012)	0.19 (ND~1.41)	1.72 (0.034~9.031)	0.003 (ND~0.038)	0.158 (ND~1.739)	1.89 (0.15~9.68)		Soils of vegetable greenhouses	[14]
Hangzhou	ND	0.59 (0.06~1.49)	0.21 (0.14~0.35)	1.48 (0.81~2.20)	0.05 (0.03~0.16)	0.14 (0.10~0.25)	2.47 (1.90~4.36)	2.75 (Σ_{11}PAEs)	Soils of vegetable greenhouses	[15]
Guangzhou, Shenzhen	(ND~0.68)	(ND~1.77)	(3.75~18.45)	(2.82~25.11)	(ND~1.48)	(ND~0.92)	(3.00~45.67)		vegetable Soils	[16]
Zhangjiang	0.02 (ND~0.45)	0.09 (ND~1.06)	0.23 (ND~7.65)	0.15 (ND~6.38)	0.05 (ND~2.83)	0.03 (ND~0.32)	0.56 (0.01~9.30)		Vegetable soil	[50]
Dongguan	0.02 (ND~0.86)	0.05 (ND~1.60)	0.49 (ND~17.51)	0.17 (ND~4.22)	0.13 (ND~5.89)	0.06 (ND~1.12)	0.92 (0.01~25.99)		Paddy soil	
Zhongshan	0.03 (ND~0.12)	0.19 (ND~2.50)	0.41 (ND~4.13)	0.12 (ND~2.69)	0.02 (ND~0.26)	0.02 (ND~0.08)	0.81 (0.05~5.92)		Banana soil	
Zhuhai	0.03 (ND~0.18)	0.06 (ND~0.44)	0.30 (ND~1.77)	0.07 (ND~0.26)	0.01 (ND~0.03)	0.02 (ND~0.07)	0.49 (ND~2.10)		Sugarcane soil	
Shunde	0.02 (ND~0.06)	0.06 (ND~0.40)	0.20 (ND~0.56)	0.11 (ND~0.99)	0.01 (ND~0.06)	0.02 (ND~0.15)	0.43 (0.04~1.38)		Orchard soil	
Leizhou Peninsula	0.02	0.02	0.45	0.01	0.01	0.02	0.53	1.11 (0.02~5.45)	Sugarcane soil	[17]
	0.01	0.01	0.24	0.24	0.01	0.02	0.53	0.86 (ND~2.78)	Paddy soil	
	0.02	0.01	0.27	0.12	0.01	0.02	0.45	0.61 (0.02~1.87)	Vegetable soil	
	0.03	0.01	0.28	0.10	0.01	0.01	0.44	0.60 (0.28~1.05)	Orchard soil	
Huizhou	0.004 (ND~0.03)	0.01 (ND~0.22)	0.15 (ND~0.39)	0.09 (ND~0.44)	0.002 (ND~0.04)	0.01 (ND~0.06)	0.31 (0.09~0.75)	0.60 (0.18~2.04)	Agricultural soil	[51]
							0.28 (0.06~0.64)	0.59 (0.18~2.04)	Vegetable soil	
							0.24 (0.08~0.64)	0.65 (0.43~1.21)	Paddy soil	

Table 3. Cont.

Area	DMP	DEP	DnBP	DEHP	BBP	DnOP	Σ_6PAEs	Σ_{16}PAEs	Type of soils	References
							0.22	0.51	Orchard soil	
							(0.08~0.64)	(0.38~0.63)		

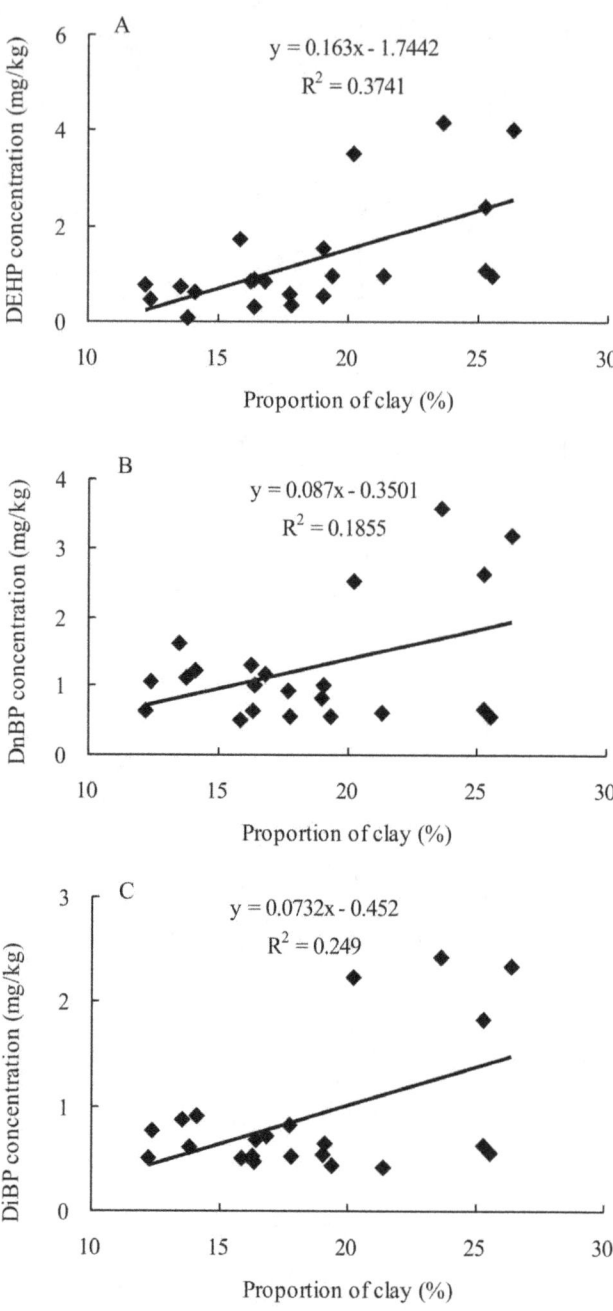

Figure 4. Correlations of the concentrations of (A) DEHP, (B) DnBP and (C) DiBP with the proportions of clay in 0–10 cm soils of vegetable greenhouses from Shandong Peninsula.

layers; by contrast, DEP, DMPP, BBP, and DPhP accounted for only approximately 1% to 5%, suggesting a small proportion.

Discussion

Potential sources of PAEs in soils from vegetable greenhouses

Soils that are being used or were used for vegetable greenhouses had higher PAE concentrations (Table 3), which suggests that PAEs are widespread in soils from vegetable greenhouses. Various PAE sources exist in soils from vegetable greenhouses. The plastic film used in vegetable greenhouses is a major source of PAEs. The

Table 4. Ratio of PAEs in samples exceeding allowable and cleanup concentrations in soils from vegetable greenhouses in Shandong Peninsula.

Soil depth (cm)	Ratio of samples exceeding allowable concentrations (%)						Ratio of samples exceeding cleanup concentrations (%)					
	DMP	DEP	DnBP	BBP	DEHP	DnOP	DMP	DEP	DnBP	BBP	DEHP	DnOP
0~10	94.6	89.2	97.3	0	2.7	45.9	0	0	0	0	0	0
10~20	100	94.6	94.6	2.7	5.4	27	0	0	13.5	0	0	0
20~30	89.2	97.3	89.2	0	0	21.6	0	0	2.7	0	0	0

maximum value of Σ_{16}PAEs in soils at 0 cm to 10 cm and 10 cm to 20 cm appeared in Weifang (Fig. 2), which has a long history of vegetable production and is famous for extensive greenhouse cultivation. Even more remarkably, the plastic film used in greenhouse cultivation in Shandong Peninsula is replaced annually, which may result in a higher concentration of PAEs in soils from vegetable greenhouses than in other soils. In addition, PAEs are found in organic fertilizers in China [26] and in compost of sewage sludge with rice straw [27]. The amount of fertilizers used for vegetable planting in greenhouses is more than that used for field crops, and the proportion of organic fertilizers has increased since 2007 [28]. Moreover, PAEs are found in the groundwater in China [29–31], and groundwater is used for irrigation in vegetable greenhouse, which may result in the buildup of PAEs in vegetable greenhouse soils. More importantly, Zeng et al. [32] found a declining trend of PAEs in agricultural soils that were far from urban centers. The highest PAE contents are found in soils close to architectural markets, where plastic materials are sold, and those close to large chemical manufacturing factories. Most vegetable greenhouses in this study are near industrialized cities with large populations. Over 300 plastic manufacturers that produced 0.3 million tons/yr of plastic existed in Shandong Peninsula by 2003. All these factors may have resulted in the high concentration of PAEs in vegetable greenhouses in Shandong Peninsula.

Among 16 PAEs, DEHP and DnBP are found to be the two most abundant PAEs in agricultural soils in Guangzhou, Shenzhen, Leizhou Peninsula, and Huizhou (Table 3). Moreover, DiBP is found to be abundant in Guangzhou agricultural soils [32], whereas DnOP is abundant in the soils of vegetable greenhouses in Nanjing [14]. Similarly, DEHP and DnOP are the two most abundant PAEs in soils of vegetable greenhouses in Shandong Peninsula, followed by DnBP and DiBP (Fig. 3). The relative contribution of PAEs in agricultural soils is consistent with that in sediment [33], air [5,34], and waters [35]. The global consumption of PAEs is about 6.0 million tones/yr, mainly as plasticizers in the plastic industry. Among plasticizers of PAEs, DEHP, DnOP, and DiBP/DnBP are widely used. It is found that DEHP and DnBP are two dominant PAE components in white and black mulch film used in vegetable production systems, ranging from 48.0~115.6 mg/kg and 2.3~3.2 mg/kg, respectively [14]. We also found that besides DEHP, DnOP and DiBP were two dominant PAEs in polyvinyl chloride (PVC) plastics mainly used in vegetable greenhouses in Shandong Peninsula, accounting for 20% and 10% of total of 16 PAEs, respectively. DMP and DEP are also detected in the plastics film, though their contents are low. Therefore, the plastics film may be a major potential source of some PAEs. Furthermore, PAEs are found in fertilizer and manure. DEHP, DnBP, DMP and DEP are the major organic pollutants in fertilizers, with contents more than

2.5 mg/kg [27]. Similarly, six PAEs (DEHP, DnBP, DnOP, DMP, DEP and BBP) are found in chicken, pig, cow and duck manure, in the range of 2.24~6.84 mg/kg [14]. These potential sources may lead to the high detection rates of DMP, DEP DnBP, DiBP, DEHP and DnOP (Table 2).

Relationship between PAEs and soil properties or age of vegetable greenhouses

Soil properties, such as pH, organic matter, texture, and redox potential, have a certain effect on the migration of hydrophobic organic compounds (HOCs) in soil [36–37]. A positive correlation between HOCs and TOC is found in several research [32,38]; however, it is not in this study. Katsoyiannis [39] reported that no correlation can be found between HOCs and TOC if continuous sources of HOCs exist in soils. In this study, several continuous inputs of PAEs in vegetable greenhouse soils, including plastic film, fertilizers, and irrigation, may hinder the achievement of equilibrium between PAEs and TOCs.

The relationship among the major PAEs, including DEHP, DnBP and DiBP, with the proportions of clay in 0 cm to 10 cm soils was analyzed (Fig. 4). DiBP, DnBP, and DEHP have a significantly positive correlation with the proportion of clay ($r = 0.431~0.611$, $p < 0.05$). A similar relationship of HOCs, such as PAEs, organic chlorinated pesticides, and PAHs, with clay is also found in soils [32,40–41]. The clay of sediment or soil shows stronger capability to adsorb HOCs than sand and silt, due to small granulometry but high specific surface area [42–43]. Besides, the aging of organic matter, such as humic material, distributes around clay complexes, resulting in the formation of films of organic material [44]. These films of organic material are very difficult to remove, and so organic matter builds up and becomes a permanent part of the clay complexes. The clay-organic complexes supply rich reactive sites for the adsorption of organic pollutants [45].

A positive correlation between Σ_{16}PAE concentration and age of vegetable greenhouses was found in this study ($r = 0.294$, $p < 0.01$), suggesting that PAEs in soils may be related with the cumulative use of potential PAE sources over years in greenhouse vegetable cultivation. However, the correlation coefficient is not high. Studies demonstrate the biodegradability of some PAEs in aerobic condition [46–47], though PAEs are resistant to degradation through hydrolysis and photolysis. The biodegradability of PAE congeners is different. Shanker et al found the degradation rates of DMP and DBP were greater than that of DEHP under aerobic conditions [48]. Additionally, PAEs migrates deeper in soils profiles, and the TOC of soil and volume of leaching water can affect the migration of PAEs [49]. These factors may result in the low correlation coefficient between PAE contents and age of vegetable greenhouses.

Comparison of PAE concentrations in the different soils in China

Comparisons of PAE contents in agricultural soils from China are presented in Table 3. The average Σ_6PAE (DMP, DEP, DnBP, DEHP, BBP, and DnOP) contents in the soils from vegetable greenhouses in Shandong Peninsula, Nanjing, and Hangzhou are approximately 2 mg/kg to 6 mg/kg, higher than other types of soils. High Σ_6PAE concentrations are also found in vegetable soils from Guangzhou and Shenzhen, where soils are previously used to plant greenhouse vegetables. In comparison, the Σ_6PAE concentrations in vegetables, paddy, banana, sugarcane, or orchard soil are low, ranging from 0.2 mg/kg to 1 mg/kg.

Potential risk assessment of soils of vegetable greenhouses from Shandong Peninsula

PAEs have a variety of toxic effects. Long term exposure to PAEs results in decreased fertility in females, fetal defect, altered hormone levels, uterine damage and male reproduction abnormalities such as reduced sperm production and motility, cell damage, cell tumors, etc [52–55]. According to human health based levels that correspond to excess lifetime cancer risks and human health based levels for systemic toxicant calculated from reference doses, allowable soil concentrations and cleanup levels of PAEs have been recommended in New York, USA [56]. The allowable soil concentrations are 0.02, 0.071, 0.081, 1.125, 4.35, and 1.20 mg/kg for DMP, DEP, DnBP, BBP, DEHP, and DnOP, respectively; and soil cleanup levels are 2, 7.1, 8.1, 50, 50, and 50 mg/kg, respectively [56]. PAEs exceeding allowable and cleanup concentrations may be a menace to human health. According to these criteria, the ratios of PAE concentration in this study exceeding allowable and cleanup concentrations are presented in Table 4. The ratios of DMP, DEP, and DnBP exceeding allowable concentrations are 90% to 100% at different soil depths, suggesting high PAE pollution. Moreover, the ratios of DnOP exceeding allowable concentrations are also high, particularly in soils at 0 cm to 10 cm; however, the ratios of BBP and DEHP are low. Similarly, in agricultural soils around Guangzhou, DMP, DEP, and DnBP also exceed allowable concentrations [32]. Notably, DnBP in some samples is approximately twice to thrice higher than the recommended cleanup concentration. These soil samples are mostly from the vegetable greenhouses with ages of approximately 10 years, suggesting that long-term application of plastic film or manure in vegetable greenhouses may increase environmental and health risks.

The cultivated vegetables can uptake and accumulate PAEs, but the difference is found in accumulated amount of PAE congeners by vegetables. Compared with DEHP, more DBP in soils is accumulated in stalk and leaf of carrot, cucumber and tomato [24]. The physical and chemical properties, such as molecular weight and octanol/water partition coefficient (K_{ow}), have effects on the accumulation of PAEs by plants. Due to the smaller molecular weight and lower K_{ow}, DBP is more easily absorbed and transported by vegetables than DEHP. Furthermore, several studies report a positive correlation between accumulated PAE amount by vegetables and contents in soils [24,57]. Thus, mitigation of PAEs in soils is important to lower the risks of PAEs to human health.

Author Contributions

Conceived and designed the experiments: CC HC YS. Performed the experiments: CC HC. Analyzed the data: HC WG. Contributed reagents/materials/analysis tools: WG DM. Wrote the paper: CC HC YS.

References

1. Staples CA, Peterson DR, Parkerton TF, Adams WJ (1997) The environmental fate of phthalate esters, a literature review. Chemosphere 4: 667–749.
2. Xie Z, Ebinghaus R, Temme C, Lohmann R, Caba A, et al. (2007) Occurrence and air-sea exchange of phthalates in the Arctic. Environmental Science & Technology 41: 4555–4560.
3. Lin C, Lee C, Mao W, Nadim F (2009) Identifying the potential sources of di-(2-ethylhexyl) phthalate contamination in the sediment of the Houjing River in southern Taiwan. Journal of Hazardous Materials 161: 270–275.
4. Buszka PM, Yeskis DJ, Kolpin DW, Furlong ET, Zaugg SD, et al. (2009) Waste-indicator and pharmaceutical compounds in landfill-leachate-affected ground water near Elkhart, Indiana, 2000–2002. Bulletin of Environmental Contamination and Toxicology 82: 653–659.
5. Wang P, Wang SL, Fan CQ (2008) Atmospheric distribution of particulate- and gas-phase phthalic esters (PAEs) in a Metropolitan City, Nanjing, East China. Chemosphere 72(10):1567–1572.
6. Zhang LF, Dong L, Ren LJ, Shi SX, Zhou L, et al. (2012) Concentration and source identification of polycyclic aromatic hydrocarbons and phthalic acid esters in the surface water of the Yangtze River Delta, China. Journal of Environmental Sciences 24(2): 335–342.
7. Bauer MJ, Herrmann R (1997) Estimation of the environmental contamination by phthalic acid esters leaching from household wastes. Science of the Total Environment 208(1–2): 49–57.
8. Amir S, Hafidi M, Merlina G, Hamdi H, Jouraiphy A, et al. (2005) Fate of phthalic acid esters during composting of both lagooning and activated sludges. Process Biochemistry 40(6): 2183–2190.
9. Hens GA, Caballos AMP (2003) Social and economic interest in the control of phthalic acid esters. Trends in Analytical Chemistry 22, 847–857.
10. United States Environmental Protection Agency (USEPA) (2013) Electronic Code of Federal Regulations, Title 40-Protection of Environment, Part 423-Steam Electric Power Generating Point Source Category. Appendix A to Part 423–126, Priority Pollutants. http://www.ecfr.gov/cgi-bin/text-idx?c¼ecfr&SID¼b960051a53c9015d817718d71f1617b7&rgn¼div5&view¼text&node¼40,30.0.1.1.23&idno¼40#40:30.0.1.1.23.0.5.9.9
11. Li ZH, Wang GZ, Qi F (2012) Current Situation and Thinking of Development of Protected Agriculture in China. Chinese Agricultural Mechanization 239(1): 7–10(in Chinese with English abstract).
12. Costa JM, Heuvelink E (2004) Protected cultivation rising in China. Fruit & Vegetable Technology 4: 8–11.
13. Zou ZR (2002) Facility Horticulture Science. China agriculture press, Beijing (in Chinese).
14. Wang J, Luo YM, Teng Y, Ma WT, Christie P, et al. (2013) Soil contamination by phthalate esters in Chinese intensive vegetable production systems with different modes of use of plastic film. Environmental Pollution 180: 265–273.
15. Chen YS, Luo YM, Zhang HB, Song J (2011) Preliminary study on PAEs pollution of greenhouse soils. ACTA Pedologica Sinica 48(3): 516–523(in Chinese with English abstract).
16. Cai QY, Mo CH, Li YH, Zeng QY, Wang BG, et al. (2005) The study of PAEs in soils from typical vegetable fields in areas of Guangzhou and Shenzhen, South China. ACTA Ecologica Sinica 25(2): 283–288(in Chinese with English abstract).
17. Guan H, Wang JS, Wan HF, Li PX, Yang GY (2007) PAEs Pollution in soils from typical agriculture area of Leizhou Peninsula. Journal of Agro-Environment Science 26(2): 622–628(in Chinese with English abstract).
18. Mariko M, Mutsuko HK, Makoto E (2008) Potential adverse effects of phthalic acid esters on human health: A review of recent studies on reproduction. Regulatory Toxicology and Pharmacology 50(1): 37–49.
19. Chou K, Robert OW (2006) Phthalates in food and medical devices. Journal of Medical Toxicology 2: 126–135.
20. Bascomb CL (1964) Rapid method for the determination of cation-exchange capacity of calcareous and non-calcareous soils. Journal of the Science of Food and Agriculture 15(12): 821–823.
21. Soil Survey Division Staff. Soil Survey Manual (1993) In: Agriculture Handbook, Revised Edition, vol. 18. United States Department of Agriculture, Washington DC.
22. Schwartz V (1995) Fractionated combustion analysis of carbon in forest soils - new possibilities for the analysis and characterization of different soils. Fresenius' journal of analytical chemistry 351(7): 629–631.
23. Flowers TH, Bremner JM (1991) A rapid dichromate procedure for routine estimation of total nitrogen in soils. Communications in Soil Science and Plant Analysis 22(13–14): 1409–1416.
24. Wang ML (2007) Research on analytical method and environmental behavior of PAEs in vegetable greenhouse. Shandong Agricultural University. Doctor thesis (in Chinese with English abstract).
25. Zeng F, Cui KY, Xie ZY, Wu L, Luo DL, et al. (2009) Distribution of phthalate esters in urban soils of subtropical city, Guangzhou, China. Journal of Hazardous Materials 164: 1171–1178.

26. Cai QY, Mo CH, Wu QT, Zeng QY, Katsoyiannis A (2007) Quantitative determination of organic priority pollutants in the composts of sewage sludge with rice straw by gas chromatography coupled with mass spectrometry. Journal of Chromatography A 1143: 207–214.

27. Mo CH, Cai QY, Li YH, Zeng QY (2008) Occurrence of priority organic pollutants in the fertilizers, China. Journal of Hazardous Materials 152: 1208–1213.

28. Liu ZH, Jiang LH, Zhang WJ, Zheng FL, Wang M, et al. (2008) Evolution of fertilization rate and variation of soil nutrient contents in greenhouse vegetable cultivation in shandong. ACTA Pedologica Sinica 45 (2): 296–303 (in Chinese with English abstract).

29. Zhang D, Liu H, Liang Y, Wang C, Liang HC, et al. (2009) Distribution of phthalate esters in the groundwater of Jianghan plain, Hubei, China. Frontiers of Earth Science in China 3(1): 73–79.

30. Xiong PX, Gong X, Deng L (2008) Analysis of PAE Pollutants in Farm Soil and Water Samples in Nanchang City. Chemistry 8: 636–640 (in Chinese with English abstract).

31. Wang C, Liu H, Cai HS, Liang Y, Liang HC, et al. (2009) Source Analysis and Detection of Trace Phthalate Esters in Groundwater in Wuhan. Environmental Science & Technology 32(10): 118–123 (in Chinese with English abstract).

32. Zeng F, Cui KY, Xie ZY, Wu LN, Liu M, et al. (2008) Phthalate esters (PAEs): Emerging organic contaminants in agricultural soils in peri-urban areas around Guangzhou, China. Environmental Pollution 156: 425–434.

33. Liu H, Liang HC, Liang Y, Zhang D, Wang C, et al. (2010) Distribution of phthalate esters in alluvial sediment: A case study at JiangHan Plain, Central China. Chemosphere 78(4): 382–388.

34. Zeng F, Lin YJ, Cui KY, Wen JX, Ma YQ, et al. (2010) Atmospheric deposition of phthalate esters in a subtropical city. Atmospheric Environment 44(6): 834–840.

35. He W, Qin N, Kong XZ, Liu WX, He QS, et al. (2013) Spatio-temporal distributions and the ecological and health risks of phthalate esters (PAEs) in the surface water of a large, shallow Chinese lake. Science of The Total Environment 461–462: 672–680.

36. Hitch RK, Day HR (1992) Unusual persistence of DDT in some western USA soils. Bulletin of Environmental Contamination and Toxicology 48: 259–264. http://www.dec.ny.gov/docs/remediation_hudson_pdf/cpsoil.pdf.

37. Cousins IT, Bondi G, Jones KC (1999) Measuring and modelling the vertical distribution of semivolatile organic compounds in soils. I: PCB and PAH soil core data. Chemosphere 39: 2507–2518.

38. Jiang YF (2009) Preliminary study on composition, distribution and source indentification of persistent organix pollutants in soil of Shanghai. Shandong University. Doctor thesis (in Chinese with English abstract).

39. Katsoyiannis A (2006) Occurrence of polychlorinated biphenyls (PCBs) in the Soulou stream in the power generation area of Eordea, northwestern Greece. Chemosphere 65: 1551–1561.

40. Wang L (2013) Pollution characteristics of organochlorine pesticides in Daling River estuary. Dalian Maritime University. Master thesis (in Chinese with English abstract).

41. Chen J, Wang XJ, Tao S (2005) The Influences of soil total organic carbon and clay contents on PAHs vertical distributions in soils in Tianjin area. Research of Environmental Sciences 18(4): 79–83 (in Chinese with English abstract).

42. Amellal N, Portal JM, Berthelin J (2001) Effect of soil structure on the bioavailability of polycyclic aromatic hydrocarbons within aggregates of a contaminated soil. Applied Geochemistry 16: 1611–1619.

43. Benlahcen KT, Chaoui A, Budzinski H, Bellocq J, Garrigues Ph (1997) Distribution and sources of polycyclic aromatic hydrocarbons in some Mediterranean coastal sediments. Marine Pollution Bulletin 34(5): 298–305.

44. Gjessing ET (1976) Physical & Chemical Characteristics of Aquatic Humus. Ann Arbor Science Publishers Inc. (Ann Arbor), Mich.

45. Evans KM, Gill RA, Robotham PWJ (1990) The PAH and organic content of sediment particle size fractions. Water, Air, and Soil Pollution 51: 13–31.

46. Shelton DR, Boyd SA, Tiedje JM (1984) Anaerobic biodegradation of phthalic acid esters in sludge. Environmental Science & Technology 18: 93–97.

47. Ejlertsson J, Meyerson U, Svensson BH (1996) Anaerobic degradation of phthalic acid esters during digestion of municipal solid waste under landfilling conditions. Biodegradation 7: 345–352.

48. Shanker R, Ramakrishna C, Seth PK (1985) Degradation of some phthalic acid esters in soil. Environmental Pollution Series A, Ecological and Biological 39(1):1–7.

49. Wan TT, He GX, Zhang ZH, Zhu L (2013) Simulation on soil column leaching of oxygen nonhydrocarbon migration in soil profiles. Acta Scientiae Circumstantiae 33(10): 2795–2806(in Chinese with English abstract).

50. Yang GY, Zhang TB, Gao ST, Huo ZX, Wan HF, et al. (2007) Distribution of phthalic acid esters in agricultural soils in typical regions of Guangdong Province. Chinese Journal of Applied Ecology 18(10): 2308–2312(in Chinese with English abstract).

51. Tan Z, Li CH, Mo CH (2012) Distribution of Phthalic Acid Esters in Agricultural Soils of Huizhou City. Environmental Science and Management 37(5): 120–123(in Chinese with English abstract).

52. Biscardi D, Monarca S, De Fusco R, Senatore F, Poli P, et al. (2003) Evaluation of the migration of mutagens/carcinogens from PET bottles into mineral water by Tradescantia/micronuclei test, Comet assay on leukocytes and GC/MS. Science of the Total Environment 302:101–108.

53. Sharpe RM, Fisher JS, Millar MM, Jobling S, Sumpter JP (1995) Gestational and lactational exposure of rats to xenoestrogens results in reduced testicular size and sperm production. Environmental Health Perspectives 103:1136–1143.

54. Jones HB, Garside DA, Liu R, Roberts JC (1993) The influence of phthalate-esters on leydig-cell structure and function in-vitro and in-vivo. Experimental and Molecular Pathology 58:179.

55. Giuseppe L, Alberto V, Claudio DF (2004) Di-2-ethylhexyl phthalate and endocrine disruption: A review. Current Drug Targets-Immune, Endocrine & Metabolic Disorders 4: 37–40.

56. Department of Environmental Conservation, New York, USA (1994) Determination of soil cleanup objectives and cleanup levels (TAGM 4046). http://www.dec.ny.gov/regulations/2612.html.

57. Chiou CT, Sheng GY, Manes M (2001) A partition-limited model for the plant uptake of organic contaminants from soil and water. Environmental Science Technology 35: 1437–1444.

6

Impacts of Agricultural Management and Climate Change on Future Soil Organic Carbon Dynamics in North China Plain

Guocheng Wang, Tingting Li*, Wen Zhang, Yongqiang Yu

State Key Laboratory of Atmospheric Boundary Layer Physics and Atmospheric Chemistry, Institute of Atmospheric Physics, Chinese Academy of Sciences, Beijing, China

Abstract

Dynamics of cropland soil organic carbon (SOC) in response to different management practices and environmental conditions across North China Plain (NCP) were studied using a modeling approach. We identified the key variables driving SOC changes at a high spatial resolution (10 km×10 km) and long time scale (90 years). The model used future climatic data from the FGOALS model based on four future greenhouse gas (GHG) concentration scenarios. Agricultural practices included different rates of nitrogen (N) fertilization, manure application, and stubble retention. We found that SOC change was significantly influenced by the management practices of stubble retention (linearly positive), manure application (linearly positive) and nitrogen fertilization (nonlinearly positive) – and the edaphic variable of initial SOC content (linearly negative). Temperature had weakly positive effects, while precipitation had negligible impacts on SOC dynamics under current irrigation management. The effects of increased N fertilization on SOC changes were most significant between the rates of 0 and 300 kg ha^{-1} yr^{-1}. With a moderate rate of manure application (i.e., 2000 kg ha^{-1} yr^{-1}), stubble retention (i.e., 50%), and an optimal rate of nitrogen fertilization (i.e., 300 kg ha^{-1} yr^{-1}), more than 60% of the study area showed an increase in SOC, and the average SOC density across NCP was relatively steady during the study period. If the rates of manure application and stubble retention doubled (i.e., manure application rate of 4000 kg ha^{-1} yr^{-1} and stubble retention rate of 100%), soils across more than 90% of the study area would act as a net C sink, and the average SOC density kept increasing from 40 Mg ha^{-1} during 2010s to the current worldwide average of ~55 Mg ha^{-1} during 2060s. The results can help target agricultural management practices for effectively mitigating climate change through soil C sequestration.

Editor: Ben Bond-Lamberty, DOE Pacific Northwest National Laboratory, United States of America

Funding: This work was jointly supported by the Ministry of Science and Technology of China (Grant No. 2010CB950603), the CAS Strategic Priority Research Program (Grant No. XDA05050507), the National Natural Science Foundation of China (Grant No. 41021004), and the State Key Program of National Natural Science of China (Grant No. 41130104). The funders had no role in study design, data collection and analysis, decision to publish, or preparation of the manuscript.

Competing Interests: The authors have declared that no competing interests exist.

* E-mail: litingting@mail.iap.ac.cn

Introduction

Cultivation generally leads to dramatic changes in soil organic carbon (SOC) by influencing the processes involved with soil C production and decomposition [1,2]. Changes in agricultural SOC are characterized by dynamic exchange processes which is strongly linked to environmental conditions such as air temperature, precipitation, soil pH and texture; and agronomic management such as crop rotation, stubble retention, tillage regimes, and application of chemical fertilizers and animal manure [3,4]. The complex interactions between management practices and environmental conditions, as well as the lack of spatiotemporal continuity in SOC monitoring data over meaningfully large areas, hamper our ability to accurately predict the regional cropland SOC change and identify the factors that control the SOC dynamics.

A modeling approach has advantages in estimating spatiotemporal changes in SOC under various management practices and environmental conditions [5]. A number of process-based carbon models such as DNDC [6], RothC [7], and CENTURY [8] have been used to simulate SOC dynamics in agricultural systems at both national and continental scales [9–11]. Recently, Huang et al. [12] developed a biogeophysical model, Agro-C, which has been validated and further used to assess the long-term agricultural SOC changes on a national scale in China [5] and Australia [2].

The North China Plain (NCP) is one of the most important agricultural production areas in China, covering a total cropland area of ~18 Mha, with a typical continuous winter wheat-summer maize cropping system (Figure 1). During the past several decades, significant changes in both climate and agricultural management practices have taken place in NCP. For example, Ding et al. [13] reported that both temperature and precipitation had shown an increasing trend across the northern part of China during the past 50 years. Improved management practices such as crop residue retention and application of mineral fertilizers and/or organic manure have been promoted as strategies to both increase crop productivity and optimize soil fertility [14]. Although an overall net increase in SOC across the croplands of NCP during the past 30 years has been reported [5], there remains large uncertainties in future cropland SOC dynamics due to the unclear future changes in both climate and management practices. For instance, it is generally recognized that the current global warming will increase the rates of organic matter decomposition, thereby accelerating climate change through C cycle feedbacks [15]. However, in areas with relatively low mean annual temperature and without water and nutrient deficiencies (e.g., the North China Plain, typically with irrigation and N fertilization to support crop production), increasing air temperature could potentially increase crop productivity and result in more input C into soils, thereby offsetting the loss in SOC through soil respiration. More recently, Yu et al. [16] simulated the Chinese cropland SOC changes during the next 40 years, by simply designing three levels of carbon input based on the historic changing trend in annual amounts of crop residues and manure. However, the quantity of input C is strongly linked to management such as stubble retention and N fertilization, which needs to be considered during a comprehensive assessment of long-term SOC dynamics, and determined to provide strategies in effectively managing the farming systems. Moreover, since it takes several decades or even a century to reach a new equilibrium of SOC following a change in management practice [17,18], the role of soils to sequester C (source or sink) as affected by environmental variables and management practices needs to be identified on a long time scale, e.g., around 100 years.

In this study, we quantified the influence of both climatic variables and management practices on SOC across croplands of North China Plain. The SOC was simulated under 900 combinations of climate, N fertilization, stubble retention, and manure application scenarios at a daily time step, from 2011 to 2100, and at a high spatial resolution, using the pre-calibrated and validated Agro-C model. The impacts of management practices and climatic variables on SOC dynamics were then statistically analyzed. This would not only provide a better understanding of the factors and processes regulating C cycling and balance in the agro-ecosystems, but would also provide insight into the effectiveness of methods for enhancing soil quality and mitigating climate change.

Materials and Methods

1.1 Agro-C

Agro-C [12] is a biogeophysical model, and it consists of two submodels, Crop-C and Soil-C. The Crop-C submodel simulates processes involved with crop photosynthesis, autotrophic respiration, and net primary production (NPP); and the Soil-C submodel simulates soil heterotrophic respiration via the decomposition of both input C (e.g., crop residues, roots and manure) and SOC. Changes in SOC are then determined by balancing the loss of SOC with the gain of input C. The input C is split into two components of labile-C and resistant-C, and the SOC pool is divided into two sub-pools named light-C and heavy-C. The light-C sub-pool is hypothesized to be more biologically reactive, while the heavy-C sub-pool is much more resistant to decomposition. Decomposition of each component and sub-pool is described by first-order kinetics at predefined potential rates (d^{-1}) of 2.6×10^{-2}, 8.4×10^{-4}, 2.5×10^{-4} and 1.8×10^{-5} for labile-C, resistant-C, light-C and heavy-C, respectively. The actual decomposition rate of each component and sub-pool is further modified by soil parameters such as temperature, moisture, texture and pH [12]. The C flow between different pools was determined based on the following assumptions: (i) the decomposition of both the labile-C and the resistant-C converts a fraction of the C into the light-C pool, (ii) a fraction of the decomposed light-C is transferred into

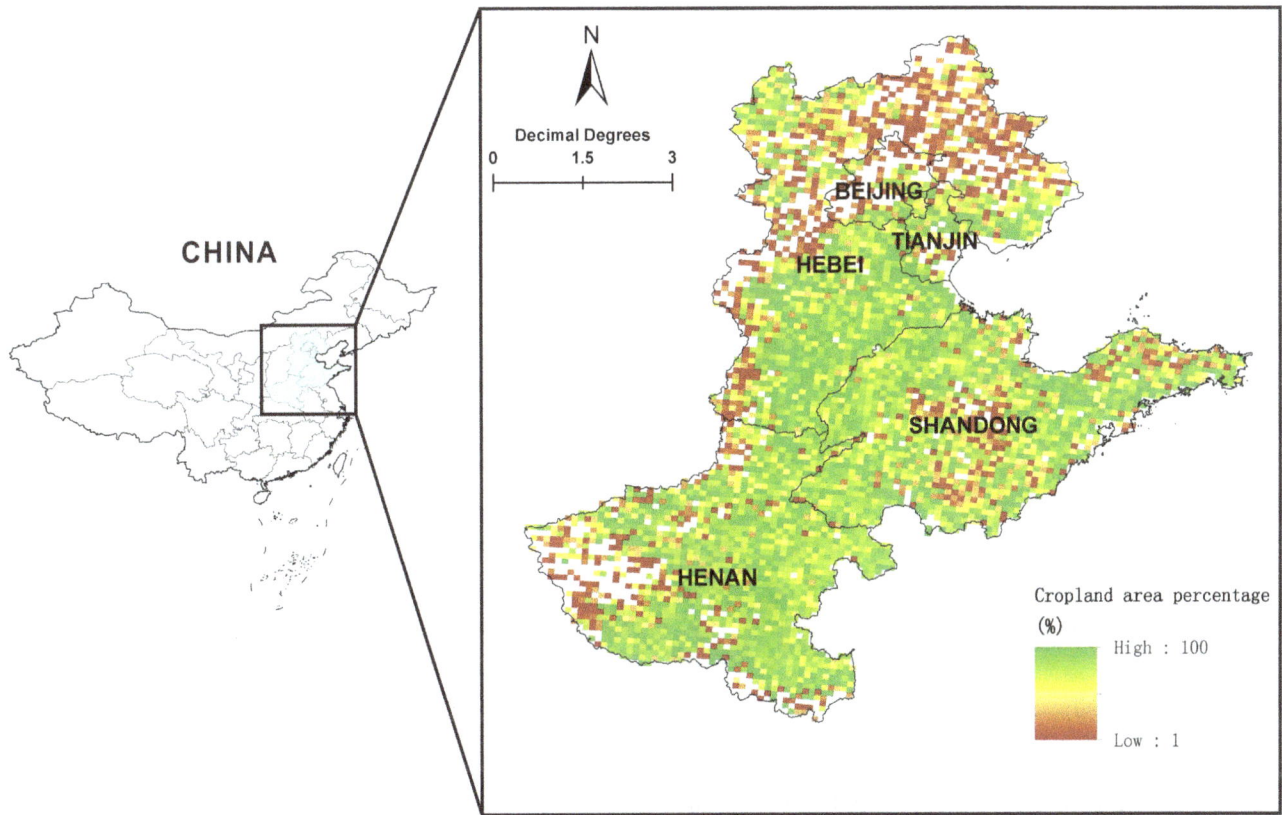

Figure 1. Spatial distribution of croplands in the North China Plain.

the heavy-C pool, and (iii) the decomposition of the heavy-C pool only produces CO_2. More detailed description of the Agro-C model was shown in Yu et al. [5] and Huang et al. [12].

In our previous studies [5,12], Agro-C has already been parameterized using observed data of several field measurements, and validated against independent datasets across a vast area representing different cropping systems, soil and climate conditions in China. The results indicate that the calibrated Agro-C model can reasonably capture observed aboveground biomass and changes in SOC under various agricultural management practices at different sites across china [5,12]. In the present study, we used the previously calibrated Agro-C model to simulate regional SOC dynamics of agro-ecosystems in North China Plain (NCP).

1.2 Spatial data

The changes in SOC under different climatic and management scenarios were simulated with a daily step from 2011 to 2100. Upscaling of the Agro-C model was accomplished by first rasterising the model inputs (e.g., climate data, initial soil properties and agricultural management information) with 10 km×10 km gridded datasets across the North China Plain and then running the model grid by grid across the study area.

Input soil properties in Agro-C include the concentrations of organic carbon and total nitrogen, bulk density, clay and sand fraction, and pH in the topsoil to 30 cm depth. These soil properties, with a 10 km×10 km spatial resolution, were firstly obtained from China Soil Scientific Database [19], representing more than 7000 soil profile measurements across China made in the late 1970s/early 1980s. Using Agro-C, we then simulated these soil properties grid by grid on a daily step from 1981 to 2010, based on the historical climate data and recorded agricultural management practices [5]. The model outputs of soil properties, including the amount of C in different pools, in the end of 2010 were further compiled as the initial model input of soil properties of the present study.

Gridded daily climate data such as maximum and minimum temperature, precipitation, and relative humidity from 2010 to 2100 were derived from the projections of the climate model FGOALS (Document S1). Due to the coarser spatial resolution (2.8° longitude×2.8° latitude) of outputs from the FGOALS model, the projections for the future GHG concentration scenarios [i.e., a set of four representative concentration pathways (RCPs) named RCP2.6, RCP4.5, RCP6.0 and RCP8.5 respectively; Figure S1 and Figure S2] of the FGOALS were therefore statistically downscaled [20,21] to a resolution of 10 km×10 km (Document S1). Additionally, as a driving variable in Agro-C, atmospheric carbon dioxide (CO_2) concentration affects the process of crop photosynthesis [12], which under each future climate scenario was obtained from Clarke et al. [22] and used as model inputs in this study.

The spatial distribution of croplands was computed based on the National Land Cover Data Sets (NLCD) of China that was developed from Landsat TM digital images around 2000 [23]. Due to a lack of yearly cropland data with sufficient spatial distribution, we assumed that the area of croplands in North China Plain did not significantly change over the study period.

1.3 Model initialization

In Agro-C model, the half-life residence time for labile-C, resistant-C, light-C and heavy-C were 0.1 y, 2.3 y, 7.6 y, and 105.4 y [5], respectively. To obtain initial fractions for the different C pools in 1981, a 'spin-up' procedure was performed by first setting the initial fractions of light-C and heavy-C as 0.25 and 0.75 [5], respectively, and then running the model until a

steady state of SOC pool was achieved. In this, an equilibrium run was not stopped until the difference of SOC pools between two successive years was less than 0.1% of the pools themselves. The length of spin-up procedure differed across different places, and generally lasted about 100 years. The average time duration of the spin-up seemed relatively short but is reasonable. Most of Chinese croplands, particularly in North China Plain, have generally experienced a very long history of cultivation. When the spin-up procedure was performed, the average amount of crop stubble retention and manure amendment over the period between 1980 and 1989 was used as organic C input, due to the unavailability of organic C input before 1980. The average of 1960–1990 climate were used as driving data, because of the unavailability of climate data before 1960. The other driving datasets such as nitrogen fertilizer and irrigation were set to 1980–1989 average values. We then calculated the initial values of different C pools based on the total SOC information obtained from the above mentioned soil database and the spin-upped fractions of different sub-C pools. The dynamics of different C pools in each grid from 1981 to 2010 on a daily time step were then modeled based on the historical climate data and recorded agricultural management practices [5]. The simulated values of above-mentioned four C pools in the end of 2010 were further compiled as the model inputs of the present study.

1.4 Agricultural management scenarios

Nitrogen (N) fertilization, application of manure, and crop stubble retention through tillage are the major management practices that affect SOC dynamics in croplands of North China Plain (NCP). A database containing the county-level yearly cultivation acreage and total amount of synthetic N application (Chinese Academy of Agricultural Sciences [24]), shows that N fertilization rates in 2010 across croplands of NCP varied from 0 to 800 kg ha^{-1} (Figure S3A). The environmental pressures that are related to excessive mineral N fertilization in Chinese croplands decrease the possibility of increasing its use in the future [25]. Another province-level database [26] shows that farm manure application rates during the past 30 years varied from 1000 to 4000 kg ha^{-1} across the studied area (Figure S3B). In the croplands of NCP, the average stubble retention rate over the period of 2000–2010 varied from ∼40% to ∼60% across space [27]. In general, an increasing fraction of crop residues were being left in the fields instead of being taken away for household fuel and animal feed in China over the past several decades [25], which was mainly attributed to the financial motivations of the government policies. To quantify the effects of both climate change and agricultural management practices on SOC, we simulated SOC changes within the winter wheat-summer maize systems under 225 combinations of N fertilization, manure application and stubble retention rates, under each of the above-mentioned four climate scenarios.

In this study, management practices included 9 nitrogen application rates (0–800 kg N ha^{-1} yr^{-1} in 100 kg N ha^{-1} yr^{-1} increments, i.e., N:0, N:100, N:200, N:300, N:400, N:500, N:600, N:700, and N:800). Apart from N:0, we specified that 55% of the N fertilizers were applied during summer-maize growing seasons, with another 45% applied during winter-wheat growing seasons. For each growing season, 40% of the N fertilizers were applied at crop emergence, another 40% applied at tillering, and the rest applied at heading for wheat and silking for maize. We also specified five manure application rates (0–4000 kg ha^{-1} yr^{-1} in 1000 kg ha^{-1} yr^{-1} increments, i.e., M:0, M:1000, M:2000, M:3000, and M:4000), and five stubble retention rates (0–100% in 25% increments, i.e., R:0, R:25, R:50, R:75, and R:100). Apart from M:0, 72% of the manure was applied before sowing of summer maize, with the rest applied before sowing of winter wheat. Stubble retention rates denoted the percentage of aboveground straw and leaf biomass incorporated into the soil by tillage after harvesting, with the rest directly removed from the system. Combinations of management practices are abbreviated in this study as M:x, N:x, R:x. The M:0, N:0, R:0 denotes the management practice with no N fertilization, no manure application, and no stubble retention.

Because of a lack of detailed crop cultivar and phenology data on the regional scale, we simply assumed that the same wheat and maize cultivars were used throughout the studied area, and crop parameters were set as default values of the pre-calibrated Agro-C model. Sowing dates of maize and wheat were based on local crop calendars. From the south to north in NCP, the sowing date ranged from 6 June to 15 June for maize, and ranged from 5 October to 11 November for wheat [28,29]. Crop growth stops at physiological maturity as determined by the accumulated temperature. In addition, according to the local agronomic management, we assumed that the soil moisture would arrive at field capacity using irrigation module when the modeled soil moisture in the 30 cm depth drops to 70% of the field capacity during the growing seasons [5].

In total, we ran 4,086,000 (225 management scenarios ×4 climate scenarios ×4540 grids) Agro-C simulations. Each simulation quantified SOC content in the top 30 cm of soil from 2011 to 2100 on a daily step.

1.5 Identifying controls on SOC dynamics

In the present study, we assessed the impacts of management practices, climatic and soil variables on SOC change using Spearman's rank correlation coefficient (rho). Selected climatic variables included mean annual temperature (hereafter simply denoted as temperature) and mean annual precipitation (hereafter simply denoted as precipitation) since these two variables have been reported uncorrelated and significantly represent the spatial variation in a range of climate patterns [30]. For correlation analysis, the long-term daily climate variables were summarized to mean annual values under each climate scenario for each grid. Selected soil parameters included the main model edaphic inputs, e.g., pH, initial SOC, clay and total nitrogen content. Change in SOC of the top 30 cm (ΔSOC) is calculated as the difference in SOC between 2100 and 2011. Spearman's rank correlation coefficient was then calculated between ΔSOC and the management practices and environmental variables across the full set of Agro-C simulations. The sign of rho, positive or negative, indicates the direction of association between the independent and dependent variables. The absolute magnitude of rho, between 0 and 1, suggests the strength of correlation between the two variables.

We also investigated the effects of the initial SOC content and each of the three management practices on ΔSOC by producing boxplots characterizing the influence of each variable on ΔSOC including the variance calculated across agricultural management practices, grids, and the four climate scenarios. All analyses were performed using statistical and graphical software R 3.0.1 [31].

Results

2.1 Correlations between environmental variables, management practices and ΔSOC

Initial SOC was strongly but negatively correlated (median rho = −0.87) with ΔSOC (Figure 2). Initial soil clay content showed a weak negative correlation (median rho = −0.21) with ΔSOC

(Figure 2). The climatic variables, e.g., temperature and precipitation, displayed a weak positive (median rho = 0.21) and negligible correlation with ΔSOC, respectively (Figure 2). Agricultural management such as stubble retention (median rho = 0.45) and manure application (median rho = 0.81) both showed a strong and positive correlation, whereas N fertilization showed a negligible correlation with ΔSOC (Figure 2).

Figure 3 presented the impacts of initial SOC content, manure application, N fertilization and stubble retention, respectively, on ΔSOC. On average, initial SOC content seemed to linearly and negatively correlate with ΔSOC (Figure 3A), whereas manure application and stubble retention had a relatively linear positive effect (Figure 3B and D) on ΔSOC. N fertilization increased ΔSOC most significantly ($P<0.05$) at rates between 0 and 300 kg ha^{-1} yr^{-1}, with a limited influence at higher fertilization rates. Across the sets involved with different climate scenarios, the median ΔSOC increased slightly, although not significantly ($P>0.05$), from RCP2.6 to RCP8.5 (Figure 4).

2.2 Spatiotemporal effects of agricultural practices on SOC

On average of the four climate scenarios, the response of ΔSOC to N fertilization, manure application and stubble retention varied over the study area, and cropland soils in Hebei and northern part of Shandong Province would accumulate more carbon than the rest of the study area if agricultural practices were optimized (Figure 5 and 6). Increasing the manure application rate led to soil across more areas turning into a net C sink, regardless of the other two management practices (Figure 5 and 6). N fertilization could not significantly halt the declining SOC trend unless more crop residues were incorporated into soils. For example, when stubble retention rate increased from 0 to 100% under M:2000, SOC increased in 11.4%, 27.7%, 40.7%, 53.9%, and 63.9% of the study area under N:0 (Figure 5), whereas increased in 13.5%, 44.5%, 62.5%, 71.3%, and 76.6% of the study area under N:300 (Figure 6). If the amount of manure application rate reached 4000 kg ha^{-1} yr^{-1} and all stubble incorporated into soils (i.e., M:4000 and R:100), SOC increased in

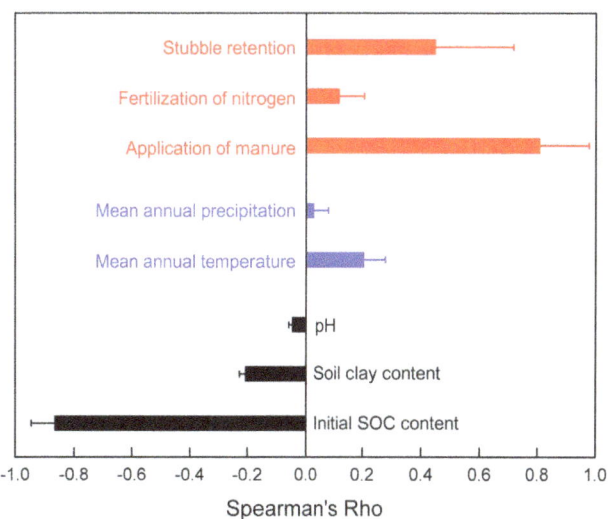

Figure 2. Spearman's rank correlation coefficients between ΔSOC (2011–2100, Mg C ha^{-1}) and climate (blue), soil (black), and management (red) variables. All the tests were significant ($P<0.001$).

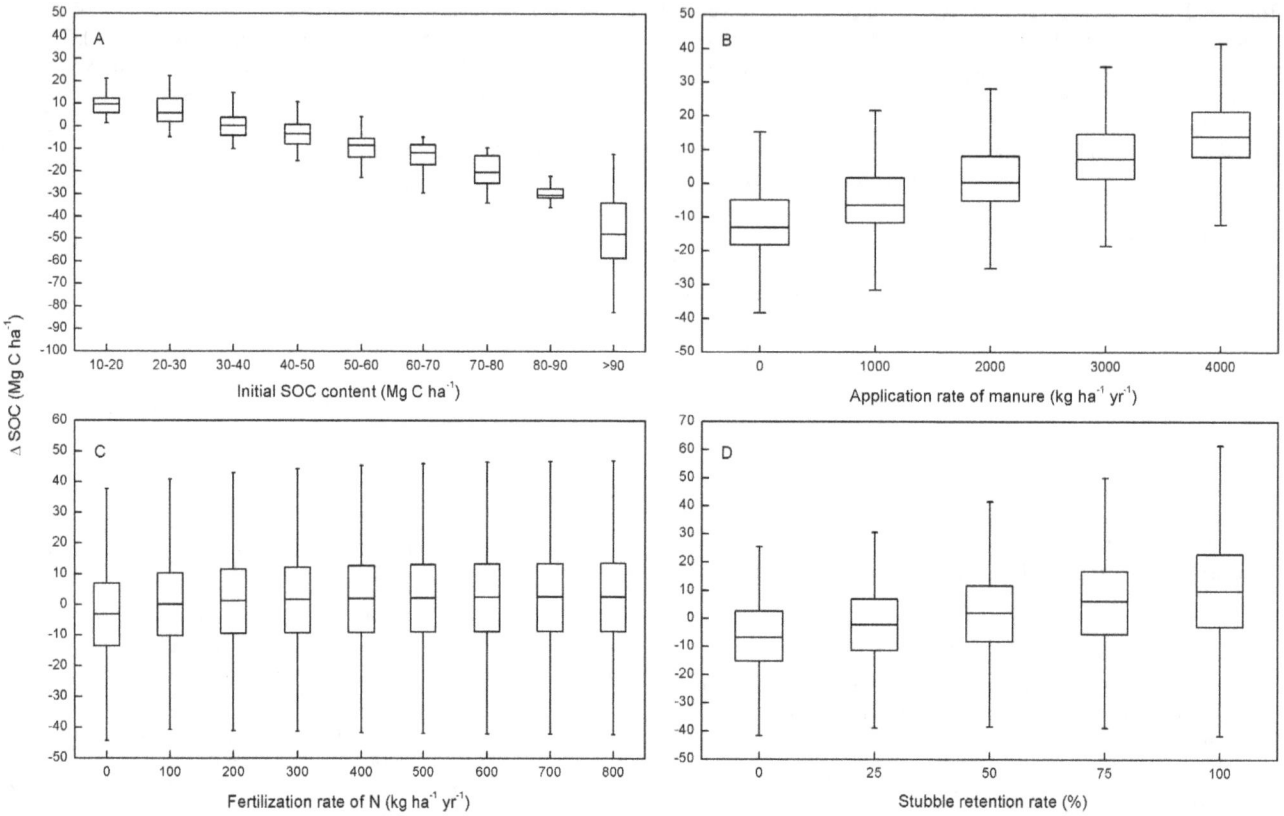

Figure 3. Response of ΔSOC (2011–2100, Mg C ha^{-1}) to initial SOC content (A), manure application (B), N fertilization (C), and stubble retention (D). Boxplots show the median and interquartile range, with whiskers extending to the most extreme data point within 1.5×(75–25%) data range.

more than 90% of the study area under both N:0 and N:300 (Figure 5 and 6).

Under M:0, N:0, R:0, the average SOC content across croplands of NCP kept decreasing from 35.7 Mg ha^{-1} during

the 2010s to 16.1 Mg ha^{-1} during the 2090s, on average of the four climate scenarios (Figure 7). However, adopting a moderate management practice (e.g., M:2000, N:100, R:50) could maintain the SOC level during the next 90 years (Figure 7). Under M:4000, N:300, R:100, the average SOC density kept increasing from 39.5 Mg ha^{-1} during the 2010s to 54.6 Mg ha^{-1} during the 2060s, with a relatively lower increasing rate afterward (Figure 7).

Discussion

3.1 Interpretation and implication of the results

Initial SOC content was identified as strongly and negatively influencing SOC change (Figure 2). This is because, in model initialization, organic C input were taken to be the average of crop stubble and manure from 1980 to 1989. However, in the following model simulations with various kinds of management scenarios, great land use changes took place and the equilibrium of soil C pools got changed. In this, under otherwise similar environmental and managed conditions, soils with higher initial SOC content would experience greater SOC loss, and vice versa [32]. This negative correlation between ΔSOC and initial SOC content has also been documented in other studies [33–35]. Although higher soil clay content has been reported to benefit the stabilization of soil organic matter (SOM) in many studies [36–38], we identified a weak negative correlation between ΔSOC and soil clay content in the present study (Figure 2). This is because, other than the benefits of SOM stabilization, higher soil clay content could also reduce the amount of C inputs to soil [12], thereby showing a negative impact on total SOC variations in the present study.

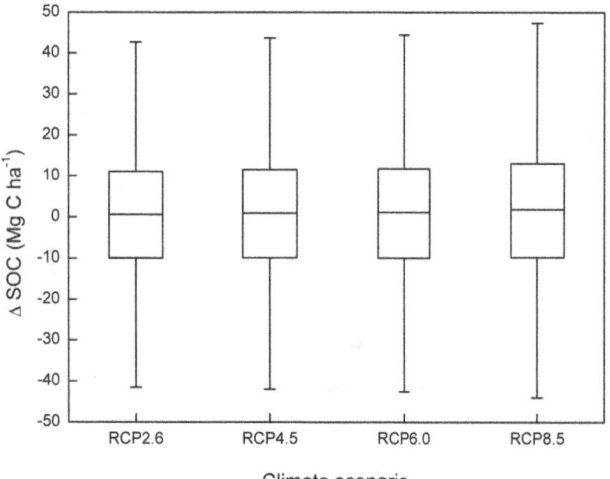

Figure 4. Response of ΔSOC (2011–2100, Mg C ha^{-1}) to different climate scenarios. Boxplots show the median and interquartile range, with whiskers extending to the most extreme data point within 1.5×(75–25%) data range.

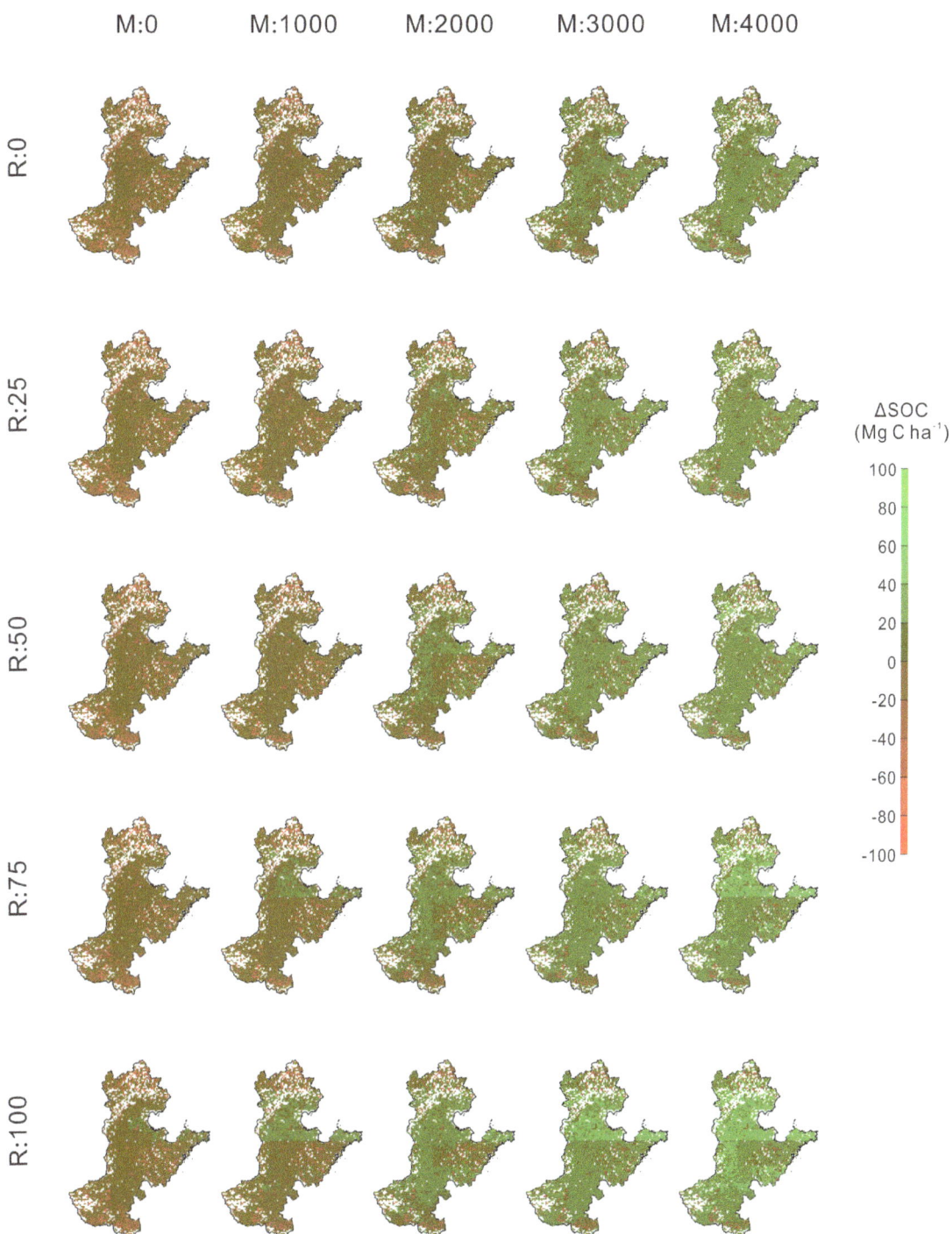

Figure 5. Spatial pattern of the effects of manure application and stubble retention on ΔSOC (2011–2100, Mg C ha^{-1}) under N:0. For a certain combination of management practices, ΔSOC in each grid represents the mean value of the results under four climate scenarios.

In Agro-C, high soil temperature and soil water content increase the SOC decomposition rate which underpins the negative relationship of SOC accumulation with both temperature and rainfall [12]. However, in NCP, we found that the mean annual temperature and precipitation were not strongly correlated with ΔSOC (Figure 2), and higher temperature seemed to benefit SOC accumulation (Figure 4 and Figure S1). This is because the influences of temperature and precipitation on SOC dynamics depend on other factors such as soil and management practices

[39]. For example, artificial irrigation, which was adopted across all sets of scenarios in this study, would certainly overshade the possible impacts of precipitation on SOC change. And the elevated atmospheric CO_2 concentration accompanied with higher temperature increased the crop productivity, thereby compensating the elevated amount of decomposed SOC induced by climate warming [40]. Furthermore, Zhang and Huang [41] reported a positive correlation between air temperature and the production of cereal crops in China during the past 30 years. This

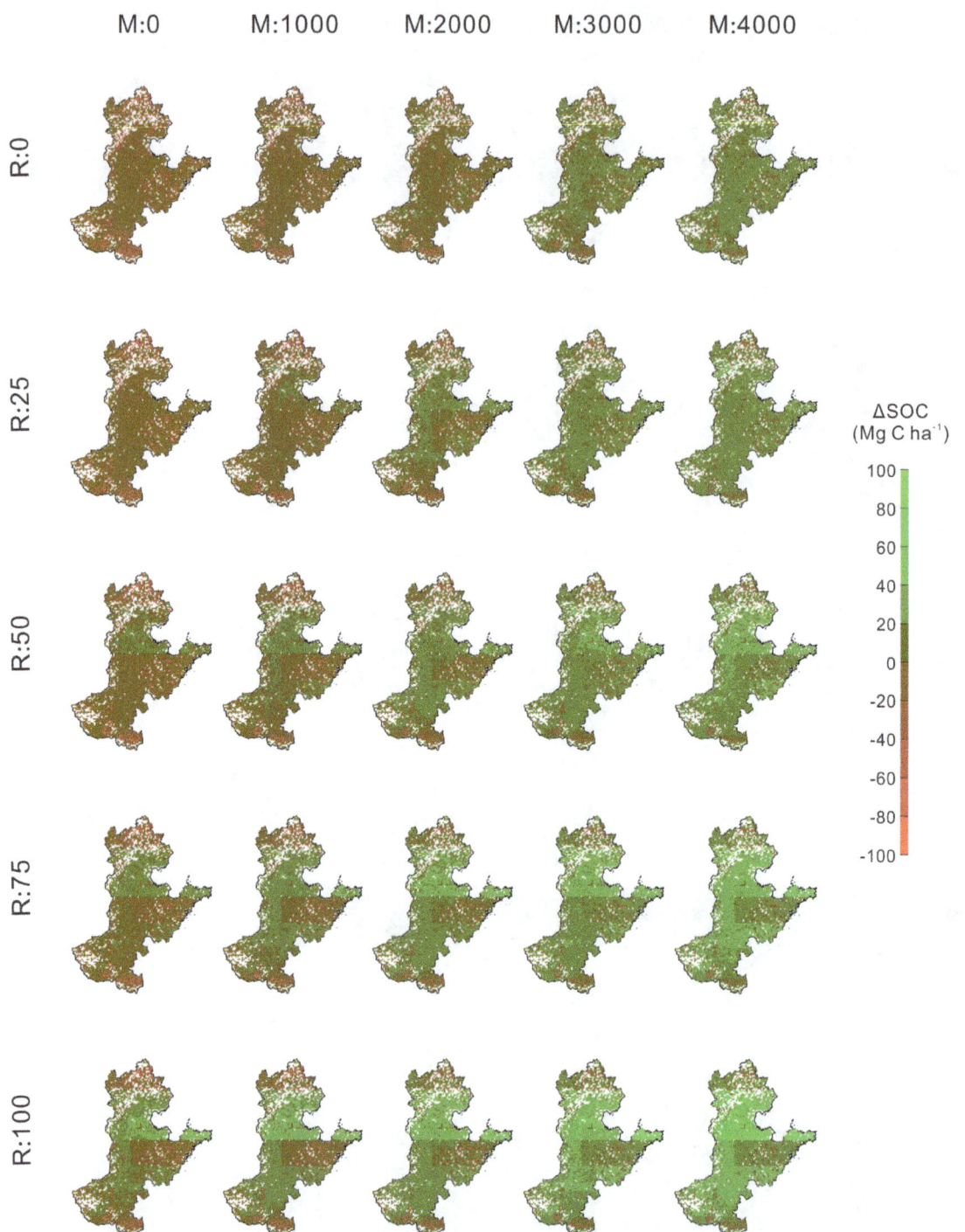

Figure 6. Spatial pattern of the effects of manure application and stubble retention on ΔSOC (2011–2100, Mg C ha^{-1}) under N:300. For a certain combination of management practices, ΔSOC in each grid represents the mean value of the results under four climate scenarios.

indicated that climate warming could potentially cause a higher amount of C input from crop residues to soils, with which our findings was consistent.

Manure application and stubble retention directly add C into soil, and were identified as the most two predominant agricultural management practices driving SOC (Figure 2 and 3). Although it has been reported that the manure application levels has been decreasing during the past 50 years, and the changing trends might continue in the future across Chinese croplands [25]. The

increasing amount of incorporated crop residues, caused by both enhanced stubble retention rate [27] and elevated amount of crop NPP [42], could still potentially override the weakening manure application rates and increase SOC.

N fertilization enhances crop residue and root biomass production (Figure 8), thus increases the amount of C incorporated into soil. However, the impact of N fertilization depends on whether nutrient is limiting. In this study, we found that N fertilization showed a negligible correlation with changes in SOC

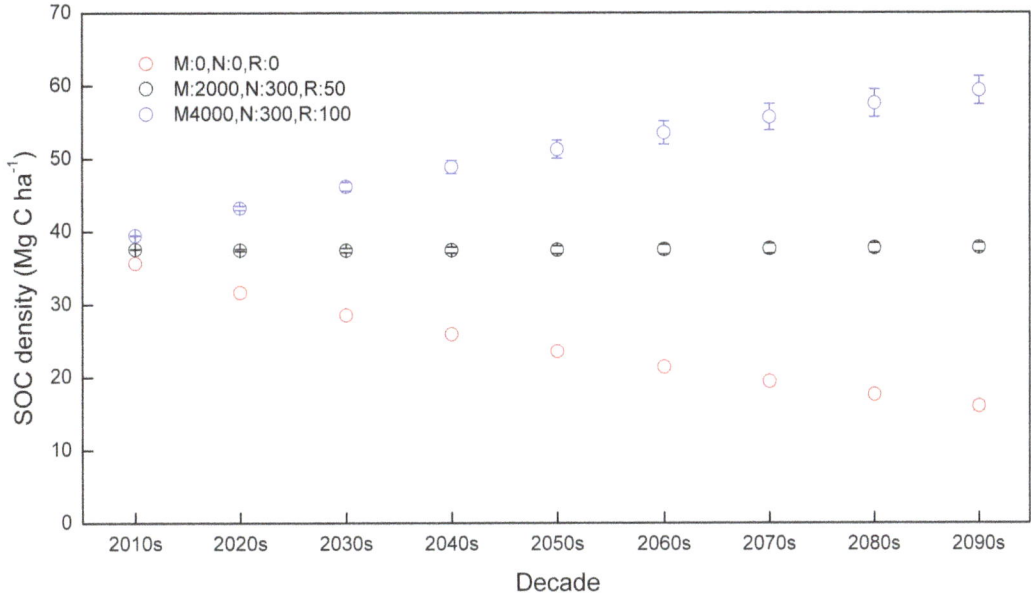

Figure 7. Temporal change of SOC (Mg C ha^{-1}) under three combinations of manure application, nitrogen fertilization and stubble retention. Open circles represent mean values of SOC under four climate scenarios, with standard deviations represented by error bars on the circles.

(Figure 2). This is mainly because changes in SOC is dominated by manure application and stubble retention in most of the scenarios. However, under scenarios with high stubble retention rates and low manure application rates, effects of N fertilization could be more obvious (Figure 5 and 6). Additionally, our results also showed that although N fertilization had a positive effect on SOC accumulation, the effect did not further increase if N fertilization rate exceeds 300 kg ha^{-1} yr^{-1} (Figure 3C). This is because with a N fertilization rate higher than 300 kg ha^{-1} yr^{-1}, nutrient was not a limiting factor of crop growth, and enhancing N inputs could not significantly elevate the amount of soil C inputs from crop residues and roots (Figure 8). Additionally, excessive use of N fertilizer is currently very common in NCP [43], which has aroused concerns about its negative impacts on ecosystems and environments [44]. We suggest that a N fertilization rate of about 300 kg ha^{-1} yr^{-1} should be adopted to both maintain SOC accumulation and reduce the environmental risks caused by excessive use of N fertilizers. As the benefit of N fertilization on SOC sequestration could be offset by N$_2$O emissions from the soil and greenhouse gas (GHG) emissions from fossil fuel combustion during the processes involving fertilizer production and transport [45], there remains a need to further investigate the proper rates and timing of N fertilization under different climate and soil conditions.

With a high (M:4000, N:300, R:100) C input management, the temporal change in SOC generally experienced a fast to low increase, with the most gain of soil C happened in the first 50 years of simulations (Figure 7). This is consistent with a number of other studies [1,46], which suggest that soil C would reach a new steady state after 20–50 years of cultivation if management practices were significantly changed. Moreover, our results revealed that under M:4000, N:300, R:100, the average SOC density of NCP croplands could achieve 55 Mg ha^{-1} during 2060s, which is similar to the current worldwide average agricultural SOC density [47]. However, the average SOC density would keep decreasing under M:0, N:0, R:0, and the "new" steady state of SOC was not attained even after 90 years of simulation (Figure 7). This could be attributed to the limited C input to balance the SOC loss by

decomposition. With a much lower decomposition rate, the

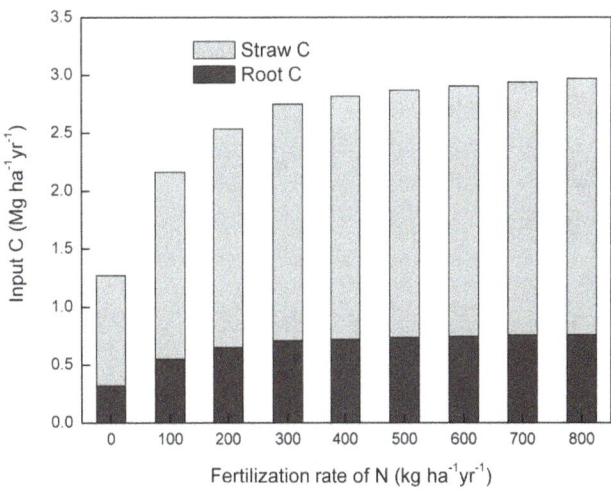

Figure 8. Response of mean annual C inputs from crop residues and roots (Mg C ha^{-1}) to different N fertilization rates.

proportion and absolute amount of heavy-C would have significant impact on long-term SOC balance.

3.2 Uncertainties and limitations

Several uncertainties and limitations should be noticed in interpreting the simulation results in this study. First, several management scenarios simulated in this study represent the most extreme conditions and may not happen in the real world. For example, zero inputs of N and C to soil, because it is improbable to extract all roots and residues from the fields, or prevent atmospheric N deposition. Second, the possible improvement in future crop varieties, which could potentially affect the quality and quantity of soil C inputs, was not considered in this study. For example, if the crop variety improvement increased the production of either crop residues or roots, this could lead to an underestimation of our results in the future SOC accumulation. Third, the effects of conservation tillage on SOC were not simulated in this study, due to a lack of the detailed information on annual conservation tillage practice. However, it could still be hypothesized that our results could represent most parts of the studied areas, because the average fraction of the area with conservation tillage practices in Chinese croplands is only about 9% in 2009 [5]. Moreover, a meta-analysis of global data on the effects of tillage on SOC change [48] indicates that conversion from conventional tillage to conservation tillage only redistributed more C into the top layer and less C into the deeper layer, and the total SOC stock in the upper 40 cm soil profile was not significantly changed. Fourth, the rainfed condition was not simulated because artificial irrigation has been traditionally adopted to support crop production in NCP. However, the current irrigation-supported agricultural production might be affected by future resource and environment related pressures [49]. In this, other environmental factors such as precipitation and temperature may play a more important role in regulating both crop productivity and SOC dynamics [50]. Last but not least, certain issues related to state-of-the-art model limitations should be mentioned. For instance, in a first-order decay model (e.g., Agro-C), there is a general linear relation between C input and SOC variation, which usually contradicts the fact that increase of C inputs by crop residues may have variable effects on SOC change in the real world [51]. Future efforts are needed from model developers to implement currently uncertain mechanisms, such as SOC saturation, priming, and different sensitivity of SOC pools to temperature, which will enable a better understanding the influence of environment on agricultural SOC dynamics.

Supporting Information

Figure S1 Climate change in North China Plain (NCP) from 2011 to 2100 under the RCP2.6 (A), RCP4.5 (B), RCP6.0 (C) and RCP8.5 (D) scenarios. The solid circles represent the annual temperature, and the open triangles represent annual precipitation.

Figure S2 Spatial distribution of mean annual temperature and precipitation increments in North China Plain (NCP) between 2011 and 2100 under four climate scenarios.

Figure S3 The range and density of N fertilization (A) and manure application (B) rates in North China Plain (NCP). The N fertilization data represent the year of 2010 on a county scale, and the manure application data represent that from 1981 to 2010 on a provincial scale. The datasets were derived from online database and published literatures. n shows the sample size.

Acknowledgments

Many thanks to Dr. Yongqiang Yu from the State Key Laboratory of Numerical Modeling for Atmospheric Sciences and Geophysical Fluid Dynamics (LASG), Institute of Atmospheric Physics, Chinese Academy of Sciences, for providing the access to the FGOALS modeling result datasets.

Author Contributions

Conceived and designed the experiments: GW TL. Performed the experiments: GW TL. Analyzed the data: GW TL. Contributed reagents/materials/analysis tools: TL WZ YY. Wrote the paper: GW.

References

1. Lal R (2004) Soil carbon sequestration impacts on global climate change and food security. Science 304: 1623–1627.
2. Wang G, Huang Y, Wang E, Yu Y, Zhang W (2013) Modeling Soil Organic Carbon Change across Australian Wheat Growing Areas, 1960–2010. PloS one 8: e63324.
3. Brady NC, Weil RR (2008) The nature and properties of soils (14th ed). Upper Saddle River: Pearson Education Publication.
4. Van Wesemael B, Paustian K, Meersmans J, Goidts E, Barancikova G, et al. (2010) Agricultural management explains historic changes in regional soil carbon stocks. Proceedings of the National Academy of Sciences 107: 14926–14930.
5. Yu Y, Huang Y, Zhang W (2012) Modelling soil organic carbon change in croplands of China, 1980–2009. Global and Planetary Change 82–83: 115–128.
6. Li C, Frolking S, Harriss R (1994) Modeling carbon biogeochemistry in agricultural soils. Global biogeochemical cycles 8: 237–254.
7. Jenkinson D, Andrew S, Lynch J, Goss M, Tinker P (1990) The Turnover of Organic Carbon and Nitrogen in Soil. Philosophical Transactions: Biological Sciences 329: 361–368.
8. Parton W, Scurlock J, Ojima D, Gilmanov T, Scholes R, et al. (1993) Observations and modeling of biomass and soil organic matter dynamics for the grassland biome worldwide. Global biogeochemical cycles 7: 785–809.
9. Li CS, Zhuang Y, Frolking S, Galloway J, Harriss R, et al. (2003) Modeling soil organic carbon change in croplands of China. Ecological Applications 13: 327–336.
10. Falloon P, Smith P, Bradley R, Milne R, Tomlinson R, et al. (2006) RothCUK– a dynamic modelling system for estimating changes in soil C from mineral soils at 1-km resolution in the UK. Soil Use and Management 22: 274–288.
11. Ogle SM, Breidt F, Easter M, Williams S, Killian K, et al. (2010) Scale and uncertainty in modeled soil organic carbon stock changes for US croplands using a process-based model. Global Change Biology 16: 810–822.
12. Huang Y, Yu Y, Zhang W, Sun W, Liu S, et al. (2009) Agro-C: A biogeophysical model for simulating the carbon budget of agroecosystems. Agricultural and Forest Meteorology 149: 106–129.
13. Ding Y, Ren G, Shi G, Gong P, Zheng X, et al. (2006) National assessment report of climate change (I): climate change in China and its future trend. Advances in Climate Change Research 2: 3–8 (in Chinese).
14. Huang Y, Sun W (2006) Changes in topsoil organic carbon of croplands in mainland China over the last two decades. Chinese Science Bulletin 51: 1785–1803.
15. Bond-Lamberty B, Thomson A (2010) Temperature-associated increases in the global soil respiration record. Nature 464: 579–582.
16. Yu Y, Huang Y, Zhang W (2013) Projected changes in soil organic carbon stocks of China's croplands under different agricultural managements, 2011–2050. Agriculture, Ecosystems & Environment 178: 109–120.
17. West TO, Post WM (2002) Soil organic carbon sequestration rates by tillage and crop rotation: A global data analysis. Soil Science Society of America Journal 66: 1930–1946.
18. Luo Z, Wang E, Sun OJ, Smith CJ, Probert ME (2011) Modeling long-term soil carbon dynamics and sequestration potential in semi-arid agro-ecosystems. Agricultural and Forest Meteorology 151: 1529–1544.
19. Yu D, Shi X, Wang H, Sun W, Chen J, et al. (2007) Regional patterns of soil organic carbon stocks in China. Journal of Environmental Management 85: 680–689.
20. Bertacchi Uvo C, Olsson J, Morita O, Jinno K, Kawamura A, et al. (2001) Statistical atmospheric downscaling for rainfall estimation in Kyushu Island, Japan. Hydrology and Earth System Sciences 5: 259–271.
21. Kidson JW, Thompson CS (1998) A comparison of statistical and model-based downscaling techniques for estimating local climate variations. Journal of Climate 11: 735–753.
22. Clarke L, Edmonds J, Krey V, Richels R, Rose S, et al. (2009) International climate policy architectures: Overview of the EMF 22 International Scenarios. Energy Economics 31: S64–S81.
23. Liu J, Liu M, Zhuang D, Zhang Z, Deng X (2003) Study on spatial pattern of land-use change in China during 1995–2000. Science in China Series D: Earth Sciences 46: 373–384.

24. Huang Y, Tang Y (2010) An estimate of greenhouse gas (N_2O and CO_2) mitigation potential under various scenarios of nitrogen use efficiency in Chinese croplands. Global Change Biology 16: 2958–2970.

25. Zhang W, Yu Y, Huang Y, Li T, Wang P (2011) Modeling methane emissions from irrigated rice cultivation in China from 1960 to 2050. Global Change Biology 17: 3511–3523.

26. National Development and Reform Commission (2007) The People's Republic of China, Initial National Communication on Climate Change (In Chinese). China Planning Press, Beijing, China.

27. Gao L, Ma L, Zhang W, Wang F, Ma W, et al. (2009) Analysis on the Quantities and Utilization of Crop Straw and its Nutrient in Huang-Huai-Hai Region. Chinese Agricultural Science Bulletin 25: 186–193 (in Chinese with English summery).

28. Wu D, Yu Q, Lu C, Hengsdijk H (2006) Quantifying production potentials of winter wheat in the North China Plain. European Journal of Agronomy 24: 226–235.

29. Wu D, Yu Q, Wang E, Hengsdijk H (2008) Impact of spatial-temporal variations of climatic variables on summer maize yield in North China Plain. International Journal of Plant Production 2: 71–88.

30. Bryan BA (2003) Physical environmental modeling, visualization and query for supporting landscape planning decisions. Landscape and urban planning 65: 237–259.

31. R Development Core Team (2013) R: A language and environment for statistical computing. Vienna, Austria: R Foundation for Statistical Computing.

32. Sanderman J, Baldock JA (2010) Accounting for soil carbon sequestration in national inventories: a soil scientist's perspective. Environmental Research Letters 5: 034003.

33. Saby N, Arrouays D, Antoni V, Lemercier B, Follain S, et al. (2008) Changes in soil organic carbon in a mountainous French region, 1990–2004. Soil Use and Management 24: 254–262.

34. Goidts E, Wesemael Bv, Van Oost K (2009) Driving forces of soil organic carbon evolution at the landscape and regional scale using data from a stratified soil monitoring. Global Change Biology 15: 2981–3000.

35. Zhao G, Bryan BA, King D, Luo Z, Wang E, et al. (2013) Impact of agricultural management practices on soil organic carbon: simulation of Australian wheat systems. Global change biology 19: 1585–1597.

36. Oades J (1988) The retention of organic matter in soils. Biogeochemistry 5: 35–70.

37. Amato M, Ladd J (1992) Decomposition of ^{14}C-labelled glucose and legume material in soils: Properties influencing the accumulation of organic residue C and microbial biomass C. Soil Biology and Biochemistry 24: 455–464.

38. Burke I, Yonker C, Parton W, Cole C, Schimel D, et al. (1989) Texture, climate, and cultivation effects on soil organic matter content in US grassland soils. Soil Science Society of America Journal 53: 800–805.

39. Giardina CP, Ryan MG (2000) Evidence that decomposition rates of organic carbon in mineral soil do not vary with temperature. Nature 404: 858–861.

40. Smith J, Smith P, Wattenbach M, Zaehle S, Hiederer R, et al. (2005) Projected changes in mineral soil carbon of European croplands and grasslands, 1990–2080. Global Change Biology 11: 2141–2152.

41. Zhang T, Huang Y (2012) Impacts of climate change and inter-annual variability on cereal crops in China from 1980 to 2008. Journal of the Science of Food and Agriculture 92: 1643–1652.

42. Huang Y, Zhang W, Sun W, Zheng X (2007) Net primary production of Chinese croplands from 1950 to 1999. Ecological Applications 17: 692–701.

43. Liu X, Ju X, Zhang F, Pan J, Christie P (2003) Nitrogen dynamics and budgets in a winter wheat–maize cropping system in the North China Plain. Field Crops Research 83: 111–124.

44. Richter A, Burrows JP, Nüβ H, Granier C, Niemeier U (2005) Increase in tropospheric nitrogen dioxide over China observed from space. Nature 437: 129–132.

45. Knorr W, Prentice I, House J, Holland E (2005) Long-term sensitivity of soil carbon turnover to warming. Nature 433: 298–301.

46. West TO, Six J (2007) Considering the influence of sequestration duration and carbon saturation on estimates of soil carbon capacity. Climatic Change 80: 25–41.

47. Batjes NH (1999) Management options for reducing CO_2-concentrations in the atmosphere by increasing carbon sequestration in the soil: National Institute of Public Health and the Environment.

48. Luo Z, Wang E, Sun OJ (2010) Can no-tillage stimulate carbon sequestration in agricultural soils? A meta-analysis of paired experiments. Agriculture, Ecosystems & Environment: 224–231.

49. Wang E, Yu Q, Wu D, Xia J (2008) Climate, agricultural production and hydrological balance in the North China Plain. International Journal of Climatology 28: 1959–1970.

50. Piao S, Ciais P, Huang Y, Shen Z, Peng S, et al. (2010) The impacts of climate change on water resources and agriculture in China. Nature 467: 43–51.

51. Powlson DS, Glendining MJ, Coleman K, Whitmore AP (2011) Implications for soil properties of removing cereal straw: results from long-term studies. Agronomy Journal 103: 279–287.

7

Zinc, Iron, Manganese and Copper Uptake Requirement in Response to Nitrogen Supply and the Increased Grain Yield of Summer Maize

Yanfang Xue[1], Shanchao Yue[1], Wei Zhang[1], Dunyi Liu[1], Zhenling Cui[1], Xinping Chen[1], Youliang Ye[2], Chunqin Zou[1]*

1 Center for Resources, Environment and Food Security, China Agricultural University, Beijing, China, 2 College of Resources and Environmental Sciences, Henan Agricultural University, Zhengzhou, China

Abstract

The relationships between grain yields and whole-plant accumulation of micronutrients such as zinc (Zn), iron (Fe), manganese (Mn) and copper (Cu) in maize (*Zea mays* L.) were investigated by studying their reciprocal internal efficiencies (RIEs, g of micronutrient requirement in plant dry matter per Mg of grain). Field experiments were conducted from 2008 to 2011 in North China to evaluate RIEs and shoot micronutrient accumulation dynamics during different growth stages under different yield and nitrogen (N) levels. Fe, Mn and Cu RIEs (average 64.4, 18.1and 5.3 g, respectively) were less affected by the yield and N levels. ZnRIE increased by 15% with an increased N supply but decreased from 36.3 to 18.0 g with increasing yield. The effect of cultivars on ZnRIE was similar to that of yield ranges. The substantial decrease in ZnRIE may be attributed to an increased Zn harvest index (from 41% to 60%) and decreased Zn concentrations in straw (a 56% decrease) and grain (decreased from 16.9 to 12.2 mg kg^{-1}) rather than greater shoot Zn accumulation. Shoot Fe, Mn and Cu accumulation at maturity tended to increase but the proportions of pre-silking shoot Fe, Cu and Zn accumulation consistently decreased (from 95% to 59%, 90% to 71% and 91% to 66%, respectively). The decrease indicated the high reproductive-stage demands for Fe, Zn and Cu with the increasing yields. Optimized N supply achieved the highest yield and tended to increase grain concentrations of micronutrients compared to no or lower N supply. Excessive N supply did not result in any increases in yield or micronutrient nutrition for shoot or grain. These results indicate that optimized N management may be an economical method of improving micronutrient concentrations in maize grain with higher grain yield.

Editor: Paul A. Cobine, Auburn University, United States of America

Funding: This research was supported by the National Natural Science Foundation of China (31272252), China Agriculture Research System (CARS-02), the 973 project (No. 2009CB18606) and the Innovative Group Grant of NSFC (31121062). The funders had no role in study design, data collection and analysis, decision to publish, or preparation of the manuscript.

Competing Interests: The authors have declared that no competing interests exist.

* E-mail: zcq0206@cau.edu.cn

Introduction

Maize (*Zea mays* L.), as one of the world's leading cereal grains along with rice and wheat, is very popular due to its diverse functionality as a food source for both humans and animals [1]. It is estimated that maize together with rice and wheat provide at least 30% of the food calories to more than 4.5 billion people in 94 developing countries [2]. Increased maize production is also required to meet the demands for animal feed and biofuel [3].

In China, maize accounts for more than one-third of Chinese cereal production and China is responsible for 20.9% of global maize output from 2009 to 2011 [4]. Pursuing high grain yields in China has been the top priority in policy. However, a 71% increase in total annual grain production from 1977 to 2005 was accompanied by a 271% increase in nitrogen (N) fertilizer input, which resulted in serious environmental problems such as eutrophication, greenhouse gas emissions and soil acidification in China [5–8]. Intensive cultivation of high-yielding cultivars with over-applications of N, phosphorous (P) and potassium (K) fertilizers also leads to micronutrient, especially zinc (Zn) and iron (Fe) deficiencies in many countries [9]. For example, in

China, approximately 40% of the soils are Zn and Fe deficient, and about 30% are manganese (Mn) and copper (Cu) deficient [10]. Therefore, improving N management and developing related policies are major issues in crop production and environmental protection in China. Several studies have reported that increasing N fertilization has little or no effect on grain micronutrient concentrations, such as Zn, Fe and Cu in maize [11–13]. Therefore, it is necessary to investigate whether optimal N management by decreasing excessive N supply, would result in negative impacts on micronutrient nutrition of maize plants and their allocation into grains under field conditions.

Maize grain yield has steadily increased over the last century through both conventional breeding and agronomic practices [14]. Increased grain yield and biomass production may be associated with greater uptake of total plant nutrients, such as N, P and K [15,16]. A double peak and a higher N and P accumulation rates in maize with higher yield have been found compared to lower yielding maize [15]. Little information is available about the accumulation dynamics of micronutrients such as Zn, Fe, Mn and Cu in response to an increased grain yield or increased biomass production. However, the patterns of micronutrient accumulation during different growth stages have been quantified [17,18]. It has been suggested that further investigation is needed to understand the timing of micronutrient uptake in response to yields, which varies between crop species, environments and management practices [19]. Knowledge of the dynamics of micronutrient accumulation associated with yield–trait relationships in crops would provide an efficient tool to synchronize micronutrient demand and supply, thus improving nutrient management efficiencies (e.g. Zn fertilizer) and benefiting sustainable production without harming the environment.

The study of the nutrient reciprocal internal efficiencies (RIEs, g of nutrient requirement in plant dry matter per Mg grain yield) could be used to determine the relationships between grain yield and whole-plant nutrient accumulation [19]. Some studies have reported large variations in the relationship between maize grain yield and nutrient accumulation in response to nutrient supplies, grain yields, genotypes, and environments [20–23]. For example, N reciprocal internal efficiency (N requirement per Mg of grain yield) decreased from 19.8 to 16.9 kg with increasing grain yield while it was higher with N application than no N application [22]. These studies mainly focused on N, P and K. Little information is available on micronutrient accumulation (such as Zn, Fe, Mn and Cu) in response to different grain yields, N fertilization rates or different cultivars. Understanding the relationships between micronutrient accumulation, especially Zn, and grain yield with different N management practices should provide valuable information leading to improvements in micro-fertilizer management practices. It is reported that crop recovery efficiency of micronutrients ranged from only 5% to 10% [24].

Recent studies have shown that increased maize grain yield decreases grain N concentration [14,22]. A negative correlation between grain yield and concentrations of Zn and Fe in maize grain has also been reported [25]. Therefore, it is important to investigate the concentrations of Zn and Fe, together with Mn and Cu, in maize grain in response to increased grain yield.

The objectives of this study were to (i) quantify RIEs (ZnRIE, FeRIE, MnRIE and CuRIE) in relation to increased grain yield, different cultivars and N management practices, (ii) estimate shoot Zn, Fe, Mn and Cu accumulation dynamics in response to increased grain yields and different N management practices, and (iii) investigate the effects of improved grain yields, different cultivars and N management practices on the concentrations of Zn, Fe, Mn and Cu in maize grain.

Materials and Methods

Experimental design

Field experiment I was conducted in four consecutive years (2008–2011) in Quzhou county (36°53′60″ N, 115°0′ E) in Hebei province, on a calcareous alluvial soil typical of the North Plain of China (NCP), pH 8.3 (1:2.5 w/v in water) and 1.4% organic matter [26]. One cultivar (Zhengdan958) of maize was planted in a randomized complete block design with four replicates (300 m² plot⁻¹) in 2008 and 2009 while three maize cultivars (Zhengdan958, Xianyu335 and Yedan13, recorded as ZD, XY, YD, respectively) were planted in 2010 and 2011 in split-plots (N fertilizer as the main plots and cultivars as sub-plots) with three or four replicates. Nitrogen rates were applied as urea as follows: no N application (recorded as N-0), 40–70% of optimized N treatment (recorded as N-low), optimized N treatment (240, 150, 105 and 193 kg N ha⁻¹ in four consecutive years, respectively, recorded as N-opt) based on the in-season root-zone N-management approach, as previously reported [3,27], and 130 or 150% of

optimized N treatment and farmer's nitrogen practice (250 kg N ha⁻¹, recorded as N-over). Nitrogen fertilizer was applied in split doses as described in Table S1. Before sowing, 45 kg ha⁻¹ of P_2O_5, 45 kg ha⁻¹ of K_2O, and 30 kg ha⁻¹ of $ZnSO_4 \cdot 7H_2O$ (without application of $ZnSO_4 \cdot 7H_2O$ in 2011) were broadcast and incorporated into the upper 0–15 cm of the soil by rotary tillage. Another 45 kg ha⁻¹ of K_2O was applied by hand as a top dressing at the V10 (ten-leaf) growth stage. The climate conditions for the experiment have been recently reported by Zhang et al. [28].

Field experiment II was conducted in 2009 in Wenxian (34°52′42″ N, 112°57′29″ E), Henan province in the middle of the NCP. One maize cultivar (Fengyu 4, recorded as FY) was planted in a randomized complete block design with three replicates (20 m² plot⁻¹). Nitrogen rates were applied as urea as follows: no N application (recorded as N-0), 50% and 75% of optimized N application (recorded as N-low), optimized N application (240 kg N ha⁻¹, recorded as N-opt) and 150% of optimized N application (recorded as N-over). Nitrogen fertilizer was applied in split doses as described in Table S1. In addition, 90 kg ha⁻¹ of P_2O_5 and 90 kg ha⁻¹ of K_2O were applied by hand at the V5 (five-leaf) growth stage.

The two experiments were conducted in a winter wheat-summer maize rotation system. Nitrogen fertilizer, P_2O_5 and K_2O were applied as urea, superphosphate and potassium chloride, respectively. Weeds were well controlled, and no obvious water or pest stress was observed during the maize growing season. No specific permissions were required for these locations. The field studies did not involve endangered or protected species.

Sampling and nutrients analysis

6-plant (experiment I) and 2-plant (experiment II) samples were collected at V6 (six-leaf), V12 (twelve-leaf), R1 (silk emerging), R3 (milk stage) and R6 (physiological maturity, divided into straw and grain) growth stages. All plant samples were collected at V6, V12, R1, R3 and R6 stages in experiment I in 2008 and 2009. In 2010, plant samples were only collected at R1 and R3 stages with three N treatments (no N application, optimized N application and farmer's nitrogen practice) and at R6 with five N treatments. In 2011, plant samples were only collected at V6 with one cultivar (ZD) and N treatments excluding 130% of optimized N treatment, at R1 with three cultivars (ZD, XY and YD) and three N treatments (no N application, optimizer N application and farmer's nitrogen practice) and R6 stages with three cultivars (ZD, XY and YD) and five N treatments. All plant samples were collected at V6, V12, R1, R3 and R6 stages in experiment II. The shoots were rapidly washed with deionized water and then oven-dried at 70°C to determine dry weight. Plant samples were ground with a stainless steel grinder (RT-02B, Taiwan, China) and digested with HNO_3-H_2O_2 in a microwave accelerated reaction system (CEM, Matthews, NC, USA). The concentrations of Zn, Fe, Mn and Cu in the digested solutions were determined by inductively coupled plasma atomic emission spectroscopy (ICP-AES, OPTIMA 3300 DV, Perkin-Elmer, USA). IPE556 grain and IPE883 straw (Wageningen University, The Netherlands) were used as reference materials.

To estimate grain yields, ears in the central part of 60 or 180 m² (experiment I) and 10 m² (experiment II) areas were harvested at maturity. At harvest, sub-samples of 6 plants in the two experiments were collected and divided into grain and straw parts to determine above-ground biomass and the harvest index (HI) after shoots were oven-dried at 70°C.

Data analysis

According to grain yield, all data collected from 2008 to 2011 were divided into four yield ranges: <7.5, 7.5–9, 9–10.5 and >10.5 Mg ha^{-1}. According to these yield ranges, maize biomass and shoot micronutrient (Zn, Fe, Mn and Cu) accumulation as well as micronutrient RIEs (ZnRIE, FeRIE, MnRIE and CuRIE) were analysed. For each of the four different cultivars (YD, XY, ZD and FY), RIEs were analysed. According to N management practices, all data were also divided into four groups: N-0 (no N application), N-low (30–75% of optimized N treatment), N-opt (optimized N treatment) and N-over (130% or 150% of optimized N treatment as well as farmer's nitrogen practice). The number of observations for shoot samples at different growth stages with different yield ranges and N levels was shown in Table S2. All the above data analysis referred to others [22,29,30]. The following parameters were calculated: Micronutrient harvest indices (ZnHI, FeHI, MnHI and CuHI) = grain micronutrient accumulation/ shoot micronutrient accumulation (1); Grain micronutrient accumulation = grain micronutrient concentration x grain dry weight (2); Shoot micronutrient accumulation = shoot micronutrient concentration x biomass (3); Micronutrient RIE = shoot micronutrient accumulation/grain yield = grain micronutrient accumulation/(grain yield x micronutrient harvest index) [16,31] (4).

The Pearson correlation procedure was used to evaluate the correlations among the measured parameters of all data using SAS software (SAS 8.0, USA). Means of different yield ranges and N levels were compared using one-way ANOVA at a 0.05 level of probability followed by Duncan test of SPSS 13.0 for Windows. All results were expressed on a dry weight basis with an exception of grain yields. Grain yields were reported at standard moisture of 15.5%.

Results

Grain yields, yield components, biomass production and micronutrient (Zn, Fe, Mn and Cu) accumulation in shoot and grain

Overall, maize grain yield (n = 149) averaged 8.0 Mg ha^{-1} with a range from 3.9 to 12.8 Mg ha^{-1} (Table 1). The total biomass production at physiological maturity averaged 13.2 Mg ha^{-1}, with a harvest index (HI) of 51% (Table 2). The Zn, Fe, Mn and Cu RIEs averaged 31.0, 64.4, 18.1, and 5.3 g, respectively. The concentrations of grain Zn, Fe, Mn and Cu averaged 16.3, 15.8, 3.2 and 1.4 mg kg^{-1}, respectively. The harvest indices of Zn, Fe, Mn and Cu averaged 48%, 23%, 16% and 24%, respectively (Table 2). Yield was significantly negatively associated with grain Zn concentration (r = −0.36***) but positively associated with grain Mn concentration (r = 0.22**). However, yield was not correlated with grain concentrations of Fe and Cu (Table 3).

Overall, seasonal biomass production and shoot Zn accumulation continued to increase throughout the growing season. Seasonal accumulation of shoot Fe and Cu showed a slight decrease from R3 to R6 while shoot Mn accumulation showed a large decrease from R3 to R6 (Figure 1 and Figure S1). At silking stage (R1), more than three quarters of shoot Zn, Fe and Cu accumulation, and more than 100% of shoot Mn accumulation, had occurred compared to only half of the biomass accumulation (Figure 1 and Figure S1). Generally, throughout the growing season, the biomass and shoot micronutrient accumulation, as a percentage of the total (at maturity) decreased in the order Mn>Cu>Fe≥Zn>biomass (Figure S1).

During the growth stages, the highest accumulation rates of biomass and shoot Fe and Zn consistently occurred between V12 and R1. During this period, biomass accumulated 3.5 Mg ha^{-1} with an average growth rate of 251 kg ha^{-1} d^{-1}; shoot Zn accumulated 98 g ha^{-1} with an average accumulation rate of 7.0 g ha^{-1} d^{-1}; shoot Fe accumulated 215 g ha^{-1} with an average accumulation rate of 15.3 g ha^{-1} d^{-1}. In contrast, the highest accumulation rates of shoot Mn (83 g ha^{-1} in total, with an average accumulation rate of 6.0 g ha^{-1} d^{-1}) and Cu (15 g ha^{-1} in total, with an average accumulation rate of 1.1 g ha^{-1} d^{-1}) occurred between V6 and V12 (Figure 1).

Relationships between yield and shoot micronutrient accumulation in response to different yield ranges

To understand the relationship between yield and shoot micronutrient accumulation in relation to grain yield, all the data in this study was grouped into 4 yield ranges as follows: <7.5 Mg ha^{-1} (the number of observations, n = 58, mean yield 5.9 Mg ha^{-1}, recorded as GY1), 7.5–9.0 Mg ha^{-1} (n = 45, mean yield 8.2 Mg ha^{-1}, recorded as GY2), 9.0–10.5 Mg ha^{-1} (n = 26, mean yield 9.7 Mg ha^{-1}, recorded as GY3) and >10.5 Mg ha^{-1} (n = 20, mean yield 11.4 Mg ha^{-1}, recorded as GY4) (Table 1). As shown in Table 2, the increase in grain yield from GY1 to GY2 was mainly attributed to thousand grain weight (an increase of 12%) and grains per ear (an increase of 11%). The increase in grain yield from GY2 to GY3 was mainly attributed to grains per ear (an increase of 14%) and HI (which increased from 50% to 53%). The further increase in grain yield from GY3 to GY4 was mainly attributed to HI (which increased from 53% to 55%).

With the increase of grain yield from GY1 to GY4, ZnRIE decreased gradually from 36.3, 32.4, 26.6 to 18.0 g. However, FeRIE (ranging from 57.4 to 68.6 g), MnRIE (ranging from 17.5 to 19.0 g) and CuRIE (ranging from 4.9 to 5.6 g) were not significantly affected by yields (Table 2). With increasing yield from GY1 to GY3, grain Zn concentration (average 16.9 mg kg^{-1}) was not affected, but it was significantly decreased to 12.2 mg kg^{-1} when grain yield increased from GY3 to GY4. Similar results were found for grain Fe concentration. Straw Zn concentration showed a decreasing trend from 24.4 to 10.7 mg kg^{-1} while straw Mn concentration showed an increasing trend from 17.3 to 23.9 mg kg^{-1} with the increasing yield from GY1 to GY4. In contrast, straw Fe and Cu concentrations were less affected by yields. Grain Mn concentration was significantly higher in GY2 and GY3 than GY1 and GY4. Similarly, grain Cu concentration was the highest in GY3 than the other three yield ranges (Table 2). With an increasing grain yield from GY1 to GY4, ZnHI increased gradually from 41%, 48%, 54% to 60% while FeHI (ranging from 21% to 27%), MnHI (ranging from 14% to 18%) and CuHI (ranging from 22% to 29%) were the highest in GY3 than the other three yield ranges (Table 2).

The response of ZnRIE to increasing grain yield could be classified into two response stages. When grain yield was increased from GY1 to GY3, ZnRIE decreased from 36.3 g to 26.6 g (a decrease of 27%) because of increasing ZnHI (which increased from 41% to 54%) and significantly declining straw Zn concentration (a decrease of 30%) coupled with the relatively constant grain Zn concentration (average 16.9 mg kg^{-1}) (Table 2). With the increasing yield from GY3 to GY4, ZnRIE further declined from 26.6 to 18.0 g (a decrease of 32%) mainly because of a further increasing ZnHI from 54% to 60% and the decreasing Zn concentrations in both straw (which decreased by 38%) and grain (which decreased by 26%) (Table 2). Furthermore, across all the grain yield ranges, ZnRIE was significantly positively correlated with Zn concentrations of grain (r = 0.54***) and especially straw (r = 0.90***) but negatively associated with grain yield (r = −0.58***) and especially ZnHI (r = −0.85***) (Table 3).

Table 1. Descriptive statistics of yield for total samples, four yield ranges (15.5% moisture), four different cultivars and four nitrogen (N) levels.

	n[a]	Mean	SD[b]	Minimum	25% Q[c]	Median	75% Q	Maximum
				------Mg ha^{-1}------				
Total	149	8.0	2.1	3.9	6.4	8.1	9.5	12.8
Yield ranges (Mg ha^{-1})								
<7.5	58	5.9	1.0	3.9	5.2	5.9	6.7	7.4
7.5–9	45	8.2	0.4	7.5	7.8	8.3	8.5	9.0
9–10.5	26	9.7	0.4	9.0	9.2	9.6	10.0	10.4
>10.5	20	11.4	0.6	10.5	11.0	11.3	11.8	12.8
Cultivars								
YD	30	5.6	1.1	3.9	4.7	5.4	6.5	8.1
XY	30	8.0	1.5	3.9	7.3	8.4	9.1	10.8
ZD	75	8.5	1.9	5.0	7.2	8.3	9.8	12.8
FY	14	10.2	1.2	8.1	9.5	10.4	11.0	12.0
N levels								
N-0	30	6.1	1.5	3.9	5.1	5.8	6.7	9.5
N-low	32	8.3	2.1	4.3	7.3	8.2	9.8	11.5
N-opt	30	8.9	2.1	4.4	7.4	9.2	10.4	12.3
N-over	57	8.3	1.8	4.5	7.2	8.3	9.3	12.8

n[a]: number of observations.
SD[b]: standard deviation.
Q[c]: quartile.

Similarly, FeRIE, MnRIE and CuRIE were consistently positively associated with their respective straw concentrations and straw yields (for MnRIE and CuRIE) but negatively correlated with their respective harvest index and grain yield (for FeRIE and CuRIE) (Table 3).

With the increase in grain yield from GY1 to GY4, the biomass production showed increasing trends, especially during the reproductive development (e.g. R3 and R6) (Figure 2A). Similar results were also found for shoot Mn and Cu accumulation due to the similar or lower shoot Mn and Cu concentrations in GY1 than the other three yield ranges, where shoot Mn and Cu concentrations were similar during the reproductive development (Figure 2D, E and Figure 3C, D). Shoot Fe accumulation among the four yield ranges were similar at V12 and R3 and tended to increase at V6 and especially at R6 with the increasing grain yield (Figure 2C). However, shoot Zn accumulation in GY4 was slightly (e.g. V6, R3 and R6 stages) or even significantly lower than the other two or three (e.g. V12 and R1 stages) yield ranges (Figure 2B).

In agreement with the overall trends shown in Figure S1, seasonal biomass production and shoot Zn accumulation continued to accumulate throughout the growing season irrespective of yield ranges (Figure 2B and Figure S2). Shoot Fe accumulation showed a decreasing trend from R3 to R6 for GY1, GY2 and GY3 but continued to accumulate throughout the growing stages for GY4 (Figure 2C and Figure S2). Shoot Mn accumulation decreased to a great extent from R3 to R6 irrespective of yields (Figure 2D and Figure S2). Similarly, shoot Cu accumulation showed a decreasing trend from R3 to R6 except GY1 (Figure 2E and Figure S2).

According to the increased grain yield from GY1 to GY4, nearly 75%, 86%, 91% and 66% of pre-silking shoot Zn accumulation, 81%, 95%, 76% and 59% of pre-silking shoot Fe accumulation, 62%, 90%, 81% and 71% of pre-silking shoot Cu accumulation and more than 100% of pre-silking shoot Mn accumulation occurred compared to 52%, 61%, 54% and 42% of pre-silking biomass accumulation (Figure 2 and Figure S2).

RIEs in response to different cultivars

To understand micronutrient RIEs in relation to different cultivars, all data were divided into four groups: YD (the number of observations, n = 30, mean yield 5.6 Mg ha^{-1}), XY (the number of observations, n = 30, mean yield 8.0 Mg ha^{-1}), ZD (the number of observations, n = 75, mean yield 8.5 Mg ha^{-1}) and FY (the number of observations, n = 14, mean yield 10.2 Mg ha^{-1}) (Table 1). Grain yield was in the order YD<XY<ZD<FY.

The effects of cultivars on ZnRIE were similar to those of yield ranges (described above). ZnRIE decreased gradually from 41.0, 36.1, 26.6 to 22.0 g for YD, XY, ZD and FY, respectively. The response of ZnRIE to cultivars could also be classified into two response stages. ZnRIE decreased from 41.0 g for YD to 26.6 g for ZD (a decrease of 35%) because of significantly increasing ZnHI (from 37% to 55%) and declining straw Zn concentration (a decrease of 45%) coupled with the relatively constant grain Zn concentration (average 16.7 mg kg^{-1}) (Table 2). ZnRIE further decreased from 26.6 g for ZD to 22.0 g for FY (a decrease of 17%) mainly because of decreasing Zn concentrations in both grain (which decreased by 17%) and straw (which decreased by 14%). FeRIE (ranging from 54.4 to 97.4 g) and MnRIE (ranging from 16.5 to 25.9 g) were in the order XY≤ZD≤YD<FY. The

Table 2. Biomass production, yield components (including ears number, grains per ear and thousand grain weight (TGW)), harvest index (HI), micronutrient reciprocal internal efficiencies (ZnRIE, FeRIE, MnRIE and CuRIE), grain micronutrient concentrations (GZnC, GFeC, GMnC and GCuC), straw micronutrient concentrations (SZnC, SFeC, SMnC and SCuC) and micronutrient harvest index (ZnHI, FeHI, MnHI and CuHI) of summer maize as affected by different yield ranges, cultivars and N levels.

Parameters	Total	Yield ranges (Mg ha^{-1})				Cultivars				N levels			
		<7.5	7.5–9	9–10.5	>10.5	YD	XY	ZD	FY	N-0	N-low	N-opt	N-over
	n^a 149	58	45	26	20	30	30	75	14	30	32	30	57
Biomass (Mg ha^{-1})	13.2	10.2d	14.0c	14.9b	17.5a	9.7c	14.2b	13.5b	16.4a	10.3b	13.9a	14.5a	13.6a
Ears number (10^4 ha^{-1})	7.4	7.3a	7.5a	7.5a	7.5a	6.8b	7.8a	7.9a	5.4c	7.5a	7.3a	7.3a	7.6a
Grains per ear	425	379c	420b	477a	500a	429b	352c	416b	616a	352b	444a	450a	439a
TGW (g)	275	257b	288a	283a	292a	233c	314a	277b	277b	268a	276a	277a	278a
HI (%)	51	50c	50c	53b	55a	49b	48b	53a	53a	50a	51a	52a	51a
ZnRIE (g)	31.0	36.3a	32.4a	26.6b	18.0c	41.0a	36.1b	26.6c	22.0d	28.9a	28.4a	31.2a	33.4a
GZnC (mg kg^{-1})	16.3	16.7a	17.4a	16.6a	12.2b	17.6a	16.0a	16.4a	13.6b	15.2b	15.2b	16.6ab	17.3a
SZnC (mg kg^{-1})	20.0	24.4a	20.2b	17.1b	10.7c	29.0a	23.7b	16.1c	13.8c	18.2ab	17.6b	20.9ab	21.8a
ZnHI (%)	48	41d	48c	54b	60a	37b	40b	55a	53a	48a	49a	49a	48a
FeRIE (g)	64.4	68.6a	64.6a	59.7a	57.4a	66.8a	54.4c	61.2bc	97.4a	65.8a	69.0a	59.9a	63.4a
GFeC (mg kg^{-1})	15.8	15.9a	16.2a	16.6a	13.8b	17.6a	13.9b	15.9a	16.0a	13.4c	15.5b	16.0ab	17.2a
SFeC (mg kg^{-1})	62.6	63a	62a	61a	66a	58.6b	46.6c	61.9b	109.4a	63.3a	69.6a	58.3a	60.6a
FeHI (%)	23	21b	23b	27a	22b	23a	22a	24a	14b	19b	21ab	25a	25a
MnRIE (g)	18.1	17.8a	18.3a	17.5a	19.0a	17.8b	16.5b	17.3b	25.9a	16.8a	18.3a	18.0a	18.7a
GMnC (mg kg^{-1})	3.2	2.9b	3.5a	3.4a	3.0b	2.7c	3.7a	3.1b	3.3b	2.95c	3.03bc	3.35a	3.28ab
SMnC (mg kg^{-1})	19.0	17.3b	18.4b	19.9b	23.9a	17.4b	14.4c	19.3b	30.2a	16.8b	19.4ab	19.1ab	19.8a
MnHI (%)	16.0	15bc	17ab	18a	14c	14c	20a	16b	11d	16a	15a	17a	16a
CuRIE (g)	5.3	5.6a	5.4a	5.1a	4.9a	6.7a	5.5b	4.6c	6.0ab	4.5b	5.4a	5.4a	5.7a
GCuC (mg kg^{-1})	1.4	1.39b	1.43b	1.68a	1.26b	1.61a	1.38b	1.40ab	1.34b	1.24b	1.27b	1.57a	1.55a
SCuC (mg kg^{-1})	5.0	4.9a	5.0a	4.9a	5.5a	6.0a	4.7b	4.5b	6.4a	4.0b	5.2a	5.1a	5.4a
CuHI (%)	24	23b	23b	29a	22b	21b	22b	27a	19b	26a	21b	26a	24ab

n^a: number of observations. Means in a row followed by different lowercase letters are significantly different at different yield ranges, cultivars and N levels (P<0.05).

significantly lower FeRIE in XY than FY is due to a lower grain Fe concentration (13.9 and 16.0 mg kg^{-1} for XY and FY, respectively) and a higher FeHI (22% and 14% for XY and FY, respectively). CuRIE was in the order ZD<XY≤FY≤YD (Table 2).

Relationship between yield and shoot micronutrient accumulation in response to different N rates

To understand the relationship between yield and shoot micronutrient accumulation in relation to N management practices, all data were divided into four groups: N-0 (the number of observations, n = 30, mean yield 6.1 Mg ha^{-1}), N-low (the

Table 3. Correlative coefficients (r) among the measured parameters at maturity (n = 149).

	ZnRIE		FeRIE		MnRIE		CuRIE		GY
GY	−0.58***	GY	−0.20*	GY	ns	GY	−0.18*	GZnC	−0.36***
SY	ns	SY	ns	SY	0.41***	SY	0.30***	GFeC	ns
GZnC	0.54***	GFeC	ns	GMnC	ns	GCuC	0.29***	GMnC	0.22**
SZnC	0.90***	SFeC	0.89***	SMnC	0.78***	SCuC	0.76***	GCuC	ns
ZnHI	−0.85***	FeHI	−0.77***	MnHI	−0.79***	CuHI	−0.58***	—	—

GY: grain yield based on 15.5% of moisture (Mg ha^{-1}); SY: straw yield based on dry weight (Mg ha^{-1}); ZnRIE, FeRIE, MnRIE and CuRIE: reciprocal internal efficiencies (g micronutrient requirement per Mg grain yield); GZnC, GFeC, GMnC and GCuC: grain micronutrient concentrations (mg kg^{-1}); SZnC, SFeC, SMnC and SCuC: straw micronutrient concentrations (mg kg^{-1}); ZnHI, FeHI, MnHI and CuHI: micronutrient harvest indices (%).
*Significant at P<0.05.
**Significant at P<0.01.
***Significance at P<0.001.
ns: not significant at P<0.05.

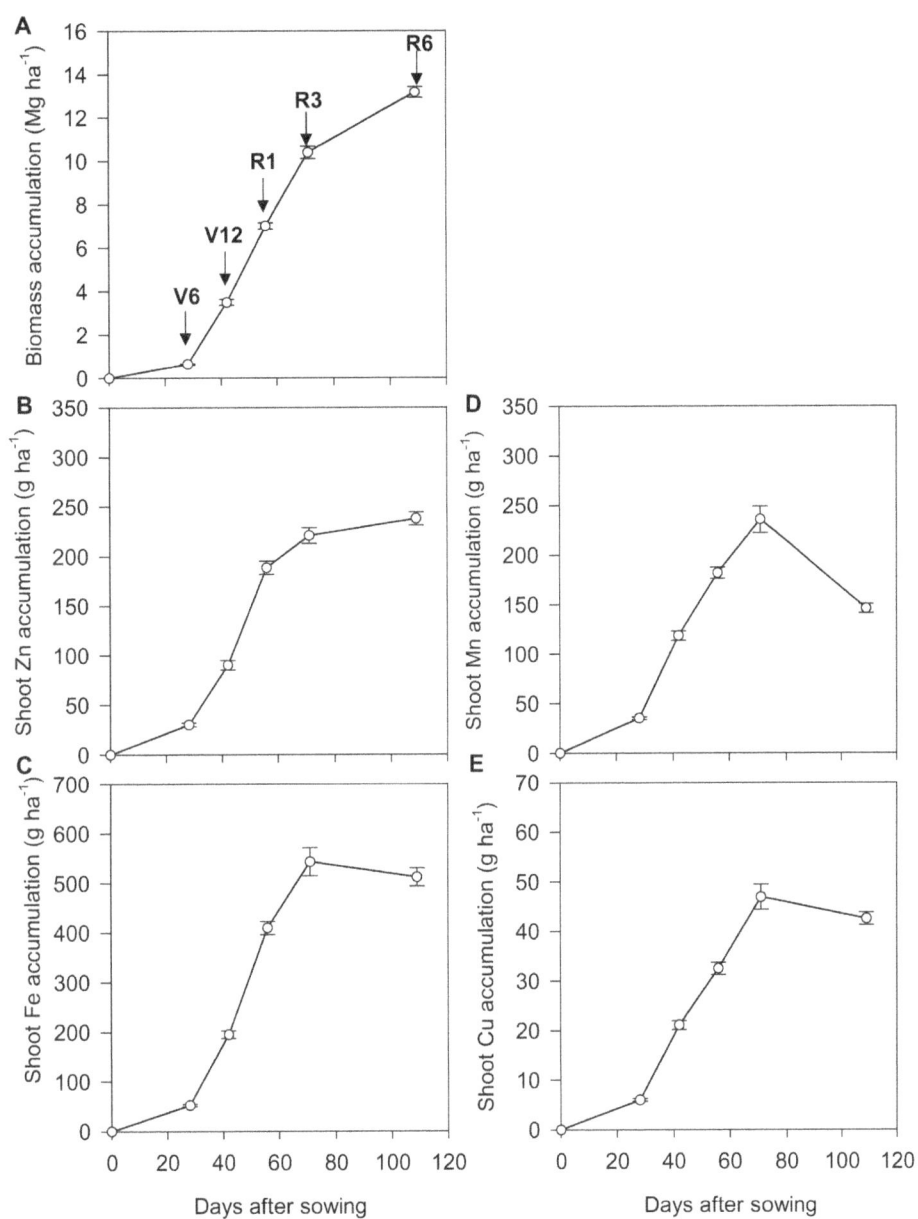

Figure 1. Dynamics of biomass (A), shoot Zn accumulation (B), shoot Fe accumulation (C), shoot Mn accumulation (D) and shoot Cu accumulation (E) of summer maize at V6 (six-leaf stage, n = 70), V12 (12-leaf stage, n = 54), R1 (silk emerging, n = 115), R3 (milk stage, n = 81) and R6 (physiological maturity, n = 149) stages, respectively. The bars represent the standard error of the mean.

number of observations, n = 32, mean yield 8.3 Mg ha^{-1}), N-opt (the number of observations, n = 30, mean yield 8.9 Mg ha^{-1}) and N-over (the number of observations, n = 57, mean yield 8.3 Mg ha^{-1}) (Table 1 and Table S1). The N-0 application rate resulted in significantly lower grain yield than the other three N application rates, while the N-opt treatment resulted in the highest grain yield. Compared to N-0 treatment, the higher grain yields with N application were mainly attributed to more biomass production and grains per ear. However, average ear numbers, thousand grain weight and HI were less affected by N rates (Table 2). Similarly, ZnRIE, FeRIE and MnRIE were also less affected by N rates. CuRIE was significantly lower in N-0 treatment compared to the other three N treatments. Grain concentrations of Zn (increased from 15.2 to 17.3 mg kg^{-1}), Fe (increased from 13.4 to 17.2 mg kg^{-1}), Mn (increased from 2.95 to 3.35 mg kg^{-1}) and Cu (increased from 1.24 to 1.57 mg kg^{-1}) showed increasing trends with the increase of N applied, although there were no significant differences between the N-opt and N-over treatments (Table 2). ZnHI and MnHI were less affected by N rates. FeHI increased from 19% to 25% with the increase of N levels from N-0 to N-opt while a further increase of N from N-opt to N-over treatments did not affect it. CuHI ranged from 21% to 26% and was inconsistently affected by N treatments (Table 2).

Before silking (R1) stage, with the increase of N levels from N-0 to N-opt treatment, the biomass production showed progressive enhancements. However, a further increasing N from N-opt to N-over treatment did not make an extra contribution to growth (Figure 4A). During the reproductive growth stages (e.g. R3 and R6), biomass production in the N-0 treatment was significantly lower than the other three N rates where there were no significant

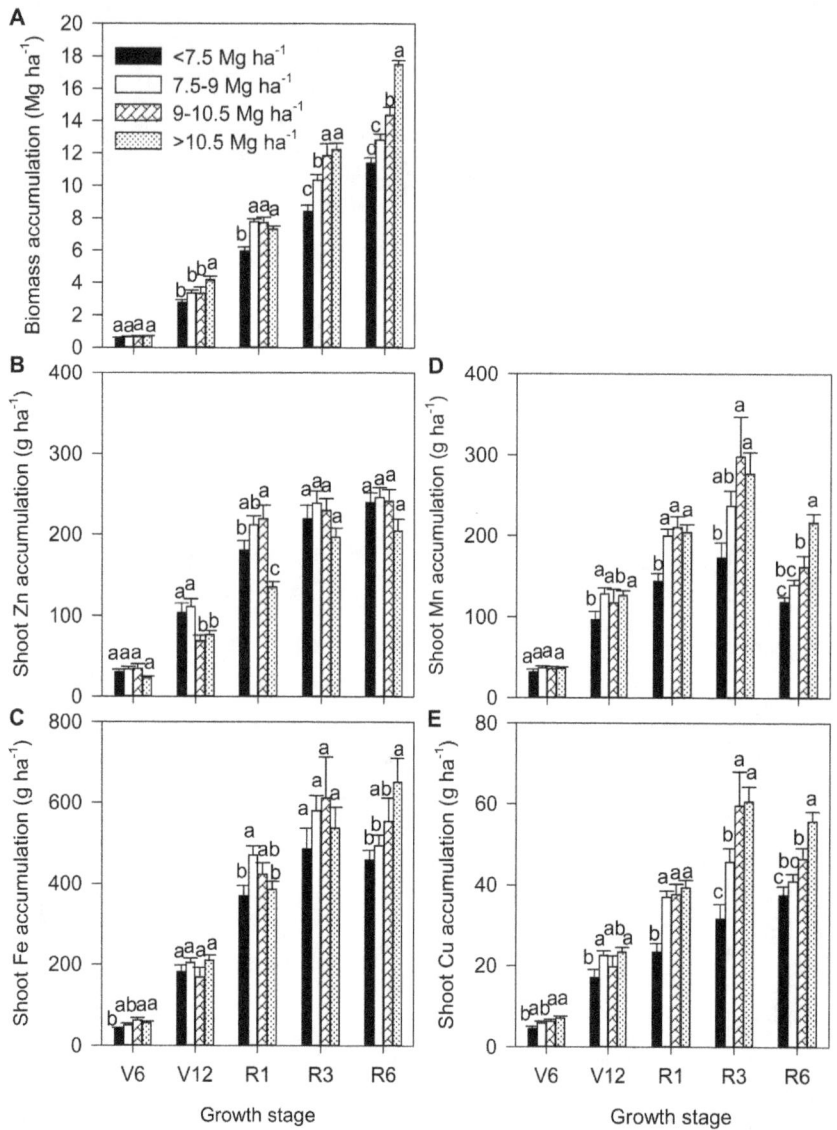

Figure 2. Dynamics of biomass (A), shoot Zn accumulation (B), shoot Fe accumulation (C), shoot Mn accumulation (D) and shoot Cu accumulation (E) of summer maize at V6 (six-leaf stage), V12 (12-leaf stage), R1 (silk emerging), R3 (milk stage) and R6 (physiological maturity) stages, respectively, with different yield ranges. The number of observations at each stage was shown in Table S2. The bars represent the standard error of the mean. Bars with different lowercase letters are significantly different at different yield ranges (P<0.05).

differences in biomass production (Figure 4A). Similar results were found in shoot Fe accumulation, as there were similar shoot Fe concentrations for each N rate at most of the growth stages (Figure 4C and Figure 5B). Similar results were also found for shoot Mn and Cu accumulation (Figure 4D, E). Shoot Mn and Cu concentrations were significantly lower in the N-0 treatment than the other three N treatments where shoot Mn and Cu concentrations were similar during the reproductive growth stages (Figure 5C, D). Shoot Zn accumulation tended to increase gradually with the increase of N from N-0 to N-opt throughout the growth stages, but there was a lower shoot Zn concentration in the N-low treatment compared to N-0 and N-opt treatments (Figure 4B and Figure 5A). A further increase in N from the N-opt to the N-over treatment did not affect shoot Zn accumulation as there was a similar biomass and shoot Zn concentration in both treatments (Figure 4B and Figure 5A).

With the increase of N rates from the N-0 to the N-over treatment, nearly 74%, 67%, 86% and 84% of pre-silking shoot Zn accumulation, 74%, 77%, 84% and 87% of pre-silking shoot Fe accumulation, 59%, 83%, 78% and 84% of pre-silking shoot Cu accumulation and more than 100% of pre-silking shoot Mn accumulation occurred compared to 48%, 54%, 54% and 57% of pre-silking biomass accumulation (Figure 4).

Discussion

Overall, maize grain yield (n = 149) averaged 8.0 Mg ha^{-1} with a range from 3.9 to 12.8 ha^{-1} (Table 1), which was 49% higher than the average yields in China (5.4 Mg ha^{-1}) and 61% higher than the world (5.0 Mg ha^{-1}) during 2006 to 2009 [4,32].

The relationships between grain yield and whole-plant micronutrient accumulation can be explained by studying the RIEs [19]. In this study, the average Zn, Fe, Mn and Cu RIEs were lower

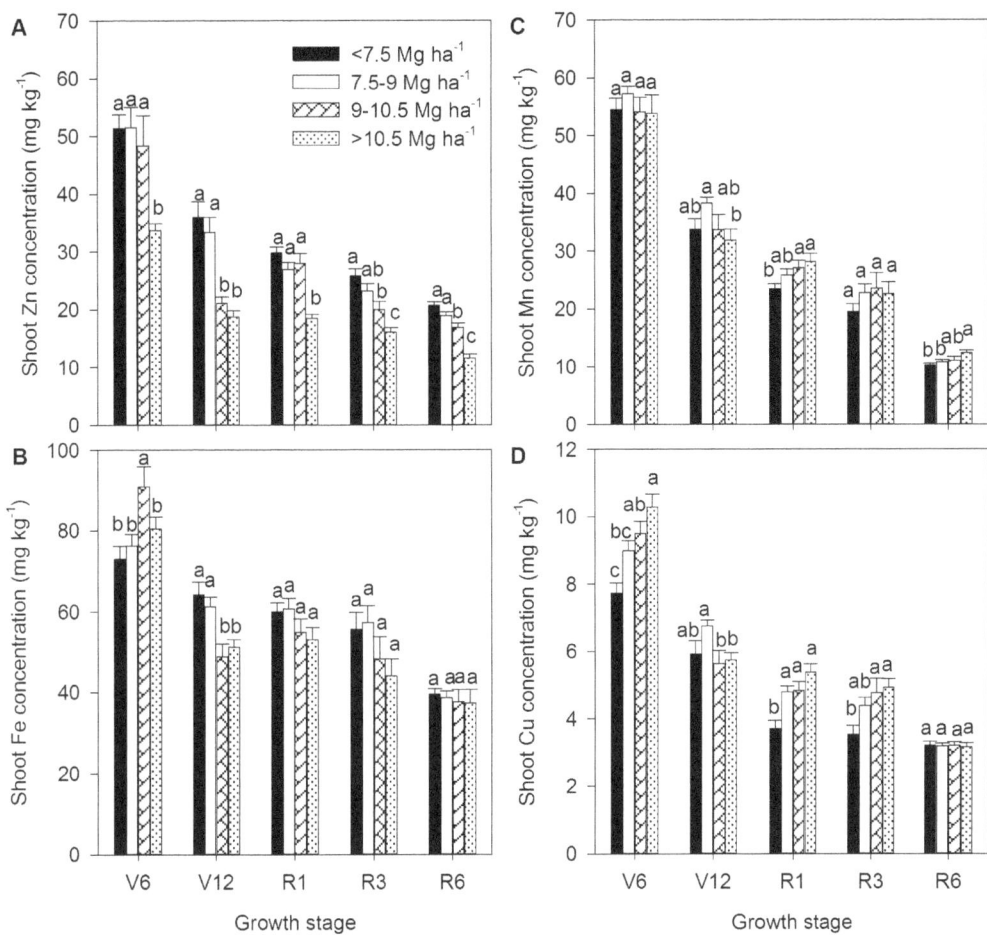

Figure 3. Dynamics of shoot Zn concentration (A) shoot Fe concentration (B), shoot Mn concentration (C) and shoot Cu concentration (D) of summer maize at V6 (six-leaf stage), V12 (12-leaf stage), R1 (silk emerging), R3 (milk stage) and R6 (physiological maturity) stages, respectively, with different grain yield ranges. The number of observations was shown in Table S2.The bars represent the standard error of the mean. Bars with different lowercase letters are significantly different at different yield ranges (P<0.05).

than corresponding values [17,19], but ZnRIE was comparable to that from Potarzycki [33] and Bender et al. [18], indicating that RIEs varied among different environments, cultivars, management regimes (such as N management) and yields [20,22,23]. With the increased N rate from N-0 to N-over treatments, there was 15% (non significant) increase in ZnRIE (Table 2). Potarzycki [33] also found a slight increase in ZnRIE with an increased supply of N from 115 to 175 kg N ha^{-1}. However, there was a significant 50% decrease in ZnRIE with the increase of grain yield from GY1 to GY4, although the data for GY4 excluded N-0 treatment (Tables 1 and 2). Similarly, with an increasing grain yield in the order YD<XY<ZD<FY, there was a significant 46% decrease in ZnRIE (Table 2). Furthermore, a significant negative relationship (r = −0.58***) between grain yield and ZnRIE was found (Table 3). Similarly, N reciprocal internal efficiency has also been found to decrease with increasing grain yields in spring maize and in winter wheat [22,29]. These results indicate yields and cultivars, indeed, exert a greater influence on ZnRIE than N management practices. The substantial decrease in the ZnRIE with the increase in grain yield was predominantly associated with an increase in ZnHI and a decrease of Zn concentrations in grain and especially in straw. Cultivars showed a similar effect on ZnRIE (Table 2). An increased ZnHI with increased grain yield was also reported [19]. Behera et al. [34] reported that ZnHI was only around 30% with

6.5 Mg ha^{-1} grain yield. These results suggest grains provide a large sink for Zn translocation.

Increased biomass production may be a driving force for the uptake and assimilation of mineral nutrients such as N, P and K [15]. Similarly, our results showed, with the increases of yield and biomass production, there was a concomitant increase in the accumulation of shoot Fe, Mn and Cu at physiological maturity (Figure 2C, D, E). Furthermore, with the increase in yield from GY2 to GY4, the declining proportions of pre-silking Fe (which decreased from 95% to 59%), Cu (which decreased from 90% to 71%) and biomass production (which decreased from 61% to 42%) were consistently found (Figure 2A, C, E and Figure S2). These results indicate the significance of increasing proportions of post-silking shoot Fe and Cu accumulation, possibly because of the significantly enhanced shoot biomass production by photosynthesis during reproductive development (which increased from 39% to 58%) with increasing yields. Due to the stable or decreasing shoot Fe and Cu accumulation between R3 and R6 (Figure 2C, E and Figure S2), more supplies of Fe and Cu during the vegetative stages (V12 to R1) and reproductive stages (R1 to R3) are necessary to maximize yields. Although shoot Mn accumulation at physiological maturity increased with the increase of yield, at least 94% of pre-silking shoot Mn accumulation (Figure 2D and Figure S2) indicates that more Mn should be applied before silking,

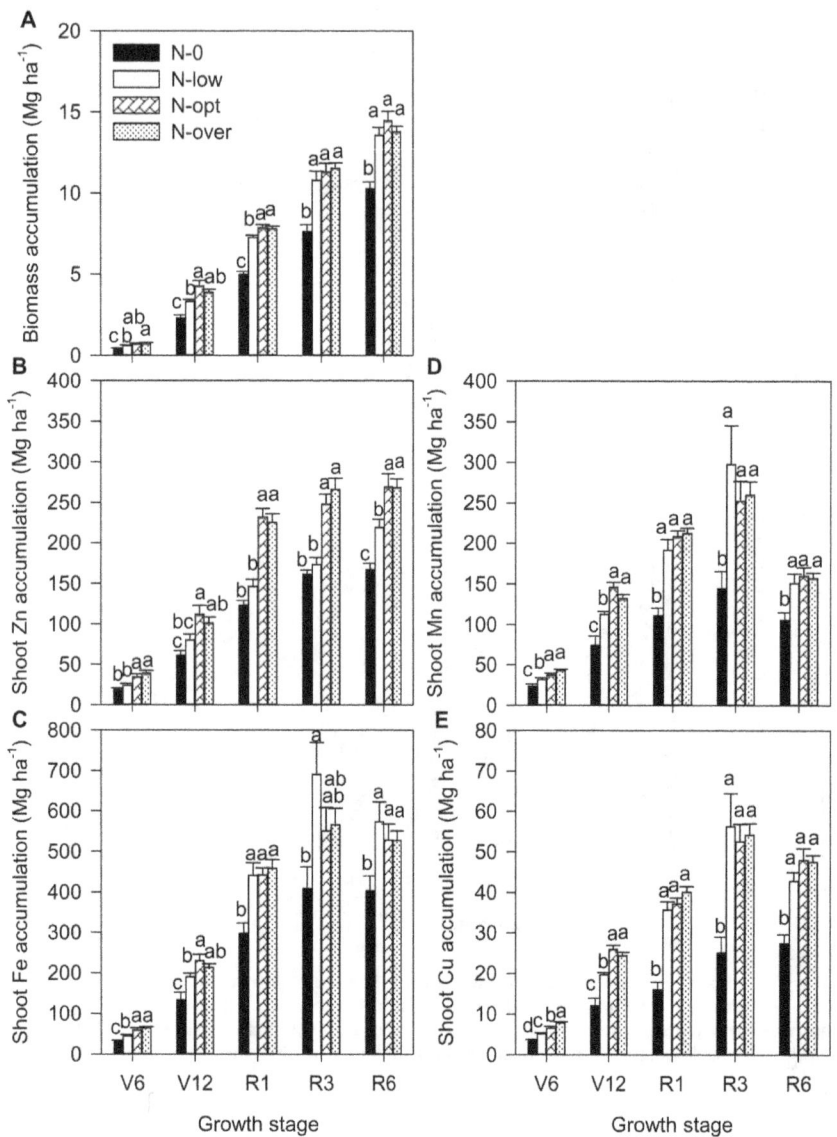

Figure 4. Dynamics of biomass (A), shoot Zn accumulation (B), shoot Fe accumulation (C), shoot Mn accumulation (D) and shoot Cu accumulation (E) of summer maize at V6 (six-leaf stage), V12 (12-leaf stage), R1 (silk emerging), R3 (milk stage) and R6 (physiological maturity) stages, respectively, with different N levels. The number of observations was shown in Table S2.The bars represent the standard error of the mean. Bars with different lowercase letters are significantly different at different N levels (P<0.05).

especially during V6 to V12 when the highest accumulation rate of shoot Mn generally occurred. Similarly, it has been reported that maximum Mn and Cu accumulation was achieved at stage R3, possibly due to losses during leaf senescence, leaf leaching and the low uptake rates resulting from root senescence during late-reproductive stages [19].

The similar or lower shoot Zn accumulation in GY4 compared to the other three yield ranges during all growth stages was at least in part due to the significantly lower shoot Zn concentration in GY4 compared to the other three yield ranges at each growth stage (Figure 2B and Figure 3A), probably due to a dilution effect as a result of the increased biomass production. Additionally, the shoot Zn concentration in GY4 was below the critical Zn-deficient range of 15–20 mg kg^{-1} [35], at almost all of the growing stages (Figure 3A), which indicates a potential Zn deficiency. Therefore, enhanced grain yield and biomass production may be involved in increasing Zn internal utilization efficiency (expressed by de-

creased ZnRIE) and Zn remobilization efficiency (expressed by increased ZnHI) at the cost of decreased Zn concentrations in straw and grain under potentially Zn deficient conditions. The higher plant Zn internal utilization efficiency in maize compared to legumes and rice was previously reported [36]. The higher Zn internal efficiency of maize with a greater yield and biomass production may be related to higher activity of carbonic anhydrase (CA) and Cu/Zn superoxide dismutase. Rengel [37] suggested that Zn-efficient wheat genotypes had an ability to maintain greater CA activity under Zn deficiency that could help maintain higher photosynthesis rates and dry matter production.

Although shoot Zn accumulation at physiological maturity was not significantly affected by the increased yield, the decreasing proportions of pre-silking shoot Zn accumulation (which decreased from 91% to 66%) together with the continued accumulation of shoot Zn throughout the growth stages with the increase of grain yield (Figure 2B and Figure S2) indicate the high reproductive-

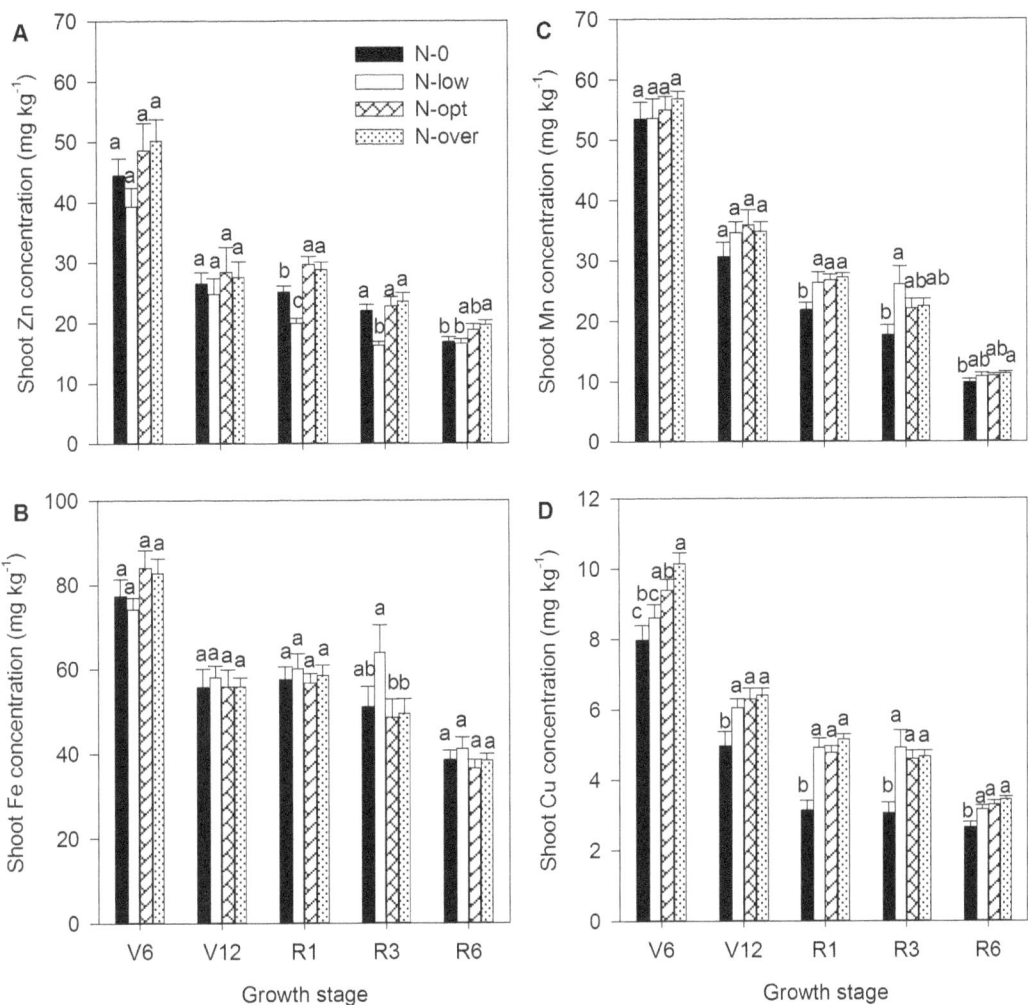

Figure 5. Dynamics of shoot Zn concentration (A) shoot Fe concentration (B), shoot Mn concentration (C) and shoot Cu concentration (D) of summer maize at V6 (six-leaf stage), V12 (12-leaf stage), R1 (silk emerging), R3 (milk stage) and R6 (physiological maturity) stages, respectively, with different N levels. The number of observations was shown in Table S2.The bars represent the standard error of the mean. Bars with different lowercase letters are significantly different at different N levels (P<0.05).

stage Zn demand. Further decreasing pre-silking accumulation of shoot Zn (48%) and Cu (45%) was found with 14.2 Mg ha^{-1} (15.5 of moisture) grain yield [18]. Grzebisz [38] also reported that vegetative tissues were only minor Zn sources while post-silking shoot Zn uptake from the soil during grain filing was the major source for growing kernels. Therefore, in order to gain higher grain yield and avoid a potential Zn deficiency, more Zn and season-long supply of Zn is necessary for maximum yields, especially during the V12 to R1 growth stages when the highest Zn accumulation rate occurs. The higher Zn and Fe accumulation rates during V12 to R1 may be because ear size and number of ovules are being established while the stalk and leaves are accumulating more photosynthates [15].

Excessive N fertilization in intensive agricultural areas of China has resulted in serious environmental problems. Currently, improving N management and developing related policies are major issues in crop production and environmental protection in China. In this study, increasing N supply from the N-0 to N-opt treatment tended to increase not only grain Zn, Fe, Mn and Cu concentrations, but also grain yields while further increased N

from N-opt to N-over treatment did not result in an increase in grain concentrations but resulted in a decrease in yield (Tables 1 and 2). Similarly, there were also not any differences between N-opt to N-over treatments in shoot concentrations and the accumulation of micronutrients during any of the growth stages (Figures 4 and 5). These results indicate that optimized N management by reasonably decreasing N input was able to maintain plant nutrition of micronutrients for maximum yields as well as grain concentrations. In agreement with our results, Oktem et al. [39] previously reported the positive effects of N on Zn, Fe and Cu, but not Mn, concentrations in maize grain. Very recently, the positive effects of N on grain micronutrient concentrations (less pronounced for Zn) were also found [19]. It has been reported that application of Zn-enriched urea results in higher productivity and grain Zn concentrations in a rice–wheat cropping system than the same rate of ZnSO$_4$ and urea applied separately [40,41]. It is also convenient to apply a split application of Zn later to maintain season-long Zn supply for maize by overcoming the rapid fix on a calcareous soil because urea is often split applied. Therefore, an optimized Zn-enriched urea management strategy may be an economical method of improving Zn concentration in maize and maintaining sufficient Zn nutrition for higher grain yield.

Conclusions and Remarks

With increases in yield and biomass production, maize shoots contained more Fe, Mn and Cu at maturity but the pre-silking proportions of shoot Fe and Cu decreased. These results indicate that with increasing yield, more Fe and Cu would be needed, not only during the vegetative stages, but also reproductive stages (e.g. R1 to R3) for maximum yields. In contrast, more Mn should be applied before silking, especially during V6 to V12 when the highest accumulation rate of shoot Mn generally occurred. Shoot Zn accumulation was non-significant among the yield ranges possibly due to a dilution effect or a potential Zn deficiency, or because of the improvements in both Zn utilization efficiency (e.g. a decrease in ZnRIE) and Zn remobilization efficiency (e.g. an increase in ZnHI). Furthermore, the substantial decrease in ZnRIE with the increase of yield was largely associated with an increase in ZnHI, at the cost of the decreased Zn concentrations in grain and especially in straw. Cultivars had the similar effects on ZnRIE. Increasing N supply generally increased micronutrient concentrations of maize grain [19,39]. However, optimizing N management by reasonably decreasing N input did not result in any negative effects on plant and grain nutrition of micronutrients, while achieving the highest yield. These results showed that optimized N management is an applicable strategy to improve micronutrient nutrition for maximum yield.

Supporting Information

Figure S1 Changes in biomass and micronutrient (Zn, Fe, Mn and Cu) accumulation expressed as biomass and micronutrient accumulation at each stage divided by their corresponding values at maturity. V6: six-leaf stage; V12: 12-leaf stage; R1: silk emerging; R3: milk stage; R6: physiological; the number of observations was 70, 54, 115, 81 and 149 at V6, V12, R1, R3 and R6, respectively, as shown in Table S2.

Figure S2 Changes in biomass and micronutrient (Zn, Fe, Mn and Cu) accumulation expressed as biomass and micronutrient accumulation at each stage divided by their corresponding values at maturity for (A) yield <7.5 Mg ha^{-1}, (B) yield between 7.5 to 9 Mg ha^{-1}, (C) yield between 9 to 10.5 Mg ha^{-1}, (D) yield >10.5 Mg ha^{-1}. V6: six-leaf stage; V12: 12-leaf stage; R1: silk emerging; R3: milk stage; R6: physiological; the number of observations was 70, 54, 115, 81 and 149 at V6, V12, R1, R3 and R6, respectively, as shown in Table S2.

Table S1 N split supply as urea during the vegetative period of summer maize from 2008 to 2011based on determination of soil mineral N (N$_{min}$) of 0–90 cm at sowing, V5, V6, V10 and V12 stages in the field.

Table S2 Number of observations at different growth stages with different yield ranges and N levels.

Acknowledgments

The authors are grateful to Meng-Long Qiu and Zhong-Xiang Li from China agricultural university for their assistance with this work. The authors are also grateful to Tom Sizmur of Rothamsted Research, UK for his improvements to the English in the manuscript.

Author Contributions

Conceived and designed the experiments: CZ XC ZC. Performed the experiments: YX SY WZ DL YY. Analyzed the data: YX CZ. Contributed reagents/materials/analysis tools: CZ XC ZC YY. Wrote the paper: YX CZ.

References

1. Nuss ET, Tanumihardjo SA (2010) Maize: a paramount staple crop in the context of global nutrition. Compr Rev Food Sci F 9: 417–436.
2. Shiferaw B, Prasanna B, Hellin J, Bänziger M (2011) Crops that feed the world 6. Past successes and future challenges to the role played by maize in global food security. Food Sec 3: 307–327.
3. Chen XP, Cui ZL, Vitousek PM, Cassman KG, Matson PA, et al. (2011) Integrated soil-crop system management for food security. Proc Natl Acad Sci USA 108: 6399–6404.
4. FAO FAOSTAT–Agriculture Database. Available: http://faostat3.fao.org/faostat-gateway/go/to/download/Q/*/E. Accessed 2013 Sep 13.
5. Ju XT, Xing GX, Chen XP, Zhang SL, Zhang LJ, et al. (2009) Reducing environmental risk by improving N management in intensive Chinese agricultural systems. Proc Natl Acad Sci USA 106: 3041–3046.
6. Guo JH, Liu XJ, Zhang Y, Shen JL, Han WX, et al. (2010) Significant acidification in major Chinese croplands. Science 327: 1008–1010.
7. Le C, Zha Y, Li Y, Sun D, Lu H, et al. (2010) Eutrophication of lake waters in China: cost, causes, and control. Environ Manage 45: 662–668.
8. Zheng XH, Han SH, Huang Y, Wang YS, Wang MX (2004) Re-quantifying the emission factors based on field measurements and estimating the direct N$_2$O emission from Chinese croplands. Global Biogeochemical Cycles 18.
9. Cakmak I (2002) Plant nutrition research: priorities to meet human needs for food in sustainable ways. Plant Soil 247: 3–24.
10. Yang XE, Chen WR, Feng Y (2007) Improving human micronutrient nutrition through biofortification in the soil–plant system: China as a case study. Environ Geochem Health 29: 413–428.
11. Losak T, Hlusek J, Martinec J, Jandak J, Szostkova M, et al. (2011) Nitrogen fertilization does not affect micronutrient uptake in grain maize (*Zea mays* L.). Acta Agric Scand B Soil Plant Sci 61: 543–550.
12. Yu WT, Zhou H, Zhu XJ, Xu YG, Ma Q (2011) Field balances and recycling rates of micronutrients with various fertilization treatments in Northeast China. Nutr Cycl Agroecosyst 90: 75–86.
13. Feil B, Moser SB, Jampatong S, Stamp P (2005) Mineral composition of the grains of tropical maize varieties as affected by pre-anthesis drought and rate of nitrogen fertilization. Crop Sci 45: 516–523.
14. Ciampitti IA, Vyn TJ (2012) Physiological perspectives of changes over time in maize yield dependency on nitrogen uptake and associated nitrogen efficiencies: A review. Field Crops Res 133: 48–67.
15. Karlen DL, Flannery RL, Sadler EJ (1987) Nutrient and dry matter accumulation rates for high yielding maize. J Plant Nutr 10: 1409–1417.
16. Ciampitti IA, Camberato JJ, Murrell ST, Vyn TJ (2013) Maize nutrient accumulation and partitioning in response to plant density and nitrogen rate: I. Macronutrients. Agron J 105: 783–795.
17. Karlen DL, Flannery RL, Sadler EJ (1988) Aerial accumulation and partitioning of nutrients by corn. Agron J 80: 232–242.
18. Bender RR, Haegele JW, Ruffo ML, Below FE (2013) Nutrient uptake, partitioning, and remobilization in modern, transgenic insect-protected maize hybrids. Agron J 105: 161–170.
19. Ciampitti IA, Vyn TJ (2013) Maize nutrient accumulation and partitioning in response to plant density and nitrogen rate: II. Calcium, magnesium, & micronutrients. Agron J. 105:1645–1657.
20. Setiyono TD, Walters DT, Cassman KG, Witt C, Dobermann A (2010) Estimating maize nutrient uptake requirements. Field Crops Res 118: 158–168.
21. Hirel B, Le Gouis J, Ney B, Gallais A (2007) The challenge of improving nitrogen use efficiency in crop plants: towards a more central role for genetic variability and quantitative genetics within integrated approaches. J Exp Bot 58: 2369–2387.
22. Hou P, Gao Q, Xie RZ, Li SK, Meng QF, et al. (2012) Grain yields in relation to N requirement: optimizing nitrogen management for spring maize grown in China. Field Crops Res 129: 1–6.
23. Zhang Y, Hou P, Gao Q, Chen XP, Zhang FS, et al. (2012) On-farm estimation of nutrient requirements for spring corn in North China. Agron J 104: 1436–1442.
24. Mortvedt JJ (1994) Needs for controlled-availability micronutrient fertilizers. Fertilizer Res 38: 213–221.
25. Banziger M, Long J (2000) The potential for increasing the iron and zinc density of maize through plant-breeding. Food Nutr Bull 21: 397–400.

26. Xue YF, Yue SC, Zhang YQ, Cui ZL, Chen XP, et al. (2012) Grain and shoot zinc accumulation in winter wheat affected by nitrogen management. Plant Soil 361: 153–163.

27. Cui ZL, Zhang FS, Chen XP, Miao YX, Li JL, et al. (2008) On-farm evaluation of an in-season nitrogen management strategy based on soil Nmin test. Field Crops Res 105: 48–55.

28. Zhang SS, Yue SC, Yan P, Qiu ML, Chen XP, et al. (2013) Testing the suitability of the end-of-season stalk nitrate test for summer corn (*Zea mays* L.) production in China. Field Crops Res 154:153–157.

29. Yue SC, Meng QF, Zhao RF, Ye YL, Zhang FS, et al. (2012) Change in nitrogen requirement with increasing grain yield for winter wheat. Agron J 104: 1687–1693.

30. Meng QF, Yue SC, Chen XP, Cui ZL, Ye YL, et al. (2013) Understanding dry matter and nitrogen accumulation with time-course for high-yielding wheat production in China. PLoS One 8: e68783.

31. Sadras VO (2006) The N:P stoichiometry of cereal, grain legume and oilseed crops. Field Crops Res 95: 13–29.

32. Wang TY, Ma XL, Li Y, Bai DP, Liu C, et al. (2011) Changes in yield and yield components of single-cross maize hybrids released in China between 1964 and 2001. Crop Sci 51: 512–525.

33. Potarzycki J (2010) The impact of fertilization systems on zinc management by grain maize. Fertilizers Fertilization. 39: 78–89.

34. Behera SK, Shukla AK, Wanjari RH, Singh MV (2011) Influence of different sources of zinc fertilizer on yield and zinc nutrition of maize (*Zea mays* L.). 3rd International Zinc Symposium: improving crop production and human health; October 2011; India. Available: http://www.Zinccrops2011.Org/presentations/2011-zinccrops2011-behera-1-abstract.pdf.

35. Marschner (2011) Mineral nutrition of higher plants, 3rd edn. Academic, London.

36. Fageria NK, Barbosa MP, Santos AB (2008) Growth and zinc uptake and use efficiency in food crops. Commun Soil Sci Plant Anal 39: 2258–2269.

37. Rengel Z (1995) Carbonic anhydrase activity in leaves of wheat genotypes differing in Zn efficiency. J Plant Physiol 147: 251–256.

38. Grzebisz W (2008) Effect of Zinc foliar application at an early stage of maize growth on patterns of nutrients and dry matter accumulation by the canopy. Part II. Nitrogen uptake and dry mattern accumulation patterns. J Elementol 13: 17–39.

39. Oktem A, Oktem AG, Emeklier HY (2010) Effect of nitrogen on yield and some quality parameters of sweet corn. Commun Soil Sci Plant Anal 41: 832–847.

40. Shivay Y, Kumar D, Prasad R (2008) Effect of zinc-enriched urea on productivity, zinc uptake and efficiency of an aromatic rice–wheat cropping system. Nutr Cycl Agroecosyst 81: 229–243.

41. Shivay Y, Kumar D, Prasad R, Ahlawat IPS (2008) Relative yield and zinc uptake by rice from zinc sulphate and zinc oxide coatings onto urea. Nutr Cycl Agroecosyst 80: 181–188.

8

Long-Term Effect of Agricultural Reclamation on Soil Chemical Properties of a Coastal Saline Marsh in Bohai Rim, Northern China

Yidong Wang[1], Zhong-Liang Wang[1,2]*, Xiaoping Feng[1,3], Changcheng Guo[1], Qing Chen[1]

1 Tianjin Key Laboratory of Water Resources and Environment, Tianjin Normal University, Tianjin, China, **2** State Key Laboratory of Environmental Geochemistry, Institute of Geochemistry, Chinese Academy of Sciences, Guiyang, China, **3** College of Urban and Environmental Sciences, Tianjin Normal University, Tianjin, China

Abstract

Over the past six decades, coastal wetlands in China have experienced rapid and extensive agricultural reclamation. In the context of saline conditions, long-term effect of cultivation after reclamation on soil chemical properties has not been well understood. We studied this issue using a case of approximately 60-years cultivation of a coastal saline marsh in Bohai Rim, northern China. The results showed that long-term reclamation significantly decreased soil organic carbon (SOC) (-42.2%) and total nitrogen (TN) (-25.8%) at surface layer (0–30 cm) as well as their stratification ratios (SRs) (0–5 cm:50–70 cm and 5–10 cm:50–70 cm). However, there was no significant change in total phosphorus (TP) as well as its SRs under cultivation. Cultivation markedly reduced ratios of SOC to TN, SOC to TP and TN to TP at surface layer (0–30 cm) and their SRs (0–5 cm:50–70 cm). After cultivation, electrical conductivity and salinity significantly decreased by 60.1% and 55.3% at 0–100 cm layer, respectively, suggesting a great desalinization. In contrast, soil pH at 20–70 cm horizons notably increased as an effect of reclamation. Cultivation also changed compositions of cations at 0–10 cm layer and anions at 5–100 cm layer, mainly decreasing the proportion of Na^+, Cl^- and SO_4^{2-}. Furthermore, cultivation significantly reduced the sodium adsorption ratio and exchangeable sodium percentage in plow-layer (0–20 cm) but not residual sodium carbonate, suggesting a reduction in sodium harm.

Editor: Jin-Song Zhang, Institute of Genetics and Developmental Biology, Chinese Academy of Sciences, China

Funding: This research was supported by the National Science & Technology Pillar Program (2012BAC07B02), the Program for New Century Excellent Talents in University (NCET-10-0954), the National Natural Science Foundation of China (31300381), and the Opening Fund of Tianjin Key Laboratory of Water Resources and Environment (YF11700102). The funders had no role in study design, data collection and analysis, decision to publish, or preparation of the manuscript.

Competing Interests: The authors have declared that no competing interests exist.

* E-mail: wangzhongliang@vip.skleg.cn

Introduction

Coastal wetlands provide essential ecosystem services to people and environment including flood protection, water supply and purification, food productivity, erosion control, wave attenuation, shoreline stabilization, wildlife habitat, biodiversity, climate regulation and amenity [1,2]. Over the past century, natural coastal wetlands all over the world have been rapidly shrunk due to intensive anthropogenic activities [3,4,5,6]. In China, approximately 51% (2.2×10^4 km^2) of coastal natural wetlands were lost or degraded since the 1950s, primarily due to agricultural reclamation [7]. Soil is one of the foundations of ecosystem services of wetland and cropland. Therefore, it is important to study the influence of agricultural cultivation after reclamation on soil properties of the coastal wetlands.

Some studies have evaluated the reclamation-induced changes of soil properties in coastal wetlands. Conversion of coastal wetlands to croplands has been reported to cause radical changes in soil chemical properties because of alterations of hydrology and agricultural activities. First, hydrologic alterations such as ditch drainage and diking led to increase in aeration and decrease in salinity [8,9,10,11]. Aeration accelerated soil organic matter decay [4,12,13] and affected substances' characteristics [5] and mobility through changing redox conditions [8,14]. Changes in salinity

further influenced substances' cycle in soils [15,16]. Second, application of fertilizer affected nutrient contents such as carbon [17], nitrogen and phosphorus [14,18]. In addition, agricultural managements including planting, harvesting and tillage also influenced balances of substances inputs and outputs in soils [19,20,21]. Most of these studies were focusing on short-term scales (e.g. [5,8,11,13,14]). In contrast, only few studies were available at long-term scales [4,9,10].

Stratification of soil chemical properties with soil depth is common in many natural ecosystems [4,22]. Different soil horizons of coastal wetlands may have distinct responses to agricultural reclamation. The surface soil is the vital interface that receives intense impact from human disturbance. In support of this concept, chemical characteristics of the surface soil in coastal wetlands were demonstrated to be more sensitive to agricultural reclamation [4,18]. However, quantitative expression of the horizon-induced differences in response of coastal wetland to agricultural reclamation is rare. Stratification ratio (SR) is a quantitative indicator to represent changes of soil profile feature [22]. It is hypothesized that the SR can be used as an indicator of the responses of different soil horizons to human disturbance.

In Bohai Rim of northern China, coastal marsh geogenesises have been realized naturally by land-sea interactions since the middle-late Holocene. Since the founding of New China, natural coastal marshes have been largely reclaimed for agriculture to alleviate the pressure of the increasing population. For instance, approximately 55% of coastal natural wetlands in Tianjin were disappeared since the 1950s. However, it is unclear that how agricultural reclamation affects soil chemical properties and their SRs of the coastal saline marshes in Bohai Rim, northern China.

The objective of this study was to evaluate a long-term (approximately 60-years) impact of agricultural reclamation on soil chemical properties as well as their SR features of a coastal saline marsh in Bohai Rim, northern China.

Materials and Methods

Study site

This study was conducted at a coastal permanent marsh and a cropland cultivated from the marsh in Qilihai (39°18′ N, 117°29′ E, elevation 2 m) of Bohai Rim, northern China. No specific permits were required for the described field studies in the research site. The field studies did not involve endangered or protected species. The study site is flat and low-lying. The region was characterized by a warm temperate semi-humid monsoon climate. Annual mean air temperature was 11.2 °C; total precipitation amount was 500–600 mm year^{-1}; evapotranspiration was approximately 1900 mm year^{-1}; frost-free period was 180–194 d; and sunshine duration hours were 2600–2800 h. Rainfall occurred synchronously with the heat, with high values in summer (June–August).

The permanent marsh was formed from the land-sea interactions since middle-late Holocene. The permanent marsh was mainly covered by reed (*Phragmites australis*) and generally had a water level of 0–40 cm above surface recently. Croplands were cultivated from the *P. australis* marshes over approximately 60 years. During the first ~20 years (1950s–1960s), paddy was cultivated. Afterwards, the paddy lands converted to dry-lands mainly dominated by corn, cotton and sorghum in rotation. Organic (manure and straw) and chemical fertilizer applications during 1950s–1990s [23] were listed in Table 1. Since the new century, organic and chemical fertilizer applications [24] were presented in Table 2. The current water level of the dry-lands was about 80–100 cm below surface.

Soil sampling and chemical analysis

The *P. australis* marsh and the cropland were adjacent (about 500 m) and had similar altitudes. Soil samples were randomly collected using a soil drill (Eijkelkamp, Netherlands) at three locations in the permanent marsh and cropland in May 2012, respectively. In both marsh and cropland, the distance between any two sampling locations was more than 15 m. In each sampling location, two adjacent soil columns (0–100 cm) were collected and

then mixed into one column according to 7 corresponding horizons: 0–10, 10–20, 20–30, 30–50, 50–70, and 70–100 cm.

Soil samples were brought to laboratory and air-dried naturally. The air-dried soil samples were sieved with a 2 mm mesh screen to remove the roots and coarse fraction before chemical analysis. Dilute HCl (0.5 mol L^{-1}) was used to remove soil inorganic carbon. Then deionized water was used to rinse off excess HCl and ensure the neutral pH of approximate 7. The contents of soil organic carbon (SOC) and total nitrogen (TN) were determined by an Elementar (Vario EL III, Elementar, Germany). The concentration of total phosphorus (TP) was determined by sulphuric and perchloric acids digestion and Mo-Sb Anti spectrophotometric method. Soil pH was measured with a pH meter (Star A 420C-01A, Thermo Orion, United States) using a soil:water ratio of 1:2.5 (g g^{-1}). Electrical conductivity (EC) was measured with a EC meter (Star A 420C-01A, Thermo Orion, United States) using a soil:water ratio of 1:5 (g g^{-1}). Soil major soluble cations (Na$^+$, Ca^{2+}, K$^+$ and Mg^{2+}) and anions (Cl$^-$, SO$_4^{2-}$, HCO$_3^-$ and CO$_3^{2-}$) were measured using a soil:water ratio of 1:5 (g g^{-1}). Cation concentrations were determined by an Atomic Absorption Spectrophotometer (TAS-990, Beijing Purkinje General Instrument Co., Ltd., China). Concentrations of Cl$^-$ and SO$_4^{2-}$ were determined by an Ion Chromatography (ICS-2100, Dionex Corporation, Sunnyvale, California, United States). Concentrations of CO$_3^{2-}$ and HCO$_3^-$ were measured using neutralization titration method using phenolphthalein and methyl orange.

Data analysis

Sodium adsorption ratio (SAR), an easily measured property that gives information on the comparative concentrations of Na$^+$, Ca^{2+}, and Mg^{2+} in soil solutions, was calculated using equation (1) [25]. Exchangeable sodium percentage (ESP) was computed using equation (2) [26,27]. Residual sodium carbonate (RSC) was estimated using equation (3) [28]. Comparisons of mean SOC, TN, TP, pH, EC and salinity between the marsh and cropland were analyzed using one-way ANOVA of SPSS 13.0 (SPSS Inc., Chicago, Illinois, United States). Figures were drawn using Origin 8.0 (OriginLabs Corporation, Northampton, Massachusetts, United States) and CorelDraw 9 (Corel Corporation, Canada).

$$SAR = \frac{Na^+}{\sqrt{(Ca^{2+} + Mg^{2+})/2}} \quad (1)$$

$$ESP = 100 \times \frac{(-0.0126 + 0.01475SAR)}{1 + (-0.0126 + 0.01475SAR)} \quad (2)$$

Table 1. Organic and chemical fertilizer applications during 1950s–1990s (source from the reference [23]).

	Organic fertilizer (manure and straw)		Chemical fertilizer	
	Application rate (m^3 ha^{-1})	Proportion (%)	Application rate (kg ha^{-1})	Proportion (%)
1950s–1960s	30	93	53	7
1970s	40	61	600	39
1980s	20	40	690	60
1990s	20	29	1156	71

Table 2. Nitrogen and phosphorus inputs of organic and chemical fertilizer applications in 2007 (source from the reference [24]).

	Organic fertilizer (manure and straw)		Chemical fertilizer	
	Nitrogen (kg ha^{-1})	Phosphorus (kg ha^{-1})	Nitrogen (kg ha^{-1})	Phosphorus (kg ha^{-1})
2007	66	41	269	149

$$RSC = \left(CO_3^{2-} + HCO_3^-\right) - \left(Ca^{2+} + Mg^{2+}\right) \qquad (3)$$

Results

Soil organic carbon, total nitrogen and phosphorus as well as their stratification ratios

Long-term agricultural reclamation significantly reduced the concentrations of SOC at surface soil (0–30 cm) but not lower soil layers (30–100 cm) (Fig. 1A). The weighted density of SOC at surface soil (0–30 cm) decreased by 10.1 g kg^{-1} (−42.2%). Similarly, the contents of TN of upper soil layers (0–50 cm) but not deeper soil layers (50–100 cm) were also evidently decreased under long-term agricultural cultivation (Fig. 1B). The weighted density of TN at surface soil (0–50 cm) reduced by 0.7 g kg^{-1} (−31.8%). In contrast, there were no significant effects of long-term agricultural reclamation on the contents of TP at the whole soil profile (0–100 cm) (Fig. 1C). The ratios of SOC/TN, SOC/TP and TN/TP were all notably decreased after agricultural reclamation at surface soil (0–30 cm, except for 10–20 cm of SOC/TN) but not lower soil layers (30–100 cm) (Fig. 2).

Stratification ratios of 0–5 cm:50–70 cm and 5–10 cm:50–70 cm of SOC were 12.3 and 3.7 for the marsh and were 3.4 and 3.0 for the cropland (Fig. 3A). Stratification ratios of 0–5 cm:50–70 cm and 5–10 cm:50–70 cm of TN were 4.0 and 1.9 for the marsh and were 1.5 and 1.4 for the cropland (Fig. 3A). Compared to the marsh, SRs of SOC and TN of the cropland were both remarkably dropped, suggesting greater decreases of SOC and TN

in the upper soil layers (Fig. 3A and B). In contrast, there was no significant impact of reclamation on the SR of TP (Fig. 3C). The SRs (0–5 cm:50–70 cm) of ratios of SOC/TN, SOC/TP and TN/TP were all significantly decreased after agricultural reclamation (Fig. 4).

Soil EC, salinity and pH

Soil profile EC of the marsh ranged from 1026 to 2365 μs cm^{-1}, with a weighted average of 1488 μs cm^{-1}. Soil profile EC of the cropland varied from 392 to 854 μs cm^{-1}, with a weighted average of 594 μs cm^{-1} (Fig. 5A). Soil profile salinity of the marsh was from 3.2 to 6.9 g kg^{-1} dry weight, with a weighted average of 4.7 g kg^{-1} dry weight. Soil profile salinity of the cropland ranged from 1.3 to 2.5 g kg^{-1} dry weight, with a weighted average of 2.1 g kg^{-1} dry weight (Fig. 5B). Thus, after long-term reclamation, EC and salinity of the whole soil profile (0–100 cm) decreased by 60.1% and 55.3%, respectively. In support of these results, concentrations of water-extractable ions showed great declines after reclamation. In contrast, soil pH (20–70 cm) significantly increased after reclamation.

Compositions of cation and anion

Even though the cations of marsh and cropland were both dominated by Na$^+$, the proportion of Na$^+$ (0–10 cm) was significantly decreased after reclamation ($p<0.05$) (Fig. 6A). However, the proportion of Ca^{2+} (0–10 cm) and Mg^{2+}+K$^+$ (5–20 cm) were significantly increased after reclamation ($p<0.05$) (Fig. 6A). Contrasted to cation, the composition of anion changed greater at 5–100 cm after reclamation, with significant decreases

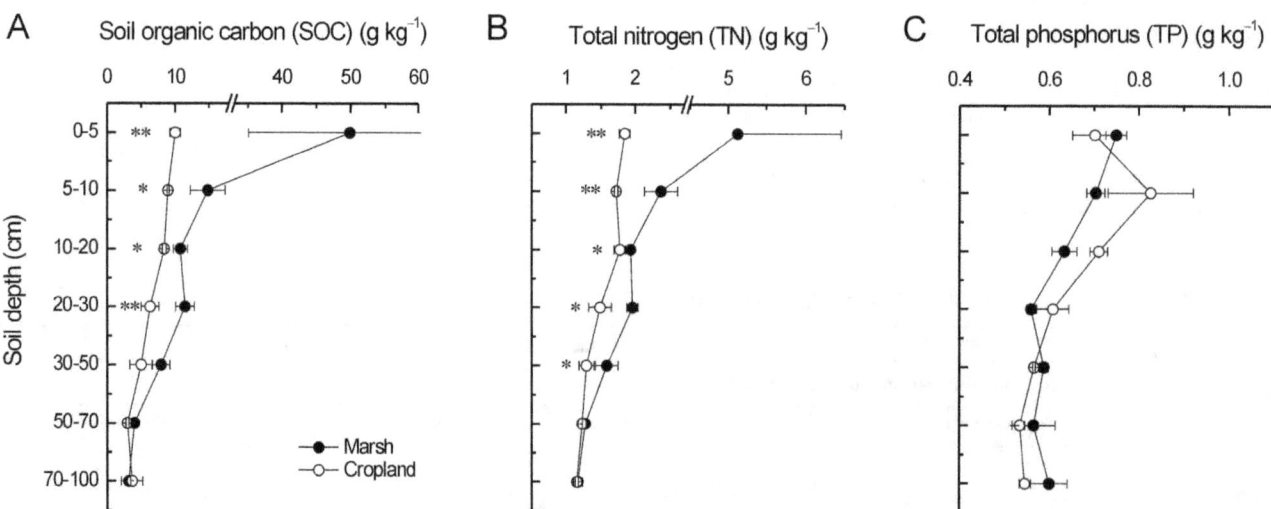

Figure 1. Long-term effects of reclamation on A) SOC, B) TN, and C) TP of the coastal marsh. * and ** indicate significant difference at $p<0.05$ and $p<0.01$ level, respectively.

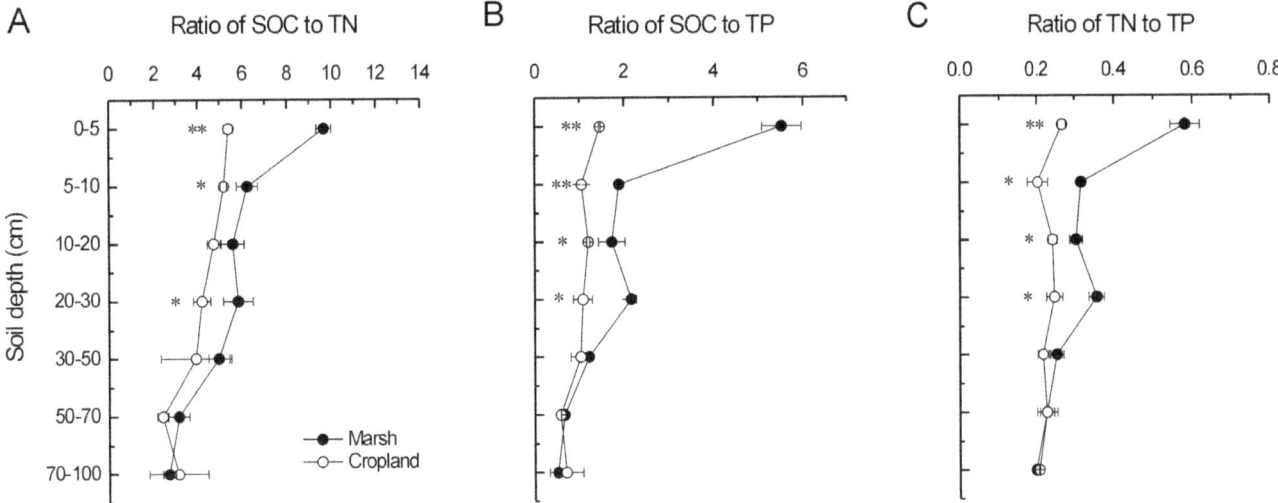

Figure 2. Long-term effects of reclamation on ratios of A) SOC/TN, B) SOC/TP, and C) TN/TP. * and ** represent significant difference at $p<0.05$ and $p<0.01$ level, respectively.

in Cl^- and SO_4^{2-} and increases in $HCO_3^-+CO_3^{2-}$ ($p<0.05$) (Fig. 6B).

Characteristics of alkalization

The SAR and ESP in plow-layer (0–20 cm) significantly reduced after reclamation ($p<0.05$) (Fig. 7A and B). This result suggested a reduction in soil alkalization and sodium harm. However, there was no significant difference in RSC under long-term cultivation (Fig. 7C).

Discussion

SOC, TN and TP as well as their stratification ratios

The soil storages of SOC, TN and TP are counterbalances of inputs (e.g. productivity and fertilizer) and outputs (e.g. loss of mineralization and harvest) during a certain period. The rates of inputs and outputs in soils have been widely recognized to be

sensitive to environmental changes (e.g. [29,30,31]) and human disturbance (e.g. [19,20,32]). Agricultural reclamation of wetland substantially altered the soil environmental conditions, thus should change soil stocks of SOC, TN and TP through breaking the balance of inputs and outputs. In our study, even though many organic (manure and straw) and chemical fertilizers were applied for a long-term time (Table 1 and Table 2), SOC and TN decreased strongly in surface soil after reclamation (Fig. 1), suggesting that the outputs were much larger than the inputs. Our result was in agreement with the results in the Yangtze Estuary [10] and the Hangzhou Bay of China [14]. Four reasons may explain the above results. First, the mineralization and accumulation of soil organic matter were both regulated by hydrologic conditions [12,33]. Submerged soils of the marsh were under anaerobic conditions, which had lessened and incomplete decompositions of organic materials and decreased humifications of organic matter. In contrast, cropland soils lacked anaerobiosis and

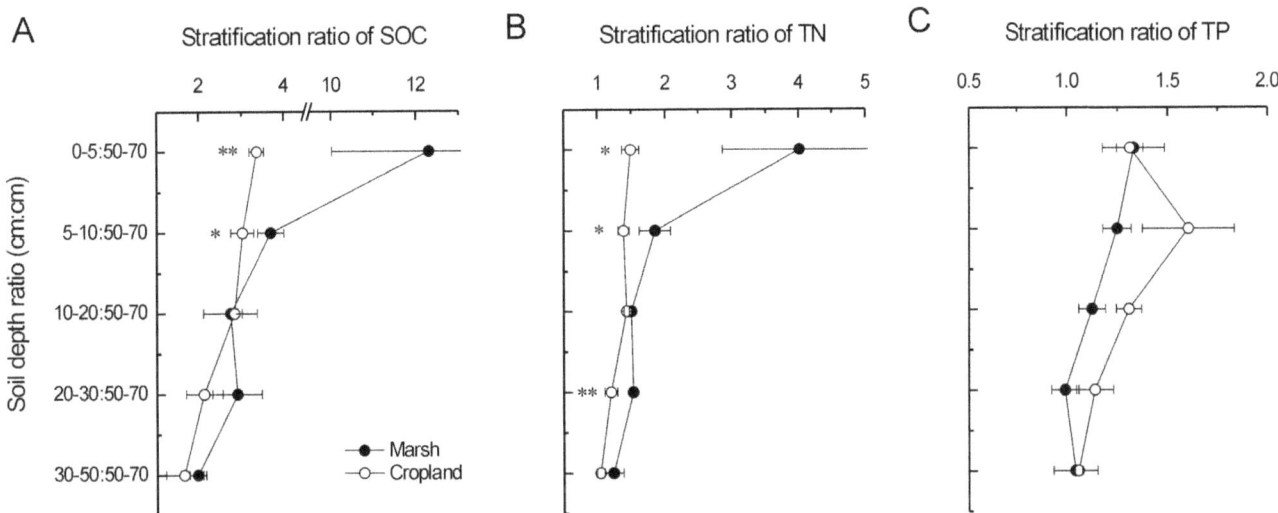

Figure 3. Long-term influences of reclamation on stratification ratios of A) SOC, B) TN, and C) TP. * and ** indicate significant difference at $p<0.05$ and $p<0.01$ level, respectively.

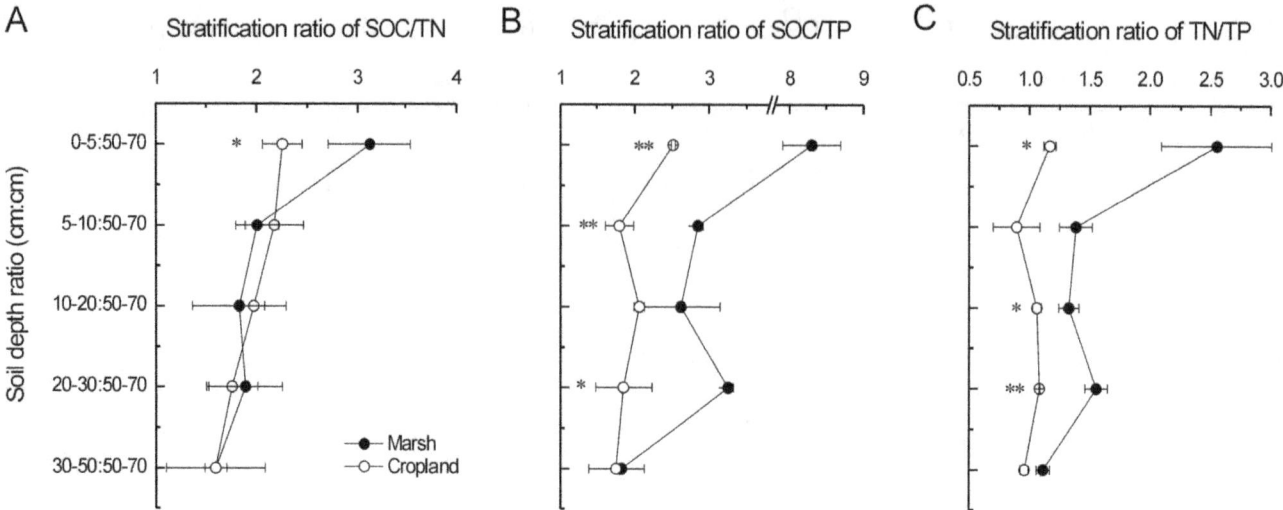

Figure 4. Long-term impacts of reclamation on stratification ratios of A) SOC/TN, B) SOC/TP, and C) TN/TP. * and ** indicate significant difference at $p<0.05$ and $p<0.01$ level, respectively.

had high rates of decomposition due to the aerobic conditions. Compared to cropland soils, there were preferential accumulations of organic matter in submerged soils [12,33]. Second, crop harvesting took lots of organic materials away from farmland ecosystem. Moreover, the net primary productivity of marshes is higher than cropland ecosystems [33]. In addition, decrease in soil salinity caused by agricultural reclamation might accelerate the mineralization of organic matters [16]. In our study, however, there was no significant change of SOC and TN in 50–100 cm soil under long-term reclamation (Fig. 1), indicating limited effects of reclamation on deeper soil horizons. This result was consistent with the results in the Sanjiang Plain [34], the Xingkai Lake [35] and the Hangzhou Bay of China [18]. Compared to SOC and TN, there was no significant variation in TP in whole soil profile after long-term reclamation (Fig. 1). However, Huang et al. [13] found significant decreases in TP caused by conversions of marshes to drylands. In marsh and cropland ecosystems,

biogeochemical cycle of phosphorus is different from carbon and nitrogen. Phosphorus has poor mobility and is easy to be fixed in the soil. The input and output of phosphorus in the soil is mainly as fertilizer application and crop harvesting, respectively [36]. Thus, reclamation-induced impacts on TP may be explained by the balance of fertilizer application and crop harvesting.

Stratification ratio is a quantitative indicator to represent the responses of soil profile feature to anthropogenic disturbance [22], which was mainly applied in farmlands (e.g. [19,20]). However, the application of SR is not common in costal wetlands. Our results showed that SRs (0–5 cm:50–70 cm and 5–10 cm:50–70 cm) of SOC and TN decreased remarkably after long-term agricultural reclamation. In support of this result, chemical properties of the surface soil were reported to be more sensitive to reclamation than that of the deeper layers in two coastal wetlands [4,18]. These results are reasonable because the surface

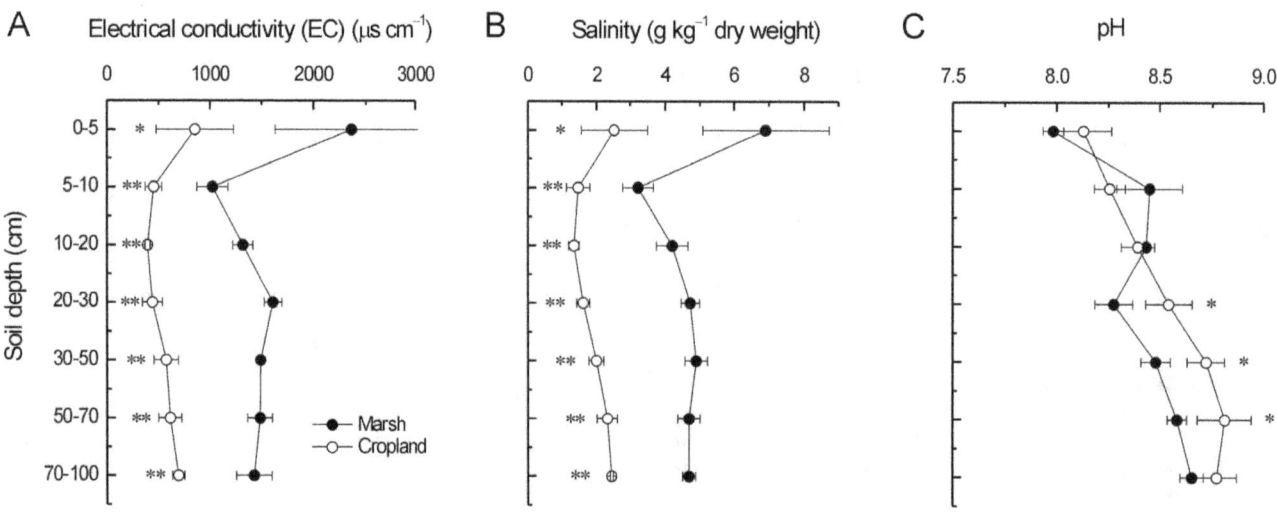

Figure 5. Long-term effects of reclamation on A) EC, B) salinity, and C) pH of coastal soils. * and ** indicate significant difference at $p<0.05$ and $p<0.01$ level, respectively.

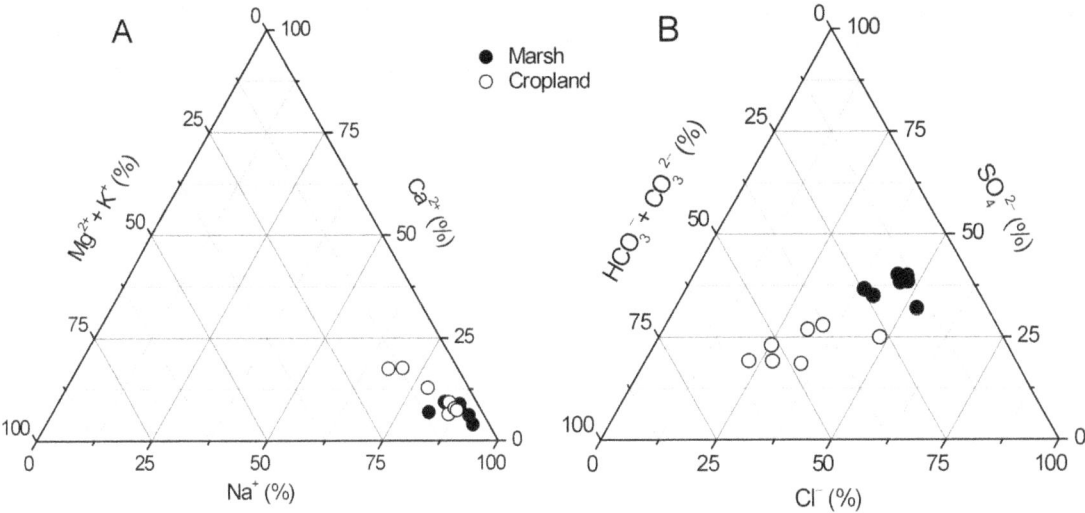

Figure 6. Long-term effects of agricultural cultivation on the compositions of A) cations and B) anions.

soils are the critical zones that receive intensive influences from anthropogenic activities.

Salinity, ion composition and alkalization

Our results showed that soil salinity and EC across the whole profile (0–100 cm) dropped significantly after agricultural reclamation (Fig. 5 A and B). This result was in agreement with the results in the Yellow River Delta [13], the Jiangsu province [11], the Yangtze Estuary [10], the Hangzhou Bay of China [14] and the Biscay Bay of Spain [9]. A reliable interpretation for this result is the effective leaching of the soil exchangeable ions induced by dewatering after reclamation. However, each of the major ions (anions and cations) in soil solutions has some unique properties which affect its production and mobility [37]. For example, the production of bicarbonate (HCO_3^-), one of the most common anions, is regulated by soil CO_2 pressure and pH. The sulfate (SO_4^{2-}) is involved in both biological and inorganic chemical reactions, whereas the chloride (Cl^-) is relatively uninvolved in

these reactions and thus has high mobility [37]. The cations are held by the negatively charged clay and organic matter particles in the soil through electrostatic forces. Therefore, ion compositions were proposed to be sensitive to disturbance due to the specific mobility of each ion. In support of this concept, we found that the proportion of Na^+ (0–10 cm), Cl^- and SO_4^{2-} (5–100 cm) decreased significantly after reclamation. However, the proportion of Ca^{2+} (0–10 cm), $Mg^{2+}+K^+$ (5–20 cm) and $HCO_3^-+CO_3^{2-}$ (5–100 cm) increased significantly after reclamation (Fig. 6). Knowing these reclamation-induced properties of the major ions, it is possible to protect and manage coastal saline soils of the wetlands and croplands as well as their ecosystem services. For example, reclamation-induced decreases in water table and soil salinity were beneficial to crop cultivation. High mobility of Na^+ further reduced the harm of salt ions after reclamation. Even though overlying water or saturated soil are basic properties of coastal wetlands, reclamation-induced hydrologic alterations such as ditch

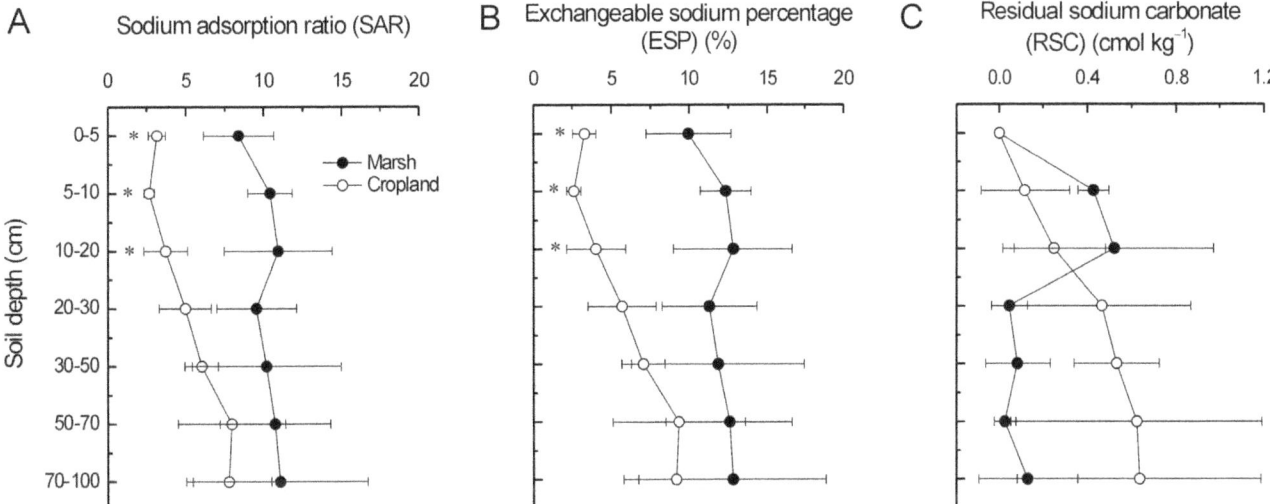

Figure 7. Long-term effects of reclamation on A) SAR, B) ESP, and C) RSC of the coastal marsh. * indicates significant difference at $p < 0.05$.

drainage and diking result in decreases in salinity, which promotes plant activities and crop yields.

The SAR, ESP and RSC are three indicators of the degree of soil alkalization. The SAR of a soil extract takes into consideration that the adverse effect of Na^+ is moderated by the presence of Ca^{2+} and Mg^{2+}. When the SAR rises above 12 to 15, serious physical soil problems arise and plants have difficulty absorbing water [38]. In our study, all SARs of the marsh and cropland were below 12. Our results showed that agricultural reclamation significantly decreased the SAR and ESP but not RSC in the plow-layer (0–20 cm), suggesting a reduction in Na^+ harm (Fig. 7). These results were consistent with those in the Cape Cod, Massachusetts of USA [8] and the Hangzhou Bay of China [14]. Moreover, significant decreases of SAR and ESP in the upper soil horizons further supported the rapid desalinization.

Conclusions

In coastal marshes of northern China, long-term agricultural reclamation significantly reduced SOC and TN at surface soil (0–30 cm) as well as their SRs (0–5 cm:50–70 cm and 5–10 cm:50–70 cm). However, there was no significant change in TP as well as its SRs. Reclamation notably decreased ratios of SOC/TN, SOC/TP and TN/TP at surface soil (0–30 cm) as well as their SRs (0–5 cm:50–70 cm). A significant desalinization of whole soil profile occurred during cultivation. Soil pH (20–70 cm) markedly increased as an impact of reclamation. Compositions of cations (0–10 cm) and anions (5–100 cm) changed mainly as decreases in proportion of Na^+, Cl^- and SO_4^{2-}. Furthermore, reclamation significantly reduced the sodium adsorption ratio and exchangeable sodium percentage in plow-layer (0–20 cm) but not residual sodium carbonate, suggesting a reduction in sodium harm.

Acknowledgments

The authors thank Mr. Yuntao Shang for his help in laboratory analysis. The authors also appreciate Dr. Jin-Song Zhang and an anonymous reviewer for their helpful comments that improved this paper.

Author Contributions

Conceived and designed the experiments: ZW YW. Performed the experiments: YW XF CG QC. Analyzed the data: YW XF. Contributed reagents/materials/analysis tools: ZW YW. Wrote the paper: YW.

References

1. Costanza R, d'Arge R, de Groot R, Farber S, Grasso M, et al. (1997) The value of the world's ecosystem services and natural capital. Nature 387: 253–260.
2. Shepard CC, Crain CM, Beck MW (2011) The Protective Role of Coastal Marshes: A Systematic Review and Meta-analysis. PLoS ONE 6(11): e27374. doi:10.1371/journal.pone.0027374.
3. Cashin GE, Dorney JR, Richardson CJ (1992) Wetland alteration trends on the North Carolina coastal plain. Wetlands 12: 63–71.
4. Ellis S, Atherton JK (2003) Properties and development of soils on reclaimed alluvial sediments of the Humber estuary, eastern England. Catena 52: 129–147.
5. Santín C, de la Rosa JM, Knicker H, Otero XL, Álvarez MÁ, et al. (2009) Effects of reclamation and regeneration processes on organic matter from estuarine soils and sediments. Org Geochem 40: 931–941.
6. An S, Tian Z, Cai Y, Wen T, Xu D, et al. (2013) Wetlands of Northeast Asia and High Asia: an overview. Aquat Sci 75: 63–71.
7. An S, Li H, Guan B, Zhou C, Wang Z, et al. (2007) China's natural wetlands: past problems, current status, and future challenges. Ambio 36: 335–342.
8. Portnoy JW (1999) Salt marsh diking and restoration: biogeochemical implications of altered wetland hydrology. Environ Manage 24: 111–120.
9. Fernández S, Santín C, Marquínez J, Álvarez MA (2010) Saltmarsh soil evolution after land reclamation in Atlantic estuaries (Bay of Biscay, North coast of Spain). Geomorphology 114: 497–507.
10. Cui J, Liu C, Li Z, Wang L, Chen X, et al. (2012) Long-term changes in topsoil chemical properties under centuries of cultivation after reclamation of coastal wetlands in the Yangtze Estuary, China. Soil Till Res 123: 50–60.
11. Jin X, Huang J, Zhou Y (2012) Impact of coastal wetland cultivation on microbial biomass, ammonia-oxidizing bacteria, gross N transformation and N2O and NO potential production. Biol Fertil Soils 48: 363–369.
12. Bridgham SD, Updegraff K, Pastor J (1998) Carbon, nitrogen, and phosphorus mineralization in northern wetlands. Ecology 79: 1545–1561.
13. Huang L, Bai J, Chen B, Zhang K, Huang C, et al. (2012) Two-decade wetland cultivation and its effects on soil properties in salt marshes in the Yellow River Delta, China. Ecol Inform 10: 49–55.
14. Iost S, Landgraf D, Makeschin F (2007) Chemical soil properties of reclaimed marsh soil from Zhejiang Province P.R. China. Geoderma 142: 245–250.
15. Weston NB, Dixon RE, Joye SB (2006) Ramifications of increased salinity in tidal freshwater sediments: geochemistry and microbial pathways of organic matter mineralization. J Geophys Res-Biogeoscience 111, G01009. doi:10.1029/2005JG000071.
16. Wong VNL, Greene RSB, Dalal RC, Murphy BW (2010) Soil carbon dynamics in saline and sodic soils: a review. Soil Use Manage 26: 2–11.
17. Lou Y, Wang J, Liang W (2011) Impacts of 22-year organic and inorganic N managements on soil organic C fractions in a maize field, northeast China. Catena 87: 386–390.
18. Wu M, Shao X, Hu F, Jiang K (2008) Effects of reclamation on soil nutrients distribution of coastal wetland in south Hangzhou Bay. Soils 40: 760–764. (in Chinese with English abstract)
19. Du Z, Ren T, Hu C (2010) Tillage and residue removal effects on soil carbon and nitrogen storage in the North China Plain. Soil Sci Soc Am J 74: 196–202.
20. Lou Y, Xu M, Chen X, He X, Zhao K (2012) Stratification of soil organic C, N and C:N ratio as affected by conservation tillage in two maize fields of China. Catena 95: 124–130.
21. Han G, Xing Q, Yu J, Luo Y, Li D, et al. (2013) Agricultural reclamation effects on ecosystem CO2 exchange of a coastal wetland in the Yellow River Delta. Agr Ecosyst Environ doi:org/10.1016/j.agee.2013.09.012.
22. Franzluebbers AJ (2002) Soil organic matter stratification ratio as an indicator of soil quality. Soil Till Res 66: 95–106.
23. Liu H, Mou S (1999) Farmland fertilization conditions and countermeasures in Tianjin. Sci Technol Tianjin Agr For 2: 43–45. (in Chinese)
24. Zhu M (2011) Study on agricultural NPS loads of Haihe Basin and assessment on its environmental impact. Beijing: Doctoral Dissertation, Institute of Agricultural Economics and Development, Chinese Academy of Agricultural Sciences. (in Chinese with English abstract)
25. Robbins CW (1984) Sodium adsorption ratio-exchangeable sodium percentage relationships in a high potassium saline-sodic soil. Irrigation Sci 5: 173–179.
26. Oster JD, Sposito G (1980) The gapon coefficient and the exchangeable sodium percentage-sodium adsorption ratio relation. Soil Sci Soc Am J 44: 258–260.
27. Shi Y, Xin D (1983) Water-salt movement and comprehensive management of droughts, floods and saline-alkali in the Huang-Huai-Hai Plain. Hebei People's Publishing House, Shijiazhuang. (in Chinese)
28. Wan H, Yang J, Yu R (1983) Investigation on calculating method of soil alkalinity in the Huang-Huai-Hai Plain. Soils 23: 319–325. (in Chinese)
29. Wang Y, Wang Z-L, Wang H, Guo C, Bao W (2012) Rainfall pulse primarily drives litterfall respiration and its contribution to soil respiration in a young exotic pine plantation in subtropical China. Can J Forest Res 42: 657–666.
30. Wang Y, Wang H, Ma Z, Dai X, Wen X, et al. (2013) The litter layer acts as a moisture-induced bidirectional buffer for atmospheric methane uptake by soil of a subtropical pine plantation. Soil Biol Biochem 66: 45–50.
31. Wang Y, Wang H, Wang Z-L, Ma Z, Dai X, et al. (2014) Effect of litter layer on soil-atmosphere N2O flux of a subtropical pine plantation in China. Atmos Environ 82: 106–112.
32. Macreadie PI, Hughes AR, Kimbro DL (2013) Loss of 'Blue Carbon' from Coastal Salt Marshes Following Habitat Disturbance. PLoS ONE 8(7): e69244. doi:10.1371/journal.pone.0069244.
33. Sahrawat KL (2003) Organic matter accumulation in submerged soils. Adv. Agron. 81: 169–201.
34. Wang L, Song C, Ge R, Song Y, Liu D (2009) Soil organic carbon storage under different land-use types in Sanjiang Plain. China Environ Sci 29: 656–660. (in Chinese with English abstract)
35. Huo L, Zou Y, Guo J, Lü X (2013) Effect of reclamation on the vertical distribution of SOC and retention of DOC. Environ Sci 34: 283–287. (in Chinese with English abstract)
36. Shen J, Yuan L, Zhang J, Li H, Bai Z, et al. (2011) Phosphorus dynamics: from soil to plant. Plant Physiol 156: 997–1005.
37. Johnson DW, Cole DW (1980) Anion mobility in soils: Relevance to nutrient transport from forest ecosystems. Environ Int 3: 79–90.
38. Munshower FF (1994) Practical handbook of disturbed land revegetation. Florida: Lewis Publishers.

9

Impact of Biotic and Abiotic Stresses on the Competitive Ability of Multiple Herbicide Resistant Wild Oat (*Avena fatua*)

Erik A. Lehnhoff[1]*, Barbara K. Keith[2], William E. Dyer[2], Fabian D. Menalled[1]

1 Department of Land Resources and Environmental Sciences, Montana State University, Bozeman, Montana, United States of America, **2** Department of Plant Sciences and Plant Pathology, Montana State University, Bozeman, Montana, United States of America

Abstract

Ecological theory predicts that fitness costs of herbicide resistance should lead to the reduced relative abundance of resistant populations upon the cessation of herbicide use. This greenhouse research investigated the potential fitness costs of two multiple herbicide resistant (MHR) wild oat (*Avena fatua*) populations, an economically important weed that affects cereal and pulse crop production in the Northern Great Plains of North America. We compared the competitive ability of two MHR and two herbicide susceptible (HS) *A. fatua* populations along a gradient of biotic and abiotic stresses The biotic stress was imposed by three levels of wheat (*Triticum aestivum*) competition (0, 4, and 8 individuals pot^{-1}) and an abiotic stress by three nitrogen (N) fertilization rates (0, 50 and 100 kg N ha^{-1}). Data were analyzed with linear mixed-effects models and results showed that the biomass of all *A. fatua* populations decreased with increasing *T. aestivum* competition at all N rates. Similarly, *A. fatua* relative growth rate (RGR) decreased with increasing *T. aestivum* competition at the medium and high N rates but there was no response with 0 N. There were no differences between the levels of biomass or RGR of HS and MHR populations in response to *T. aestivum* competition. Overall, the results indicate that MHR does not confer growth-related fitness costs in these *A. fatua* populations, and that their relative abundance will not be diminished with respect to HS populations in the absence of herbicide treatment.

Editor: Randall P. Niedz, United States Department of Agriculture, United States of America

Funding: This research was funded through grants by the U.S. Environmental Protection Agency, http://www.epa.gov, (Grant # X8-97873401-0); the Montana Wheat and Barley Committee, http://wbc.agr.mt.gov, (Grant # 408-1655); and the Montana Noxious Weed Trust Fund, http://agr.mt.gov/agr/Programs/Weeds/TrustFund, (Grant# 2011-033). The funders had no role in study design, data collection and analysis, decision to publish, or preparation of the manuscript.

Competing Interests: The authors have declared that no competing interests exist.

* E-mail: erik.lehnhoff@montana.edu

Introduction

The reliance on herbicides for weed control has posed strong selection pressure for resistant populations, and there are now nearly 400 unique cases (plant species × site of action) of herbicide resistance in 217 plant species [1]. Most of these cases involve target site mutations that confer resistance to a single herbicide or related herbicides with the same mechanism of action. However, non-target-site-based resistance has recently become more common, and in some cases the use of one herbicide mode of action may substantially increase selection for non-target-site-based resistance genes that confer resistance to other unrelated herbicides [2,3]. The physiological mechanisms of non-target-site-based resistance are usually based on enhanced herbicide metabolism or detoxification as mediated by cytochrome P450 monooxygenases [4] (hereafter P450s), glutathione *S*-transferases (GSTs) [5], and other enzymes of Phase II metabolism [6].

Ecological theory predicts that individuals with heritable resistance to an environmental stress may have an ecological disadvantage as compared to susceptible individuals in the absence of the stress [7,8]. For example, herbicide resistant biotypes are predicted to experience a fitness cost as resources are shifted to resistance mechanism(s) rather than to growth and reproduction. Such fitness costs have been associated with a number of specific gene mutations conferring resistance to herbicides (see [9] for a review). For example, fitness costs have been demonstrated for the Pro197His and Trp547Leu mutations that confer resistance to acetolactate synthase (ALS)-inhibiting herbicides in prickly lettuce (*Lactuca serriola*) [10,11] and Powell's amaranth (*Amaranthus powellii*) [12]. However, while Menchari et al. [13] determined fitness costs for the resistance-conferring Asp2078Gly and Ile2041Asn mutations in acetyl CoA carboxylase (ACCase) in slender meadow foxtail (*Alopecuris myosuroides*), they did not find fitness costs associated with the Ile1781Leu mutation. Similarly, Vila-Aiub et al. [9] found resistance costs for the Cys2088Arg mutation associated with resistance to ACCase herbicides in Wimmera ryegrass (*Lolium rigidum*), yet fitness costs were not demonstrated for the Ile1781Leu mutation. Thus, particular target site mutations may or may not be associated with fitness costs in resistant populations.

Weed populations with herbicide resistance conferred by enhanced metabolic rates may be more likely to exhibit fitness costs, due to the constitutive and/or inducible overexpression of genes involved in energetically expensive pathways like those involving P450- and GST-mediated metabolism [14]. For example, a *L. rigidum* biotype (isolated from one population) with suspected P450-mediated herbicide metabolism produced less aboveground biomass and had a lower relative growth rate (RGR)

Table 1. Soil nutrient concentrations.

Greenhouse	Nitrate (kg ha^{-1})	Phosphorus (mg kg^{-1})	Potassium (mg kg^{-1})	Organic Matter (%)
1	39.7±3.5[a]	13±0.0[a]	261.3±8.0[a]	7.2±0.7[a]
2	129.3±5.0[b]	14.7±0.6[b]	302.0±13.9[b]	7.6±0.6[a]

Initial nutrient and organic matter concentrations in greenhouse soil (mean ± SD, n = 3 per greenhouse) used in experiments to assess the impact of environmental and biological stressors on *Avena fatua* growth. Significant differences across greenhouses are indicated by different letters (P<0.05).

[15] and was a weaker competitor with *T. aestivum* [4] in the greenhouse than susceptible biotypes. Similarly, Park and Mallory-Smith [16] found that plants from a metabolically-based resistant downy brome (*Bromus tectorum*) population produced less shoot biomass, leaf area and seeds, and was less competitive than plants from an adjacent susceptible population.

A. fatua is one of the most economically important weeds across the Northern Great Plains of North America, where it competes with and reduces yields of cereal and pulse crops [17,18]. ACCase and ALS-inhibiting herbicides have been used to control *A. fatua* since the 1970s and 1980s, respectively. However, repeated use of these herbicides, as well as others such as triallate and difenzoquat, led to the evolution of herbicide resistant *A. fatua* populations in the 1980s and 1990s [19,20,21]. Recently, two *A. fatua* populations with resistance to multiple herbicides including triallate (emergence inhibitor), flucarbazone (ALS-inhibitor), imazamethabenz (ACCase-inhibitor), paraquat (membrane disruptor), and difenzoquat (membrane disruptor) were reported [22]. Resistance to some of these modes of action may be conferred by P450-enhanced metabolism [3,23,24] or by protection against oxidative stress via GST [6]. While controlling resistant weeds with herbicides is difficult, resistance via enhanced herbicide metabolism is especially problematic because weeds do not need to be exposed to a herbicide in order to be resistant to it [6].

Few studies have evaluated fitness costs for herbicide resistance in *A. fatua*. O'Donovan et al. [25] showed that shoot weight in triallate/difenzoquat-resistant *A. fatua* populations was generally greater than in susceptible ones, but this difference did not translate into competitive advantages. Similarly, Lehnhoff et al. [22] demonstrated that in non-competitive conditions, the MHR *A. fatua* populations utilized in this study had photosynthetic capacities and relative growth rates similar to susceptible populations, and while the MHR populations reached anthesis quicker, they ultimately produced fewer tillers and seeds.

Understanding the effects of herbicide resistance on individual fitness is crucial to the effective management of MHR weed biotypes. Specifically, assessment of fitness costs is important for the prediction of resistant biotype population demographics after the cessation of herbicide use [26,27]. Three issues highlight the importance of assessing fitness costs of the MHR *A. fatua* populations. First, the observations that herbicide resistance fitness costs are not universal [9] indicates that each case should be examined individually. Second, the increasing incidence of MHR populations and their spread has profound implications for their prevention and management [28]. And finally, the fact that the MHR populations studied here are the first with confirmed constitutively elevated and inducible P450 expression levels (Keith et al. unpublished data) provides an important genetic resource in which to test the resource-based allocation theory.

The purpose of this research was to evaluate the competitive performance of two MHR and two herbicide susceptible (HS) *A. fatua* populations along a gradient of biotic and abiotic stresses. The biotic stress was provided by growing *A. fatua* plants in competition with three planting densities of *T. aestivum*, and the abiotic stress was a gradient of resource availability using three rates of N fertilization. Our objectives were to compare (1) total above-ground biomass production, and (2) relative growth rates of the HS and MHR *A. fatua* populations across a factorial combination of these abiotic and biotic gradients.

Materials and Methods

Plant Materials

Two herbicide susceptible (HS1 and HS2) *A. fatua* populations were used as controls. Population HS1 was obtained originally from seeds collected from a field adjacent to where MHR populations were collected, and subsequently confirmed to be 100% HS. All seeds were collected with private landowner permission, and no endangered or protected species were involved in this research. The second susceptible population (HS2, technically a biotype) was the nondormant inbred SH430 biotype used for seed dormancy research [29,30]. This biotype exhibited similar growth to HS1 and the MHR populations in previous research [22].

MHR populations (MHR3 and MHR4) used for this research were derived from seeds collected from two *A. fatua* populations not controlled by 60 g a.i. ha^{-1} pinoxaden (Axial, Syngenta Crop Protection, Inc., Greensboro, NC, USA; ACCase inhibitor) from two fields in Teton County, Montana, USA in 2006. To ensure a 100% resistant population, these seeds (initially about 90% resistant to 60 g a.i. ha^{-1} pinoxaden) were subjected to two generations of recurrent group selection with 50 plants per generation by spraying with the same pinoxaden dose. Because we maintained some genetic diversity by using 50 random seeds in each generation, we refer to MHR3 and MHR4 as populations rather than biotypes. Therefore this work assessed the population level effects of MHR [22] and did not explicitly assess fitness costs of resistance as would be done via isogenic lines of MHR biotypes [31].

Prior to conducting this study, MHR3 and MH4 populations were determined to be resistant to field use rates of difenzoquat, flucarbazone, imazamethabenz, and tralkoxydim as compared to HS1 and HS2 [22]. Resistance to triallate and paraquat was subsequently determined for both MHR populations (Keith et al. unpublished data). We use the MHR acronym to describe these populations because they are resistant to members of five different mode of action families, and we suspect the presence of different physiological mechanisms (Keith et al. unpublished data).

Plant Growth

This study was conducted as a complete randomized block design with four blocks of four *A. fatua* populations, three N fertilization rates, three levels of *T. aestivum* (Reeder hard red spring wheat) competition, and two harvest times, for a total of 288 experimental units (plastic pots, 17.8 cm dia. ×15.2 cm deep) per trial. It was conducted twice, simultaneously in two different

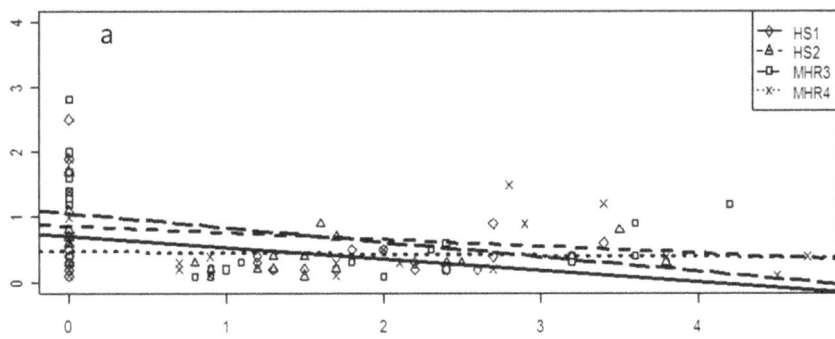

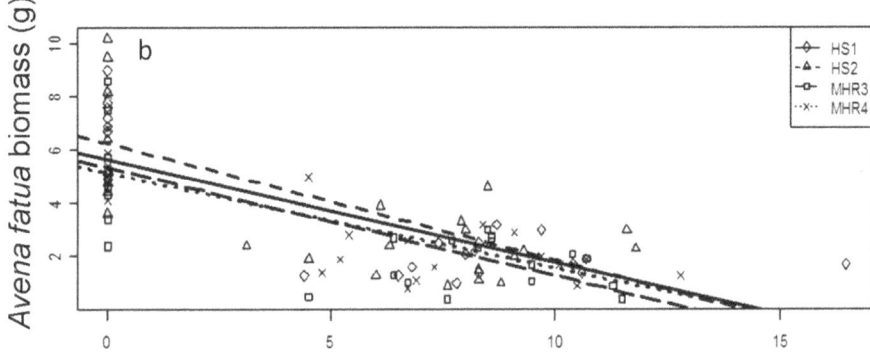

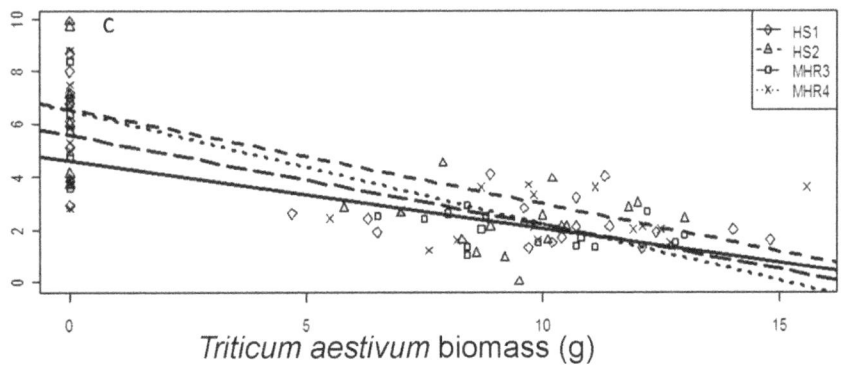

Figure 1. Effects of *Triticum aestivum* **competition and nitrogen on** *Avena fatua* **biomass.** Nitrogen fertilizer rates (N) are (a) 0, (b) 50 and (c) 100 kg N ha^{-1}. HS1 and HS2 are herbicide susceptible *A. fatua* populations and MHR3 and MHR4 are multiple herbicide resistant populations. n = 8 for each N, *T. aestivum* competition, and *A. fatua* combination.

greenhouses at the Montana State University Plant Growth Center under a 16-hr photoperiod of natural sunlight supplemented with mercury vapor lamps (165 μE m^{-2} sec^{-1}) at 25±4 C. Pots were filled with a mixture of 1:1:1 [by vol] sphagnum moss, sand, and Sunshine Mix #1 (Sun Gro Horticulture, Inc., Bellevue, WA) and leached to reduce background N concentrations by draining four pot volumes of water through the soil over two days. Because of changes in greenhouse soil availability, different batches of soil were used in each greenhouse. Three soil samples, each composited from ten pots, were collected from each greenhouse and analyzed to determine initial concentrations of nitrate (NO$_3^-$), Olsen phosphorus (P), potassium (K$^+$), and organic

matter. The pots were then fertilized with ammonium sulfate at rates of 0, 50, or 100 kg N ha^{-1}. In each pot, two seeds of one of the four *A. fatua* populations (HS1, HS2, MHR3, or MHR4) were planted 2 cm deep in the center of the pot, and *T. aestivum* seeds (0, 5, or 9) were planted evenly spaced around the circumference of the pot approximately 2 cm from the edge. *A. fatua* seedlings were thinned to the single largest individual per pot and *T. aestivum* seedlings were thinned to create densities of 0, 4, and 8 individuals per pot at 7 days after planting (DAP).

At 32 DAP, one half of the pots from each treatment were randomly chosen and *A. fatua* and *T. aestivum* plants were harvested separately by cutting them at the root crown, followed by drying at

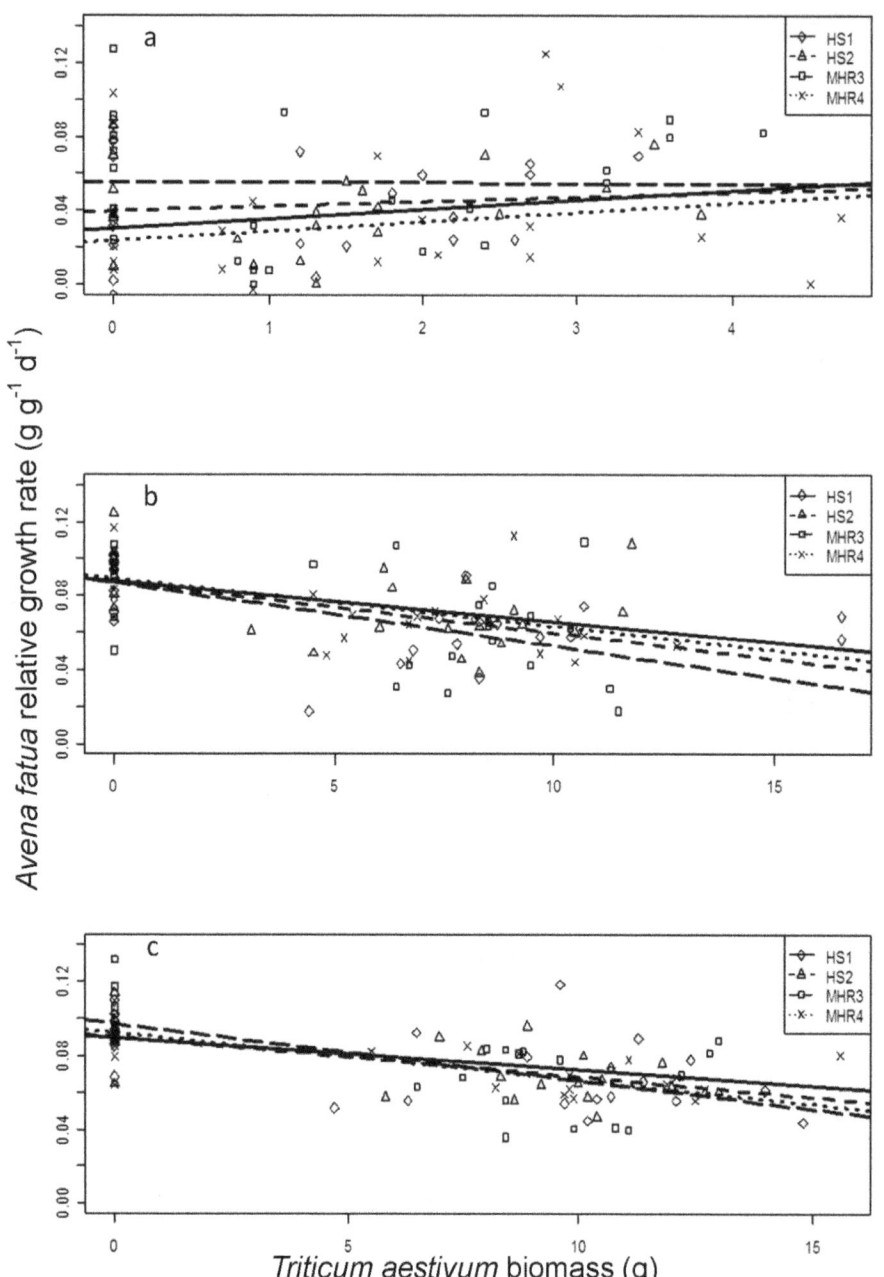

Figure 2. Effects of *Triticum aestivum* **competition on** *Avena fatua* **relative growth rate.** Nitrogen fertilizer rates (N) are (a) 0, (b) 50 and (c) 100 kg N ha^{-1}. HS1 and HS2 are herbicide susceptible *A. fatua* populations and MHR3 and MHR4 are multiple herbicide resistant populations. n = 8 for each N, *T. aestivum* competition, and *A. fatua* combination.

40 C for 10 days, and weighing. The pots receiving 50 and 100 kg N ha^{-1} were then fertilized at these rates bi-weekly and weekly, respectively, for the next four weeks. The remaining plants were harvested at 61 DAP, and plant materials were dried and weighed as described above.

Data Analysis

Differences in initial soil nutrient contents between greenhouse soil sources were evaluated by Student's t-test and were found to be different for concentrations of NO_3^- (P<0.001), P (P = 0.038), and K^+ (P = 0.019), but there were no differences in percent organic matter (P = 0.484) (Table 1). The effects of initial nutrient content on *A. fatua* biomass and RGR were evaluated via ANOVA with greenhouse as a random error term, and initial nutrient differences between greenhouses did not translate into differences in final *A. fatua* biomass ($F_{1,283} = 1.09$, P = 0.297) or relative growth rate ($F_{1,283} = 3.03$, P = 0.083), so data were combined for analysis.

A. fatua RGRs were calculated following Hunt [32] (equation 1):

$$RGR = \frac{(\ln W_2 - \ln W_1)}{(t_2 - t_1)} \quad (1)$$

where ln is the natural log, W_2 and W_1 are final and first harvest

Table 2. Models predicting Avena fatua biomass and growth rate.

Model variables	k	Biomass			Relative growth rate		
		AIC	Δ (AIC)	w(AIC)	AIC	Δ (AIC)	w(AIC)
Intercept only	3	1296	237	0	−1158	62	0
N	4	1215	156	0	−1210	10	0.01
T. aestivum biomass	4	1287	228	0	−1143	77	0
A. fatua population	4	1300	241	0	−1145	75	0
N + T. aestivum biomass	5	1076	17	0	**−1220**	**0**	**0.97**
N + A. fatua population	5	1219	160	0	−1197	23	0
T. aestivum biomass + A. fatua population	5	1291	232	0	−1130	90	0
N × T. aestivum biomass	6	**1059**	**0**	**1**	−1212	8	0.02
N × A. fatua population	6	1231	172	0	−1177	43	0
T. aestivum biomass × A. fatua population	6	1298	239	0	−1114	106	0
N + T. aestivum biomass + A. fatua population	6	1081	22	0	−1206	14	0
N × T. aestivum biomass × A. fatua population	10	1096	37	0	−1139	81	0

Akaike information criterion (AIC) scores for mixed-effects models of Avena fatua biomass and relative growth rate, where k is the number of parameters predicted by intercept only, nitrogen fertilization rate (N), competition from Triticum aestivum (T. aestivum biomass), A. fatua population, and their interactions. Bold values indicate the best-fit models with the lowest AIC scores. Δ(AIC) is the change in AIC with respect to the best candidate model and wAIC is the AIC weight. N rate, T. aestivum biomass, and A. fatua population were fixed effects and greenhouse was a random effect included in every model. n = 8 for each N, T. aestivum competition, and A. fatua combination.

biomasses, respectively, and t_2 and t_1 are final and first harvest times in days. The biomass and RGR responses of the A. fatua populations to N rates and T. aestivum competition were analyzed via linear mixed effects-models, with N rate, T. aestivum biomass, and A. fatua population as fixed effects, and the greenhouse in which experiments were conducted as a random effect to account for initial soil differences, using the 'lme4' package in R (version 2.12.1, The R Foundation for Statistical Computing). Best models predicting A. fatua responses were determined based on the lowest Akaike Information Criterion (AIC) values. Change in AIC (ΔAIC) values between the best candidate model and all other models, and the models' AIC weights (wAIC, higher weight indicates better model), were also assessed for model comparison. For individual N rates, the response of each A. fatua population to T. aestivum was modeled via linear regression in R.

Results and Discussion

The best candidate models predicting A. fatua biomass and RGR based on AIC scores and weights were N fertilization rate × T. aestivum biomass and N fertilization rate + T. aestivum biomass, respectively (Table 2). The second best candidate models for A. fatua biomass and RGR based on ΔAIC were N + T. aestivum biomass and N × T. aestivum biomass, respectively. The best models that included A. fatua population as a predictor had ΔAIC values of 22 and 14 for A. fatua biomass and RGR, respectively; however, these models had w(AIC) values of zero, indicating a poor fit to the data. At all N rates, the biomass of A. fatua decreased as T. aestivum biomass increased (P<0.05) (Figure 1) while the RGR of all A. fatua populations decreased with increasing T. aestivum biomass for the two higher N fertilization rates, but not at the zero N rate (P = 0.213) (Figure 2).

As expected, A. fatua biomass declined with increasing T. aestivum competition, although this relationship was not significant in the zero N treatments, where both A. fatua and T. aestivum biomasses were quite low compared to those in the 50 and 100 kg N ha^{-1} treatments. The minimal decrease of A. fatua biomass in

the unfertilized pots can be explained by the apparent lack of competition from T. aestivum, as both species were stunted in these treatments (Lehnhoff, personal observation). The similar pattern observed with respect to A. fatua RGR can also be attributed to the low T. aestivum biomass in the unfertilized pots compared to fertilized ones. These results are generally in accordance with those reported by Blackshaw and Brandt [33] who showed that A. fatua competition with T. aestivum was unaffected by fertilizer rate.

More importantly, there were no differences in the response of HS and MHR populations to T. aestivum competition or N stress, suggesting that there were no growth-related fitness costs for multiple herbicide resistance. This is similar to our previous findings [22], where MHR populations did not experience growth-related fitness costs in the absence of competition. Because these MHR populations exhibit constitutively higher levels of P450 expression (Keith et al., unpublished data), the resource allocation theory would predict that the resulting resource diversion would be associated with a fitness cost. These results based on A. fatua growth therefore contrast with previous studies reporting various fitness costs associated with enhanced rates of herbicide metabolism [4,15,16].

The current results suggest that if competitiveness is positively related to reproductive fitness, the frequency of resistant alleles in the A. fatua MHR populations should not decline due to fitness costs. However, we previously showed that, in the absence of competition, one MHR population produced fewer seeds than a HS population in the greenhouse [22]. Thus, it is possible that if the selection pressure is interrupted, the proportion of MHR to HS seeds in the seedbank will decline, ultimately leading to a reduction in MHR populations, sensu Maxwell et al. [34] who predicted a decrease in resistant populations after the cessation of herbicide use. The relationship between A. fatua competitiveness (i.e., biomass production and RGR) and reproductive fitness is unknown, and will need to be examined under realistic field conditions to determine the actual implications for field management of MHR A. fatua.

While competitive ability provides a potential indication of the fitness costs associated with herbicide resistance [4], other traits that contribute to individual success through the entire life cycle, such as seed germination, seedling survival, and seed production should be included in a comprehensive analysis [14,35]. In this context, this study provides useful information that can be integrated with future field research on MHR and HS demography. Such investigations will be required in order to develop successful alternative management options for herbicide resistant weed populations.

Finally, it should be noted that our study examined a small number of populations (two HS and two MHR). As Cousens et al. [36] discussed, a non-significant result from a small number of populations may not be enough to support a hypothesis that there is no fitness cost of resistance. Additional studies comparing the

MHR populations to more HS populations could provide valuable insight into potential fitness costs. Alternatively, fitness costs could be assessed through the generation and testing of HS and MHR isogenic biotypes, and this process is currently underway in our laboratory.

Acknowledgments

We thank Mark Boyd and Ethan Mayes for assistance with greenhouse work.

Author Contributions

Conceived and designed the experiments: EAL FDM WED. Performed the experiments: EAL BKK. Analyzed the data: EAL. Wrote the paper: EAL BKK WED FDM.

References

1. Heap IM (2013) International survey of herbicide resistant weeds. Weed Science Society of America website. Available: http://www.weedscience.org. Accessed 2013 Mar 1.
2. Delye C, Gardin JAC, Boucansaud K, Chauvel B, Petit C (2011) Non-target-site-based resistance should be the centre of attention for herbicide resistance research: Alopecurus myosuroides as an illustration. Weed Research 51: 433–437.
3. Beckie HJ, Tardif FJ (2012) Herbicide cross resistance in weeds. Crop Protection 35: 15–28.
4. Vila-Aiub MM, Neve P, Powles SB (2009) Evidence for an ecological cost of enhanced herbicide metabolism in Lolium rigidum. Journal of Ecology 97: 772–780.
5. Reade JPH, Milner LJ, Cobb AH (2004) A role for glutathione S-transferases in resistance to herbicides in grasses. Weed Science 52: 468–474.
6. Powles SB, Yu Q (2010) Evolution in Action: Plants Resistant to Herbicides. In: Merchant SBWROD, editor. Annual Review of Plant Biology, Vol 61. 317–347.
7. Herms DA, Mattson WJ (1992) The dilemna of plants – to grow or defend. Quarterly Review of Biology 67: 283–335.
8. Bazzaz FA, Chiariello NR, Coley PD, Pitelka LF (1987) Allocating resources to reproduction and defense. Bioscience 37: 58–67.
9. Vila-Aiub MM, Neve P, Powles SB (2009) Fitness costs associated with evolved herbicide resistance alleles in plants. New Phytologist 184: 751–767.
10. Alcocer-Ruthling M, Thill DC, Mallorysmith C (1992) Monitoring the occurrence of sulfonylurea-resistant prickly lettuce (Lactuca serriola). Weed Technology 6: 437–440.
11. Alcocer-Ruthling M, Thill DC, Shafii B (1992) Differential competitiveness of sulfonylurea resistant and susceptible prickly lettice (Lactuca serriola). Weed Technology 6: 303–309.
12. Tardif FJ, Rajcan I, Costea M (2006) A mutation in the herbicide target site acetohydroxyacid synthase produces morphological and structural alterations and reduces fitness in Amaranthus powellii. New Phytologist 169: 251–264.
13. Menchari Y, Chauvel B, Darmency H, Delye C (2008) Fitness costs associated with three mutant acetyl-coenzyme A carboxylase alleles endowing herbicide resistance in black-grass Alopecurus myosuroides. Journal of Applied Ecology 45: 939–947.
14. Vila-Aiub MM, Neve P, Steadman KJ, Powles SB (2005) Ecological fitness of a multiple herbicide-resistant Lolium rigidum population: dynamics of seed germination and seedling emergence of resistant and susceptible phenotypes. Journal of Applied Ecology 42: 288–298.
15. Vila-Aiub MM, Neve P, Powles SB (2005) Resistance cost of a cytochrome P450 herbicide metabolism mechanism but not an ACCase target site mutation in a multiple resistant Lolium rigidum population. New Phytologist 167: 787–796.
16. Park KW, Mallory-Smith CA (2005) Multiple herbicide resistance in downy brome (Bromus tectorum) and its impact on fitness. Weed Science 53: 780–786.
17. Evans RM, Thill DC, Tapia L, Shafii B, Lish JM (1991) Wild oat (Avena fatua) and spring barley (Hordeum vulgare) density affect spring barley grain yield. Weed Technology 5: 33–39.
18. Beckie HJ, Francis A, Hall LM (2012) The Biology of Canadian Weeds. 27. Avena fatua L. (updated). Canadian Journal of Plant Science 92: 1329–1357.

19. O'Donovan JT, Sharma MP, Harker KN, Maurice D, Baig MN, et al. (1994) Wild oat (Avena fatua) populations resistant to triallate are also resistant to difenzoquat. Weed Science 42: 195–199.
20. Heap IM, Murray BG, Loeppky HA, Morrison IN (1993) Resistance to aryloxyphenoxypropionate and cyclohexanedione herbicides in wild oat (Avena fatua). Weed Science 41: 232–238.
21. Somody CN, Nalewaja JD, Miller SD (1984) Wild oat (Avena fatua) and Avena sterilis morphological characteristics and response to herbicides. Weed Science 32: 353–359.
22. Lehnhoff EA, Keith BK, Dyer WE, Peterson RKD, Menalled FD (in press) Multiple herbicide resistance in wild oat (Avena fatua) and its impacts on physiology, germinability, and seed production. Agronomy Journal.
23. Beckie HJ, Warwick SI, Sauder CA (2012) Basis for herbicide resistance in Canadian populations of wild oat (Avena fatua). Weed Science 60: 10–18.
24. Maneechote C, Preston C, Powles SB (1997) A diclofop-methyl-resistant Avena sterilis biotype with a herbicide-resistant acetyl-coenzyme A carboxylase and enhanced metabolism of diclofop-methyl. Pesticide Science 49: 105–114.
25. O'Donovan JT, Newman JC, Blackshaw RE, Harker KN, Derksen DA, et al. (1999) Growth, competitiveness, and seed germination of triallate/difenzoquat-susceptible and -resistant wild oat populations. Canadian Journal of Plant Science 79: 303–312.
26. Holt J, Thill DC (1994) Growth and productivity of resistant plants. In: J.A.M PSBaH, editor. Herbicide Resistance in Plants: Biology and Biochemistry. Boca Raton, FL: Lewis. 299–316.
27. Maxwell BD, Mortimer AM (1994) Selection for Herbicide Resistance. In: Powles SB, Holtum JAM, editors. Herbicide Resistance in Plants: Biology and Chemistry. Boca Raton, Florida: CRC Press, Inc. 353.
28. Mortensen DA, Egan JF, Maxwell BD, Ryan MR, Smith RG (2012) Navigating a Critical Juncture for Sustainable Weed Management. Bioscience 62: 75–84.
29. Naylor JM, Jana S (1976) Genetic adaptation for seed dormancy in Avena fatua. Canadian Journal of Botany-Revue Canadienne De Botanique 54: 306–312.
30. Johnson RR, Cranston HJ, Chaverra ME, Dyer WE (1995) Characterization of CDNA clones for differentially expressed genes in embryos of dormant and nondormant Avana fatua caryopses. Plant Molecular Biology 28: 113–122.
31. Vila-Aiub MM, Neve P, Roux F (2011) A unified approach to the estimation and interpretation of resistance costs in plants. Heredity 107: 386–394.
32. Hunt R (1982) Plant Growth Curves: The functional approach to plant growth analysis. London: Edward Arnond. 248 p.
33. Blackshaw RE, Brandt RN (2008) Nitrogen fertilizer rate effects on weed competitiveness is species dependent. Weed Science 56: 743–747.
34. Maxwell BD, Roush ML, Radosevich SR (1990) Predicting the evolution and dynamics of herbicide resistance in weed populations. Weed Technology 4: 2–13.
35. Délye C, Menchari Y, Michel S, Cadet É, Corre VL (2013) A new insight into arable weed adaptive evolution: mutations endowing herbicide resistance also affect germination dynamics and seedling emergence. Annals of Botany 111: 481–691.
36. Cousens RD, Gill GS, Speijers EJ (1997) Comment: Number of sample populations required to determine the effects of herbicide resistance on plant growth and fitness. Weed Research 37: 1–4.

10

Evaluation of the CENTURY Model Using Long-Term Fertilization Trials under Corn-Wheat Cropping Systems in the Typical Croplands of China

Rihuan Cong[1,2], Xiujun Wang[3,4], Minggang Xu[1]*, Stephen M. Ogle[5], William J. Parton[5]

1 Ministry of Agriculture Key Laboratory of Crop Nutrition and Fertilization, Institute of Agricultural Resources and Regional Planning, Chinese Academy of Agricultural Sciences, Beijing, China, **2** College of Resources and Environment, Huazhong Agricultural University, Wuhan, China, **3** State Key Laboratory of Desert and Oasis Ecology, Xinjiang Institute of Ecology and Geography, Chinese Academy of Sciences, Urumqi, China, **4** Earth System Science Interdisciplinary Center, University of Maryland, College Park, Maryland, United States of America, **5** Natural Resource Ecology Laboratory, Colorado State University, Fort Collins, Colorado, United States of America

Abstract

Soil organic matter models are widely used to study soil organic carbon (SOC) dynamics. Here, we used the CENTURY model to simulate SOC in wheat-corn cropping systems at three long-term fertilization trials. Our study indicates that CENTURY can simulate fertilization effects on SOC dynamics under different climate and soil conditions. The normalized root mean square error is less than 15% for all the treatments. Soil carbon presents various changes under different fertilization management. Treatment with straw return would enhance SOC to a relatively stable level whereas chemical fertilization affects SOC differently across the three sites. After running CENTURY over the period of 1990–2050, the SOC levels are predicted to increase from 31.8 to 52.1 Mg ha^{-1} across the three sites. We estimate that the carbon sequestration potential between 1990 and 2050 would be 9.4–35.7 Mg ha^{-1} under the current high manure application at the three sites. Analysis of SOC in each carbon pool indicates that long-term fertilization enhances the slow pool proportion but decreases the passive pool proportion. Model results suggest that change in the slow carbon pool is the major driver of the overall trends in SOC stocks under long-term fertilization.

Editor: Julio Vera, University of Erlangen-Nuremberg, Germany

Funding: Financial supports are from the National Science Foundation of China (41171239) and the National Basic Research Program (2011CB100501). The funders had no role in study design, data collection and analysis, decision to publish, or preparation of the manuscript.

Competing Interests: The authors have declared that no competing interests exist.

* E-mail: mgxu@caas.ac.cn

Introduction

Soil organic carbon (SOC) is one of the most important terrestrial pools for C storage. It is estimated that the total soil carbon pool is around 1400–1500 Pg C, which is approximately three times greater than the atmospheric pool (750 Pg C) [1,2]. The SOC pool represents a dynamic equilibrium resulting from changes in gains and losses. Even small changes in SOC at a site may potentially add up to significant changes in large-scale carbon cycling across a region [3]. Furthermore, SOC is relatively dynamic and can be greatly influenced by agricultural practices. Increases in SOC storage in cropland soils would benefit soil productivity and environmental health [4,5], and so alternative farming management practices have been evaluated to identify their potentials for increasing SOC in the agroecosystems [4–7].

Long-term experiments are crucial for determining fundamental crop, soil and ecological processes and their impacts on the environment [6–9]. Data from long-term experiments provide a unique resource to investigate long-term influences of climate, crop rotation and crop residue management on soil fertility [6–12]. However, SOC change is affected by complex interactions that vary across space and time depending on the environmental conditions and agricultural management practices. A weakness of long-term experiments is that they are typically restricted to small subset of the entire set of environmental conditions and management practices that exists [13].

Process-based models are an effective way to evaluate SOC changes across a broader set of environmental conditions and management practices [14]. In recent decades, the development and evaluation of soil organic matter models has improved the understanding of factors controlling SOC dynamics, and thus increased our ability to predict future SOC trends. A number of SOC models have been developed, but applying these models requires adequate evaluation with measured SOC trends from experimental for different environmental conditions and management practices [15]. For example, the CENTURY model [16] has been widely used to simulate SOC changes under different management conditions in long-term experiments (e.g., [17], [18] and [19]). With the development of CENTURY, the model has been successfully employed in long-term fertilizer, irrigation, pest management, and site-specific farming applications [20,21]. In China, CENTURY model has been used in grassland [22], forest [23], and regional farmland [24]. However, CENTURY modeling research was still limited in farmland especially under the double cropping rotations and in the acidic soil.

Here, we evaluate the CENTURY with data from three long-term experiments with wheat-corn cropping rotations and different fertilization practices. Specifically, our objectives were

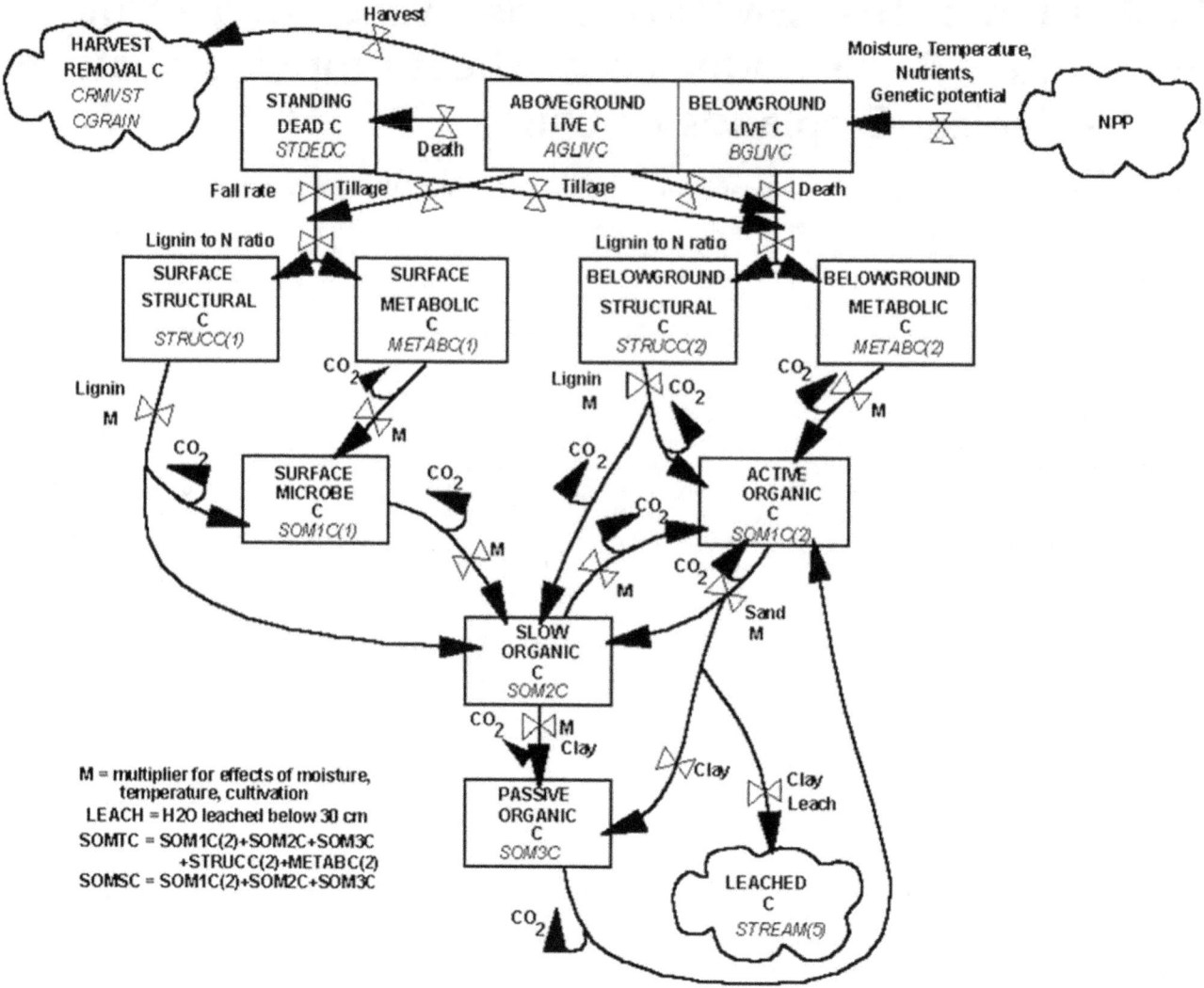

Figure 1. The pools and flows of carbon in the CENTURY model. The diagram showed the major factors which control the flows [29].

Materials and Methods

Long-term Experiment

(i) to evaluate the performance of CENTURY with evaluation of modeled SOC stocks for different fertilizations and under acidic soil; (ii) to study the effect of fertilization practices on different SOC pools in the modeling framework; and (iii) to predict soil carbon potential under long-term fertilization.

Three long-term experiments were utilized for this study, which were located at Changping (40°13′N, 116°15′E), Yangling (34°17′N, 108°00′E) and Qiyang (26°45′N, 111°52′E) in China. Climate conditions varied from semi-humid (Changping site) to

Table 1. Land management practices for the long-term experimental sites during the different blocks and periods used in the CENTURY model simulations.

No.	Periods	Management practices	Repeating sequence
1	up to 1900 (4000 yrs)	Native grassland, low-intensity grazing	1 year
2	1901–1960	Corn, low crop yield cultivar, ploughing, remove straw, applying manure	1 year
3	1961–1988	Wheat-corn rotation, low crop yield cultivar, ploughing, remove straw, applying chemical fertilizers and manure	2 years
4	1989–1989	Uniform tillage to set up long-term experiments	1 year
5	1990–2050	Long-term fertilization experiments	2 years

Table 2. Carbon input (kg ha^{-1}yr^{-1}) from manure and straw residue in each period used in the CENTURY model at Changping, Yangling, and Qiyang sites.

Sites	1901–1960	1961–1988	1989–1989	1990–2050 (long-term experiment)		
				NPKM	hNPKM	NPKS
Changping	500	500	–	3150	4725	1000
Yangling	500	500	–	3327	4991	1998
Qiyang	500	500	–	5838	8757	1052

humid warm-temperate (Yangling sites) to humid subtropical climate (Qiyang site). Annual mean temperature was 13.1°C at the Changping site, 14.9°C at the Yangling site, and 18.1°C at the Qiyang site. Annual precipitation was generally low at Changping (515 mm) and Yangling (525 mm) sites but 1445 mm at Qiyang. However, annual evaporation was much higher, varying from 993 mm to 1470 mm [25].

The experimental sites had double cropping systems, i.e., winter wheat and summer corn. Winter wheat was seeded in early November and harvested in early May at the Qiyang site. For the other two sites, winter wheat was seeded around October 20th, and harvested around June 1st. The wheat seeding rates ranged from 165 to 225 kg ha^{-1}. Summer corn was sown around June 10th, with the rate of 63 000–75 000 seeds per hectare, and harvested around October 1st at most sites. For the Qiyang site, summer corn was sown in holes between the wheat strips in early April, and harvested in the middle of July. Thus, there was a short period of inter-cropping for wheat and corn at the Qiyang site. However, in the CENTURY modeling, wheat was harvested in April and corn was sown in May at the Qiyang site since inter-cropping could not be fitted in the model. We believed that there would be little impact on the belowground biomass of wheat and corn, and thus to the soil organic carbon turnover.

Seven treatments were selected in this study: (i) control (no fertilizer); (ii) mineral nitrogen (N); (iii) mineral nitrogen and phosphorus combination (NP); (iv) mineral nitrogen, phosphorus and potassium combination (NPK); (v) NPK combinations with livestock or farmyard (i.e., livestock manure mixed with soil and/or crop residue) manure (NPKM); (vi) 1.5 times' application rate of NPKM (hNPKM); and (vii) mineral NPK combined with crop straw (NPKS). The total nitrogen applied (i.e., mineral plus organic) was equal (i.e., nitrogen balanced) in each of the fertilizer treatments (i.e., N, NP, NPK, NPKM and NPKS treatments). The NPKM and hNPKM treatments had 30% of total N applied from mineral fertilizer and the remaining 70% from organic manure. The treatment plots were initially randomized and isolated by 100-cm-cement baffle plates. There was no replicate for the treatments at these sites due to field availability. However, the plot size was relative large (196 m^2–400 m^2) at each study site. The durations of experiments were from 1990 to 2005 at the Changping site, from 1990 to 2008 at the Qiyang sites, and from 1990 to 2009 at the Yangling site.

Urea, calcium superphosphate, and potassium chloride were used as sources of mineral N, P, and K, respectively. All P and K fertilizers and nearly half of the mineral N fertilizer were applied as basal fertilizers prior to seeding. The remaining mineral N fertilizer was applied as top dressing during the growing season. Manure was generally applied once each year before wheat sowing. However, at the Qiyang site, 30% of manure was applied before wheat seeding and remaining 70% was applied before corn seeding. The sources of organic manure were farmyard manure mixed with crop residue at Changping site, cattle manure at Yangling site, and pig manure at Qiyang site [25]. For the NPKS treatment, half amount (2000 kg ha^{-1} yr^{-1}) of both wheat and

Table 3. Soil classification and initial physical and chemical properties (0–20 cm) in 1989.

Properties		Changping	Yangling	Qiyang
Soil classification (FAO)		Haplic Luvisol	Calcaric Regosol	Eutric Cambisol
Parent material		Diluvial Alluvium	Loess	Quaternary Red Clay
Clay mineral type		Hydromica, Montmorillonite	Hydromica, Montmorillonite	Kaolinite
Bulk density	(g cm^{-3})	1.58	1.30	1.19
Total porosity	(%)	40.4	49.6	51.7
Field capacity	(%)	24.8	21.2	23.7
Texture		silt loam	silt loam	clay
Clay (<2 μm)	(%)	10.2	16.8	40.9
Silt (2–50 μm)	(%)	72.0	51.6	27.7
Sand (50–2000 μm)	(%)	16.2	31.6	31.4
Soil pH		8.7	8.6	5.7
SOC	(g kg^{-1})	7.1	6.3	8.6
TN	(g kg^{-1})	0.80	0.83	1.07
C/N ratio		8.9	7.6	8.0

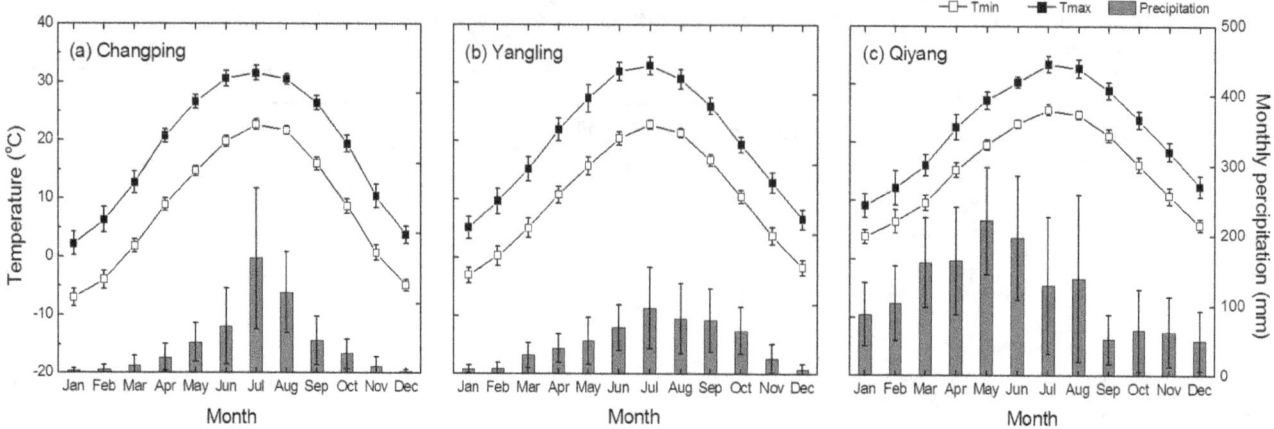

Figure 2. Average monthly precipitation, maximum (Tmax) and minimum (Tmin) temperatures from 1990 to 2010 at (A) Changping, (B) Yangling, and (C) Qiyang. Error bars are standard deviations for the mean values from 1990–2010.

corn straw was incorporated at the Qiyang site, whereas 2250 kg ha^{-1} yr^{-1} and 4500 kg ha^{-1} yr^{-1} of corn straw were incorporated at the Changping and Yangling sites, respectively [26]. Carbon and nutrient contents of manure are measured in 2009 and 2010. Nutrient contents of straw are measured annually over the period of experiment.

There was no irrigation at the Qiyang site since sufficient precipitation occurs during the growing season (i.e., from March to August). However, at the other two sites, irrigations were applied 2–3 times during wheat growing season: 5 mm before seeding, 4 mm for the over wintering stage if needed, and 5 mm at

jointing stage. Irrigation was applied once during corn growing season, with 5 mm before seeding.

Plots were ploughed to a depth of 20 cm twice a year, usually in early June at most sites after wheat harvest and early October after corn harvest. In contrast, at the Qiyang site, the whole plot area was ploughed (20 cm depth) in late October before wheat seeding. Then, area between the wheat belts was ploughed (20 cm depth) in early April before the corn seeding.

Aboveground biomass was removed with negligible stubble left in the field. Thus, organic carbon input was mainly from manure application (i.e., NPKM and hNPKM treatments), residue

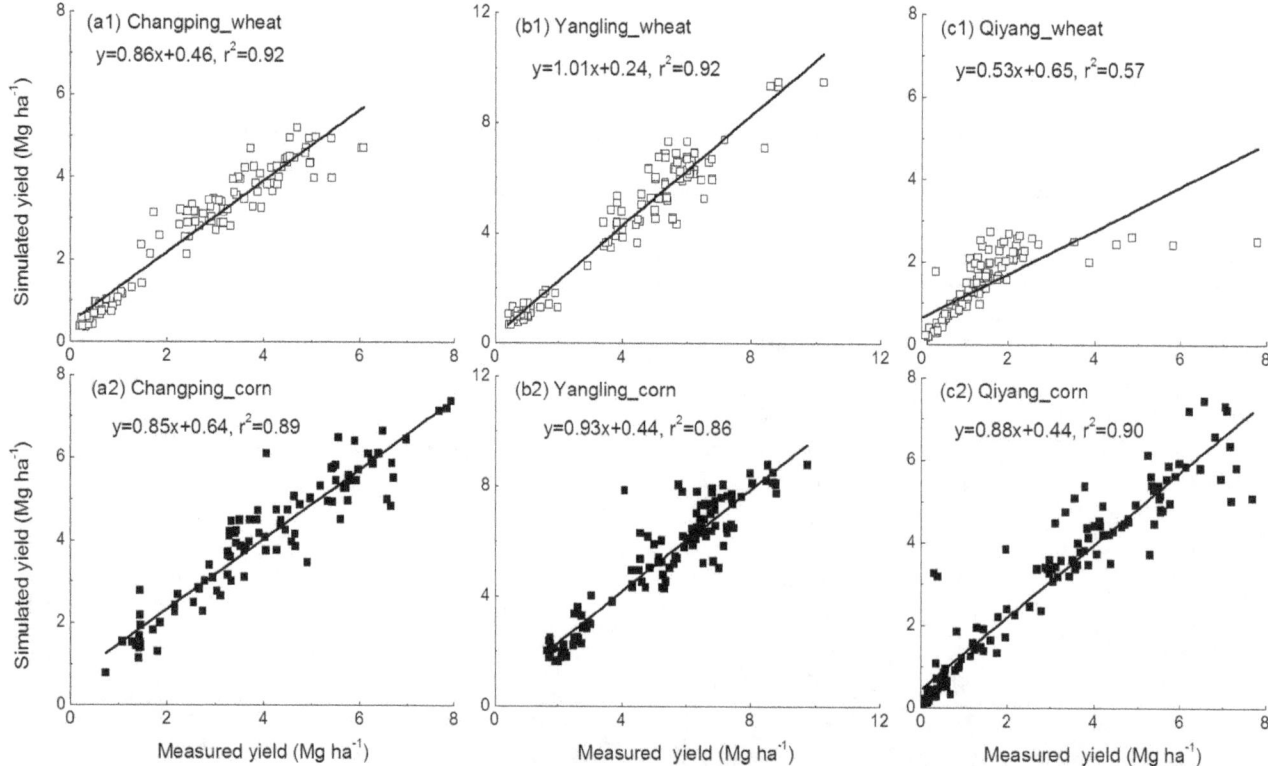

Figure 3. Correlationship between simulated and observed crop (i.e., □wheat and ■corn) grain yield data under the control, N, NP, NPK, NPKM, hNPKM and NPKS treatments at (A) Changping, (B) Yangling, and (C) Qiyang sites.

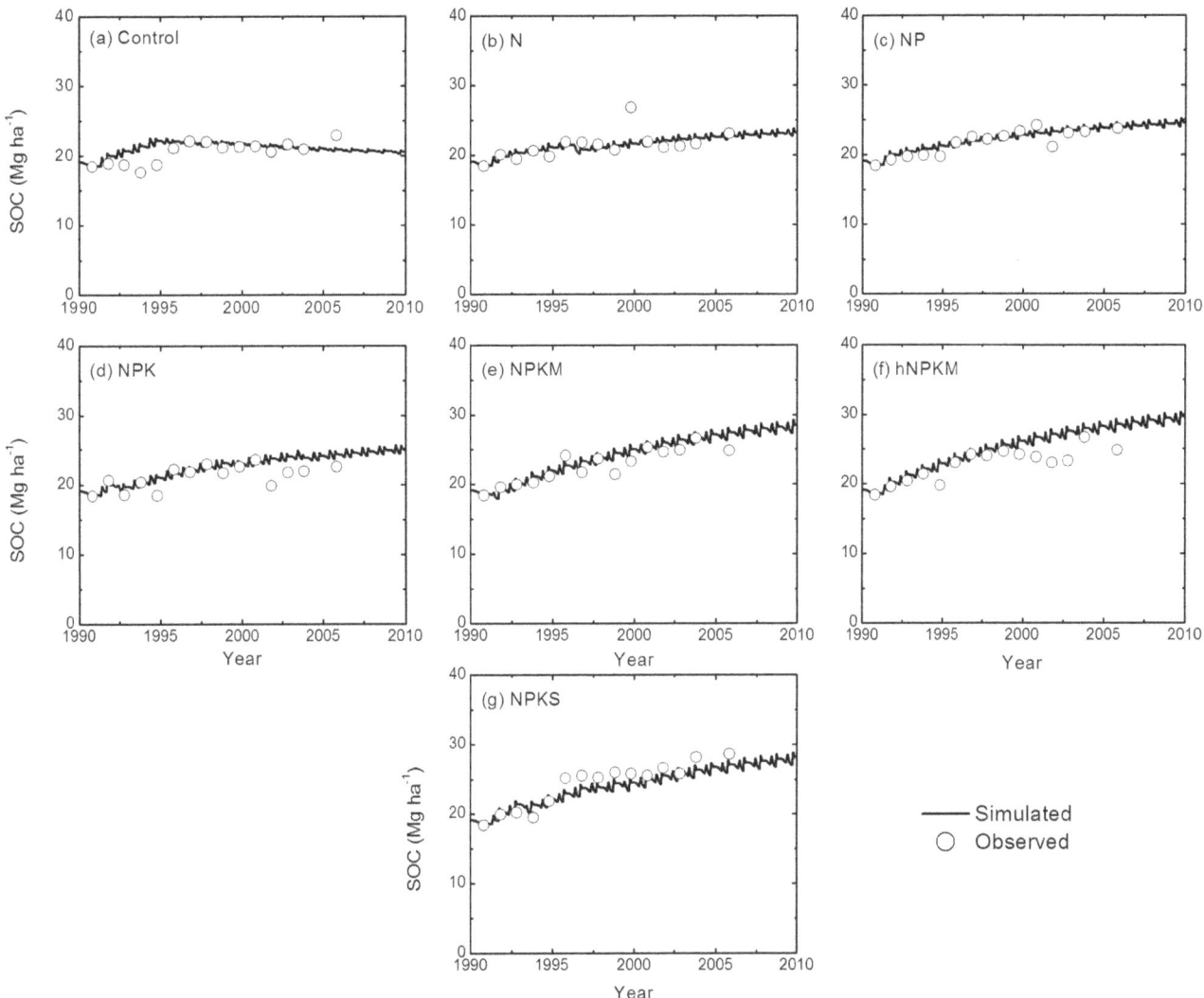

Figure 4. Simulated and measured soil organic carbon (SOC) stocks (0–20 cm) under the control, N, NP, NPK, NPKM, hNPKM and NPKS treatments at Changping.

addition (i.e., NPKS treatment) and from root system during crop growing season. All harvested biomass was removed from the plots for determining crop grain and residue yields. The grain and residue samples were air-dried, threshed, oven-dried at 70°C to a uniform moisture level, and then weighed separately.

During the experimental periods, surface soil (0–20 cm) was collected in each plot annually approximately 15 days after corn harvest. Five cores of soil from each plot were randomly taken, and soils were mixed thoroughly, and air dried for seven days after removing discrete plant residues and visible soil organism. Air-dried soil was sieved through 2 mm screen to determine pH (1:1w/v water). Representative subsamples were crushed to 0.25 mm for measurements of SOC [27]. Soil bulk density was measured in situ once every five or ten years. Surface soil SOC stock expressed as Mg ha^{-1} was calculated by multiplying the SOC content (g kg^{-1}) by averaged soil bulk density (g cm^{-3}) and depth (20 cm).

Model Description and Application

The simulation of SOC dynamics was performed with the CENTURY model (version 4.5) which was described by Parton

and Rasmussen [28]. As shown in Fig. 1 [29], the arrows showing the CO_2 evolved in the transformations was indicative of the microbial growth efficiencies. The first-order decay rates for each of the pools corresponded to turnover times of roughly 3 and 0.5 years for the structural and metabolic components; 1.5 years for the active fraction, 25 years for the slow pool, and 1000 years for the passive pool [28]. The actual turnover times of each carbon pool was a function of the maximum turnover time of specific carbon pool and DEFAC. The value of DEFAC was calculated by multiplying the soil moisture factor (function of precipitation and stored soil water) and the soil temperature factor (function of the average monthly soil surface temperature). The turnover rate of active carbon pool was also a function of soil texture (higher for sandy soils), while the stabilization of active carbon into slow carbon was a function of the silt plus clay content [28].

Model simulations were set-up based on the historical crop rotations and farm practices' investigation from the local farmers at these sites. Five distinct periods were modeled (Table 1), including (1) an initialization period, i.e., 4000 years of native vegetation to reach an equilibrium; (2) the first cultivation period, i.e., plowing of the native grassland in 1901, and planting of corn

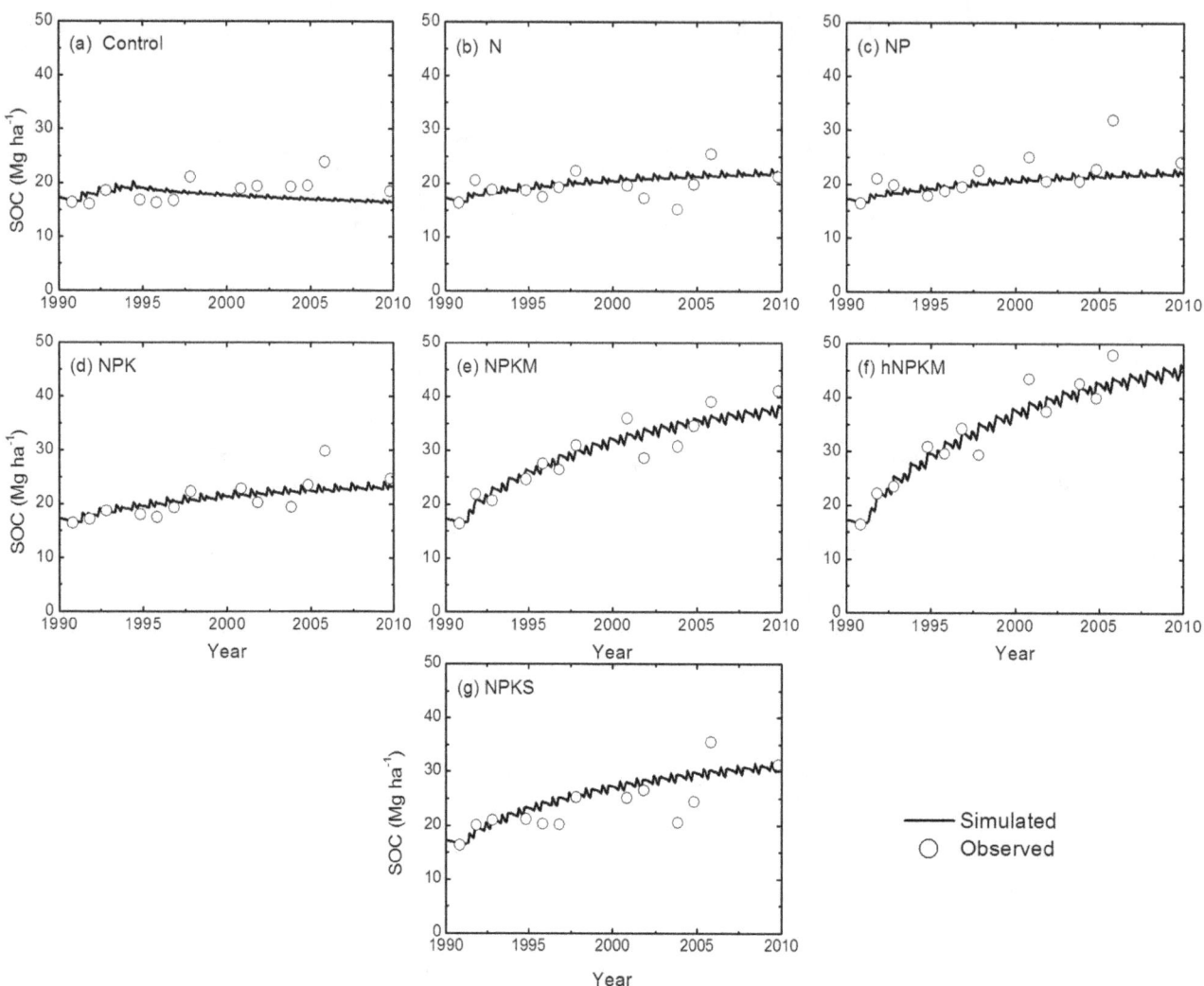

Figure 5. Simulated and measured soil organic carbon (SOC) stocks (0–20 cm) under the control, N, NP, NPK, NPKM, hNPKM and NPKS treatments at Yangling.

with removal of straw until 1960; (3) a second cultivation period, i.e., corn-wheat cropping with removal of straw from 1961 to 1988; (4) pre-experimental treatment period, i.e., uniform tillage with no crop in 1989 prior to establishing the long-term experiment; and (5) experiment period, i.e., the long-term corn-wheat fertility experiment from 1990 to 2050. Management during the experiment (1990–2010) was repeated for remainder of the experimental period until the end time of simulation to evaluate future trends. As shown in Table 2, carbon inputs from manure application were around 500 kg C ha^{-1} before the long-term experiment. Then carbon input varied based on measured data at different sites.

Weather data from 1990–2010 were obtained from China meteorological sharing service system (http://cdc.cma.gov.cn/). For the future simulations (i.e., 2011–2050), we used monthly mean climate data from the 1990–2010 period. Soil properties parameters were soil texture (sand, silt and clay content), soil pH, and bulk density (Table 3). Soils from the Changping and Yangling sites had high soil pH (~8.5) and silt loam texture with similar clay minerals (e.g., Hydromica and Montmorillonite). On the contrary, soil at the Qiyang site, developed from Quaternary

red clay, had a lower soil pH (5.7) and heavy texture with the main clay mineral of Kaolinite.

As shown in Fig. 2, Changping and Yangling sites had similar climate condition whereas temperature and precipitation were much higher at the Qiyang site. The humid and warm climate at Qiyang site would accelerate decomposition of soil organic matter [28]. After setting up all the information (e.g., climate condition, soil texture and etc.) and modeling crop growth successfully, we modeled soil organic carbon pool by calibrating the dec4 value (i.e., maximum decomposition rate of soil organic matter with slow turnover) in the fix.100 file. By raising the parameter 10% higher for every test compared with the measured SOC, we finally got the sound fixed parameter as 0.0045 at the Qiyang site but 0.0023 (i.e., default value in the CENTURY model) at the other two sites [28].

For the plant production submodel, there were pools for live shoots and roots, and standing dead plant material. The potential production was a function of a genetic maximum (PRDX(1)) defined for each crop and 0–1 scalars depending on soil temperature, moisture status, shading by dead vegetation, and seedling growth. The maximum potential production of a crop, unlimited by temperature, moisture or nutrient stresses, was

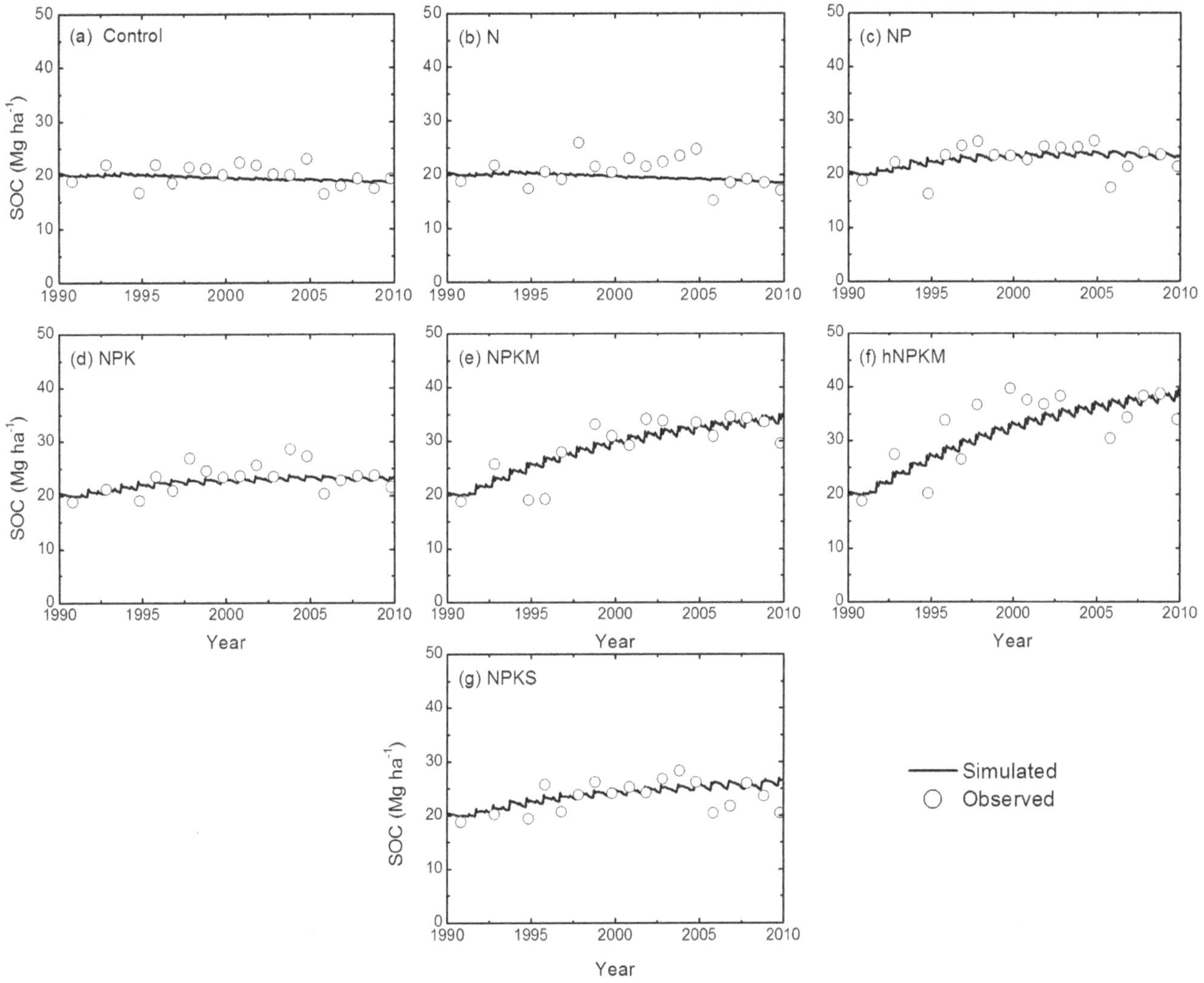

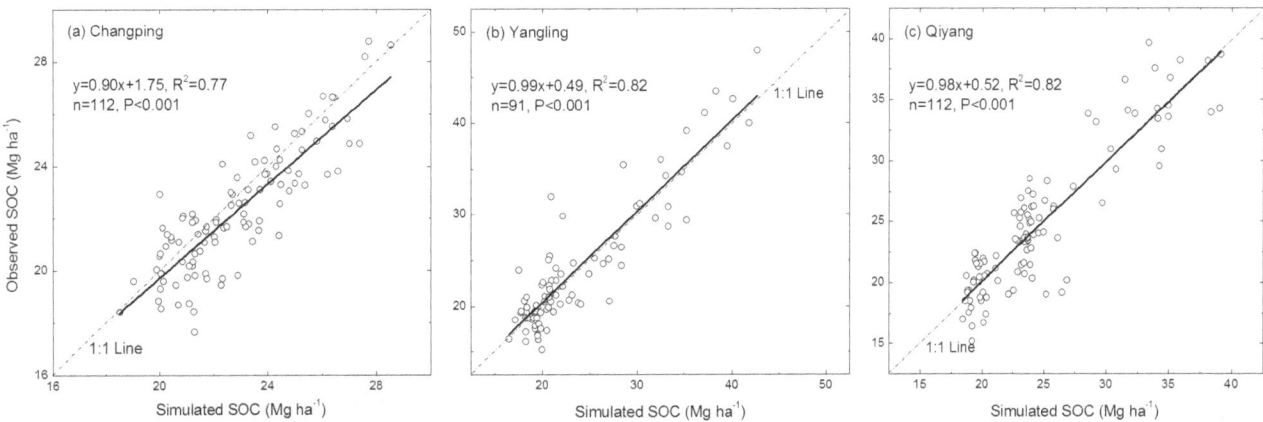

Figure 6. Simulated and measured soil organic carbon (SOC) stocks (0–20 cm) under the control, N, NP, NPK, NPKM, hNPKM and NPKS treatments at Qiyang.

Figure 7. Correlation between simulated SOC and measured SOC under the control, N, NP, NPK, NPKM, hNPKM and NPKS treatments at (A) Changping, (B) Yangling, and (C) Qiyang sites.

Table 4. Statistics comparing simulated and measured SOC stocks at the three long-term experimental sites.

Site	Parameter[a]	Treatment						
		Control	N	NP	NPK	NPKM	hNPKM	NPKS
Changping	n-RMSE	6%	5%	5%	6%	7%	4%	6%
(n = 16×7)	d	0.57	0.81	0.9	0.81	0.91	0.97	0.91
	R^2	0.01	0.76	0.83	0.46	0.80	0.90	0.76
Yangling	n-RMSE	9%	12%	9%	12%	9%	9%	14%
(n = 13×7)	d	0.5	1.00	0.83	0.77	0.96	0.97	0.85
	R^2	0.39	0.19	0.43	0.59	0.89	0.88	0.61
Qiyang	n-RMSE	8%	7%	6%	9%	10%	11%	7%
(n = 16×7)	d	0.37	0.51	0.75	0.61	0.89	0.88	0.82
	R^2	0.07	0.16	0.47	0.34	0.69	0.64	0.61

[a]Note: n-RMSE, the normalized-root mean square err; d, the index of agreement; R^2, the sample correlation coefficient.

primarily determined by the level of photosynthetically active radiation, the maximum net assimilation rate of photosynthesis, the efficiency of conversion of carbohydrate into plant constituents, and the maintenance respiration rate [30]. Thus, the parameter for maximum potential production (PRDX(1)) had both genetic and environmental components. However, in CENTURY, the seasonal distribution of production was primarily controlled by the temperature response function rather than the seasonal variation in photosynthetically active radiation, so the maximum potential production parameter would reflect aboveground crop production in optimal summer conditions. This parameter would frequently be used to calibrate the predicted crop production for different environments, species, and varieties [31].

The CENTURY model had no function for pH effects on aboveground growth. Based on the 20 years of field measurement at the Qiyang site, soil pH had no impact on crop growth as long as the value remains above pH 5.5, and would completely stop growth in wheat and corn if the value declined to less than pH 4.0 [32]. We added the following soil pH factor (f_{pH}) in the crop growth sub-model to account for this effect:

$$\begin{cases} f_{pH} = 0 & \text{if } pH \leq 4.0 \\ f_{pH} = \dfrac{(pH - \delta)}{(\delta - \beta)} + 1 & \text{if } 4.0 \leq pH \leq 5.5 \\ f_{pH} = 1 & \text{if } pH \geq 5.5 \end{cases} \quad (1)$$

where δ and β were the maximum (i.e., $\delta = 5.5$) and minimum (i.e., $\beta = 4.0$) pH value that effected plant growth, respectively.

Model Evaluation

We compared the simulated SOC (0–20 cm) with field measured values. Both visual examination of graphic output and several statistical tests were used to evaluate the CENTURY model performance. Four statistical parameters were selected [33–36] for the evaluation, including (i) the sample coefficients of determination (R^2), (ii) the normalized-root mean square error (n-RMSE, equation 2), and (iii) the index of agreement (d, equation 3), as a descriptive measure of the average relative error [37].

$$n - RMSE = \frac{1}{\bar{M}} \sqrt{\sum_{i=1}^{n} (S_i - M_i)^2 / n} \quad (2)$$

where M_i were the measured values, S_i were the simulated values, $\bar{M}$ was the mean of the measured data and n was the number of the paired values. The index of agreement is computed with the following equation:

$$d = 1 - \frac{\sum_{i=1}^{n} (M_i - S_i)^2}{\sum_{i=1}^{n} (|S_i'| + |M_i'|)^2} \quad (3)$$

where $S_i' = S_i - \bar{M}$ and $M_i' = M_i - \bar{M}$.

The classical R^2 statistic ($0 < R^2 < 1$) provided the percentage of data variance that was accounted for by the model. The RMSE evaluated the difference between observed and modeled values in the original units of the data, while the normalized-RMSE (n-RMSE, equation 2) removed the influence of the units, and placed the results on a percentage scale for comparison of model performance among variables with different units. For an ideal fit,

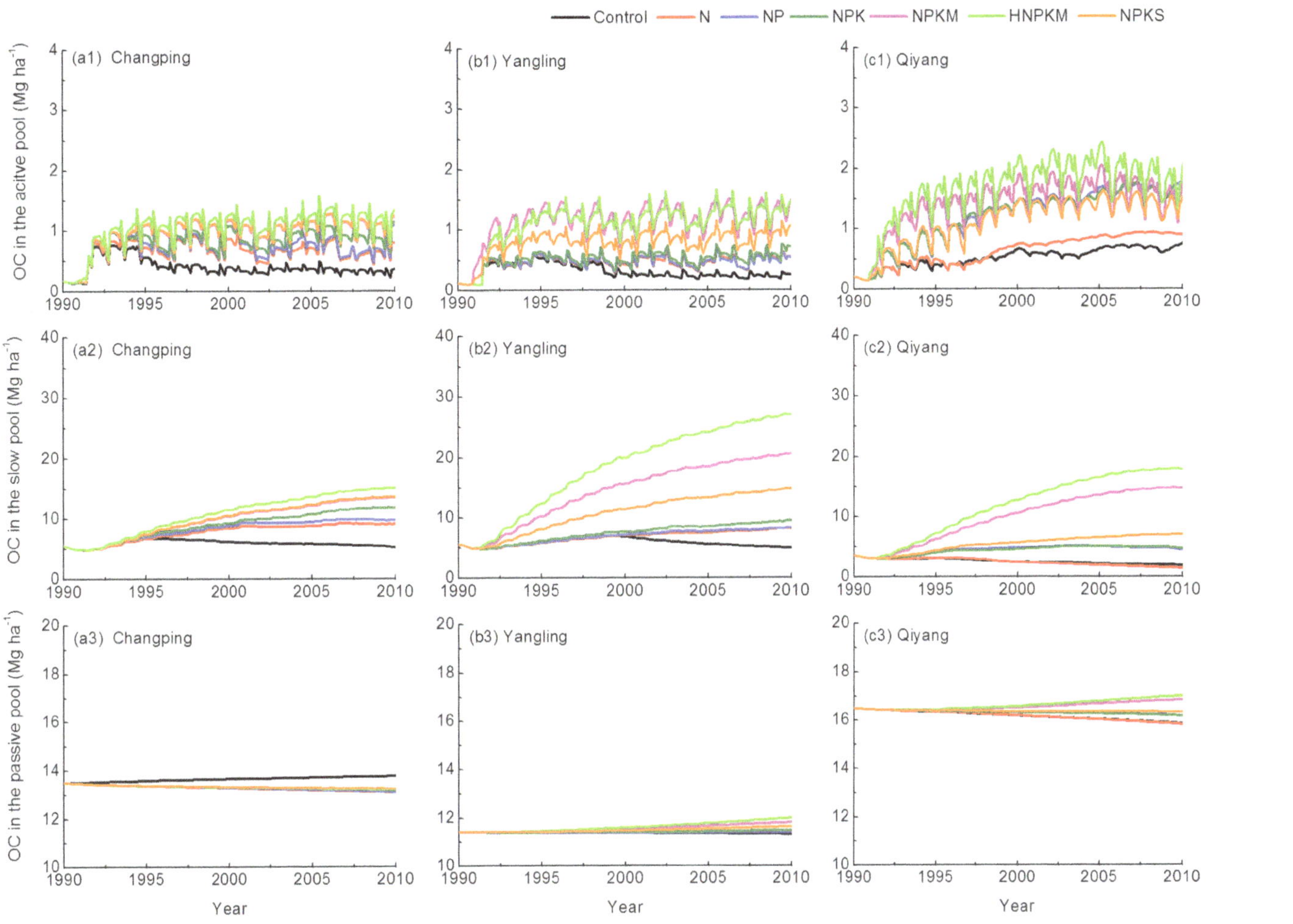

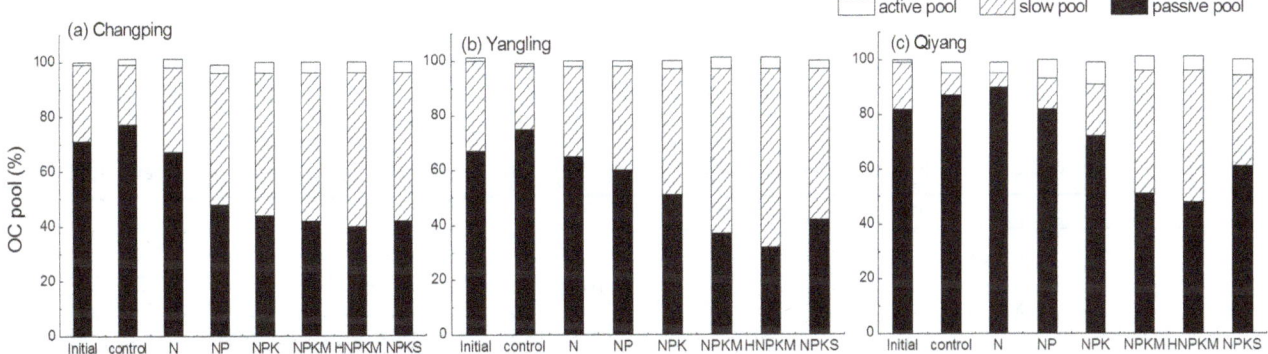

Figure 8. Soil organic carbon in the (1) active, (2) slow, and (3) passive pool under the control, N, NP, NPK, NPKM, hNPKM and NPKS treatments at (A) Changping, (B) Yangling, and (C) Qiyang sites.

R^2 would equal 100%, $RMSE$ would equal zero and d would equal 1. In our study, $RMSE < 15\%$, $d > 0.5$ and significant R^2 were used to bound the best case parameter sets for both calibration and validation simulations [38].

Results

Model Evaluation

We evaluated the model results for both grain yields and the SOC stock predictions. We were focusing on the SOC stock predictions of the model. In summary for the grain yield results (Fig. 3), we found that the model had correlation coefficients (R^2)

Figure 9. Proportion of changed SOC between the 1990 (initial) and 2010 in each soil organic matter pool under the control, N, NP, NPK, NPKM, hNPKM and NPKS treatments at (A) Changping, (B) Yangling, and (C) Qiyang sites.

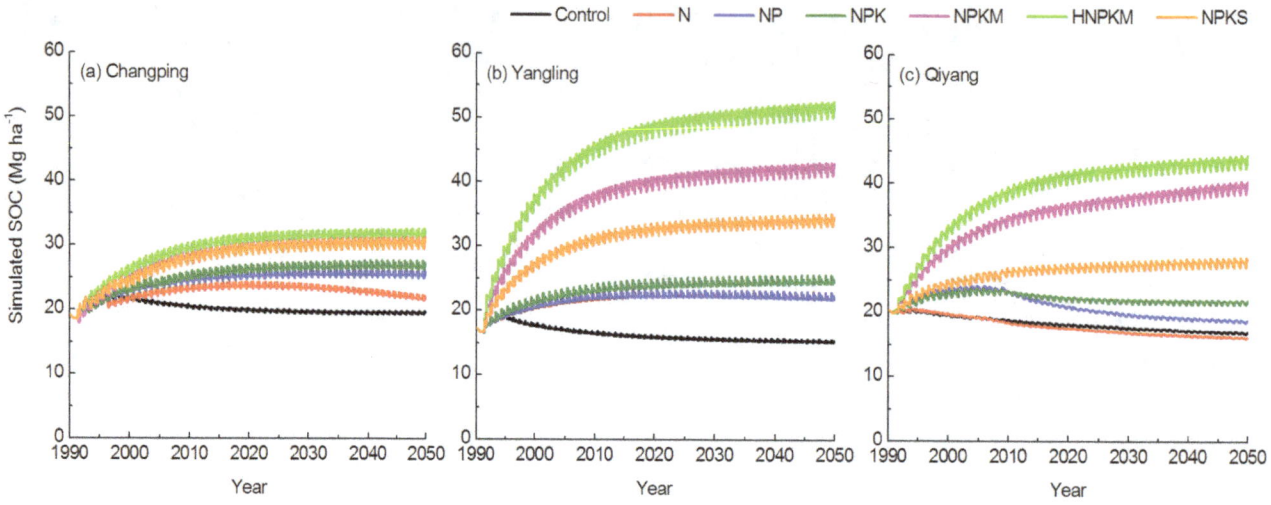

Figure 10. Soil organic carbon dynamics under the control, N, NP, NPK, NPKM, hNPKM and NPKS treatments at (A) Changping, (B) Yangling, and (C) Qiyang sites.

ranging from 0.57 to 0.92 ($p<0.05$), indicating that CENTURY model simulates the grain yields with reasonable accuracy.

The modeled SOC follows the trend of the measurements reasonably well in all treatments across the sites (Figs. 4, 5, and 6). Model results and measured SOC demonstrated that carbon levels decrease in the control treatment at all sites but increase under the fertilization treatments, except N only. For the N treatment, SOC is maintained at a stable level at most sites but did decline at the Qiyang site. The reason was related to a low soil pH that caused a decline in crop production and thus decline in residue input to the soil [25].

Linear regressions of observed vs. simulated SOC stocks (Fig. 7) are all highly significant ($P<0.001$) under various fertilization treatments. Approximately 77%–84% of the variability in the observed SOC stocks can be explained by the simulation. Therefore, in our experiment, the CENTURY model performs well and is suitable for predicting SOC dynamics for different types of N fertilizer (i.e., manure *versus* mineral) and N rates. The coefficients of determination (R^2) are higher at the Yangling and Qiyang sites than the Changping site.

RMSE is useful for evaluation of precision in model results [33]. Several studies have emphasized the need to estimate the precision in model predictions to evaluate model uncertainty [14,39,40]. In this study, we used several parameters for quantitative statistical analysis of measured and simulated SOC (Table 4). The *n-RMSE* ranges from 4% to 7% at the Changping site, 9%–14% at the Yangling site, and 6%–11% at the Qiyang site, respectively.

Similarly, Fallon and Smith [41] calculated *n-RMSE* ranging from 1.8% to 16.4% when they modeled SOC in several long-term experiments in Europe. Also Álvaro et al. [20] had a low *n-RMSE* (below 6%) when they used the CENTURY model to evaluate tillage effects on SOC. Overall, the *n-RMSE* value is less than 15% for all the treatments in our study, demonstrating that the model is relatively precise in predicting the trends for different fertilization practices that are common in Chinese agricultural systems.

The *d* values and coefficients of determination (R^2) are highest in the NPKM, hNPKM and NPKS treatments at most sites. However, the R^2 value was much lower in the control (~0.07) or N (~0.19) treatments. The reason was that inter-annual fluctuations in crop yield were larger in the control and N treatments due to the long-term nutrient deficiency [25]. Therefore, carbon input from biomass was fluctuated and thus the simulation was not satisfied as other treatments.

Fertilization Effects on Organic Carbon Pools

Long-term fertilization has a variety of effects on SOC pools (Fig. 8 and 9). Generally, the active pool is the smallest carbon pool [16], and our results are consistent with past studies with an active pool containing only 1–3 Mg ha^{-1} across the sites (first row in Fig. 8). By 2010, the largest active pool is at the Qiyang site (Fig. 8_c1), followed by Changping (Fig. 7_a1) and Yangling (Fig. 8_b1) sites.

Long-term manure fertilization significantly increased the proportion of slow soil organic matter at all sites (Fig. 8). By the

Table 5. Estimated carbon sequestration potential (0–20 cm) under long-term hNPKM and NPKS fertilization treatments.

Site	SOC in 1990 (Mg ha^{-1})	SOC in 2050 (Mg ha^{-1})		Carbon sequestration potential (Mg ha^{-1})	
		NPKS	hNPKM	NPKS	hNPKM
Changping	22.4	30.5	31.8	8.1	9.4
Yangling	16.4	34.3	52.1	17.9	35.7
Qiyang	20.5	28.2	44.2	7.7	23.7

end of 2010, the slow pool accounted for ~42% of the total SOC at the Changping site, ~65% at the Yangling site, and ~48% at the Qiyang site (Fig. 9). In addition, trends in the slow pool are consistent with the overall change in total SOC stock at each of the sites (Figs. 4, 5, and 6). These results suggest that the carbon dynamics associated with the slow pool are key drivers of the trend in total SOC.

The percentage of SOC in the passive soil organic matter is 75%–87% under the control treatment but only 32%–48% under the hNPKM treatment (Fig. 9). Our modeling results suggest that a higher proportion of the organic carbon is contained in the passive pool with the slowest turnover rate if no fertilizer is added. However, long-term fertilization generally has a smaller effect on the passive pool compared to the slow pool according to the experimental data (third row in Fig. 8). The reason would be corresponded to turnover times of 25 years for the slow pool but 1000 years for the passive pool in the CENTURY [28].

Soil Carbon Dynamics and Carbon Sequestration Potential

Long-term fertilization has different effects on soil organic carbon across the three sites (Fig. 10). For example, the SOC stock (0–20 cm) increases and then decreases to a relative stable level in the control and N treatments at Changping site (Fig. 10a). Plots with balanced fertilization or manure application and straw return (i.e., NPK, NPKM, hNPKM and NPKS) have an increase in SOC and then remain relatively stable. By 2050, the SOC stocks reach 30.5–31.8 Mg ha^{-1} with organic fertilization and 26.8 Mg ha^{-1} with chemical fertilization (NPK).

For the Yangling site (Fig. 10b), SOC contents increase over time in the NPK, NPKM, hNPKM, and NPKS treatments. By the end of simulated period, SOC reaches 42.6 Mg ha^{-1} and 52.1 Mg ha^{-1} under the NPKM and hNPKM treatments, respectively. With straw return, the SOC level reaches 34.3 Mg ha^{-1}. Balanced chemical fertilization (NPK) also yields relatively high SOC contents at 24.6 Mg ha^{-1} by 2050. There is little difference in SOC stocks between N and NP treatments. SOC increases to 20.1 Mg ha^{-1} in 1994, then declines to 15.3 Mg ha^{-1} in 2050 for the control treatment.

Similar to the Yangling site, SOC levels increase through the entire period of simulation under the NPKM, hNPKM, and NPKS treatments for the Qiyang site (Fig. 10c). By 2050, SOC stocks are 40.2 Mg ha^{-1}, 44.2 Mg ha^{-1}, and 28.2 Mg ha^{-1} for the NPKM, hNPKM, and NPKS treatments, respectively. For all chemical fertilization treatments, we assume no decrease in pH after year 2010. Therefore, soil pH has not negative impact on carbon input due to a reduction in crop production associated with pH, and applying NPK fertilizers maintain SOC levels at 21.7 Mg ha^{-1}.

The SOC trends are relatively stable by the end of 2050 in the different treatments (Fig. 10). In order to estimating the carbon sequestration potential (i.e., the SOC increment by the end of simulation) with manure addition and regular straw incorporation, we further evaluated the SOC change under the hNPKM and NPKS treatments in the period of 1990–2050 (Table 5). The initial SOC level is highest at the Changping site (22.4 Mg ha^{-1}) and lowest at the Yangling site (16.4 Mg ha^{-1}). In contrasts, the highest simulated SOC level in 2050 happens at the Yangling site and lowest at the Changping site. Compared with the NPKS treatment, carbon sequestration potential is greater with hNPKM management, ranging from 9.4 Mg ha^{-1} to 35.7 Mg ha^{-1}. For the NPKS treatment, SOC increases to 17.9 Mg ha^{-1} at the Yangling site but only 7.1 Mg ha^{-1}–8.1 Mg ha^{-1} at the other two sites.

Discussion

Studies has proved that optimizing fertilization managements can raise crop yield as well as biomass, which would enhance biomass input into the soil from crop straw and roots [28,42,43]. Halvorson et al [44] reported that increasing nitrogen fertilizer rate would increase SOC levels in the 0–7.5 cm soil depth after 11 years cropping. In our study, chemical fertilization (expect N treatment) would also increase crop yield and thus increase SOC in the first 20 years of crop rotation at two of three sites. However, most chemical fertilization treatment would only maintain the SOC level in the long-term run (Fig. 10).

Straw incorporation would also increase carbon input to the soil. Li et al [45] found that cropland soil lost 1.6% of SOC due the less (25%) aboveground residue's returning. Using the DNDC model, Tang et al. [46] predicted that the soil carbon loss in China's cropland would be partly reversed if straw return and no tillage practices were expanded widely. In our study, averaged carbon sequestration rate would reach to 118 kg ha^{-1} yr^{-1} to 298 kg ha^{-1} yr^{-1} with straw returning during the period of 1990–2050 under current returning rate of straw residue (i.e., 2000–4500 kg ha^{-1} yr^{-1}). The results are lower than the estimated rate (108–728 kg ha^{-1} yr^{-1}) in the similar region by Lu et al [47], in which the estimated straw returning rates were also higher (4530–8110 kg ha^{-1} yr^{-1}).

Moreover, there are differences in the influence of manure on SOC trends among the sites. At the Changping site, SOC stocks reach a relatively stable level (30.0 Mg ha^{-1}–31.0 Mg ha^{-1}) by 2013 (Fig. 10a). In contrast, plots from the other two sites continue to increase in SOC over the entire simulated period. The reason for the site difference in the simulated trends is related to the manure quality between the sites, and the associated decomposition rates of the organic matter [28,48]. For the Changping site, farmyard manure is livestock manure mixed with soil and/or crop residue. The lignin content in farmyard manure is lower and decompose at a faster rate [49]. When we enhanced the lignin content of farmyard manure in CENTURY, the simulation of SOC rose to the much higher level (data not shown).

Many modeling research also focus on the regional estimation of SOC sequestration. Follett [50] estimated soil carbon sequestration potential would be 11–56 Tg C yr^{-1} for residue management in the USA. In China, Yan et al [51] estimated 32.5 Tg C yr^{-1} to be sequestered in croplands if 50% of the crop residue was returned to soils, and no tillage was adopted on 50% of the arable lands. However, these results are criticized for the models that are calibrated against few or no field experimental data. Our study estimated carbon sequestration potential using long-term experiment data, providing credible data to evaluate carbon sequestration response to different fertilization managements.

Conclusion

We used the CENTURY model (version 4.5) to simulate soil organic carbon (SOC) dynamics for long-term fertilization experiments with wheat-corn cropping systems in China. Our study indicated that CENTURY can simulate fertilization effects on SOC trends for different climatic conditions and soil properties that are common for wheat-corn systems in China. For most sites, the model was more precise in predicting the SOC trends for treatments with balanced nutrient additions (NPK, NPKM, hNPKM, and NPKS) than the control and unbalanced fertilization treatments (N and NP). After simulating SOC dynamics from 1990–2050, the SOC levels increased from 31.8–52.1 Mg ha^{-1} across the four sites. We estimate that carbon sequestration potential ranges from 9.4–35.7 Mg ha^{-1} under the high manure application practice (i.e., hNPKM treatment). Analysis of the proportion of organic carbon in each of the soil organic matter pools indicates that long-term fertilization enhances the slow pool and leads to a decline in the proportion of carbon in the passive pool. Model results suggest that changes in slow organic matter drive the overall trends in SOC stocks associated with long-term fertilization.

References

1. Schlesinger WH (1997) Biogeochemistry: an analysis of global change: New York Academic Press.
2. Schuman GE, Janzen HH, Herrick JE (2002) Soil carbon dynamics and potential carbon sequestration by rangelands. Environmental pollution 116: 391–396.
3. Manzoni S, Porporato A (2009) Soil carbon and nitrogen mineralization: Theory and models across scales. Soil Biology & Biochemistry 41: 1355–1379.
4. Lal R, Follett F, Stewart BA, Kimble JM (2007) Soil carbon sequestration to mitigate climate change and advance food security. Soil Science 172: 943–956.
5. Lal R (2004) Soil Carbon Sequestration Impacts on Global Climate Change and Food Security. Science 304: 1623–1627.
6. Körschens M (2006) The importance of long-term field experimentsfor soil science and environmental research - a review. Plant, soil and environment 52: 1–8.
7. Zhang W, Xu M, Wang X, Huang Q, Nie J, et al. (2012) Effects of organic amendments on soil carbon sequestration in paddy fields of subtropical China. Journal of Soils and Sediments: 1–14.
8. Powlson D, Zhao BQ, Li XY, Li XP, Shi XJ, et al. (2010) Long-Term Fertilizer Experiment Network in China: Crop Yields and Soil Nutrient Trends. Agronomy Journal 102: 216–230.
9. Edmeades DC (2003) The long-term effects of manures and fertilisers on soil productivity and quality: a review. Nutrient Cycling in Agroecosystems 66: 165–180.
10. Thomsen IK, Christensen BT (2004) Yields of wheat and soil carbon and nitrogen contents following long-term incorporation of barley straw and ryegrass catch crops. Soil Use and Management 20: 432–438.
11. Bruun S, Christensen BT, Hansen EM, Magid J, Jensen LS (2003) Calibration and validation of the soil organic matter dynamics of the Daisy model with data from the Askov long-term experiments. Soil Biology & Biochemistry 35: 67–76.
12. Shirato Y, Yokozawa M (2005) Applying the Rothamsted Carbon Model for long-term experiments on Japanese paddy soils and modifying it by simple mining of the decomposition rate. Soil Science and Plant Nutrition 51: 405–415.
13. Yang Y, Luo Y, Finzi AC (2011) Carbon and nitrogen dynamics during forest stand development: a global synthesis. New Phytologist 190: 977–989.
14. Smith P, Smith JU, Powlson DS, McGill WB, Arah JRM, et al. (1997) A comparison of the performance of nine soil organic matter models using datasets from seven long-term experiments. Geoderma 81: 153–225.
15. Chilcott CR, Dalal RC, Parton WJ, Carter JO, King AJ (2007) Long-term trends in fertility of soils under continuous cultivation and cereal cropping in southern Queensland. IX*. Simulation of soil carbon and nitrogen pools using CENTURY model. Australian Journal of Soil Research 45: 206–217.
16. Denef K, Six J, Merckx R, Paustian K (2004) Carbon sequestration in microaggregates of no-tillage soils with different clay mineralogy. Soil Science Society of America Journal 68: 1935–1944.
17. Kelly RH, Parton WJ, Crocker GJ, Grace PR, Klir J, et al. (1997) Simulating trends in soil organic carbon in long-term experiments using the century model. Geoderma 81: 75–90.
18. Falloon P, Smith P (2002) Simulating SOC changes in long-term experiments with RothC and CENTURY: model evaluation for a regional scale application. Soil Use and Management 18: 101–111.
19. Bhattacharyya T, Pal DK, Williams S, Telpande BA, Deshmukh AS, et al. (2010) Evaluating the Century C model using two long-term fertilizer trials representing humid and semi-arid sites from India. Agriculture, Ecosystems & Environment 139: 264–272.
20. Álvaro-Fuentes J, López MV, Arrúe JL, Moret D, Paustian K (2009) Tillage and cropping effects on soil organic carbon in Mediterranean semiarid agroecosystems: Testing the Century model. Agriculture, Ecosystems & Environment 134: 211–217.
21. Cerri CEP, Easter M, Paustian K, Killian K, Coleman K, et al. (2007) Simulating SOC changes in 11 land use change chronosequences from the Brazilian Amazon with RothC and Century models. Agriculture, Ecosystems & Environment 122: 46–57.
22. Feng XM, Zhao YS (2011) Grazing intensity monitoring in Northern China steppe: Integrating CENTURY model and MODIS data. Ecological Indicators 11: 175–182.
23. Fang D, Zhou G, Jiang Y, Jia B, Xu Z, et al. (2012) Impact of fire on carbon dynamics of Larix gmelinii forest in Daxingan Mountains of Northeast China:A simulation with CENTURY model. The Journal of Applied Ecology 23: 2411–2421.
24. Wang SH, Shi XZ, Zhao YC, Weindorf DC, Yu DS, et al. (2011) Regional Simulation of Soil Organic Carbon Dynamics for Dry Farmland in East China by Coupling a 1:500 000 Soil Database with the Century Model. Pedosphere 21: 277–287.
25. Cong RH, Wang XJ, Xu MG, Zhang WJ, Xie LJ, et al. (2012) Dynamics of soil carbon to nitrogen ratio changes under long-term fertilizer addition in wheat-corn double cropping systems of China. European Journal of Soil Science 63: 341–350.
26. Cong R, Xu M, Wang X, Zhang W, Yang X, et al. (2012) An analysis of soil carbon dynamics in long-term soil fertility trials in China. Nutrient Cycling in Agroecosystems 93: 201–213.
27. Walkley A, Black IA (1934) An examination of Degtjareff method for determining soil organic matter and a proposed modification of the chromic acid titration method. Soil Science 37: 29–38.
28. Parton WJ, Rasmussen PE (1994) Long-Term Effects of Crop Management in Wheat-Fallow: II. CENTURY Model Simulations. Soil Science Society of America Journal 58: 530–536.
29. Alister K Metherell, Laura A Harding, C. Vernon Cole, Parton WJ (1993) CENTURY Soil Organic Matter Model Environment.
30. van Heemst HDJ (1986) Physiological principles. In: Keulen; HV, Wolf J, editors. Modelling of agricultural production: weather, soils and crops. Pudoc Wageningen. pp. 13–26.
31. Metherell AK, Harding LA, Cole CV, Parton WJ (1993) Plant Production Submodels. CENTURY Soil Organic Matter Model Environment Technical Documentation Agroecosystem Version 40.
32. Zhang H, Wang B, Xu M (2008) Effects of Inorganic Fertilizer Inputs on Grain Yields and Soil Properties in a Long-Term Wheat-corn Cropping System in South China. Communications in Soil Science and Plant Analysis 39: 1583–1599.
33. Smith JU, Smith P, Addiscott TM (1996) Quantitative methods to evaluate and compare soil organic matter (SOM) models. In: Powlson DS, Smith P, Smith JU, editors. Evaluation of Soil Organic Matter Models Using Existing, Long-term Datasets. Heidelberg: Springer-Verlag. pp. 181–200.
34. Cheng WX, Zhang QL, Coleman DC, Carroll CR, Hoffman CA (1996) Is available carbon limiting microbial respiration in the rhizosphere? Soil Biology & Biochemistry 28: 1283–1288.
35. Kong AYY, Six J, Bryant DC, Denison RF, van Kessel C (2005) The Relationship between Carbon Input, Aggregation, and Soil Organic Carbon Stabilization in Sustainable Cropping Systems. Soil Sci Soc Am J 69: 1078–1085.
36. Domke GM, Woodall CW, Walters BF, Smith JE (2013) From Models to Measurements: Comparing Downed Dead Wood Carbon Stock Estimates in the U.S. Forest Inventory. PLoS One 8: e59949.
37. Willmott CJ (1982) Some Comments on the Evaluation of Model Performance. Bulletin of the American Meteorological Society 63: 1309–1313.
38. Tonitto C, David MB, Drinkwater LE, Li CS (2007) Application of the DNDC model to tile-drained Illinois agroecosystems: model calibration, validation, and uncertainty analysis. Nutrient Cycling in Agroecosystems 78: 51–63.
39. Blagodatskaya E, Kuzyakov Y (2008) Mechanisms of real and apparent priming effects and their dependence on soil microbial biomass and community structure: critical review. Biology and Fertility of Soils 45: 115–131.
40. Ogle SM, Breidt FJ, Easter M, Williams S, Paustian K (2007) An empirically based approach for estimating uncertainty associated with modelling carbon sequestration in soils. Ecological Modelling 205: 453–463.
41. Falloon P, Smith P (2003) Accounting for changes in soil carbon under the Kyoto Protocol: need for improved long-term data sets to reduce uncertainty in model projections. Soil Use and Management 19: 265–269.
42. Halvorson AD, Wienhold BJ, Black AL (2002) Tillage, nitrogen, and cropping system effects on soil carbon sequestration. Soil Science Society of America Journal 66: 906–912.

Acknowledgments

The authors would like to thank Dr. Sigen Chen and Ms. Cynthia Keough for the technical support on the CENTURY model. We acknowledge all the colleagues for their unremitting efforts on the long-term experiments from these sites.

Author Contributions

Conceived and designed the experiments: RHC MGX XJW. Performed the experiments: RHC SMO WJP. Analyzed the data: RHC MGX XJW. Contributed reagents/materials/analysis tools: SMO WJP. Wrote the paper: RHC.

43. Dumanski J, Desjardins RL, Tarnocai C, Monreal C, Gregorich EG, et al. (1998) Possibilities for future carbon sequestration in Canadian agriculture in relation to land use changes. Climatic Change 40: 81–103.

44. Halvorson AD, Reule CA, Follett RF (1999) Nitrogen fertilization effects on soil carbon and nitrogen in a dryland cropping system. Soil Science Society of America Journal 63: 912–917.

45. Li C, Zhuang Y, Frolking S, Galloway J, Harriss R, et al. (2003) Modeling Soil Organic Carbon Change in Croplands of China. Ecological Applications 13: 327–336.

46. Tang HJ, Qiu JJ, Van Ranst E, Li CS (2006) Estimations of soil organic carbon storage in cropland of China based on DNDC model. Geoderma 134: 200–206.

47. Lu F, Wang XK, Han B, Ouyang ZY, Duan XN, et al. (2009) Soil carbon sequestrations by nitrogen fertilizer application, straw return and no-tillage in China's cropland. Global Change Biology 15: 281–305.

48. Paustian K, Parton W, Persson J (1992) Modeling Soil Organic Matter in Organic-Amended and Nitrogen-Fertilized Long-Term Plots. Soil Science Society of America Journal 56: 476–488.

49. Zhang WJ, Wang XJ, Xu MG, Huang SM, Liu H, et al. (2010) Soil organic carbon dynamics under long-term fertilizations in arable land of northern China. Biogeosciences 7: 409–425.

50. R.F F (2001) Soil management concepts and carbon sequestration in cropland soils. Soil and Tillage Research 61: 77–92.

51. Yan HM, Cao MK, Liu JY, Tao B (2007) Potential and sustainability for carbon sequestration with improved soil management in agricultural soils of China. Agriculture Ecosystems & Environment 121: 325–335.

11

Microbial Growth and Carbon Use Efficiency in the Rhizosphere and Root-Free Soil

Evgenia Blagodatskaya[1,2,3]*, Sergey Blagodatsky[2,4], Traute-Heidi Anderson[5], Yakov Kuzyakov[1,3]

1 Soil Science of Temperate Ecosystems, Büsgen-Institute, University of Göttingen, Göttingen, Germany, **2** Institute of Physicochemical and Biological Problems in Soil Science, Russian Academy of Sciences, Pushchino, Russia, **3** Agricultural Soil Science, Büsgen-Institute, University of Göttingen, Göttingen, Germany, **4** Institute for Plant Production and Agroecology in the Tropics and Subtropics, University of Hohenheim, Stuttgart, Germany, **5** Thünen-Institute of Climate-Smart Agriculture (vTI), Braunschweig, Germany

Abstract

Plant-microbial interactions alter C and N balance in the rhizosphere and affect the microbial carbon use efficiency (CUE)– the fundamental characteristic of microbial metabolism. Estimation of CUE in microbial hotspots with high dynamics of activity and changes of microbial physiological state from dormancy to activity is a challenge in soil microbiology. We analyzed respiratory activity, microbial DNA content and CUE by manipulation the C and nutrients availability in the soil under *Beta vulgaris*. All measurements were done in root-free and rhizosphere soil under steady-state conditions and during microbial growth induced by addition of glucose. Microorganisms in the rhizosphere and root-free soil differed in their CUE dynamics due to varying time delays between respiration burst and DNA increase. Constant CUE in an exponentially-growing microbial community in rhizosphere demonstrated the balanced growth. In contrast, the CUE in the root-free soil increased more than three times at the end of exponential growth and was 1.5 times higher than in the rhizosphere. Plants alter the dynamics of microbial CUE by balancing the catabolic and anabolic processes, which were decoupled in the root-free soil. The effects of N and C availability on CUE in rhizosphere and root-free soil are discussed.

Editor: Jeffrey L. Blanchard, University of Massachusetts, United States of America

Funding: This authors acknowledge the following: the European Commission (Marie Curie IIF program, project IIF 039907-MICROSOM) for supporting EB (http://ec.europa.eu/research/mariecurieactions/); the Alexander von Humboldt Foundation for supporting SB (http://www.humboldt-foundation.de/web/start.html); Russian Academy of Sciences (Scientific School Program 6123.2014.4, https://grants.extech.ru/grants/res/winners.php?OZ = 4&TZ = S&year = 2014); Russian Foundation for Basic Research (grant No 12-04-01170, http://www.rfbr.ru/rffi/ru/); and German Research Foundation (DFG) within project KU 1184/13-1/2 (http://www.dfg.de/en/). The funders had no role in study, design, data collection and analysis, decision to manuscript, or preparation of the manuscript.

Competing Interests: The authors have declared that no competing interests exist.

* E-mail: janeblag@mail.ru

Introduction

Analysis of microbial carbon use efficiency (CUE) and microbial turnover rates are critical for accounting of C balance in soil with the goal of correct estimation of C sequestration potential as well as for modelling the turnover of soil C and CO_2 fluxes [1–3]. The efficiency of microbial growth on a carbonaceous substrate coming with plant residues is positively related to formation rates of soil organic carbon [4]. A magnitude and dynamics of CUE is a function of numerous physical, chemical and ecological factors, e.g. soil quality [5], microbial community composition [6], [7], substrate and nutrient availability [3], [8], etc. At that the factor specific mechanisms, which control the CUE, remain uncertain [9]. This calls for the case studies under control conditions, so that the number of influencing factors can be reduced. So, preferential objects for CUE studies are the soils similar in physico-chemical characteristics but contrasting in substrate availability: e.g. rhizosphere and root-free soil. Higher microbial abundance and diversity and faster microbial growth occur in the rhizosphere soil as compared to root-free soil [10], [11] due to the high availability of C exuded by roots [1], [12]. Contrary to this, permanent limitation by available substrates in root-free soil leads to the selection of microorganisms with slower growth rates and more efficient metabolism [13]. So, rhizosphere and root-free soil can

serve as good model for an *in situ* comparison of microbial physiology and CUE in microhabitats with contrasting resource levels.

CUE has become a very popular but ambiguous term in soil science. It is often used with a broad meaning, combining the efficiency of growth and the efficiency of maintenance of soil microorganisms [3]. Here, we introduce basic terms and approaches applicable either for distinct growth or for sustaining microbial biomass.

CUE Estimation for Growing Microbial Biomass

During microbial growth, CUE is equivalent to the microbial yield coefficient (Y, g C_{mic} g^{-1} C_s), i.e. biomass-C increment per amount of substrate-C used (Eq. 1, [14]):

$$Y = -\frac{\Delta C_{mic}}{\Delta C_s} \qquad (1)$$

where ΔC_{mic} is the increase in microbial biomass-C caused by the consumption of substrate-C ΔC_s. So, for **estimation of CUE for growing microbial biomass,** we used the microbial yield coefficient (Y). In spite of wide variability of the experimental Y estimations in the range of 0.1 to 0.8 [6], [15], [16] and a maximal theoretical value of 0.62 for glucose [17], the fixed value of

$Y = 0.45$ is often assumed in soil studies and models [1], [18]. Considering very high variation (about 8 times) such a rough overall assumption of the average of 0.45 applied for different soils can distort the estimations and predictions of C stocks and fluxes [5], [18].

CUE Estimation under Steady-state Conditions

In the absence of microbial growth, the estimation of Y (Eq. 1) is not applicable. However, even without distinct exponential growth, the substrate can be used both for maintenance and for the very slow, "cryptic" growth [19], so that microbial biomass does not decrease in time. Under such steady-state conditions, the estimation of the efficiency of microbial metabolism by specific respiration (CO_2 produced per time and microbial biomass unit) can be used as a physiological characteristic.

The dormancy or maintenance state of microbial community reveals itself as a low respiration-to-biomass ratio which has been suggested as a physiological index of soil microbial communities [20]. The maintenance requirements are higher for microorganisms adapted to permanent input of available substrates than for microbial communities from nutrient-limited microhabitats [21]. The similar relationship is valid for growth expenses: the amount of respired CO_2 during growth is larger for microbial communities with a higher growth rate and comparatively less efficient metabolism [22]. So, we hypothesised that both in the presence and absence of an available substrate, microbial communities in rhizosphere soil will have higher specific respiration rates than those in root-free soil.

CUE Estimation during Shift from Dormancy to Active Stage

It is important to consider the CUE not only as a growth parameter (Y) and as a dormancy characteristic (maintenance coefficient), but also as the amount of CO_2 produced per biomass unit in the course of the famine-to-feast transition. How such a transition alters CUE dynamics under changing environmental conditions, i.e. from substrate-limited to substrate-rich microhabitats, remains unclear. In contrast to steady-state or growth conditions where CUE remains constant, the experimental estimation of CUE during the famine-to-feast microbial transition remains a challenge for environmental microbiology. This is because the application of standard methods (fumigation-extraction or substrate-induced respiration) is restricted for biomass assessment in growing microbial communities.

A strong positive correlation between DNA and microbial C in soil [5], [23–25] led us use the DNA content as a proxy of microbial biomass. The increase in microbial DNA content corresponds to the respiratory response during exponential microbial growth after substrate addition [24], [26]. Therefore, we used the CO_2/DNA ratio for comparison of the CUE by transition from dormant to active stage for microbial communities with contrasting growth strategies. Experimentally, the growth strategies can be evaluated by the maximal specific growth rate under unlimited conditions that is greater for r- than for K-strategists [27], [28]. So, we used two parameters of microbial metabolism: microbial maximal specific growth rates and CUE, to evaluate the relative abundance of slow- or fast-growing microorganisms in rhizosphere and root-free soil.

Nitrogen Effect on CUE

The efficiency of microbial metabolism depends strongly on nitrogen (N) availability [29]. Lower respiration due to higher efficiency of microbial C reutilisation has been observed in the absence of N limitation as compared to N-limited conditions [30]. Nitrogen addition reduces cumulative microbial respiration in soil amended with glucose [31] and plant litter [32] and increased the growth yield efficiency [18]. While the CUE decline under N limitation is commonly expected [3], it is unknown whether N availability affects equally microbial respiration and growth rates in microhabitats with contrast substrate availability, e.g. in root-free and rhizosphere soil [33]. Therefore, we compared the specific respiration and microbial growth kinetics in the root-free and in rhizosphere soil with different N fertilization rates. We expected to find more distinct effect of N availability in the rhizosphere where microbial activity and abundance are higher and N limitation may be more important as compared to root-free soil. We hypothesized that the increase of N availability improves CUE and decreases specific respiration, especially in the rhizosphere.

We analyzed the ratio between respiration and microbial DNA content 1) under steady state conditions (in unamended soil), 2) during microbial growth in soil amended with glucose, and 3) during transition from steady state conditions to growth. In addition, effect of N availability on microbial growth rate and CUE was determined. Three complementary indices were applied as indicators of the efficiency of microbial metabolism in the rhizosphere and in root-free soil: 1) the CO_2/DNA ratio further referred to as 'specific respiration rate', 2) the ΔCO_2/ΔDNA ratio for growing biomass, and 3) CUE during microbial growth on glucose.

Materials and Methods

Soil Sampling

Soil samples were taken from the field experimental station at the Institute of Agroecology (FAL, Braunschweig, Germany). No specific permission was required as one of the co-authors (THA) had been working in the Institute of Agroecology, and soil was regularly sampled in the course of long-term field trial described elsewhere [34]. The soil is a loamy sand Haplic Cambisol (C_{org} 1.1%; N_{tot} 0.087%; pH_{CaCl2} 6.7). The plots under sugar beet (*Beta vulgaris* subsp. *rapacea* (KOCH-DÖLL, cv. Wiebke) with full and half the recommended rate of mineral N fertiliser (126 and 63 kg N ha^{-1} year^{-1}, respectively) were chosen for analysis of the N effects on microbial communities of root-free and rhizosphere soil. Soil was sampled during harvesting the sugar beet at a mature stage (age 4.5-month). Soil samples were taken from the 0–10 cm layer from five randomly chosen replicate microsites and then mixed. Rhizosphere soil was sampled at a distance 1–5 mm adjacent to the roots (i.e. collecting the soil aggregates falling off when shaking the root system), whereas root-free soil was taken between rows of sugar beets. Fine roots and other plant debris were carefully removed during sampling. No significant differences were detected in pH, C_t or N_t content of the rhizosphere and root-free soil. The soil was stored field-fresh in aerated polyethylene bags at 4°C for 1–2 weeks. Prior to analysis the soil was sieved (< 2 mm), moistured to 60% of WHC, and preincubated at 22°C for 24 h.

Soil Respiration and Chemical Analysis

Microbial biomass (C_{mic}) was determined by the initial rate of substrate-induced respiration after soil amendment with glucose and according to the equation of Anderson & Domsch [35]:

$$C_{mic}(\mu g \cdot g^{-1} soil) = (\mu l CO_2 \cdot g^{-1} soil \cdot h^{-1}) \cdot 40.04 \quad (2)$$

Rate of basal respiration (V_{basal}) was estimated for soil without glucose as the hourly mean of 10 h of CO_2 evolution at 22°C, after 2–3 hours diminishing of the initial CO_2 flush caused by soil disturbance during sample preparation [36]. The CO_2 emission rate (V_{CO2}) was measured hourly at 22°C using an automated infrared-gas analyser system [37].

Soil organic C and total N were analysed by dry combustion (C-IR 12, Leco, and Macro-N, Hereaus, respectively). Soil pH was measured in 0.01 M $CaCl_2$ with a soil-to-solution ratio of 1:2.

Total DNA

Quantity of double-stranded DNA was determined by direct DNA isolation from the soil with mechanic and enzymatic disruption of microbial cell walls and subsequent spectrofluorimetric detection with PicoGreen [23], [24]. For rhizosphere and root-free soil from plot fertilized with 126 kg N ha^{-1} $year^{-1}$ the dsDNA determination was done at 0, 12, 15, 20, 25 and 36 hours after addition of glucose and nutrients (as described below for respiration kinetics).

The procedure of DNA isolation involved sonication of the soil suspension in Tris-EDTA buffer (TE) at pH 8, addition of aurintricarboxylic acid (a nuclease inhibitor) and sodium dodecyl sulphate. Then two cycles of quick freeze at −80°C in Deep Freezer (ProfiMaster EPF3080/N, National Lab GmbH, Mölln, Germany) for 1 h and subsequent thaw at +65°C in water bath with thermostat (Model 1002, GFL Gesellschaft für Labortechnik mbH, Burgwedel, Germany) were performed to destroy microbial cells. Enzymatic digestion was accomplished with lysozyme and Proteinase K for 1 h at 37°C. Mechanical destruction of microbial cells was implemented by shaking with sterile acid-washed glass-beads (Sigma-Aldrich, Inc.) of three sizes (710–1180, 212–300, and <106 μm) on a Vortex homogeniser at 2000 rpm. The samples were diluted with an equal volume of TE-buffer and centrifuged for 10 min at 5500 g. Half a millilitre of the diluted supernatant (1:100) was mixed with 0.5 ml of a 1:200 dilution of PicoGreen™ (Molecular Probes). After 4 min incubation, the fluorescence was measured on an SFM-25 spectrofluorimeter (Kontron, Germany) at an excitation wavelength of 480 nm and an emission wavelength of 523 nm. The dsDNA of bacteriophage lambda was used as a standard; samples for the standard curve were prepared in TE-buffer in the same way as the experimental samples.

Kinetic Parameters of Microbial Growth

Kinetics of microbial growth was determined indirectly by the rate of CO_2 emission from soil amended with glucose and mineral nutrients [38]. It has to be noted that despite substrate addition is required for the estimation of kinetic parameters (specific growth rate, active and total microbial biomass, see below), the results obtained by this approach (substrate induced growth response – SIGR) are the characteristics of the soil microbial community at the sampling instant, i.e. before substrate addition. Samples of 10 g (dry weight) soil were amended with a powder-mixture containing glucose (10 mg g^{-1}), talcum (20 mg g^{-1}) and mineral salts: $(NH_4)_2SO_4-1.9$ mg g^{-1}, $K_2HPO_4-2.25$ mg g^{-1} and $MgSO_4 \cdot 7H_2O-3.8$ mg g^{-1} [39]. These optimal concentrations of the substrates were selected in preliminary experiments and are sufficient for unlimited exponential growth of soil microorganisms at least during several hours needed for recording of respiration kinetics. Mineral salts were chosen considering the pH value and buffer capacity of the soil so that the pH was not changed more than 0.1 pH units. Soil samples were placed (in triplicate) in an ADC2250 24-channel Soil Respiration System (ADC Bioscientific, Herts, UK) at 22°C. Each sample was continuously aerated

(300 mL min^{-1}), and the rate of CO_2 production from each sample was measured every hour using an infrared detector and mass-flow meter [37].

Maximal specific microbial growth rate (μ_m) was determined by fitting the model parameters to the measured data on CO_2 production:

$$v(t) = A + B \cdot \exp(\mu_m \cdot t) \qquad (3)$$

where $v(t)$ - CO_2 evolution rate at time (t), A - initial rate of uncoupled (non-growth) respiration, B - initial rate of coupled (growth) respiration [19], [40]. Fitting was restricted to the initial phase of the curve, which corresponded to unlimited exponential growth [41]. Maximum values of statistic criteria: r^2, the fraction of total variation explained by the model were used for fitting optimisation. Further goodness of fit estimations were made and based on the Q value derived from χ^2 [42].

Activity status of the microbial biomass r_0 was calculated from the ratio of A:B [19]:

$$r_0 = \frac{B(1-\lambda)}{A + B(1-\lambda)} \qquad (4)$$

where λ may be accepted as a basic stoichiometric constant = 0.9 [19]. The total glucose-metabolizing microbial biomass (sustaining + growing; x_0) was calculated as following:

$$x_0 = \frac{B \cdot \lambda \cdot Y_{CO_2}}{r_0 \cdot \mu_m} \qquad (5)$$

where Y_{CO2} is yield of biomass C per unit of respired C-CO2.

The *growing microbial biomass* (x_0') was calculated using equation:

$$x'_0 = x_0 \cdot r_0 \qquad (6)$$

More complete theoretical background and details on equations derivation were described elsewhere [28], [38], [40].

The duration of lag-period (t_{lag}) – a period characterised by stable respiration preceding microbial growth – was defined as the time from glucose addition to the time when the increasing rate of growth-associated respiration ($B* exp(\mu_m *t)$) equalled the rate of non-growth respiration (A) [43]. The lag-period was calculated using parameters of Eq. 3:

$$t_{lag} = \ln(A/B)/\mu_m \qquad (7)$$

The ratio of *CO_2 increment-to-DNA increment* ($\Delta CO_2/\Delta DNA$) was calculated as the amount of CO_2 in μg C evolved per μg of DNA increment during the same period. The amount of respired CO_2 in soil amended with glucose was corrected for basal respiration, i.e. the corresponding amount of CO_2 respired from the unamended soil during the same period was subtracted from the CO_2 increment for glucose-amended soil.

The carbon use efficiency or CUE (in the growth phase, this is equivalent to the growth yield quotient, Y, Eq.1) was calculated as biomass C increment per amount of consumed C-substrate, which is in turn equal to biomass C increment plus CO_2 evolved:

$$CUE = \Delta C_{mic}/(\Delta C_{mic} + \Delta C_{CO_2}) \qquad (8)$$

where ΔC_{mic} is the net increase in microbial biomass C (μg C g^{-1}) and ΔC_{CO2} is the net increase in cumulative respiration (μg C g^{-1}) corrected for basal respiration. Microbial C content was calculated from mean measured DNA content found in our study (11% of dry biomass), assuming that the C content in microbial biomass is 45% [5], [44].

Statistical Analyses

The means of three replicates with standard errors are presented in tables and figures. Two-way ANOVA was applied to characterise the effects of C and N availability: 1) C availability: rhizosphere versus root-free soil, and 2) N availability: half versus full N fertilisation. When significant effects were found, a multiple comparison using the Student-Newman-Keuls test ($P<0.05$) was performed. All variables passed normality and equal variance tests.

Results

Basal Respiration Rate, DNA Content and Microbial Biomass

The basal respiration rate (V_{basal}) was significantly higher in the rhizosphere as compared to root-free soil (Fig. 1a). This rhizosphere effect amounted to 66% at the half N rate while it was only 14% at the full rate of N application. The V_{basal} in root-free soil was significantly higher at the full versus half rate of N-fertilisation (Fig. 1a). In rhizosphere soil, however, N fertilisation significantly decreased basal respiration.

Microbial DNA content was higher at the full N rate than in the corresponding treatments with the half N (Fig. 1b). Higher DNA content in rhizosphere versus root-free soil (28% at the full and 21% at the half N rate) reflects a pronounced rhizosphere effect.

Microbial respiration curves during growth on glucose were clearly different between the rhizosphere and root-free soil (Fig. 2). These differences were more pronounced under N limitation (Fig. 2). Maximal specific growth rates (μ_m) were significantly higher, while the duration of the lag-period was 1.7–1.9 h shorter in the rhizosphere than in root-free soil (Table 1).

Both the total microbial biomass C and its growing fraction were always higher in the rhizosphere as compared to root-free soil (Table 1). This rhizosphere effect was most pronounced at half versus the full N rate (Table 1) and amounted to 31% and 14% of the total microbial biomass, respectively. Actively growing microbial biomass did not exceed 0.34% of total microbial C and was much more sensitive to the presence of roots as compared to total microbial biomass. So, the rhizosphere effect for growing microbial biomass was much greater than for the total microbial biomass and amounted to 45% at full N and to 83% at the half N rate (Table 1). The direct effect of N on total microbial biomass was insignificant in rhizosphere soil, while in root-free soil significantly higher microbial biomass C was observed at the full N rate.

Two-way ANOVA confirmed the strong effects of roots of *Beta vulgaris* on all microbial parameters tested (Table 2). The portion of active microbial biomass and the lag-period were affected by roots at the largest extent: more than 90% of their variation was explained by the rhizosphere effect. The direct effect of N on the specific growth rate (μ_m) and DNA was even stronger than the effect of roots (Table 2).

We conclude that significantly higher basal respiration, DNA content and total and actively growing microbial biomass were observed in the rhizosphere versus root-free soil and this effect was more pronounced under low N fertilisation.

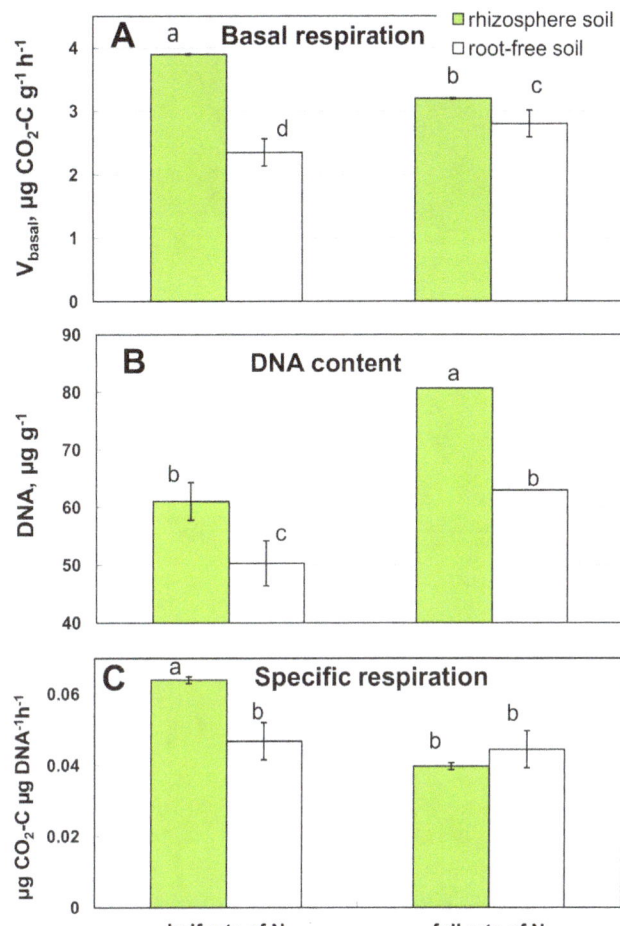

Figure 1. Respiration rate and microbial DNA in soil and rhizosphere. Basal respiration rate (a), microbial DNA content (b), and ratio of basal respiration rate (V_{basal}) to DNA content (c) of rhizosphere and root-free soil under *Beta vulgaris* at half (63 kg N ha^{-1}) and full (126 kg N ha^{-1}) rates of nitrogen fertilisation.

Respiratory Activity in Relation to DNA Content in Rhizosphere and Root-Free Soil

The CO_2/DNA ratio in the non-growing microbial community varied between 0.038 and 0.064 μg CO_2-C μg^{-1} DNA h^{-1} (Fig. 1c). The rhizosphere effect on the CO_2/DNA ratio was significant only at the half N rate (Fig. 1c). A significant N effect was observed only in rhizosphere soil: the CO_2/DNA ratio was 64% greater at the half versus the full N rate (Fig. 1c).

Respiratory Response and Microbial DNA Dynamics during Glucose-induced Growth

According to respiratory kinetics, we defined three phases of microbial growth on glucose (Fig. 2): an initial phase corresponding to the absence of microbial growth lasting in rhizosphere soil between 0 and −10.7 h (Table 1, see lag period); followed by the phase of exponential growth to 25.5 h; and by the phase of growth retardation thereafter. In root-free soil duration of corresponding microbial growth phases was for ca. 2 h (lag-phase) and even for 4 h longer than in the rhizosphere (Table 1, Fig. 2). The DNA content in the rhizosphere significantly increased within two hours after the end of the lag-period (t_{lag} 10.3 h, Tables 1, 3). Thus, the amount of DNA in the rhizosphere soil increased almost

Table 1. Biomass and kinetic parameters of the respiratory response of microorganisms growing on glucose.

Soil	N rate	Microbial biomass C			Total cell mass	Maximal growth rate (μ_m)	Lag-period (t_{lag})
		Total	Growing	Growing			
		µg C g^{-1}	µg C g^{-1}	% of total	µg g^{-1}	h^{-1}	h
Root-free	50%	221^c±1	0.486^a±0.04	0.22	491±2	0.250^b±0.003	12.2^a±0.3
Rhizosphere	50%	290^a±20	0.888^a±0.02	0.31	644±44	0.260^a±0.001	10.3^b±0.2
Root-free	100%	245^b±14	0.637^b±0.03	0.26	544±31	0.238^c±0.002	12.4^c±0.2
Rhizosphere	100%	280^a±1	0.922^a±0.05	0.33	622±2	0.246^b±0.002	10.7^b±0.5

Total cell mass was calculated assuming a C content of the microbial biomass of 45% of dry weight (Christensen et al., 1993). Small letters show significant differences within the same column (p<0.05).

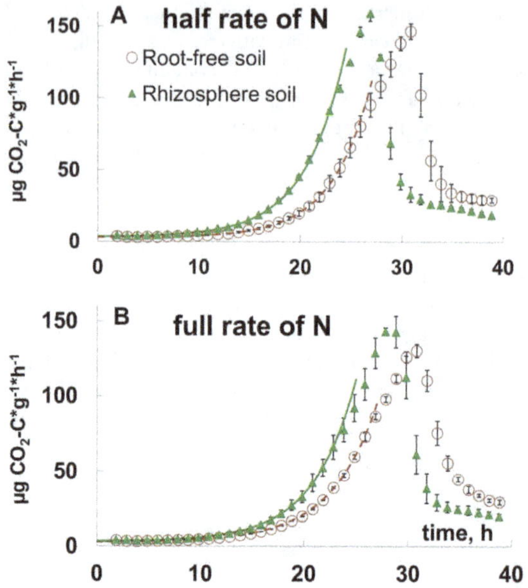

Figure 2. Dynamics of microbial respiration after glucose addition to root-free and rhizosphere soil. Glucose and nutrients induced respiration rate in root-free and rhizosphere soil under *Beta vulgaris* at half (a) and full (b) rates of N fertiliser. Experimental points and curves fitted by Eq. 3 for unlimited growth period are presented.

simultaneously with the respiration (Fig. 3a). In contrast, there were no changes in DNA content 15 hours after glucose application in root-free soil (Fig. 3b). So, contrary to the rhizosphere a time shift of at least three hours was observed between the increase of CO_2 and of DNA.

During the exponential growth, the specific rate of CO_2 emission (V_{CO2}/DNA ratio) steadily increased in both soils (Fig. 3 inserts). Despite the DNA content was significantly lower in root-free as compared to rhizosphere soil during the 35 h after glucose addition (Fig. 3), no significant differences (exception for one point at 20 h) between root-free and rhizosphere soil were found for the V_{CO2}/DNA ratio, which peaked at 25 h after glucose addition and exceeded 1 µg C µg^{-1} DNA h^{-1}. After growth retardation, the V_{CO2}/DNA ratios returned to the initial state and were close to 0.1 µg C µg^{-1} DNA h^{-1} (Fig. 3 inserts).

The quantity of CO_2 evolved per unit of newly-formed DNA ($\Delta CO_2/\Delta DNA$) from the rhizosphere soil continuously increased until the middle of the exponential growth, then stabilised until the end of incubation at 13.6±0.3 µg CO_2-C µg^{-1} DNA (Fig. 4a), indicating a proportional increase in CO_2 and DNA content. In the root-free soil however, the $\Delta CO_2/\Delta DNA$ ratio was 1.5–2 times lower than in rhizosphere during exponential growth (until 20–23 h after glucose addition) and increased only after growth retardation (Fig. 4b). The microbial respiration rate decreased in the rhizosphere after 25 h, and in the root-free soil after 30 hours (Fig. 2), but the DNA content increased for at least 10 more hours in both soils (Fig. 3,). Twice as much CO_2 was produced during exponential growth in rhizosphere versus root-free soil (Table 3), but only 8% more CO_2 was evolved from rhizosphere as compared to root-free soil during the whole incubation (36 h after glucose addition). Thus, the more efficient growth in the exponential phase (according to the $\Delta CO_2/\Delta DNA$ ratio) was counterbalanced by a less efficient metabolism after substrate exhaustion in the root-free soil.

The CUE (Eq. 8) also indicated more efficient microbial metabolism in root-free versus rhizosphere soil during exponential

Table 2. Contribution of two factors: living roots (Roots) and N fertilisation rate (N) and their interactions (Roots x N) to the variance of microbial parameters.

Factor	Basal respiration	Microbial biomass total	active	dsDNA content	Maximal growth rate, μ_m	Lag-period
Roots	67.2***	86.7***	89.8***	40.7***	30.6**	95.1***
N	0.6[ns]	1.5[ns]	6.7**	48.1***	63.8**	1.7[ns]
Roots x N	28.6**	8.5*	2.5*	7.5***	0.4[ns]	0.3[ns]
Residual	3.9	3.3	1	3.7	5.2	2.9

two-way ANOVA, % of explained variance.
***, **, * - significant effects at $P<0.001$, <0.01 and <0.05, respectively.
[ns]– not significant.

growth (Table 3). At the early stage of glucose utilization and after growth retardation, however, the efficiency of microbial metabolism was lower in root-free than in rhizosphere soil. Remarkably, CUE estimated for the whole incubation period did not differ between both soils (Table 3).

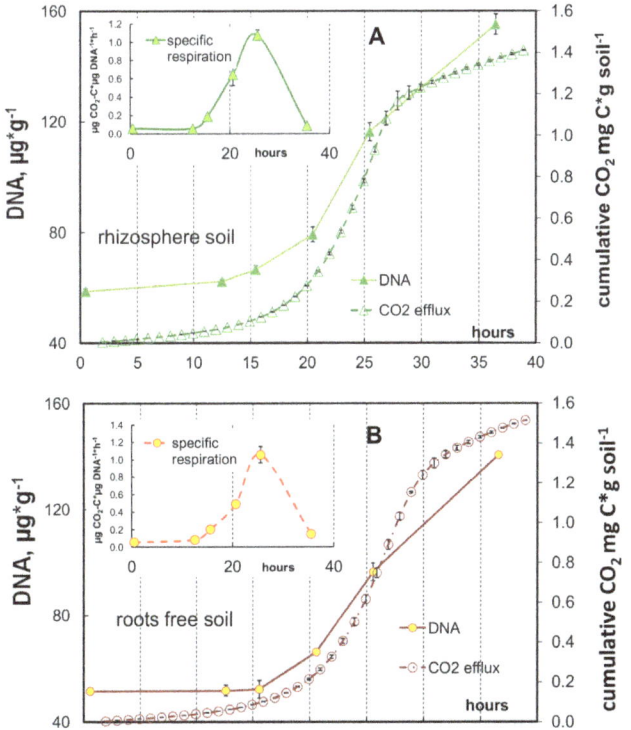

Figure 3. Microbial DNA dynamics and cumulative CO₂ production in root-free and rhizosphere soil. Dynamics of microbial DNA content and CO_2 accumulation after glucose addition in rhizosphere (a) and root-free (b) soil collected from the plot fertilized with 126 kg N ha^{-1} year^{-1}. Dynamics of specific CO_2 production (V_{CO2}-to-DNA ratio) are shown in the inserted graphs for rhizosphere and root-free soil, correspondingly.

Discussion

Microbial Biomass and DNA Content as a Basis for CUE Estimation

Assuming a C content of microbial biomass of 45% of dry weight [5], the total cell mass in soil without glucose varied from 491 to 644 µg g^{-1} soil (according to the SIR method, Eq. 2, Table 1). Therefore, the DNA content in microbial biomass amounted to 9.5–13% of dry weight which is in the upper range of the values reported for cultures extracted or isolated from soil bacteria, 5.2–13% [45] and is very close to the microbial DNA content *in situ* in soil (7–9%) when microbial biomass was assessed by a fumigation-extraction technique [26]. The comparison of several independent observations indicated that approximately 13% of the soil microbial biomass consisted of DNA [25]. However, the DNA content per biomass unit was not constant and decreased with increasing cell size from 13 to 5.2% [45] and was greater in non-growing than in growing bacterial cells. Therefore, the high DNA percentage in microbial biomass in our soil reflected the domination of small-sized cells in the non-growing microbial community.

Respiration and DNA Content under Steady-state and Unlimited Growth Conditions

Our results (Fig. 3, insert) confirm the findings of Marstorp & Witter [26] for a sandy loam soil from central Sweden, where CO_2/DNA ratios were lower than 0.1 µg CO_2-C µg^{-1} DNA h^{-1} for a non-growing microbial community. During exponential growth, however, we observed a quick increase in CO_2/DNA ratios. The CO_2/DNA ratio calculated according to Figure 1 in Marstorp & Witter [26] also increased during glucose-induced growth up to 0.5 µg CO_2-C µg^{-1} DNA h^{-1}. The CO_2/DNA ratio changed along with the physiological state of microorganisms and therefore, together with the metabolic quotient qCO_2, can be used as a valuable ecophysiological indicator reflecting the activity status of microbial biomass in soil.

A constant DNA content during the lag-period has been observed for *in situ* soil conditions [26]. We noticed, however, that the increase in DNA content in root-free soil began several hours after the increase in respiration, reflecting a period necessary for the activation of microbial metabolism (CO_2 increase) before the real growth (DNA increase) start. Such behaviour is common for *K*-strategists [46]. The delay between respiratory increase and DNA synthesis after the stimulation of microbial growth was much shorter in rhizosphere than in root-free soil, where no increase in DNA content was evident, even at the start of the exponential respiration increase. This was supported by the amount of active

Table 3. The amount of produced CO_2, DNA increment and carbon use efficiency (CUE) at different phases of microbial growth after glucose addition.

Period after glucose addition, (h)	Location	Phase of microbial growth	DNA increase during specified period	CO_2 accumulated during specified period	CUE, calculated according Eq.8, see details in text
			µg g soil^{-1}	µg C g soil^{-1}	g C g C^{-1}
0–12.5	Rhizosphere	lag-phase & initial growth	3.5±1.3	59^d±3	0.41^a±0.04
	Root-free soil	lag-phase	0.2±3.6	40^d±2	0.39^a±0.05
12.5–25.5	Rhizosphere	exponential growth	54.1±3.4	772^b±22	0.23^b±0.02
	Root-free soil	exponential growth	44.8±8.2	383^c±39	0.35^a±0.07
25.5–36.5	Rhizosphere	growth retardation growth	38.6±4.9	578^c±26	0.22^b±0.04
	Root-free soil	& growth retardation	43.5±8.5	877^b±65	0.17^b±0.06
0–36.5	Rhizosphere	all phases	96±3.8	1408^a±1	0.23^b±0.01
	Root-free soil	all phases	87.7±4.3	1300^a±24	0.23^b±0.02

Small letters show significant differences within the same column (p<0.05).

microbial biomass capable for immediate growth that was twice as large in rhizosphere as compared to root-free soil (Table 1).

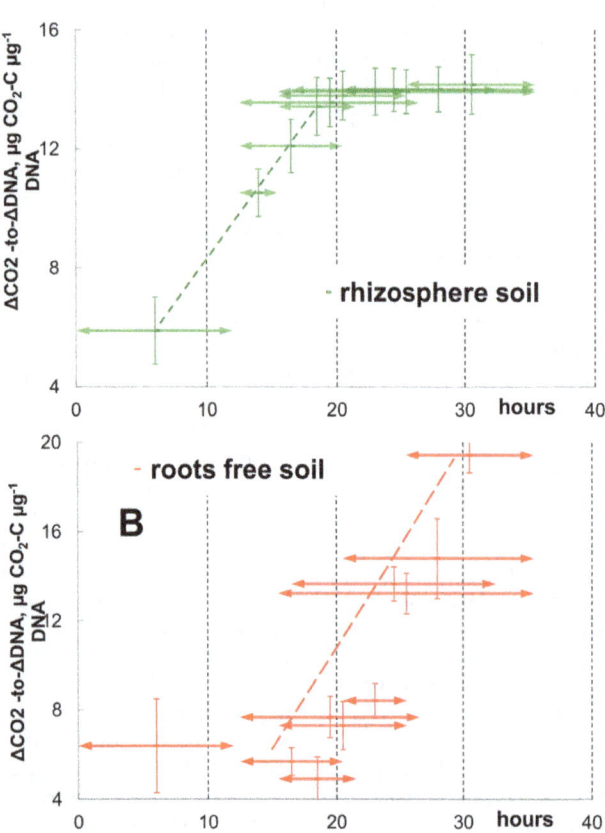

Figure 4. The ratio of CO_2 increment-to-DNA increment in rhizosphere and root-free soil. Soil was collected from the plot fertilized with 126 kg N ha^{-1} year^{-1}, (a) – rhizosphere soil, (b) – root-free soil. Horizontal arrows show the time period used for $\Delta CO_2/\Delta DNA$ ratio calculation. Vertical bars show standard deviations.

We demonstrated two kinds of physiological responses to glucose addition in microbial communities in rhizosphere and root-free soil. The DNA synthesis after glucose addition was more closely coupled with CO_2 production in rhizosphere soil as compared to root-free soil, where the dynamics of DNA synthesis and CO_2 production were decoupled both immediately after glucose addition and after its exhaustion. Microorganisms in the root-free soil persisted in a dormant state and reacted to increased substrate availability with a distinct delay between respiration response and DNA synthesis. In the rhizosphere, where the fraction of active microorganisms capable for immediate growth was two-fold larger than in root-free soil, the microbial community responded to glucose earlier in terms of both respiration and DNA synthesis (Figs. 2, 3).

Lag Period and Specific Growth Rates of Microorganisms in the Rhizosphere and Root-free Soil

The significantly greater μ_m values in rhizosphere as compared to root-free soil (Table 1) indicated a greater portion of fast growing microorganisms with r-strategy in the rhizosphere. Selective stimulation of some bacterial species in the rhizosphere (e. g. *Pseudomonas sp.*), [12], [47] with higher specific growth rates than most other soil bacteria [38] explains this phenomenon. The microbial community of the rhizosphere has a shorter lag-period and was ready for immediate growth on available substrate compared to the microbial community in root-free soil. According to Eq. 7, the duration of t_{lag} is dependent both on μ_m and on the fraction of actively growing microorganisms in the total microbial biomass. The negative correlation between lag-period and the amount of active biomass ($r^2 = -0.78$, p<0.12) was stronger compared to correlation between t_{lag} and μ_m ($r^2 = -0.49$, p<0.30). Thus, we conclude that the activity state of microbial biomass rather than such feature of the microorganisms as maximal specific growth rate (μ_m) is responsible for the duration of t_{lag}.

Basal Respiration as a Response to N Limitation in Rhizosphere versus Root-free Soil

The inverse response of basal respiration rate to N fertilization level in the rhizosphere and root-free soil (Fig. 1a) reflected the different strategies of microbial growth in soil microhabitats. Microorganisms with r-strategy dominating in rhizosphere soil increased basal respiration under N limitation. This resulted in highest values of specific respiration (maintenance efficiency) and consequently in lowest CUE. Contrary to that, the K-strategists prevailing in root-free soil even decreased basal respiration in low N treatment, thus, maintaining CUE similar to that in high N plot under steady-state. There were no differences in fine root development between the plots with full and half rate of N at time of soil sampling [34]; therefore we do not attribute the observed differences in V_{basal} to the variation in C input from roots to the soil [48]. Double limitation by C and N in the root-free soil at the half N rate decreased both microbial DNA content and basal respiration compared to root-free soil at the full N rate. However, specific respiration (maintenance efficiency) did not differ significantly between half and full rate of N fertilization in root-free soil (Fig. 1c) demonstrating stronger competitive abilities of K-strategists under N limitation. Therefore, both the level of metabolic activity and CUE should be considered when the N effect on soil respiration is estimated.

CUE in Rhizosphere and Root-free Soil: Dynamics and Proof of Estimates

Our study revealed the basic differences between microbial communities in rhizosphere and root-free soil in catabolic and anabolic processes traced by the dynamics of two fundamental microbial parameters: respiration activity (CO_2) and cell proliferation (DNA), which were used for estimation of CUE. Lower CUE during exponential growth of the r-selected rhizosphere community (Table 3) was confirmed by the two-fold higher $\Delta CO_2/\Delta DNA$ ratios in rhizosphere versus root-free soil (Fig. 4, 15–20 hours). This agrees with the negative correlation between growth rate and yield [22], [49]. Contrary to r-strategists, the K-strategists relatively more abundant in root-free soil do not mineralise glucose immediately, but can partly store it as an intracellular reserve during lag-phase and use it later after substrate exhaustion [38], [50], [51], thus maintaining their respiratory activity longer. Remarkably, distinct differences in CUE between rhizosphere and root-free soil observed during exponential growth were completely smoothed for CUE estimated for the whole incubation period. Thus, the same energy input caused different patterns of catabolic and anabolic processes in r- and K-selected communities resulting in similar energy output per unit of newly formed DNA in rhizosphere and root-free soils. This demonstrates that the shift in balance between catabolic and anabolic processes can serve as a tool for microbial community to maintain CUE independently of changing environment.

The CUE estimated during the exponential growth was 22% and 35% for rhizosphere and root-free soil, respectively. This is close to the range of 20–30% found for a cultured population of indigenous soil bacteria in the growth phase [45] and it is in the range of 14–51% observed for 8 agricultural soils [5]. However, much higher CUE has been obtained by other methods for *in situ* microbial communities growing on ^{14}C- or ^{13}C-labeled glucose (50–61%, [30]; 69–78%, [18], see for review [3]).

We used the average DNA value of 11% of total microbial biomass that was determined in soil without glucose addition [5], [45]. Considering lower DNA content in growing cells versus the cells in stationary phase [45], and that the DNA content in fungal

mycelium can be much lower than in bacterial cells [52,53] the CUE of 38% and 51% can be obtained for rhizosphere and root-free soil, respectively (based on the lowest DNA content of 5.2% of cell mass for pure cultures [45]). These CUE exactly fits to the estimates for glucose use efficiency in N-amended and in N-limited soil (Y = 0.52 and 0.38, respectively) using a balance calculation [30]. The scatter of CUE values found in the literature can be explained by the variation in growth conditions of microorganisms affecting also the DNA content in microbial cells. More experimental studies on the variability of DNA content *in situ* are needed for narrowing CUE estimates in experiments similar to ours.

Conclusions

The applied combination of approaches: analysis of the double-stranded DNA content in soil and of respiration kinetics allows quantitative distinguishing of microbial traits in the rhizosphere versus root free soil. Total microbial biomass in the rhizosphere was 14–31% higher than that in the root free soil, while the growing (active) part of microbial biomass was 45–83% higher. The higher microbial specific growth rate (μ_m) and lower CUE indicated the greater contribution of r-strategists in rhizosphere as compared with root-free soil. We partly confirmed hypotheses posed in the introduction: microbial communities in rhizosphere soil have specific respiration rate higher than microorganisms in root-free soil. This holds true under N limiting conditions but no difference was observed for fully fertilized N plot. Lower content of available N decreased microbial DNA, but increased the μ_m values. The N limitation in the rhizosphere increased microbial respiration, presumably due to lower C use efficiency confirming domination of r-selected species in rhizosphere microbial community and supporting our second hypotheses.

The $\Delta CO_2/\Delta DNA$ ratio was stable in the growing microbial community in the rhizosphere while it increased consistently in root-free soil, revealing contrasting patterns of microbial metabolism in different microhabitats. The K-strategy typical for root-free soil manifested itself by decoupling of the respiration burst after glucose addition and DNA increase, more efficient growth (high CUE) and longer persistence of respiratory activity. The r-strategy (common for rhizosphere) was exhibited as a faster and simultaneous response on substrate addition, lower growth efficiency and a shorter period of high activity following by more abrupt respiration decrease after substrate exhaustion. The CUE during exponential growth was by the factor of 1.5 higher in root-free than in rhizosphere soil indicating the necessity to consider variable Y depending on substrate availability in soil microhabitats. Further studies are necessary for the determination of the range of differences in CUE in soil microhabitats, because microbial community composition depends on multiple factors such as host plant species, soil properties, plant development stage [10], [54] and these factors will affect also the microbial physiology in rhizosphere and root-free soils.

Author Contributions

Conceived and designed the experiments: EB SB THA YK. Performed the experiments: EB SB. Analyzed the data: EB SB THA YK. Contributed reagents/materials/analysis tools: THA YK. Wrote the paper: EB SB THA YK. Obtained permission for use of soil samples: THA.

References

1. Cheng W (2009) Rhizosphere priming effect: Its functional relationships with microbial turnover, evapotranspiration, and C-N budgets. Soil Biol Biochem 41: 1795–1801.

2. Gärdenäs AI, Ågren GI, Bird JA, Clarholm M, Hallin S, et al. (2011) Knowledge gaps in soil carbon and nitrogen interactions - From molecular to global scale. Soil Biol Biochem 43: 702–717.

3. Manzoni S, Taylor P, Richter A, Porporato A, Agren GI (2013) Environmental and stoichiometric controls on microbial carbon-use efficiency in soils. New Phytol 196: 79–91.

4. Bradford M, Keiser A, Davies C, Mersmann C, Strickland M (2013) Empirical evidence that soil carbon formation from plant inputs is positively related to microbial growth. Biogeochem 113: 271–281.

5. Anderson TH, Martens R (2013) DNA determinations during growth of soil microbial biomasses. Soil Biol Biochem 57: 487–495.

6. Keiblinger KM, Hall EK, Wanek W, Szukics U, Hammerle I, et al. (2010) The effect of resource quantity and resource stoichiometry on microbial carbon-use-efficiency. FEMS Microbiol Ecol 73: 430–440.

7. Schimel J, Schaeffer SM (2012) Microbial control over carbon cycling in soil. Frontiers in Microbiology 3: 348.

8. Allison SD, Wallenstein MD, Bradford MA (2010) Soil-carbon response to warming dependent on microbial physiology. Nature Geosci 3: 336–340.

9. Paterson E (2009) Comments on the regulatory gate hypothesis and implications for C-cycling in soil. Soil Biol Biochem 41: 1352–1354.

10. Berg G, Smalla K (2009) Plant species and soil type cooperatively shape the structure and function of microbial communities in the rhizosphere. FEMS Microbiol Ecol 68: 1–13.

11. Paterson E, Midwood AJ, Millard P (2009) Through the eye of the needle: a review of isotope approaches to quantify microbial processes mediating soil carbon balance. New Phytol 184: 19–33.

12. Grayston SJ, Wang S, Campbell CD, Edwards AC (1998) Selective influence of plant species on microbial diversity in the rhizosphere. Soil Biol Biochem 30: 369–378.

13. Blagodatskaya EV, Blagodatsky SA, Anderson TH, Kuzyakov Y (2007) Priming effects in Chernozem induced by glucose and N in relation to microbial growth strategies. Applied Soil Ecology 37: 95–105.

14. Pirt SJ (1975) Principles of microbe and cell cultivation: John Wiley & Sons. 274 p.

15. Blagodatsky SA, Demyanova EG, Kobzeva EI, Kudeyarov VN (2002) Changes in the efficiency of microbial growth upon soil amendment with available substrates. Euras Soil Sci 35: 874–880.

16. Herron PM, Stark JM, Holt C, Hooker T, Cardon ZG (2009) Microbial growth efficiencies across a soil moisture gradient assessed using 13C-acetic acid vapor and ^{15}N-ammonia gas. Soil Biol Biochem 41: 1262–1269.

17. Payne JW (1970) Energy yields and growth of heterotrophs. Annual Reviews in Microbiology 24: 17–52.

18. Thiet RK, Frey SD, Six J (2006) Do growth yield efficiencies differ between soil microbial communities differing in fungal: bacterial ratios? Reality check and methodological issues. Soil Biol Biochem 38: 837–844.

19. Panikov NS, Sizova MV (1996) A kinetic method for estimating the biomass of microbial functional groups in soil. J Microbiol Meth 24: 219–230.

20. Anderson TH, Domsch KH (1985) Maintenance carbon requirements of actively metabolizing microbial populations under in situ conditions. Soil Biol Biochem 17: 197–203.

21. van Bodegom P (2007) Microbial Maintenance: A Critical Review on Its Quantification. Microbial Ecol 53: 513–523.

22. Lipson D, Monson R, Schmidt S, Weintraub M (2009) The trade-off between growth rate and yield in microbial communities and the consequences for under-snow soil respiration in a high elevation coniferous forest. Biogeochem 95: 23–35.

23. Marstorp H, Guan X, Gong P (2000) Relationship between dsDNA, chloroform labile C and ergosterol in soils of different organic matter contents and pH. Soil Biol Biochem 32: 879–882.

24. Blagodatskaya EV, Blagodatskii SA, Anderson TH (2003) Quantitative isolation of microbial DNA from different types of soils from natural and agricultural ecosystems. Mikrobiology 72: 840–846.

25. Joergensen RG, Emmerling C (2006) Methods for evaluating human impact on soil microorganisms based on their activity, biomass, and diversity in agricultural soils. J Plant Nutr Soil Sci 169: 295–309.

26. Marstorp H, Witter E (1999) Extractable dsDNA and product formation as measures of microbial growth in soil upon substrate addition. Soil Biol Biochem 31: 1443–1453.

27. Andrews JH, Harris RF (1986) r and K-selection and microbial ecology. In: Marshall KC, editor. Adv Microb Ecol. New York. 99–144.

28. Dorodnikov M, Blagodatskaya E, Blagodatsky S, Fangmeier A, Kuzyakov Y (2009) Stimulation of r- vs. K-selected microorganisms by elevated atmospheric CO2 depends on soil aggregate size. FEMS Microbiol Ecol 69.

29. del Giorgio PA, Cole JJ (1998) Bacterial growth efficiency in natural aquatic systems. Ann Rev Ecol Systematics 29: 503–541.

30. Blagodatskiy SA, Larionova AA, Yevdokimov IV (1993) Effect of mineral nitrogen on the respiration rate and growth efficiency of soil microorganisms. Euras Soil Sci 25: 85–95.

31. Blagodatsky SA, Yevdokimov IV, Larionova AA, Richter J (1998) Microbial growth in soil and nitrogen turnover: Model calibration with laboratory data. Soil Biol Biochem 30: 1757–1764.

32. Rousk J, Bååth E (2007) Fungal and bacterial growth in soil with plant materials of different C/N ratios. FEMS Microbiol Ecol 62: 258–267.

33. Kuzyakov Y, Xu X (2013) Competition between roots and microorganisms for nitrogen: mechanisms and ecological relevance. New Phytologist 198: 656–669.

34. Weigel HJ, Pacholski A, Burkart S, Helal M, Heinemeyer O, et al. (2005) Carbon turnover in a crop rotation under free air CO_2 enrichment (FACE). Pedosphere 15: 728–738.

35. Anderson JPE, Domsch KH (1978) A physiological method for the quantative measurement of microbial biomass in soils. Soil Biol Biochem 10: 215–221.

36. Anderson TH, Domsch KH (1986) Carbon assimilation and microbial activity in soil. Zeitschrift für Pflanzenernährung und Bodenkunde 149: 457–468.

37. Heinemeyer O (1989) Soil microbial biomass and respiration measurements: An automated technique based on infra-red gas analysis. Plant Soil: 191–195.

38. Panikov NS (1995) Microbial Growth Kinetics. London, Glasgow: Chapman and Hall. 378 p.

39. Blagodatsky SA, Blagodatskaya EV, Anderson TH, Weigel HJ (2006) Kinetics of the respiratory response of the soil and rhizosphere microbial communities in a field experiment with an elevated concentration of atmospheric CO_2. Euras Soil Sci 39: 290–297.

40. Blagodatsky SA, Heinemeyer O, Richter J (2000) Estimating the active and total soil microbial biomass by kinetic respiration analysis. Biol Fertil Soils 32: 73–81.

41. Wutzler T, Blagodatsky SA, Blagodatskaya E, Kuzyakov Y (2012) Soil microbial biomass and its activity estimated by kinetic respiration analysis - Statistical guidelines. Soil Biol Biochem 45: 102–112.

42. ModelMaker (1997) ModelMaker Version 3.0.3 Software. CherwellScientific Publishing Limited, Oxford.

43. Blagodatskaya EV, Blagodatsky SA, Anderson TH, Kuzyakov Y (2009) Contrasting effects of glucose, living roots and maize straw on microbial growth kinetics and substrate availability in soil. Europ J Soil Sci 60.

44. Christensen H, Bakken LR, Olsen RA (1993) Soil bacterial DNA and biovolume profiles measured by flow-cytometry. FEMS Microbiol Ecol 102: 129–140.

45. Christensen H, Olsen RA, Bakken LR (1995) Flow Cytometric Measurements of Cell Volumes and DNA Contents During Culture of Indigenous Soil Bacteria. Microbial Ecol 29: 49.

46. Panikov NS (2010) Microbial Ecology. Environmental Biotechnology. 121–191.

47. Goddard VJ, Bailey MJ, Darrah P, Lilley AK, Thompson IP (2001) Monitoring temporal and spatial variation in rhizosphere bacterial population diversity: A community approach for the improved selection of rhizosphere competent bacteria. Plant Soil 232: 181–193.

48. Gershenson A, Bader NE, Cheng W (2009) Effects of substrate availability on the temperature sensitivity of soil organic matter decomposition. Global Change Biol 15: 176–183.

49. Pfeiffer T, Schuster S, Bonhoeffer S (2001) Cooperation and competition in the evolution of ATP-producing pathways. Science (Washington D C) 292: 504–507.

50. Hill PW, Farrar JF, Jones DL (2008) Decoupling of microbial glucose uptake and mineralization in soil. Soil Biol Biochem 40: 616–624.

51. Schneckenberger K, Demin D, Stahr K, Kuzyakov Y (2008) Microbial utilization and mineralization of [14C] glucose added in six orders of concentration to soil. Soil Biology and Biochemistry 40: 1981–1988.

52. Anderson TH (2008) Assessment of DNA contents of soil fungi. Landbauforsch Volkenrode 58: 19–28.

53. Leckie SE, Prescott CE, Grayston SJ, Neufeld JD, Mohn WW (2004) Comparison of chloroform fumigation-extraction, phospholipid fatty acid, and DNA methods to determine microbial biomass in forest humus. Soil Biology and Biochemistry 36: 529–532.

54. Zachow C, Tilcher R, Berg G (2008) Sugar beet-associated bacterial and fungal communities show a high indigenous antagonistic potential against plant pathogens. Microbial Ecol 55: 119–129.

12

Effects of Nitrogen and Phosphorus Fertilization on Soil Carbon Fractions in Alpine Meadows on the Qinghai-Tibetan Plateau

Jin Hua Li[1]*, Yu Jie Yang[1], Bo Wen Li[1], Wen Jin Li[1], Gang Wang[1], Johannes M. H. Knops[2]

1 State Key Laboratory of Grassland Agro-Ecosystems, School of Life Sciences, Lanzhou University, Lanzhou, P.R. China, **2** School of Biological Sciences, University of Nebraska, Lincoln, Nebraska, United States of America

Abstract

In grassland ecosystems, N and P fertilization often increase plant productivity, but there is no concensus if fertilization affects soil C fractions. We tested effects of N, P and N+P fertilization at 5, 10, 15 $g\ m^{-2}\ yr^{-1}$ (N_5, N_{10}, N_{15}, P_5, P_{10}, P_{15}, N_5P_5, $N_{10}P_{10}$, and $N_{15}P_{15}$) compared to unfertilized control on soil C, soil microbial biomass and functional diversity at the 0–20 cm and 20–40 cm depth in an alpine meadow after 5 years of continuous fertilization. Fertilization increased total aboveground biomass of community and grass but decreased legume and forb biomass compared to no fertilization. All fertilization treatments decreased the C:N ratios of legumes and roots compared to control, however fertilization at rates of 5 and 15 $g\ m^{-2}\ yr^{-1}$ decreased the C:N ratios of the grasses. Compared to the control, soil microbial biomass C increased in N_5, N_{10}, P_5, and P_{10} in 0–20 cm, and increased in N_{10} and P_5 while decreased in other treatments in 20–40 cm. Most of the fertilization treatments decreased the respiratory quotient (qCO_2) in 0–20 cm but increased qCO_2 in 20–40 cm. Fertilization increased soil microbial functional diversity (except N_{15}) but decreased cumulative C mineralization (except in N_{15} in 0–20 cm and N_5 in 20–40 cm). Soil organic C (SOC) decreased in P_5 and P_{15} in 0–20 cm and for most of the fertilization treatments (except $N_{15}P_{15}$) in 20–40 cm. Overall, these results suggested that soils will not be a C sink (except $N_{15}P_{15}$). Nitrogen and phosphorus fertilization may lower the SOC pool by altering the plant biomass composition, especially the C:N ratios of different plant functional groups, and modifying C substrate utilization patterns of soil microbial communities. The N+P fertilization at 15 $g\ m^{-2}\ yr^{-1}$ may be used in increasing plant aboveground biomass and soil C accumulation under these meadows.

Editor: Upendra M. Sainju, Agricultural Research Service, United States of America

Funding: This research was supported by Program for New Century Excellent Talents in University (NCET-11-0210), the Fundamental Research Funds for the Central Universities (LZUJBKY-2013-K12) and National Natural Science Foundation of China (No. 30871823). The funders had no role in study design, data collection and analysis, decision to publish, or preparation of the manuscript.

Competing Interests: The authors have declared that no competing interests exist.

* Email: lijinhuap@sohu.com

Introduction

In grassland ecosystems, soils represent the largest active pool of organic C which may have global implications [1–3]. In N-limited grasslands, N deposition through precipitation and/or dust accumulation or fertilization and sometimes P fertilization can increase the productivity of the plant community but also can decrease species richness [2,4–9]. However, N and P fertilization may affect soil C fractions and the ecological function of grassland soils in different ways [2,10]. Long-term N fertilization can decrease microbial biomass carbon (MBC) of grassland soil [11,12]. Fertilization effects on C mineralization are variable [13–15]. Mack et al. [16] showed that NP fertilization contributed to decreases in soil organic C (SOC) pools in tundra ecosystems. Phosphorus additions can lead to greater C sequestration in soils [17]. In addition, N additions can significantly enhanced soil C stocks, but combined NPKMg additions may not result in soil C sequestration [9]. Thus, it is clear that in contrast to the consistent increases of productivity due to N and P fertilization, the impacts of N and P fertilization on microbial composition, activity and soil C is not consistent and differs markedly between sites.

Alpine grasslands in the Qinghai-Tibetan Plateau are estimated to hold approximately 55% of China's total grassland C, out of which 93% is present in the soil [18]. Little is known about how fertilization types and rates affect soil C fractions in alpine meadows.

Here we specifically examined how 5 years of N and/or P additions had influenced soil organic C fractions. Our objective was to develop a better understanding of the effects of N and P fertilization on changes in soil organic C dynamics. Previous studies from the same experiment have shown that N or N+P fertilization increased plant aboveground productivity and there was a yearly variation in plant biomass [19–21]. Forbs and legumes richness decreased after N and P fertilization [21]. Nitrogen concentration and N/P ratios of grasses and forbs increased significantly but were relatively constant for legumes after N addition [22,23]. In this paper, we divided species into different functional groups to address the changes in soil organic C fractions in relation to the species changes. Specifically, we evaluated the effects of N and P fertilization on the aboveground plant biomass of different functional groups and soil C fractions at 0–20 cm and 20–40 cm.

Fertilization can affect the quantity of C input to soil by alteration of the plant productivity, diversity, and stoichiometry of plant tissues [16,24], as well as soil organic matter decomposition and respiration by altering soil microbial communities and activity [11,25]. We hypothesized that a combination of N and P fertilization will increase SOC but decrease MBC, microbial functional diversity and C mineralization rates compared to N and P fertilization alone or no fertilization.

Methods

Ethics Statement

Our study area was located at the Alpine Meadow Ecosystem Research Station in Hezuo, Gansu, eastern Qinghai-Tibetan Plateau, China (N34°55', E102°53' 3,000 m above sea level), which is managed by Lanzhou University. No specific permissions were required for conducting experiment in this location, and no endangered, protected species or vertebrate species were involved with this research.

Study site

This study was conducted from 2009 to 2013 in the Research Station of Alpine Meadow and Wetland Ecosystems of Lanzhou University, located in Hezuo, Gansu, eastern Qinhai-Tibetan Plateau of China (N34°55', E102°53', 3,000 m above sea level). Hezuo has a 30-year mean annual precipitation of 550 mm, with 85% of the precipitation occurring during the growing season from June through September (Institute of Hezuo Meteorology). Mean annual temperature is 2.4°C, ranging from −8.3°C during December-February to 11.9°C during June–August periods. The soils are classified as chestnut soils or Haplic Calcisols according to the FAO classification or sub-alpine meadow soil according to the Chinese soil classification system [26]. The vegetation is a typical alpine meadow, dominated by grasses such as *Festuca ovina* Linn., *Poa poophagorum* Bor and *Elymus nutans* Griseb; sedges such as *Scirpus pumilus* Vahl and *Kobresia capillifolia* (Decne.) C.B. *Kobresia pygmaea* C.B. Clarke in Hook; forbs such as *Anemone rivularis* Buch.-Ham, *Trollius farreri* Stapf and *Anemone obtusiloba* D. Don, *Taraxacum lugubre*, *Geranium pylzowianum*, *Polygonum viviparum*, and legumes such as *Astragalus polycladus* Bur., *Oxytropis ochrocephala* and *Gueldenstaedtia verna* Georgi.

Experimental design

Soil organic C fractions resulting from the increased plant production by N and P fertilization were evaluated using the following fertilization treatments: (i) N alone; (ii) P alone; and (iii) N+P together. There were three rates (5, 10, 15 g m^{-2} yr^{-1}) of N (urea with 46% N), and P (sodium dihydrogen phosphate anhydrous with 44.6% P). A control treatment (CK) without any N or P fertilization was also included. As a result, there were ten treatments (N$_5$, N$_{10}$, N$_{15}$, P$_5$, P$_{10}$, P$_{15}$, N$_5$P$_5$, N$_{10}$P$_{10}$, N$_{15}$P$_{15}$, and CK) with five replicates for a total of 50 treatment plots. Only four replicates were used in this study. All treatments were randomly assigned in 50 plots (5×5 m area each). In addition, there were four plots without fertilization but destroyed by zokors (*Myospalax fontanieri*) in 2009, but this didn't affect the neighbor plots. All plots were separated by 1 m unfertilized buffers. Fertilizer was broadcast evenly in the plot once per year on 20th August for 5 years since 2009. Fertilizer was applied in August for two reasons: (i) In our study site, the growing season was from June through September and 85% of the rainfall occurred during the growing season. (ii) Plant biomass peaked at this time, which was reasonable to be contrasted with that before fertilization application. However, applying fertilizers in the middle of the growing

season might result in the inefficient use and wastage. Before applying fertilizers in August 2009, average SOC, soil total N and total P was 33.8±1.8 g kg^{-1}, 3.7±0.1 g kg^{-1}, 0.65±0.01 g kg^{-1} in 0–20 cm and 26.1±1.5 g kg^{-1}, 2.8±0.1 g kg^{-1}, 0.65±0.01 g kg^{-1} in 20–40 cm, respectively, across all plots. The soil had 200 g kg^{-1} sand, 600 g kg^{-1} silt, and 200 g kg^{-1} clay.

Plant sampling

In early August 2013 before applying fertilizers, four sampling quadrates of 50 cm×50 cm (selected at random from the central of each 5×5 m plot to avoid edge effects) were used to harvest plant samples at 1 cm above the ground level using scissors. Plant samples were separated into three functional groups: grasses, legumes, and forbs. At the same time, root samples were collected also within the quadrates after clipping the aboveground plant. Four soil cores (5 cm diameter by 20 cm depth) in each quadrate were collected randomly using a hand probe and composited by plot. Soil samples were washed gently with water over a 60 mesh screen until roots were separated from the soil. All above- and belowground plant samples were dried at 65°C to constant mass and ground to 1 mm prior to determination of total C and N. Aboveground plant samples were also collected from 2009 to 2012, oven dried, ground, and C and N concentrations determined as above.

Soil sampling

In addition to samples collected for root biomass, soil samples were also collected for determining C fractions. Five cores (5 cm in diameter) in 0–20 cm and 20–40 cm depths were taken randomly in each plot, and mixed into one composite sample by depth. After removing gravel, coarse fragments and roots, each soil sample was homogenized and divided into two portions. One portion was air dried and sieved to 2 mm to analyze abiotic parameters. The other portion was kept at 4°C for analysis of microbial parameters (soil MBC, functional diversity of microbial community, and C mineralization).

Plant and soil analysis

Plant C and SOC concentrations were determined using the dichromate oxidation method [27]. Plant and soil total N was determined following Kjeldehl digestion by a Nitrogen Analyzer System (KJELTEC 2300 AUTO SYSTEM II, Foss Tecator AB, Sweden).

Soil MBC was determined by using the chloroform fumigation extraction method [28]. Field-moist soils were adjusted to 50% water holding capacity, incubated at 25°C for 2 weeks for uniform rewetting and stabilizing the microbial activity after the initial disturbances. Three 25 g sub-samples were fumigated with alcohol-free CHCl$_3$ for 24 h at 25°C. Samples were then extracted for MBC by adding 100 ml of 0.5 M K$_2$SO$_4$, shaking for 60 min and filtered through Whatman No. 2 paper. Three 25 g non-fumigated soil sub-samples were processed in the same manner. Carbon concentration in the extract was determined using the same method as SOC and MBC was calculated as the difference in C concentrations between fumigated and non-fumigated samples divided by the efficiency factor 0.38 [28].

Soil incubation for C mineralization was carried out in the laboratory for 28 days. Dry weight equivalent of 30 g of the sieved field-moist soil samples were weighed into 250 ml Schott jars. Small beakers filled with 15 ml of 1 M NaOH were placed at the soil surface in the jars to trap the evolved CO$_2$. The jars were fastened airtight and incubated for 28 days at 25°C. Moisture content of the samples was periodically adjusted to a value of 30% of water holding capacity since this is generally considered to mimic the moisture content of the field conditions [29]. Constant soil water content was maintained by weighing each sample once a week and adding water to the soil. The CO_2 evolved from the soil was measured at 1, 7, 14, 21, and 28 day after incubation. After each day of incubation, the small beaker with NaOH solution was removed and replaced by a new one with fresh NaOH. The CO_2 absorbed in NaOH solution was titrated with 0.1 M HCl after the addition of $BaCl_2$. A NaOH solution without soil, incubated as above, was also titrated. Basal respiration was calculated as the amount of CO_2 evolution in the first 24-h incubation divided by the dry mass of soil. The metabolic quotient (qCO_2) was calculated as the ratio of basal respiration to MBC [30]. The cumulative C mineralization was calculated as the sum of carbon dioxide (CO_2) released during 28 days of incubation.

Bacterial functional diversity in Biolog EcoPlate

Field-moist soils were used to assess the functional diversity of the microbial community, measured by Biolog EcoPlate system [28,30–32] according to the procedure describes by Li et al [33]. Average well color development (AWCD) in each microplate was determined as described by Garland [34]. AWCD and Shannon-Wiener diversity index in Biolog EcoPlate were calculated using data of Biolog EcoPlate collected during 0–168 h and at 168 h respectively as described by Li et al [33].

AWCD = $[\sum(C-R)/31]$, C is sum of optical densities of 31 wells in a plate, R is optical density of control well in the same plate.

Shannon-Wiener diversity H = $-\Sigma P_i \ln P_i$, of which P_i is proportional color development of the ith well over total color development of all wells of a plate.

Data analysis

Data were analyzed using SPSS 16.0 and graphs were plotted using Sigma Plot 10.0. In this paper, biomass data for only in the fifth year (i.e. 2013) was analyzed because data for the first four years have already been published [20,21]. A one-way ANOVA was used to analyze data by considering all treatments (three fertilization types × three rates + unfertilized control = 10 treatments). When the analysis was significant, the control was taken out and a two-way MANOVA was used to test the overall effects of fertilization types and rates on measured parameters. The least significant difference (LSD) test was used to separate means at $P = 0.05$. Principal component analysis (PCA) was performed on AWCD data at 168 h of incubation in the BIOLOG Ecoplate as described by Zhang et al [35] to determine the carbon utilization patterns of soil microbial communities. Regression analysis was used to analyze the relationship between cumulative carbon mineralization rate at the end of incubation (28th day) and soil MBC and SOC in all treatments. Regression analysis was also used to analyze the relationship between SOC and C:N ratios of different plant functional groups, roots or soils.

Results

Fertilization effects on plant aboveground biomass

Both fertilization type and rate influenced plant community biomass, grass and legume biomass, with a significant type × rate interaction for grass and community biomass (Table 1).

Eight of nine fertilization treatments (except the 15 g m^{-2} yr^{-1} P fertilization) had higher community biomass than the control (Figure 1A). Combination of N+P at 5 g m^{-2} yr^{-1} and 10 g m^{-2} yr^{-1} had much higher community biomass than corresponding rates of N fertilization or any rates of P fertilization. Nitrogen or N+P at 15 g m^{-2} yr^{-1} had two times community biomass and

Table 1. Univariate tests: ANOVAs on plant biomass of community and three functional groups: grasses, legumes and forbs, and on soil microbial biomass C (MBC), qCO_2, cumulative C mineralization, soil organic C (SOC) in 0–20 cm and 20–40 cm for fertilization types and rates.

	Plant biomass				MBC		qCO_2		Cumulative C mineralization		SOC	
	Community	Grasses	Legumes	Forbs	0–20	20–40	0–20	20–40	0–20	20–40	0–20	20–40
Type	0.000	0.000	0.717	0.468	0.003	0.004	<0.001	0.073	<0.001	<0.001	0.049	0.016
Rate	0.018	0.001	0.663	0.009	<0.001	<0.001	0.016	0.003	0.573	0.693	0.807	0.009
Type* Rate	0.018	0.002	0.762	0.357	<0.001	<0.001	<0.001	0.118	0.031	0.503	0.095	0.052

four times grass biomass compared to the control (Figure 1A, B). Grasses comprised 70% of the total biomass, and its biomass trend with fertilization was similar to that of total biomass. All N+P rates had higher grasses biomass compared to corresponding N fertilization rates although no difference occurred between 15 g m^{-2} yr^{-1} N+P and 15 g m^{-2} yr^{-1} N. The N+P treatments at 10 and 15 g m^{-2} yr^{-1} had much higher grass biomass as compared to the corresponding P fertilization rates (Figure 1B). However, the biomass of legumes and forbs in all fertilization treatments (except forbs in 5 g m^{-2} yr^{-1} of N+P fertilization) were lower than the control (Figure 1C, D). The N+P fertilization at 15 g m^{-2} yr^{-1} had significantly lower legume and forb biomass than other treatments, except the forb biomass at 15 g m^{-2} yr^{-1} of N and P.

Fertilization effects on carbon/nitrogen ratio of plant and soil

Fertilization type influenced C:N ratios of the different plant functional groups. Fertilization rate and the fertilization type × rate interaction were significant for forb C:N ratio (Table 2). All fertilization at 5 and 10 g m^{-2} yr^{-1} decreased whereas fertilization at 15 g m^{-2} yr^{-1} increased grasses' C:N ratios compared to the control (Figure 2A). The N, P, and N+P rates decreased C: N

ratios of legume and root (Figure 2B, 2D). The 5 and 10 g m^{-2} yr^{-1} N fertilization and 10 g m^{-2} yr^{-1} P decreased while N+P at all rates increased forbs' C:N ratios (Figure 2C). Fertilization type had significant effect on soil C concentrations in both layers and fertilization rate had significant effect on soil C concentrations in 20–40 cm. However, neither fertilization type nor fertilization rate had significant effects on soil N concentrations and C:N ratios in both layers (Table 2). Nevertheless, soil C:N was lower in fertilization treatments (except in 15 g m^{-2} yr^{-1} of N or N+P in 0–20 cm and 15 g m^{-2} yr^{-1} of N+P in 20–40 cm) than the control (Figure 2E, 2F).

Effect of fertilization on soil carbon fractions

Fertilization type and rate had significant effects on soil MBC, qCO_2, cumulative C mineralization, and SOC at 0–20 cm and 20–40 cm layers, except for the effect of rate on C mineralization at both layers and SOC at 0–20 cm (Table 1). The fertilization type × rate interaction was significant for all parameters in all layers, except for qCO_2 and C mineralization at 20–40 cm.

Nitrogen and P at rates of 5 and 10 g m^{-2} yr^{-1} increased MBC compared to other treatments in 0–20 cm (Figure 3A). In 20–40 cm, 10 g m^{-2} yr^{-1} of N fertilization and 5 g m^{-2} yr^{-1} of P

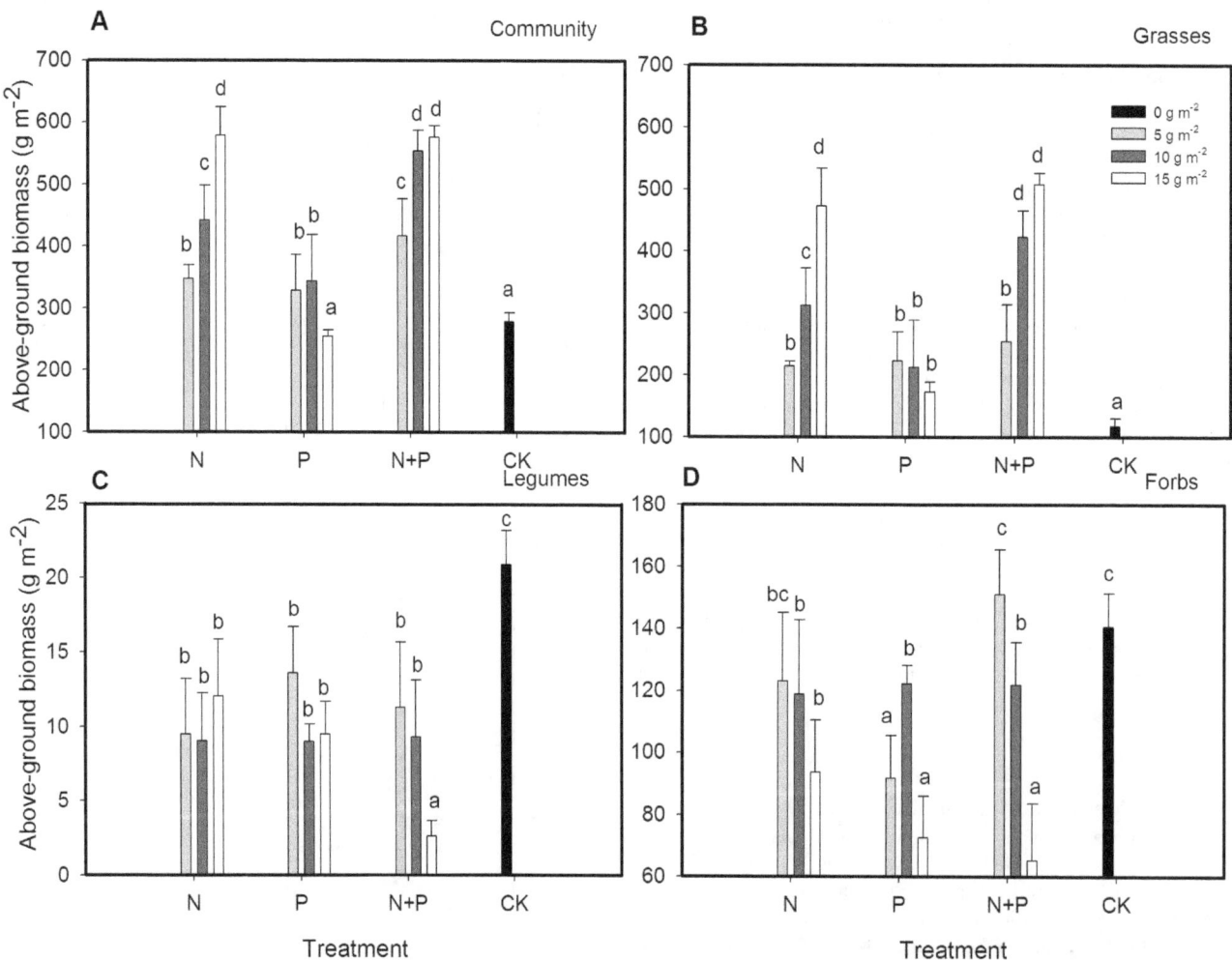

Figure 1. Mean ± SE of total aboveground plant biomass of (A) community, (B) grasses, (C) legumes, and (D) forbs as affected by N and P fertilization treatments. Different letters above bars indicate significantly different at $P = 0.05$.

Table 2. Univariate tests: ANOVAs on C:N ratios of plant (different functional groups and roots) and soil in 0–20 cm and 20–40 cm for fertilization types and rates

	Plant C:N ratio				Soil C:N ratio	
	Grasses	Legumes	Forbs	Roots	0–20 cm	20–40 cm
Type	<0.001	<0.001	<0.001	0.381	0.626	0.058
Rate	0.152	0.474	<0.001	0.695	0.807	0.169
Type* Rate	0.095	0.676	<0.001	0.390	0.890	0.346

fertilization significantly increased MBC compared to other treatments (Figure 3B). Soil MBC was significantly lower in 5 and 10 g m^{-2} yr^{-1} of N+P than those in corresponding rates of N or P fertilization in both layers (Figure 3A, B).

Nitrogen fertilization at 5 and 15 g m^{-2} yr^{-1} significantly increased $q CO_2$ while P fertilization at 5 and 15 g m^{-2} yr^{-1} and N+P at all rates significantly decreased $q CO_2$ in 0–20 cm (Figure 3C). The N+P at 5 and 10 g m^{-2} yr^{-1} decreased $q CO_2$ compared to other treatments in 20–40 cm (Figure 3D).

Most of the fertilization treatments (except the 15 g m^{-2} yr^{-1} N in 0–20 cm soil and 5 g m^{-2} yr^{-1} N in 20–40 cm soil) had lower cumulative C mineralization rates than the control. All N+P fertilization rates had lower cumulative C mineralization than other treatments (Figure 3E, F).

Soil organic C averaged across treatments was higher in 0–20 cm than 20–40 cm (Figure 4A, B). Compared to the control, 10 g m^{-2} yr^{-1} of N or P fertilization, and 15 g m^{-2} yr^{-1} of N+P fertilization significantly increased SOC in 0–20 cm (Figure 4A). The SOC at 20–40 cm was greater in 15 g m^{-2} yr^{-1} of N+P than other treatments (Figure 4B).

Effect of fertilization on functional diversity of soil microbial community

The AWCD was usually higher in 10 g m^{-2} yr^{-1} of N+P application throughout the incubation than other treatments (Figure 5). The AWCD was lower in N+P than N fertilization at 5 g m^{-2} yr^{-1}, but it was higher in N+P than N fertilization at 10 and 15 g m^{-2}. Eight of nine fertilization treatments (except 15 g m^{-2} yr^{-1} N) significantly increased the Shannon functional diversity of soil bacterial community compared to the control (Figure 6). The N+P fertilization at 10 g m^{-2} yr^{-1} had the highest diversity, followed by 10 and 5 g m^{-2} yr^{-1} N fertilization

In the PCA ordination diagram, samples with similar AWCD datasets were located close to one another, and those dissimilar were located far apart (Figure 7). The PCA results showed that the C substrate utilizing profiles were separated into four distinct groups: 1) 5 and 15 g m^{-2} yr^{-1} of P fertilization; 2) 10 g m^{-2} yr^{-1} of N or P fertilization, 15 g m^{-2} yr^{-1} of N+P fertilization; 3) 5 and 15 g m^{-2} yr^{-1} of N fertilization, 5 and 10 g m^{-2} yr^{-1} of N+P fertilization; and 4) unfertilized control. Metabolic profiles from the treatments within each group were similar to each other, but significantly different from other groups. The first two PCs (PC1 and PC2) explained 42.86% and 8.46% of the variance in AWCD data. The main loadings on the PC1 axis were carboxylic acids (0.87), polymers (0.83), amino acids (0.79), amine/amides (0.78), and carbohydrates (0.75), and on the PC2 axis were amino acids (0.77), and carboxylic acids (0.59).

Discussion

Response of plant biomass, functional group composition, plant and root C:N ratio to fertilization

Fertilization probably increased the growth and biomass of plants, especially grasses, by supplying essential nutrients, such as N and P [21]. The N and P-induced increase in plant biomass was consistent with previous study [21] and N addition experiments in other grasslands [4–9]. Previous study has shown that N and N+P fertilization significantly decreased species richness but N+P addition significantly increased species number of grasses, and species loss was mainly due to the loss of forbs and legumes after fertilization [21]. Studies from other grasslands also have shown that N addition significantly decreased species diversity [5–9]. Fertilization-induced increase in aboveground plant productivity and community coverage resulted in a decrease of light penetration in community canopy. This resulted in reduced growth of legumes and forbs under intense light competition, leading to local extinction of these species [21,36,37].

Fertilization-induced decrease in C:N ratios of all plant functional groups and roots was consistent with previous studies in alpine meadows that N concentrations of grass and forbs increased significantly thereby decreasing C/N ratios while these of legumes were relatively constant after N addition [22,23]. This might be due to the species specific ecological stoichiometry [38]. Grasses had higher N and P use efficiency than other functional groups, which resulted in increase in N concentration after fertilization [38]. The height and biomass of forbs didn't change or even decreased. Legume species were not sensitive to N additions for they could get additional N by N-fixing [22,38]. Studies from other grasslands also showed that N inputs consistently increased soil N availability, leading to tissue and litter N-enrichment in semiarid shrublands [15,39].

Response of soil microbial community to fertilization

Increase soil MBC in N or P fertilization treatments was contrary to our hypothesis and inconsistent with some studies that showed long-term N addition decreased MBC by 20–35% [25,40]. However, in our study site, N fertilization rate was much lower than those in other studies conducted in Europe and North America [25,40]. Increase in aboveground plant biomass and in turn C input to soil following N and P fertilization might be responsible for the observed increase in soil MBC. The increase in carbon substrate to microbes and alleviation of microbial N or P-limitation could increase the microbial growth [41], which was responsible for the increase in AWCD and functional diversity after fertilization in our study. Reduction in microbial biomass and activity with P and N+P fertilization might be due to the toxicity of nutrients that affected MBC and favored one species of microbe

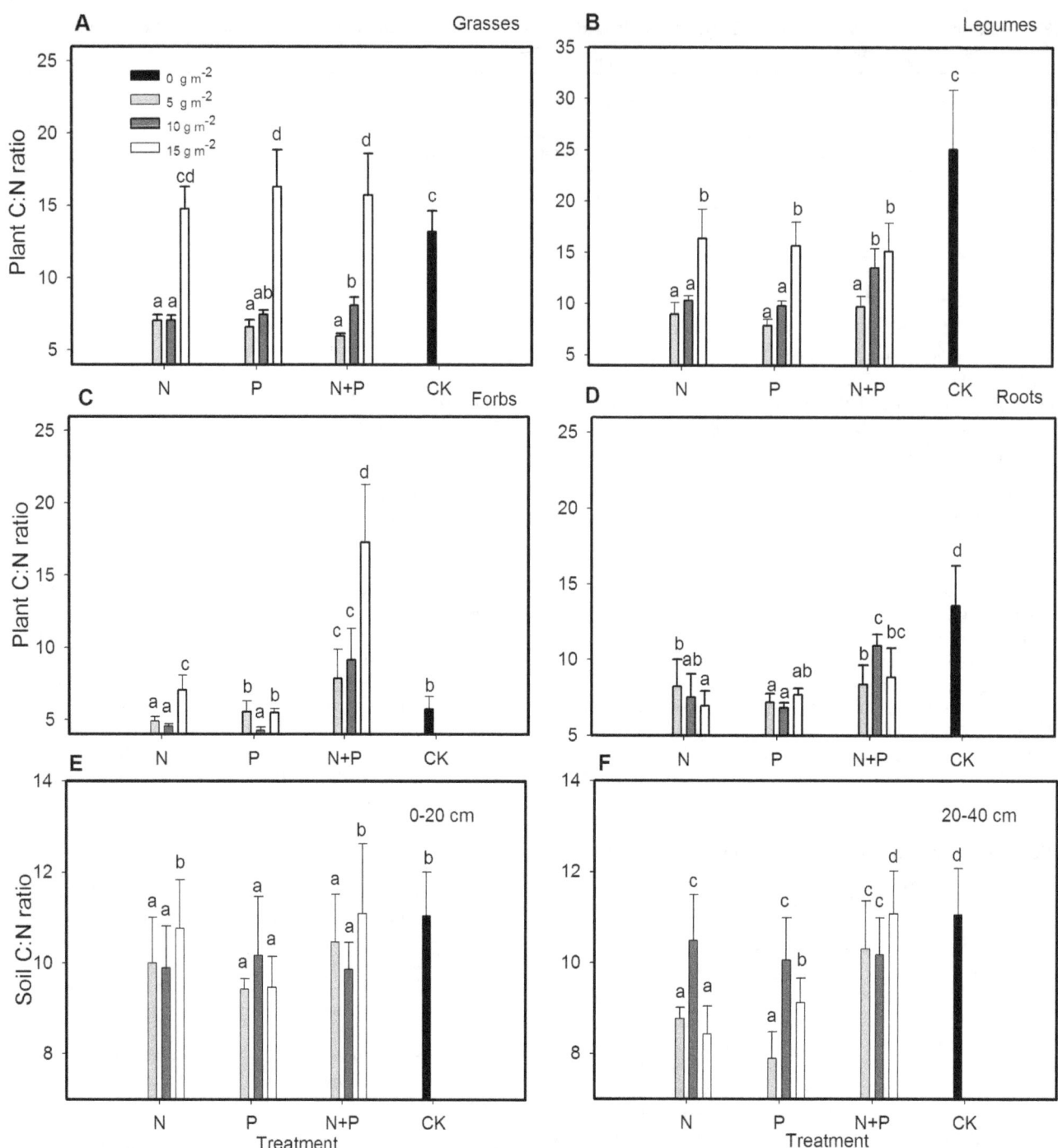

Figure 2. Mean ± SE of C:N ratios of (A) grasses, (B) legumes, (C) forbs, and (D) roots, (E) 0–20 cm soil, (F) 20–40 cm soil as affected by N and P fertilization treatments. Different letters above bars indicate significantly different at $P = 0.05$.

over another [42]. However, increased N availability could stimulate carbon-degrading enzyme activities [43,44] and microbial enzyme shifts could result in increased C mineralization and qCO_2 [46].

Our study showed similar patterns of qCO_2 and C mineralization with fertilization treatments, but inverse to the pattern of MBC. The decrease in C mineralization might be due to a decline in lignin degradation induced by reduction enzyme activity and/or fungal population densities, and modification of the soil microbe

composition following fertilization [45–47]. Ramirez et al. [40] reported that there were consistent phylum-level changes in the bacterial communities across soils with the N-amended soil. The N-induced shifts in microbial community structure should yield corresponding shifts in the functional and metabolic potentials of the communities, resulting in a change in decomposition rates [40]. In our study, we did not measure soil microbial composition or enzyme activities; however the results of C substrate utilization data based on Biolog Ecoplate indicated that N, P and N+P

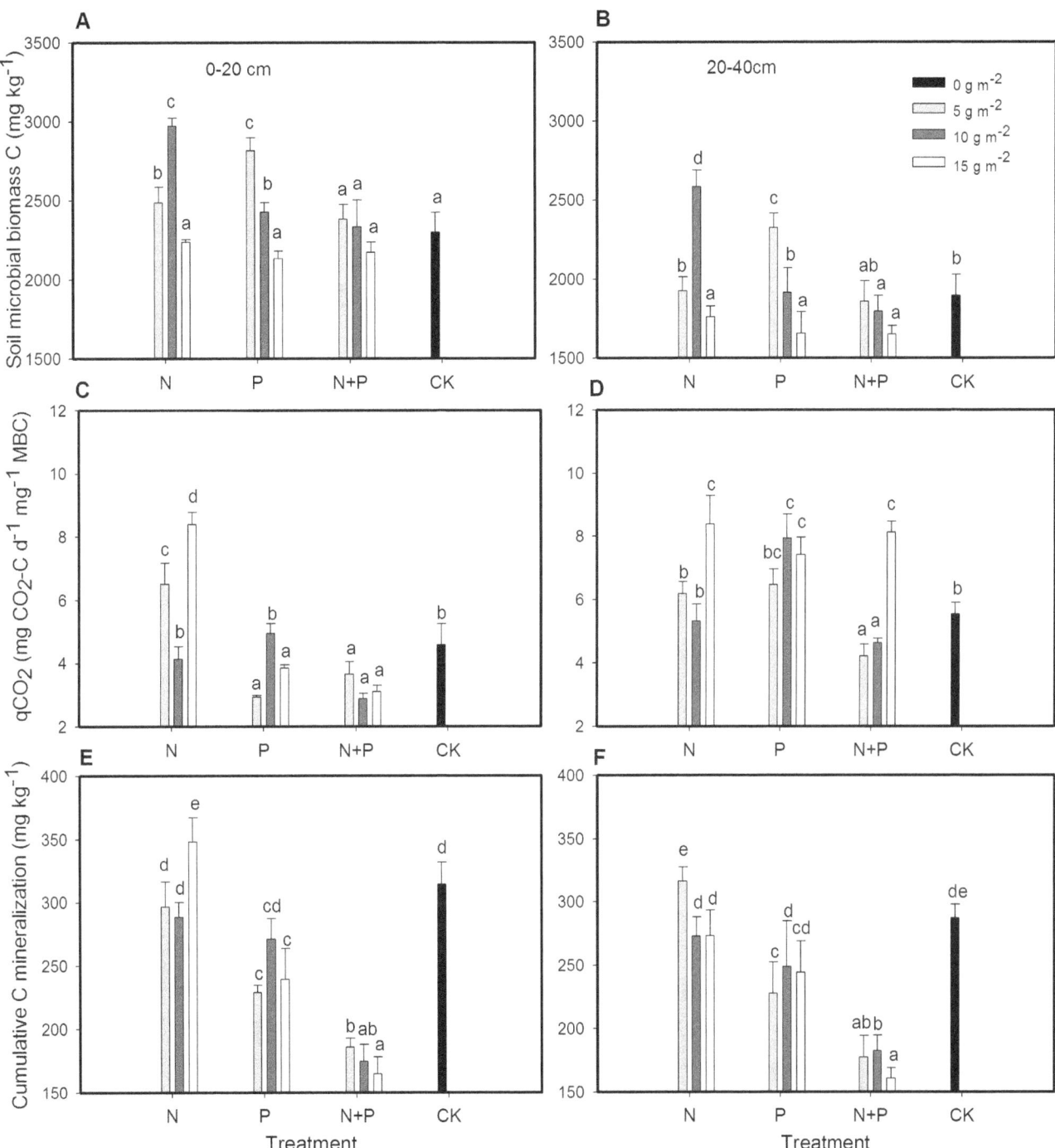

Figure 3. Mean ± SE of (A) and (B) soil microbial biomass C, (C) and (D) metabolic quotient (qCO$_2$), (E) and (F) cumulative C mineralization in 0–20 cm and 20–40 cm soil layers as affected by N and P fertilization treatments. Different letters above bars indicate significantly different at $P = 0.05$.

fertilization could increase microbial functional diversity and affect microbial carbon utilization and metabolism. Increased rates of N or P fertilization alone increased microbial biomass but reduced their activity, probably due to favorable growth of certain microbes over another due to fertilization. This was likely the negative impact on microbes by decreasing soil pH [42] and increasing toxicity [43] following N fertilization.

The combination of N+P fertilization however increased AWCD and Shannon diversity. This might be due to the following

reasons: (i) Alleviation of microbial N and P-limitation and the alteration of community composition, resulting in increased growth in certain microbial population [48]. (ii) Changes in plant community structure and composition which affected soil microbial community. (iii) Grass roots provided unique attachment sites for certain microbial populations, e.g. mycorrhizal fungi, or grass growth stimulated the formation of certain microbial communities after fertilization [49]. This was supported by principal component analysis of carbon utilization data which revealed that microbial

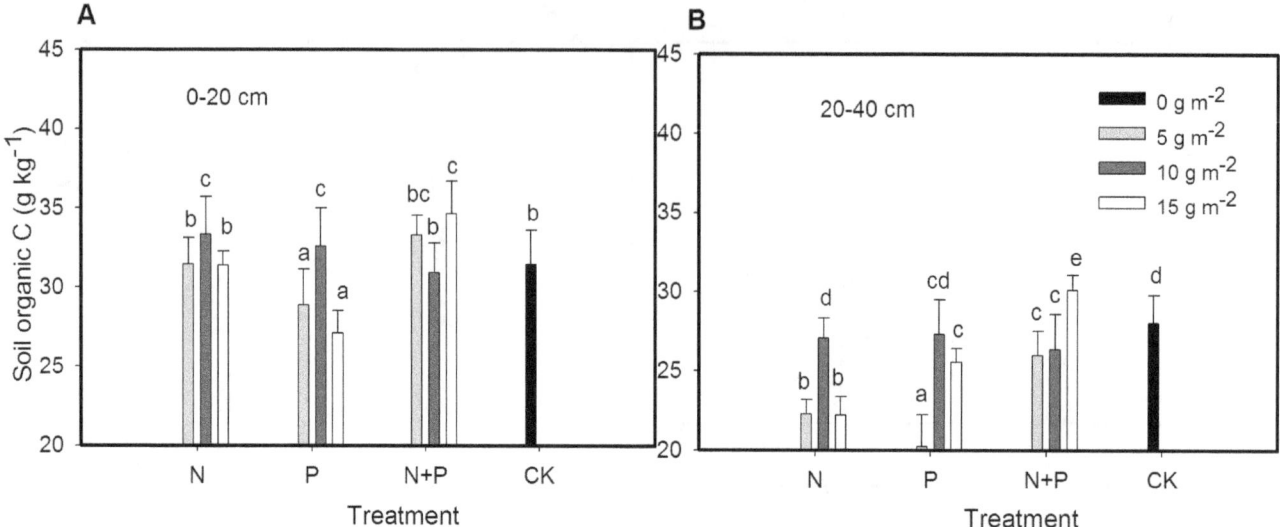

Figure 4. Mean ± SE of soil organic C in 0–20 cm (A) and 20–40 cm (B) as affected by N and P fertilization treatments. Different letters above bars indicate significantly different at $P = 0.05$.

functional diversity depended on the plant community components. These findings suggested that fertilization may alter the structure and composition of soil microbes and modify substrate utilization patterns by soil microbes in the alpine meadow, resulting in a change in C mineralization rates.

Soil organic C

Our data showed that plant community biomass was positively correlated with SOC in 0–20 cm ($R^2 = 0.115$, $P = 0.032$). The increased soil organic C only appeared in 10 g m^{-2} yr^{-1} of N, P and 15 g m^{-2} yr^{-1} of N+P compared to control. The increase in soil organic C agreed with previous studies, which showed that N

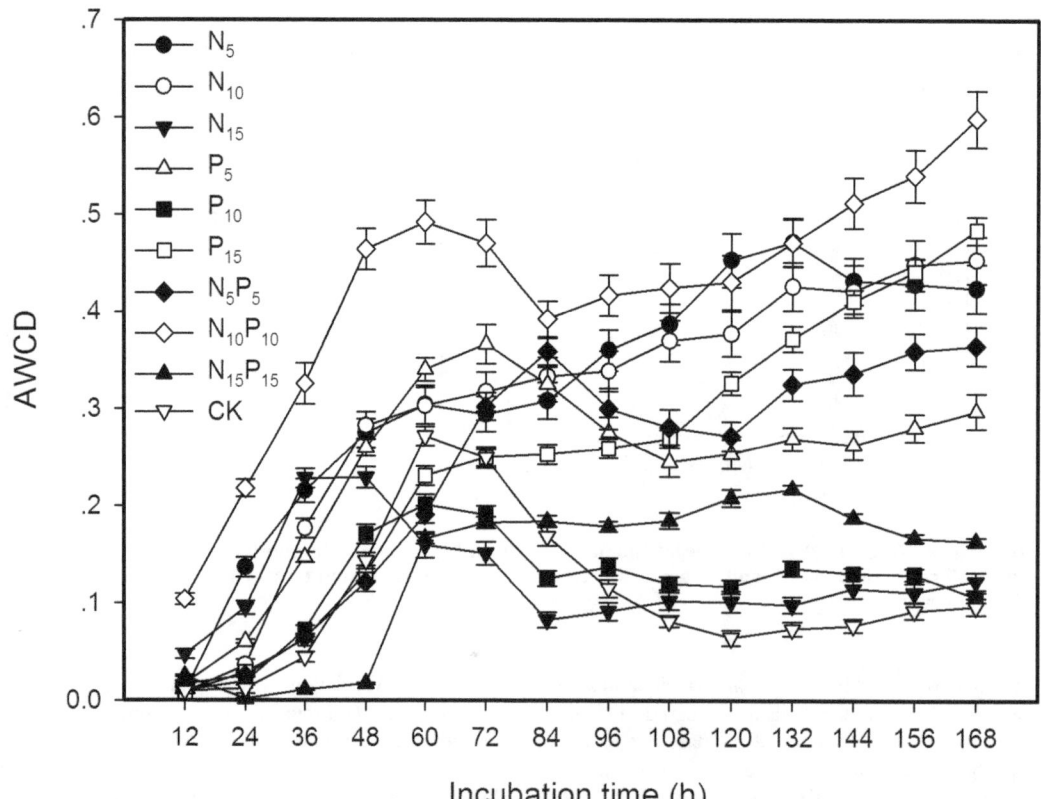

Figure 5. Average well color development (AWCD) (means ± SE) of soil microbial community in BIOLOG Ecoplate as affected by N and P fertilization treatments.

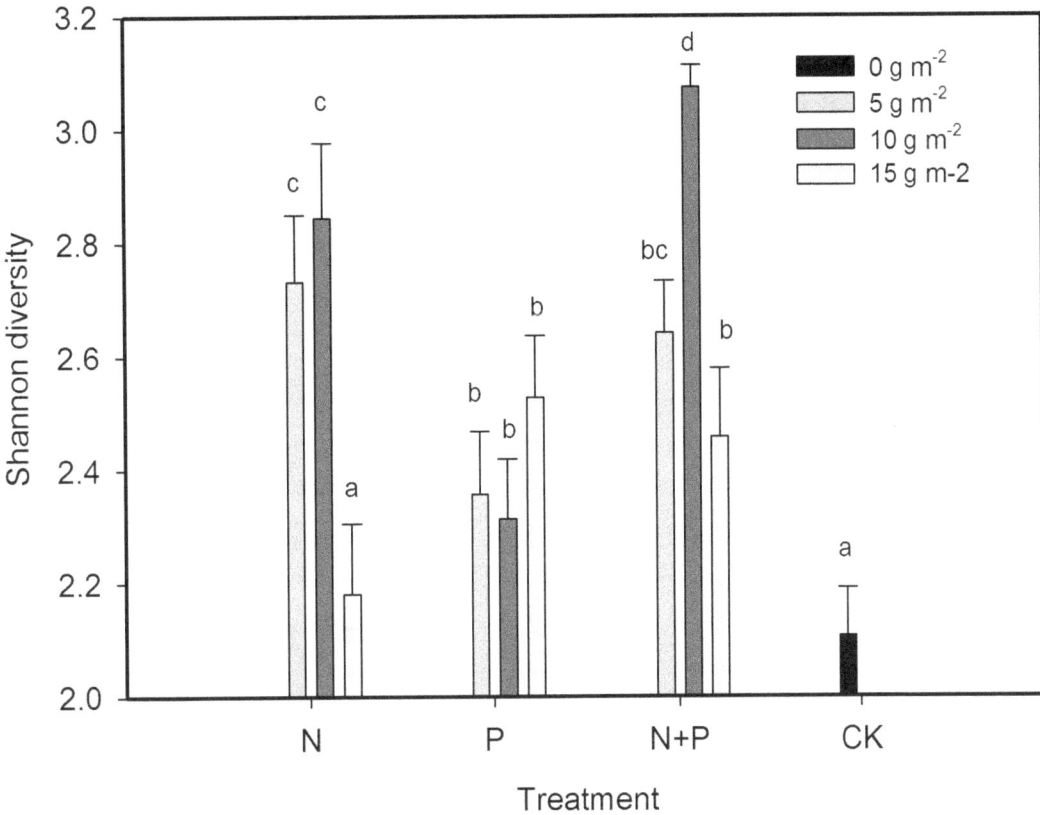

Figure 6. Functional diversity of soil microbial community in 0–20 cm as affected by N and P fertilization treatments Different letters above bars indicate significantly different at $P = 0.05$**.**

and P additions significantly enhanced C stocks and C sequestration in grassland soils [34,49–51]. In our study, plant community biomass increased and grasses accounted for 70% of the total biomass in eight of nine fertilization treatments. So, it was the amount of C input returned mostly from grasses that increased SOC. Also C:N ratio was higher in grasses than forbs in two of the three fertilization treatments. Since 15 g m^{-2} yr^{-1} of N+P fertilization increased grass biomass, the increase in SOC with this treatment was most likely to be due to increased C input and slower decomposition of grasses due to higher C/N ratio. These findings suggested that 15 g m^{-2} yr^{-1} of N+P fertilization can be used to sequester C in such alpine meadow soils. The decrease in soil organic C with 5 and 15 g m^{-2} yr^{-1} of P in 0–20 cm and most of the fertilization treatments (except 15 g m^{-2} yr^{-1} of N+P) in 20–40 cm agreed with several studies which have found that N and P fertilization contributed to decreases in soil organic C pools in tundra ecosystems [16,45]. In our study, SOC concentration was positively correlated with C:N ratios of grasses ($R^2 = 0.112$, $P = 0.035$ in 0–20 cm soil and $R^2 = 0.261$, $P = 0.001$ in 20–40 cm soil respectively) and forbs ($R^2 = 0.124$, $P = 0.026$ in 0–20 cm soil and $R^2 = 0.185$, $P = 0.006$ in 20–40 cm soil respectively), suggesting that the decrease in soil organic C concentration was partially due to changes in plant diversity and productivity of functional groups and decrease in plant C:N ratios, in particular the dominant functional groups—grasses and forbs. Given that soil represented the largest active C pool in terrestrial ecosystems, and, in particular, alpine meadows with higher soil organic C [52], a decrease in SOC and C accumulation might lead to an increase in CO_2 emission to atmosphere.

After 5 years of fertilization, decreased SOC in 5 and 15 g m^{-2} yr^{-1} of P in the upper soil layer and most of the fertilization treatments (except 15 g m^{-2} yr^{-1} of N+P) in the deeper soil layer indicated that soils with these fertilization treatments in these meadows will not be a strong C sink, and that N or P fertilization may even lower the soil organic C pool. Combined with the increase in plant community productivity and changes in the C:N ratios of plant tissues, our results also suggested that a new balance between C sequestration by vegetation and soil organic C mineralization will determine whether alpine meadow ecosystems become a C sink, source or vary across space and time. A study from European grassland showed that multi-nutrient (N, P, K, Mg) additions increased aboveground plant productivity (APP), and N+P had higher APP than N alone, but lower APP than NPKMg fertilization [9]. However, soil C sequestration was increased by N-only additions, not multi-nutrient fertilization [9]. An experiment from our study site showed that APP increased with the increasing number of added limiting resources (N, P, K, and water) and APP was higher in N+P treatment than in N+P+K treatment or N, P alone treatment [19]. Therefore, in alpine meadows, both N and P fertilization need to be applied at adequate rates (15 g m^{-2} yr^{-1}) to increase C sequestration, not N or P alone. Potassium combination with N and P might not further increase plant biomass according to study of Ren et al. [21], but further studies were needed to examine their effects on SOC in alpine meadows.

Conclusions

Nitrogen alone and N+P fertilization at rates of 15 g m^{-2} yr^{-1} doubled aboveground biomass of plant communities, driven by a

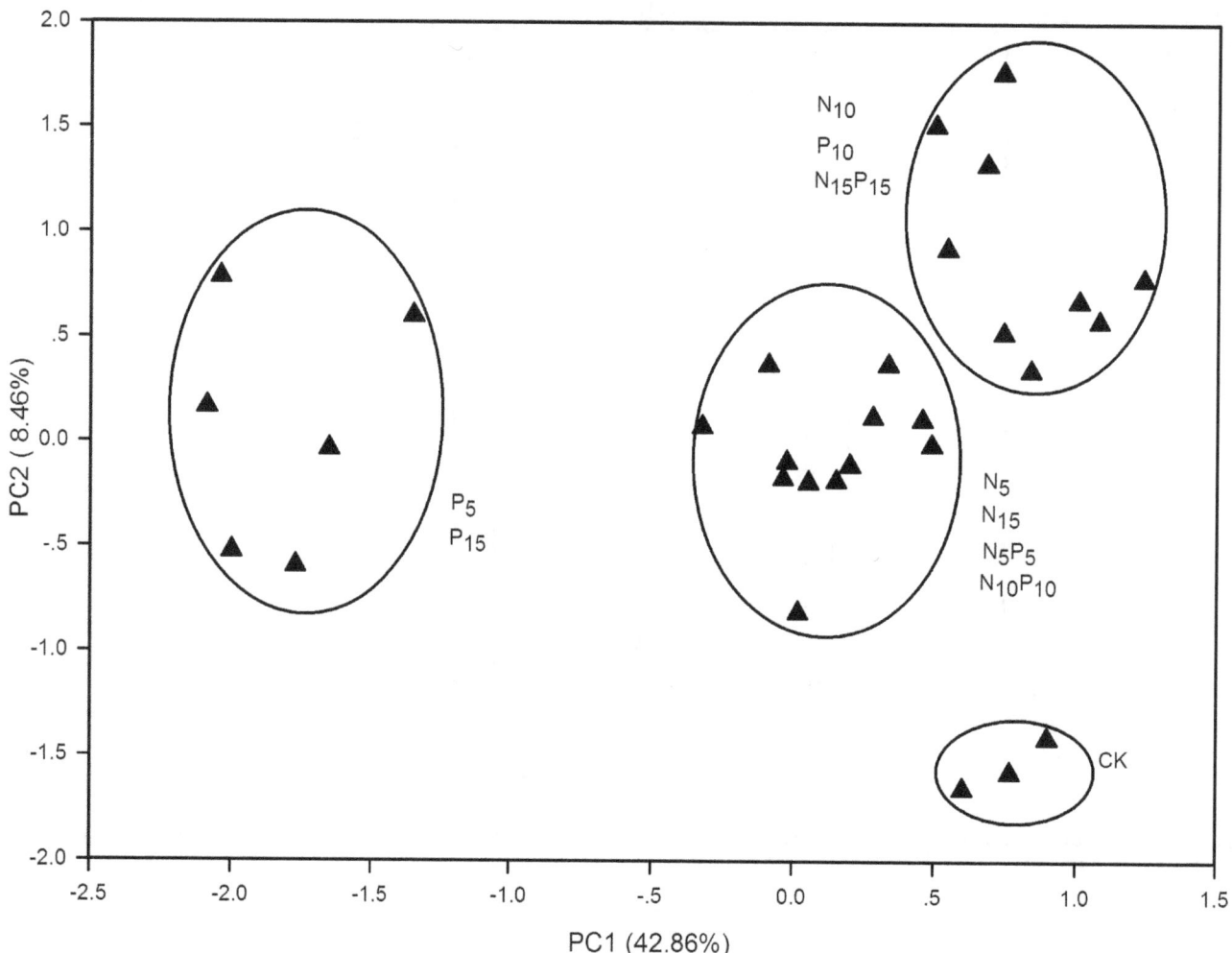

Figure 7. Principal component analysis results according to average well color development data of carbon substrate groups at 168 h in BIOLOG Ecoplates. N, nitrogen fertilization; P, phosphorus fertilization; CK, unfertilized control; subscript numbers 5, 10, 15 indicate corresponding fertilization rates (g m^{-2} yr^{-1}).

fourfold increase in grass biomass. The combination of N and P fertilizers increased SOC but reduced microbial biomass and activity and C mineralization compared to N or P alone or no fertilization. The N+P at rate of 15 g m^{-2} yr^{-1} can be used in increasing aboveground biomass of plant community and especially grass biomass and dominance, resulting in increased soil C accumulation at the surface layer in these alpine meadows.

Acknowledgments

We are grateful to Alpine Meadow Ecosystem Research Station of Lanzhou University for providing experimental site.

Author Contributions

Conceived and designed the experiments: JHL GW. Performed the experiments: YJY BWL. Analyzed the data: JHL JK. Contributed reagents/materials/analysis tools: WJL. Wrote the paper: JHL.

References

1. Post WM, Emanuel WR, Zinke PJ, Stangenberger AG (1982) Soil carbon pools and world life zones. Nature 298: 156–159.
2. Scurlock JMO, Hall DO (1998) The global carbon sink: a grassland perspective. Glob Chang Biol 4(2):229–233.
3. Yang H, Shaojie MU, Chengming SUN, Jianlong LI, Weimin JU (2011) Summary of research on estimation of organic carbon storage in grassland ecosystem. Chin J Grassland 33: 107–114.
4. Bai YF, Wu JG, Clark CM, Naeem S, Pan QM, et al. (2010) Tradeoffs and thresholds in the effects of nitrogen addition on biodiversity and ecosystem functioning: evidence from inner Mongolia Grasslands. Glob Chang Biol 16: 358–372.
5. Clark CM, Tilman D (2008) Loss of plant species after chronic low-level nitrogen deposition to prairie grasslands. Nature 45: 712–715.

6. Isbell F, Tilman D, Polasky S, Binder S, Hawthorne P (2013) Low biodiversity state persists two decades after cessation of nutrient enrichment. Ecol Lett 16: 454–460.
7. Stevens CJ, Dise NB, Mountford JD (2004) Impacts of nitrogen deposition on the species richness of grasslands. Science 303: 1676–1679.
8. Wedin DA, Tilman D (1996) Influence of nitrogen loading and species composition on the carbon balance of grasslands. Science 274: 1720–1723.
9. Fornara DA, Banin L, Crawley MJ (2013) Multi-nutrient vs. nitrogen-only effects on carbon sequestration in grassland soils. Glob Chang Biol 19: 3848–3857.
10. Wang GX, Qian J, Cheng GD, Lai YM (2002) Soil organic carbon pool of grassland soils on the Qinghai-Tibetan Plateau and its global implication. Sci Total Environ 291: 207–217.

11. Allison SD, Lu Y, Weihe C, Goulden ML, Martiny AC, et al. (2013) Microbial abundance and composition influence litter decomposition response to environmental change. Ecology 94(3):714–725.

12. Lovell RD, Jarvis SC, Bardgett RD (1995) Soil microbial biomass and activity in long term grassland: Effects of management changes. Soil Biol Biochem 27(7):969–975.

13. Li YC, Song CC, Hou CC, Song YY (2010) Effects of nitrogen input on meadow marsh soil N_2O emission and organic carbon mineralization. Chin J Ecol 29(11):2091–2096.

14. Bai JB, Xu XL, Song MH, He YT, Jiang J, et al. (2011) Effects of temperature and added nitrogen on carbon mineralization in alpine soils on the Tibetan Plateau. Ecol Environ Sci 20(5):855–859.

15. Vourlitis GL, Zorba G (2007) Nitrogen and carbon mineralization of semi-arid shrubland soil exposed to long-term atmospheric nitrogen deposition. Biol Fert Soils 43: 611–615.

16. Mack MC, Schuur EAG, Bret Harte MS, Shaver GR, Chapin III FS (2004) Ecosystem carbon storage in arctic tundra reduced by long term nutrient fertilization. Nature 431: 440–443.

17. Bradford MA, Fierer N, Jackson RB, Maddox TR, Reynolds JF (2008) Nonlinear root-derived carbon sequestration across a gradient of nitrogen and phosphorous deposition in experimental mesocosms. Glob Chang Biol 14: 1113–1124.

18. Ni J (2002) Carbon storage in grasslands of China. J Arid Environ 50: 205–218.

19. Ren ZW, Li Q, Chu CJ, Zhao LQ, Zhang JQ, et al. (2010) Effects of resource additions on species richness and *ANPP* in an alpine meadow community. J Plant Ecol 3: 25–31.

20. Xin XJ (2011) Effects of N, P additions on above/below-ground biomass allocation and plant functional types' composition in a sub-alpine meadow. Thesis, Lanzhou University.

21. Qi R (2013) Response of plant community to nitrogen and phosphorous additions in sub-alpine meadows of the Qinghai Tibetan Plateau. Thesis, Lanzhou University.

22. Chen LY (2010) Effect of N, P addition on N:P stioehiometry of different functional groups in *Potentilla fruticosa* community in a sub-alpine meadow. Thesis, Lanzhou University.

23. Zhang RY, Gou X, Bai Y, Zhao J, Chen LY, et al. (2011) Biomass fraction of graminoids and forbs in N-limited alpine grassland N:P stoichiometry. Polish J Ecol 59(1):105–114.

24. Fornara DA. Tilman D (2012) Soil carbon sequestration in prairie grasslands increased by chronic nitrogen addition. Ecology 93: 2030–2036.

25. Liu L, Greaver TL (2010) A global perspective on belowground carbon dynamics under nitrogen enrichment. Ecol Lett 13: 819–828.

26. Gong ZT (1999) Chinese Soil Taxonomy: Theories Methods and Applications. Science Press, Beijing.

27. Kalembasa SJ, Jenkinson DS (1973) A comparative study of titrimetric and gravimetric methods for the determination of organic carbon in soil. J Sci Food Agric 24: 1085–1090.

28. Vance ED, Brookes PC, Jenkinson DS (1987) An extraction method for measuring soil microbial biomass C. Soil Biol Biochem 19: 703–707.

29. Song MH, Jiang J, Cao GM, Xu XL (2010) Effects of temperature, glucose and inorganic nitrogen inputs on carbon mineralization in a Tibetan alpine meadow soil. Eur J of Soil Biol 46: 375–380.

30. Anderson TH, Domsch KH (1990) Application of ecophysiological quotients (qCO_2 and qD) on microbial biomass from soils of different cropping histories. Soil Biol Biochem 22: 251–255.

31. Garland JL, Mills AL (1991) Classification and characterization of heterotrophic microbial communities on the basis of patterns of community-level sole-carbon-source utilization. Appl Enviro Microbio 7: 2351–2359.

32. Insam H (1997) A new set of substrates proposed for community characterization in environmental samples. In: Insam H, Rangger A, editors. Microbial Communities. Berlin Heidelberg: Springer pp. 259–260.

33. Li JH, Jiao SM, Gao RQ, Bargett RD (2012) Differential effects of legume species on the recovery of soil microbial communities, and carbon and nitrogen contents in abandoned fields of the Loess Plateau, China. Environ Manage 50: 1193–1203.

34. Garland JL (1996) Analytical approaches to the characterization of samples of microbial communities using patterns of potential C source utilization. Soil Biol Biochem 28: 213–221.

35. Zhang CB, Wang J, Liu WL, Zhu SX, Ge HL, et al. (2010) Effects of plant diversity on microbial biomass and community metabolic profiles in a full-scale constructed wetland. Ecol Eng 36 (1):62–68.

36. Suding KN, Collins SL, Gough L, Clark C, Cleland EE, et al. (2005) Functional- and abundance- based mechanisms explain diversity loss due to N fertilization. PNAS 102: 4387–4392.

37. Hautier Y, Niklaus PA, Hector A (2009) Competition for light causes plant diversity loss after eutrophication. Science 324: 636–638.

38. Güsewell S (2004) N:P ratios in terrestrial plants: variation and functional significance. New Phytol 164: 243–266.

39. Vourlitis GL, Pasquini SC, Mustard R (2009) Effects of dry-season N input on the productivity and N storage of Mediterranean-type shrublands. Ecosystems 12: 473–488.

40. Ramirez KS, Craine JM, Fierer N (2012) Consistent effects of nitrogen amendments on soil microbial communities and processes across biomes. Glob Chang Biol 18: 1918–1927.

41. Stapleton LM, Crout NMJ, Sawstrom C, Marshall WA, Poulton PR, et al. (2005). Microbial carbon dynamics in nitrogen amended Arctic tundra soil: measurement and model testing. Soil Biol Biochem 37: 2088–2098.

42. Vitousek PM, Aber JD, Howarth RW, Likens GE, Matson PA, et al. (1997) Human alteration of the global nitrogen cycle: sources and consequences. Ecol Appl 7: 737–750.

43. Keeler B, Hobbie S, Kellogg L (2009) Effects of long-term nitrogen addition on microbial enzyme activity in eight forested and grassland sites: Implications for litter and soil organic matter decomposition. Ecosystems 12: 1–15.

44. Koyama A, Wallenstein MD, Simpson RT, Moore JC (2013) Carbon-Degrading Enzyme Activities Stimulated by Increased Nutrient Availability in Arctic Tundra Soils. PLoS ONE 8(10): e77212. doi:10.1371/journal.pone.0077212

45. Carreiro MM, Sinsabaugh RL, Repert DA, Parkhurst DE (2000) Microbial enzyme shifts explain litter decay responses to simulated N deposition. Ecology 81: 2359–2365.

46. Liu K, Crowley D (2009) Nitrogen deposition effects on carbon storage and fungal:bacterial ratios in coastal sage scrub soils of southern California. J Environ Qual 38: 2267–2272.

47. Bradley K, Drijber RA, Knops J (2006) Increased N availability in grassland soils modifies their microbial communities and decreases the abundance of arbuscular mycorrhizal fungi. Soil Biol Biochem 38: 1583–1595.

48. Wardle DA, Bardgett RD, Klironomos JN, Setala H, van der Putten WH, et al. (2004) Ecological linkages between aboveground and belowground biota. Science 304: 1629–1633.

49. Malhi SS, Harapiak JT, Nyborg M, Gill KS, Monreal CM, et al. (2003) Total and light fraction organic C in a thin Black Chernozemic grassland soil as affected by 27 annual applications of six rates of fertilizer N. Nutr Cycl Agroecosys 66: 33–41.

50. Biudes MS, Vourlitis GL (2012) Carbon and Nitrogen Mineralization of a Semiarid Shrubland Exposed to Experimental Nitrogen Deposition. Soil Sci. Soc. Am. J. 76: 2068–2073.

51. Li LJ, Zeng DH, Yu ZY, Fan ZP, Mao R (2010) Soil microbial properties under N and P additions in a semi-arid, sandy grassland. Biol Fert Soils 46(6):653–658.

52. Zhang JX, Cao GM, Zhou DW, Hu QW, Zhao XQ (2003) The carbon storage and carbon cycle among the atmosphere, soil, vegetation and animal in the *Kobresia humilis* alpine meadow ecosystem. Acta Ecol Sin 23(4):627–634.

Nutrient Addition Dramatically Accelerates Microbial Community Succession

Joseph E. Knelman[1,2], Steven K. Schmidt[2], Ryan C. Lynch[2], John L. Darcy[2], Sarah C. Castle[3], Cory C. Cleveland[3], Diana R. Nemergut[1,4]*

1 Institute of Arctic and Alpine Research, University of Colorado, Boulder, Colorado, United States of America, 2 Department of Ecology and Evolutionary Biology, University of Colorado, Boulder, Colorado, United States of America, 3 Department of Ecosystem and Conservation Sciences, University of Montana, Missoula, Montana, United States of America, 4 Environmental Studies Program, University of Colorado, Boulder, Colorado, United States of America

Abstract

The ecological mechanisms driving community succession are widely debated, particularly for microorganisms. While successional soil microbial communities are known to undergo predictable changes in structure concomitant with shifts in a variety of edaphic properties, the causal mechanisms underlying these patterns are poorly understood. Thus, to specifically isolate how nutrients – important drivers of plant succession – affect soil microbial succession, we established a full factorial nitrogen (N) and phosphorus (P) fertilization plot experiment in recently deglaciated (~3 years since exposure), unvegetated soils of the Puca Glacier forefield in Southeastern Peru. We evaluated soil properties and examined bacterial community composition in plots before and one year after fertilization. Fertilized soils were then compared to samples from three reference successional transects representing advancing stages of soil development ranging from 5 years to 85 years since exposure. We found that a single application of +NP fertilizer caused the soil bacterial community structure of the three-year old soils to most resemble the 85-year old soils after one year. Despite differences in a variety of soil edaphic properties between fertilizer plots and late successional soils, bacterial community composition of +NP plots converged with late successional communities. Thus, our work suggests a mechanism for microbial succession whereby changes in resource availability drive shifts in community composition, supporting a role for nutrient colimitation in primary succession. These results suggest that nutrients alone, independent of other edaphic factors that change with succession, act as an important control over soil microbial community development, greatly accelerating the rate of succession.

Editor: Jack Anthony Gilbert, Argonne National Laboratory, United States of America

Funding: This work was supported by the National Science Foundation of the USA (http://nsf.gov/) through grants for studying microbial succession following glacial retreat (DEB-0922267) to DRN, SKS, and CCC; the Alpine Microbial Observatory (MCB-0455606) to SKS; and a graduate research fellowship to JEK. The funders had no role in study design, data collection and analysis, decision to publish, or preparation of the manuscript.

Competing Interests: The authors have declared that no competing interests exist.

* Email: nemergut@colorado.edu

Introduction

Deglaciated forefields have been valuable model systems for developing and testing theories of succession and have greatly enhanced our understanding of the relationship between community structure and function during ecosystem development [1–3]. Shifts in soil nutrient pools, including increases in available nitrogen (N) and phosphorus (P), have been well documented along early primary successional chronosequences [4–6] and have been shown to correlate with changes in plant community succession [1,7,8]. Recently, studies in such systems have revealed that – like plants – microbial communities also progress through successional stages [9–11]. However, the forces that control microbial succession are not well understood.

Some evidence suggests that shifts in nutrient availability may also, in part, drive microbial community succession. For example, in primary successional ecosystems, research has corroborated relationships between natural gradients in soil nutrients and microbial community composition [12,13]. Such correlations can be difficult to interpret, however, as changes in microbial community composition could be both a cause and consequence of shifts in soil fertility. Furthermore, the mechanisms underlying correlations between standing nutrient pools and microbial communities may be temporally disconnected, in that current soil biogeochemical status may not accurately reflect the historical nutrient conditions that structured the microbial community. Thus, manipulation experiments are essential in evaluating the direct impact of nutrients and their limitations on microbial communities. Indeed, fertilizer treatments are known to elicit changes in soil microbial community structure and function in more developed ecosystems [14,15] suggesting that nutrient availability may also be important in controlling successional changes in microbial community composition.

Yet, it would be surprising if nutrients alone drove microbial community succession for several reasons. First, other edaphic properties also undergo concomitant shifts with microbial community structure and function during succession, some of which are known to more strongly correlate with microbial community structure than nutrient pools in developed soils. For example, organic carbon (C) pools and pH, which typically show dramatic changes across primary successional chronosequences [2], are key determinants of soil microbial community composition at regional to global scales [16–19]. Second, soil microbial community structure can correlate with plant community composition [20,21], which can show strong spatial gradients in early succession [3]. Third, stochastic processes can be key in shaping early successional communities where the importance of dispersal events may be accentuated, [22–24] and arrival order may influence assembly through priority effects [25]. Given the large functional and phylogenetic diversity of microbial communities, it is possible that succession is influenced by a diverse combination of such factors [26].

Thus, the extent to which nutrients themselves influence microbial community assembly outside of the myriad of factors that change over succession is unknown. To specifically isolate the effects of nutrients, we performed a full factorial N×P fertilization experiment in soils that had been exposed for ~3 years in the forefield of the Puca Glacier in Southeastern Peru. We analyzed soil bacterial communities before and one year following nutrient additions and compared them with soils sampled from three different locations over an 85-year section of the Puca Glacier chronosequence. The Puca Glacier soils constitute an autotrophic successional sequence [27], and both photosynthesis and respiration respond strongly to P additions in microcosms [28,29]. Nitrogen appears to be limiting in this system as well and N-fixation rates in 4 year old unvegetated soils are comparable to rates measured in developed soil crusts [30]. Thus, given work that demonstrates relationships between nutrients and microbial community composition, we hypothesized that fertilizer additions to early successional soils would drive communities to be compositionally different than unfertilized (control) soils. However, due to the potential influence of other edaphic (e.g. pH, organic C, soil moisture) and stochastic factors on microbial succession, we hypothesized that fertilized communities would be unique from communities found along the natural chronosequence.

Materials and Methods

Study site description, fertilization, and sampling

The study site is located in the forelands of the Puca Glacier in the Cordillera Vilcanota of Peru (13°46′24″S, 71°04′17″W, ~5,000 m.a.s.l.). No specific permits were required for our field studies and our work did not involve endangered or protected species. Mean annual precipitation is roughly 100 cm and mean annual temperature is ~5°C. Moraine rocks at this site have high quartz and calcite mineral content. Further details of this site can be found in previous work [9,30] and soil characteristics are presented in Table S2.

We established permanent plots (1 m^2) near the terminus of the glacier, in soils that had been deglaciated for approximately 3 years at the time of initial sampling. Corners were marked with long nails (approximately 15 cm shank length) to guide resampling. Sampling occurred in August 2010 (pre-treatment) and August 2011 (post-treatment). All of the plots were unvegetated and no mosses and lichens were present at the time of establishment. Each of the 16 plots was randomly chosen to receive one of three nutrient amendments (nitrogen addition (+N), phosphorus addition (+P), the combination of the two (+NP)) or to serve as controls, resulting in a total of four plots per treatment and four control plots.

Pre-weighed amounts of fertilizer were dissolved in glacier-melt stream water and fertilizer solutions were applied with handheld sprayers. Each sprayer was designated for a particular treatment to avoid cross contamination. For the +N plots, nitrogen was added

in the form of ammonium nitrate (NH_4NO_3) resulting in 15 g of NH_4NO_3 and 5.25 g of N/m^2. The +P plots received 0.5 g of phosphorus in the form of 2.2 g of potassium dihydrogen phosphate (KH_2PO_4). +NP plots received 15 g of NH_4NO_3 and 2.2 g of KH_2PO_4. For controls, stream water from the same source was sprayed onto the plots. These levels of nutrient addition were designed to result in a pulse of nutrients that would greatly overcome any possible natural limitations.

Plots were sampled prior to the application of fertilization treatment. In each plot surface soil was collected (0–5 cm) from 2 locations, and samples were composited to generate one sample per plot. Samples were obtained in the same manner one year following the fertilization treatment. Ethanol and paper towels were used to sterilize the tools before sampling each individual plot.

Samples were collected in a similar manner along three transects of varying age across the glacial forefield both years; molecular analyses were done on the samples collected in 2011. These reference soils represented advancing stages of succession: soils that had been exposed for approximately 5 years, soils with biological soil crust formation (approximately 20 years after exposure), and soils with 25–50% vegetation cover (approximately 85 years after exposure). At the field site, samples were kept in a cooler on ice for transport to Boulder, CO. Soils were sieved (to 2 mm), and then stored at 4°C for soil characterization. A subsample was immediately archived in a −80°C freezer for molecular analysis and later used for KCl extractions.

Soil Analysis

Gravimetric soil moisture and pH (using a ratio of 2 g soil to 4 mL DI H_2O) were assayed based on standard methods [9]. For total organic C analysis, carbonate (inorganic C) removal was first performed on dried, ground soils [9]. 50 mg of these processed soils were packed into tin capsules; %C and %N were determined using a Thermo Finnigan EA 1112 Series Flash Elemental Analyzer (Thermo Fisher Scientific, Inc., Waltham, Massachusetts, USA) [31]. Bio-available P concentrations were measured on air-dried and sieved soil (2 mm × 2 mm) by extracting 3–5 g of soil with 0.5 M sodium bicarbonate for 30 minutes [32]. Extracts were filtered and analyzed colorimetrically using the ammonium molybdate-malachite green method [33] adapted for microplate analysis. NH_4^+ and NO_3^-/NO_2^- extractable N were analyzed from soils using 2M KCl with 1 hour shaking and a 22 hour extraction period [34]. This analysis was performed on soils that were frozen at −80°C. Although not fresh samples, these soils typically withstand extreme fluctuations in temperature [35] and the data presented here are intended for within study comparison only. NH_4^+ and NO_3^-/NO_2^- were measured on a Lachat QuikChem 8500 Flow Injection Analyzer (Lachat Instruments, Hach Company, Loveland, CO) and BioTek Synergy 2 Multi-detection Microplate Reader (BioTek, Winooski, VT) respectively.

DNA Extractions for 454 pyrosequencing

MO BIO PowerSoil^{TM} DNA Isolation kits were used as per the manufacturer's instructions for DNA extractions of soil samples (Mo Bio Laboratories, Inc., Carlsbad, CA). PCR-amplified bacterial 16S rRNA genes from the genomic DNA of the soil samples were generated using a universal bacterial 27F and 338R primer set as described by Hamady et al. [36], and reaction conditions followed those described by Fierer et al. [37], though modified to 25 PCR cycles. Primers included a 2 bp linker, the 454 Roche Titanium A/B primer, and a unique, 12 base pair error-correcting Golay barcode for pyrosequencing as detailed by Knelman et al. [21]. 454 Life Sciences GS FLX Titanium pyrosequencing of the 16S rRNA gene amplicons was completed by the Duke Institute for Genome Sciences & Policy (Duke University, North Carolina).

Pyrosequence and statistical analysis

Using QIIME, sequences were limited to those of a sequence length of 200 to 400 base pairs, a maximum of 5 homopolymers, a minimum quality score of 25, and a maximum of ambiguous bases/primer mismatches of 0; reverse primers were removed, and all samples were then denoised using flowgram clustering in QIIME [38]. Chloroplast sequences were removed. OTUs were selected at a 97% identity level by clustering based on representative sequences via UCLUST [39]. The Ribosomal Database Classifier [40], a naïve Bayesian classifier, was employed to assign taxonomic identification to OTUs. After sequence alignments based on the NAST algorithm [41], a phylogeny was constructed with the FastTree algorithm [42]. OTU tables were rarified to the lowest number of sequences in a sample: 407 for community dissimilarity analyses of fertilization plots. Reference transects of advancing age included 6, 5, and 3 sequenced replicate samples, respectively, and were rarefied to 71 to include all of these samples. For comparison of reference samples and fertilization plots this workflow was repeated. In order to examine differences among bacterial communities, pairwise distance matrices based on weighted UniFrac, a phylogenetic distance metric, were generated for entire communities and the cyanobacterial subset of communities in fertilization plots [43,44]. The Principal Coordinate Analysis (PCoA) ordinations were constructed based on OTU tables and weighted UniFrac distance matrices for overall communities. The QIIME-generated OTU tables were used to evaluate the relative abundance of all taxa.

Primer v6 software [45] was used to perform permutational ANOVAs (PERMANOVA) to compare phylogenetic distances among bacterial communities. PERMANOVA tests were used on both UniFrac beta diversity matrices of the entire communities and cyanobacterial portions of communities. PERMANOVA analysis was also employed to assess differences among treatment-affected communities and successional reference communities. For all comparisons with reference communities, data were rarefied to the lowest sampling depth among both fertilization plot and reference plot samples.

R software [46] was used for further statistical analysis. The PERMDISP2 procedure (with permutational P-values) from the R vegan package to test homogeneity of group dispersions (variances) was also employed via QIIME in order to test for differences in community phylogenetic dispersion (UniFrac) in fertilized samples and reference successional communities [47,48]. As well, the pgirmess package in R was used to evaluate comparisons among reference chronosequence soil relative abundance data via the Kruskal Wallis test. To assess treatment vs. temporal effects underlying shifts in overall phylogenetic community composition, a Tukey's HSD post-hoc test was used to compare UniFrac distances of paired pre- and post-treatment plots with paired control plots from both years. Additionally, to assess the relative abundances of bacterial taxa, we compared the differences in paired pre- to post-treatment taxon relative abundances for each treatment with that of paired control plots via Tukey's HSD post-hoc tests. To examine the relationship between treatment-related community shifts from our fertilization experiment and reference communities across advancing stages of soil development, we examined the relationship between weighted UniFrac phylogenetic dissimilarity and time between +NP communities and reference communities via a Spearman correlation Mantel test. The Mantel test tests the null hypothesis that there is no

correlation between +NP and reference community dissimilarity and chronosequence age rank.

All relative abundance data and environmental variables were evaluated for normality. Taxon relative abundances and fertilizer plot NO_3^-/NO_2^- were square root transformed to achieve a normal distribution prior to statistical analysis. All other edaphic factors were natural log transformed. ANOVAs, Tukey's HSD, and Kruskal Wallis post-hoc tests were used to assess differences in pH, %C, P, N pools and soil moisture in fertilization plots and reference chronosequence soils. Percent N was below the detection limit in a majority of samples and thus removed from statistical evaluations.

Sequences and metadata have been deposited in FigShare and are available with the DOIs: 10.6084/m9.figshare.1050042 (metadata) and 10.6084/m9.figshare.1048992 (sequences).

Results and Discussion

Together, our analyses demonstrate that a single +NP application caused the bacterial community structure of the 3-year-old barren soils to converge with the structure of 85-year-old vegetated soils after only one year. First, paired pre- and post-treatment plot community differences (weighted UniFrac distance) were assessed among all plot categories using an ANOVA. The +NP plots showed a significant community shift in response to the treatment (Tukey's HSD; P = 0.037); no other significant differences in community structure were detected between treatments and controls (Tukey's HSD; P>0.05). A PCoA ordination (Fig. 1) revealed a successional trend in community composition across the reference chronosequence, with post-treatment +NP communities clustering with the oldest reference communities. A PERMANOVA analysis demonstrated that there were no significant differences among pre-treatment communities (Table 1). However, communities in post-treatment +NP plots were significantly different from both pre- and post-treatment controls, including the paired pre-treatment +NP plots (PERMANOVA, P<0.05, Table 1). When +NP communities were compared to reference communities across the natural chronosequence, a Mantel test of pairwise average UniFrac [43,44] distances between +NP plots and reference samples revealed significant patterns of decreasing dissimilarity: +NP communities were most similar to the 85 year old successional soils (Fig. 2, $\rho_M = -0.35$ P = 0.01). The PERMANOVA analysis also showed that +NP communities were significantly different than communities of all successional stages except those of the oldest transect (85 years old) (Table 1). These results suggest that fertilization drives community composition away from early successional stages and results in convergence with communities of older soils. Likewise, the phylogenetic dispersion [47,48] of +NP communities was significantly different from all reference communities except those in the 85 year old soils (Table 2). We note that our PERMANOVA analysis was not corrected for multiple comparisons due to the low statistical power of our study, but the general results of this analysis were nonetheless corroborated by our other statistical analyses of treatment effect (ANOVA/Tukey's HSD of pre- and post-treatment community shifts) and convergence of the +NP plots to the oldest successional soils (Mantel test of +NP community distance compared to successional reference samples over time).

Our results suggest that nutrient colimitation is an important control on microbial primary succession in this system. Because of low statistical power, it is difficult to discern whether this colimitation is simultaneous, meaning that both nutrients need to be present for a community response, or independent, meaning that each nutrient in isolation may elicit some response [49].

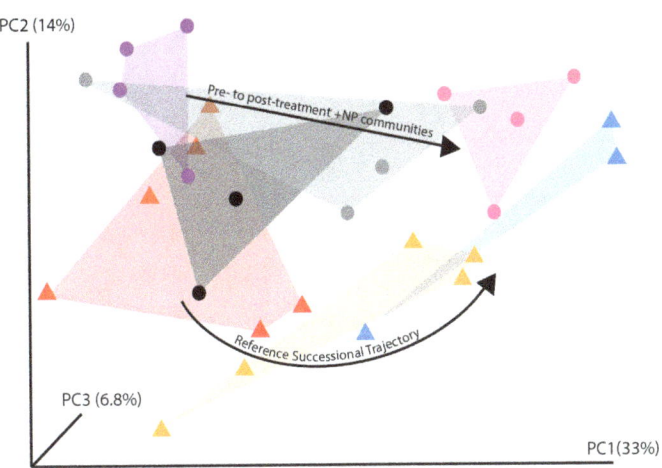

Figure 1. Principal Coordinates Analysis (PCoA) ordination plot of bacterial communities from the field fertilization experiment and bacterial communities from the successional chronosequence. Only the +NP treatment communities are shown because the +N and +P treatments did not result in significant community shifts. PCoA visually represents differences among community composition as the distance between points. Triangles represent communities from the natural chronosequence: red = 5 years old; orange = 20 years old; blue = 85 years old. Circles represent communities from the fertilization experiment: black = pre-treatment control; grey = post-treatment control; purple = pretreatment +NP; Pink = post-treatment +NP. Our analysis revealed significant community shifts over the reference chronosequence (triangles) as well as a significant response to +NP fertilization (circles). As well, the PCoA analysis demonstrates that the +NP communities (pink circles) group with the oldest soils from the chronosequence (blue triangles).

However, there is some evidence that single nutrient additions may cause a smaller response than when both nutrients are abundant. For example, our results show that post-treatment +P communities are not significantly different from post treatment + NP communities (Table 1). As well, both +N and +P plots show patterns of convergence similar to +NP plots in comparison with ongoing natural succession; by contrast, control plots do not display convergence (Table 1). Thus, +N and +P communities may represent intermediate states between control and +NP plots, but we were not able to statistically demonstrate an underlying treatment effect.

While our study is unique as we established and resampled nutrient addition plots in a remote glacial forefield, the rapidly changing nature of the Puca Glacier landscape and criteria for setting up plots on a stable and relatively homogenous surface limited replication and necessitated rarefaction of sequencing depth to 71 to include all available samples. As such, we

acknowledge the need to be circumspect in drawing conclusions as such factors curbed the statistical power of our study and potentially our ability to detect smaller magnitude treatment effects in the +N and +P additions, for example. However, we note that the patterns shown here are robust to even lower rarefaction depths (55–70); thus, it is likely that observed patterns are real. Nonetheless, our research shows the greatest, and only statistically significant treatment effect on microbial communities under +NP additions, suggesting the effect of both nutrients in tandem is important in succession.

Interestingly, standing nutrient pool analysis lends some insight into particular dynamics that may underlie nutrient colimitation in this autotrophic chronosequence. For example, +P and +NP soils both show significant increases in ammonium pools in comparison with control plot soils (Table S1), which is consistent with a body of research that demonstrates P limitation is a strong control of N-fixation [50,51], and may be particularly strong in this autotrophic

Table 1. Post-treatment +NP phylogenetic community structure was significantly different from controls and from all communities from the reference chronosequence with the exception of communities in the oldest soils (P<0.05).

Permutational MANOVA (PERMANOVA) contrast P-values				
Sample vs. Sample	post-treatment control	post-treatment +N	post-treatment +P	post-treatment +NP
pre-treatment control	0.415	0.066	**0.031**	**0.026**
post-treatment control	---	0.422	0.072	**0.036**
pre-treatment +NP-paired	0.114	0.085	**0.024**	**0.032**
post-treatment +NP plots	**0.036**	**0.023**	0.18	---
succession timepoint 1	0.124	**0.003**	**0.005**	**0.006**
succession timepoint 2	0.105	**0.009**	**0.018**	**0.022**
succession timepoint 3	0.152	0.055	0.179	0.162
Significant P-values (P<0.05) bolded.				

Controls showed no differences from any contrasts (P>0.05). Significant P-values (P<0.05) are bolded.

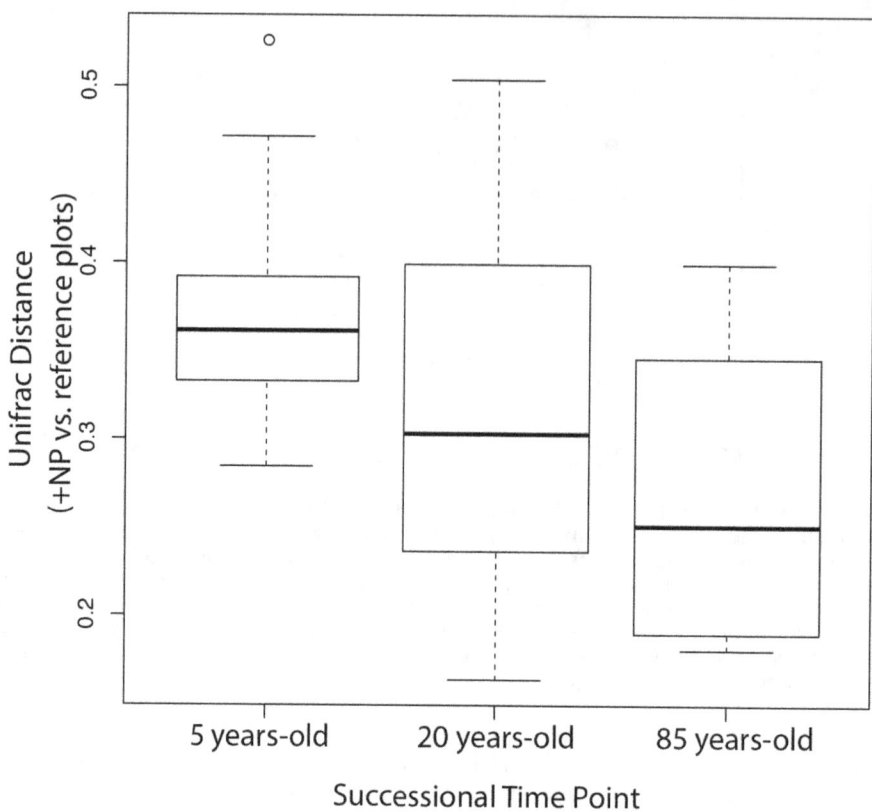

Figure 2. Relationship between +NP treatment-affected communities and reference communities. A box plot shows the average weighted UniFrac [43,44] distance between +NP-treated communities and reference communities with increasing successional time. A Mantel test demonstrates that +NP communities show decreasing dissimiliarty as compared to the reference communities over advancing stages of succession ($\rho_M = -0.35$ $P = 0.01$).

chronosequence that features cyanobacterial N-fixers [9]. Likewise, +N plots show a significant increase in bioavailable-P relative to control plots (Table S1), a pattern supported by research that shows N availability may limit the production of phosphatase enzymes [51–53]. Thus, these particular biochemical pathways lead to a coupling of nutrient cycles, which appears to be reflected in a colimitation to successional processes.

Despite the multitude of well documented changes across successional gradients including shifts in pH, C pools, plant cover and biotic historical factors, nutrient addition alone not only caused changes in early successional community structure, but induced convergence with late successional soil communities of the

natural chronosequence (Fig. 1 and 2 and Tables 1 and 2). For example, strong changes in %C, another known filter on microbial communities, were observed across the natural chronosequence but not in +NP plots (Tables S1 and S2). In other ecosystems, the effects of fertilization on microbial community structure have been attributed to changes in plant productivity or community structure [14]. However, it is important to note that while the +NP fertilization caused sparse vegetation (<15 cm tall) to colonize after one year at our site, soils were collected at least 75 cm from these small plants. Altogether, our results suggest that the effects of the +NP fertilization on microbial community succession were direct and not mediated through changes in other aspects of the

Table 2. Post-treatment +NP communities showed differences from all reference succession communities with the exception of the oldest transect (P<0.05).

Homogeneity of Dispersion (PERMDISP) P-values		
Sample vs. Sample	post-treatment +NP	post-treatment control
succession timepoint 1	**0.022**	0.508
succession timepoint 2	**0.042**	0.588
succession timepoint 3	0.555	0.997
Significant P-values (P<0.05) bolded.		

Controls showed no difference in community dispersion from communities of any of the reference succession transects (P>0.05). Significant P-values (P<0.05) are bolded.

abiotic environment or through the effects of plants on soil communities.

Our field-based fertilization experiment helps to extend existing ecological theory regarding the role of nutrient limitations in succession [4,5,54] to microbial communities present in the earliest primary successional soils, which are important for biogeochemical cycling, physical soil development, and plant colonization [9,21,30]. While it is widely acknowledged that microbes can alter soil fertility and nutrient cycling processes, and that changes in soil nutrient pools and microbial communities occur over primary succession [9,12,13,30], to what extent nutrients directly structure soil microbial communities is not clear. Our fertilization experiment allowed us to decouple the effects of changes in microbial communities on nutrient cycles and to directly demonstrate the influence of nutrient pools on microbial succession. Correlative studies are less powerful because they cannot isolate the impact of individual factors amidst the multiplicity of soil properties that change with succession, and because measured soil properties may be decoupled from microbial community composition in time.

Despite the high fertilization rate we used, the nutrient addition treatment did not push communities to an alternative or novel state, but simply accelerated succession, rapidly producing a community that was structurally most similar to the community in the 85 year old soils in the chronosequence (Fig. 1 and 2 and Tables 1 and 2). Thus, our data highlight the stability of soil microbial communities [55]. Few studies have explicitly evaluated nutrients in the context of longer-term successional reference plant communities to understand how nutrients may either drive succession or shape alternative stable states in communities. However, in a study of salt marsh vegetation, Van Wijnen and Bakker [56] observed that fertilization of young marsh communities resulted in plant communities that resembled those of older, unfertilized marshes. These results further suggest that nutrient-related mechanisms for succession may be generalizable between plant and microbial communities.

The relative abundance of cyanobacteria significantly increased in the +NP plots and the phylogenetic structure of the cyanobacterial communities in post-treatment +NP plots was significantly different from the paired pre-treatment +NP and pre-/post-treatment control plots (PERMANOVA, P<0.05, Table S3). Although not significant, cyanobacterial relative abundance nearly doubled between the oldest and youngest stages of the reference chronosequence and past work at this site has documented similar successional changes in cyanobacterial community structure (Table S2) [9,30]. Consistent with these results, a laboratory experiment evaluating microbial autotrophs from this site demonstrated that P additions resulted in significant increases in the growth rate of photoautotrophic crusts [28]. Both N fixation rates and the relative abundance of N-fixing cyanobacteria show successional trends at this site as well [30], suggesting that N availability may also limit microbial growth and activity. The current study adds to this work and demonstrates that both N and P together are important colimiting controls over community successional processes in this system (Tables 1 and 2).

The increase in the relative abundance of cyanobacteria in the + NP plots may reflect their ecological advantage in this low C environment. In a laboratory study, Drakare [57] observed that P additions enhanced cyanobacterial populations, but only in an environment where low C concentrations constrained heterotrophic growth. Incubation studies of early successional soils that found increases in heterotrophic activity in response to both N and C (but not to N alone) are also consistent with this interpretation [58,59]. These results indicate that C often limits the response of the heterotrophic community to nutrient additions, whereas cyanobacteria can readily take advantage of such nutrients to fuel photosynthesis. By extension, we argue that the observed effects of N and P additions on microbial community succession are likely to apply only to autotrophic successional sequences, and that heterotrophic succession (sensu Fierer et. al [27]) may be controlled by a different suite of resources, including C availability.

Microbes are fundamental to soil physical and chemical development and underlie ecosystem function, thus understanding the factors that drive soil microbial community succession is key to predicting and managing ecosystem development. Particularly in low nutrient environments, microbial activity has major effects on soil, plant community, and ecosystem development [9,30,60,61]. Likewise, low nutrient environments may feature more prominent nutrient colimitations [49]. As such, the results of this study have important implications for understanding nutrient controls on ecosystem development and relevant models for microbial succession. Furthermore, while early successional microbial communities may vary strongly in both composition and in terms of the specifics of resource availability (e.g., heterotrophic vs. autotrophic), our study provides evidence that nutrient colimitation may provide a generalizable mechanism for microbial community succession in autotrophic successional sequences. Our data also support recent evidence for the stability of soil microbial communities, as fertilization simply accelerated succession and did not push communities into a novel state. Overall, the details of microbial nutrient limitations presented herein are essential to understanding the factors that structure early successional microbial communities, the profound contributions they make to soil development, and the ecosystem processes they mediate.

Acknowledgments

We thank S.P. O'Neill, A.J. King, P. Sowell and K. Yager for assistance in the field and laboratory. We also thank A.R. Townsend, E.B. Graham, and J.S. Prevéy for contributing to discussions regarding this research. We are grateful for the constructive feedback provided via the review process.

Author Contributions

Conceived and designed the experiments: JEK SKS RCL CCC DRN. Performed the experiments: JEK SKS RCL JLD SCC CCC DRN. Analyzed the data: JEK JLD DRN. Contributed to the writing of the manuscript: JEK SKS RCL JLD SCC CCC DRN.

References

1. Chapin FS, Walker LR, Fastie CL, Sharman LC (1994) Mechanisms of primary succession following deglaciation at Glacier Bay, Alaska. Ecological Monographs 64: 149–175.

2. Walker LR, Moral R del (2003) Primary succession and ecosystem rehabilitation. Cambridge: Cambridge University Press. 348 p.

3. Matthews JA (1992) The ecology of recently-deglaciated terrain: a geoecological approach to glacier forelands and primary succession. Cambridge: Cambridge University Press. 409 p.

4. Walker TW, Syers JK (1976) The fate of phosphorus during pedogenesis. Geoderma 15: 1–19.

5. Crews TE, Kitayama K, Fownes JH, Riley RH, Herbert DA, et al. (1995) Changes in soil phosphorus fractions and ecosystem dynamics across a long chronosequence in Hawaii. Ecology 76: 1407–1424.

6. Ugolini FC (1968) Soil development and alder invasion in a recently deglaciated area of Glacier Bay, Alaska. In: Trappe JM, Franklin JF, Tarrant RF, Hansen GM, editors. Biology of Alder pp. 115–148 Portland: Pacific Northwest Forest and Range Experiment Station, Portland, OR. pp. 115–148.

7. Vitousek PM, Walker LR, Whiteaker LD, Matson PA (1993) Nutrient limitations to plant growth during primary succession in Hawaii Volcanoes National Park. Biogeochemistry 23: 197–215.

8. Richardson SJ, Peltzer DA, Allen RB, McGlone MS, Parfitt RL (2004) Rapid development of phosphorus limitation in temperate rainforest along the Franz Josef soil chronosequence. Oecologia 139: 267–276.

9. Nemergut DR, Anderson SP, Cleveland CC, Martin AP, Miller AE, et al. (2007) Microbial community succession in an unvegetated, recently deglaciated soil. Microbial Ecology 53: 110–122.

10. Brown SP, Jumpponen A (2013) Contrasting primary successional trajectories of fungi and bacteria in retreating glacier soils. Molecular Ecology 23: 481–497.

11. Schütte UME, Abdo Z, Bent SJ, Williams CJ, Schneider GM, et al. (2009) Bacterial succession in a glacier foreland of the High Arctic. ISME Journal 3: 1258–1268.

12. Zumsteg A, Luster J, Göransson H, Smittenberg RH, Brunner I, et al. (2012) Bacterial, archaeal and fungal succession in the forefield of a receding glacier. Microbial Ecology 63: 552–564.

13. Edwards IP, Bürgmann H, Miniaci C, Zeyer J (2006) Variation in microbial community composition and culturability in the rhizosphere of Leucanthemopsis alpina (L.) Heywood and adjacent bare soil along an alpine chronosequence. Microbial Ecology 52: 679–692.

14. Ramirez KS, Lauber CL, Knight R, Bradford MA, Fierer N (2010) Consistent effects of nitrogen fertilization on soil bacterial communities in contrasting systems. Ecology 91: 3463–3470.

15. Nemergut DR, Townsend AR, Sattin SR, Freeman KR, Fierer N, et al. (2008) The effects of chronic nitrogen fertilization on alpine tundra soil microbial communities: implications for carbon and nitrogen cycling. Environmental Microbiology 10: 3093–3105.

16. Fierer N, Jackson RB (2006) The diversity and biogeography of soil bacterial communities. Proceedings of the National Academy of Sciences 103: 626–631.

17. Deiglmayr K, Philippot L, Tscherko D, Kandeler E (2006) Microbial succession of nitrate-reducing bacteria in the rhizosphere of Poa alpina across a glacier foreland in the Central Alps. Environmental Microbiology 8: 1600–1612.

18. Noll M, Wellinger M (2008) Changes of the soil ecosystem along a receding glacier: Testing the correlation between environmental factors and bacterial community structure. Soil Biology and Biochemistry 40: 2611–2619.

19. Fierer N, Bradford MA, Jackson RB (2007) Toward an ecological classification of soil bacteria. Ecology 88: 1354–1364.

20. Jangid K, Whitman W, Condron L, Turner B, Williams M (2013) Soil bacterial community succession during long-term ecosystem development. Molecular Ecology 22: 3415–3424.

21. Knelman JE, Legg TM, O'Neill SP, Washenberger CL, González A, et al. (2012) Bacterial community structure and function change in association with colonizer plants during early primary succession in a glacier forefield. Soil Biology and Biochemistry 46: 172–180.

22. Ferrenberg S, O'Neill SP, Knelman JE, Todd B, Duggan S, et al. (2013) Changes in assembly processes in soil bacterial communities following a wildfire disturbance. ISME Journal 7: 1102–1111.

23. Cline LC, Zak DR (2013) Dispersal limitation structures fungal community assembly in a long-term glacial chronosequence. Environmental Microbiology 16:..1538–1548

24. Meola M, Lazzaro A, Zeyer J (2014) Diversity, resistance and resilience of the bacterial communities at two alpine glacier forefields after a reciprocal soil transplantation. Environmental Microbiology 16: 1918–1934.

25. Fukami T, Morin PJ (2003) Productivity–biodiversity relationships depend on the history of community assembly. Nature 424: 423–426.

26. Nemergut DR, Schmidt SK, Fukami T, O'Neill SP, Bilinski TM, et al. (2013) Patterns and processes of microbial community assembly. Microbiology and Molecular Biology Reviews 77: 342–356.

27. Fierer N, Nemergut D, Knight R, Craine JM (2010) Changes through time: integrating microorganisms into the study of succession. Research in Microbiology 161: 635–642.

28. Schmidt SK, Nemergut DR, Todd BT, Lynch RC, Darcy JL, et al. (2012) A simple method for determining limiting nutrients for photosynthetic crusts. Plant Ecology and Diversity 5: 513–519.

29. Schmidt SK, Cleveland CC, Nemergut DR, Reed SC, King AJ, et al. (2011) Estimating phosphorus availability for microbial growth in an emerging landscape. Geoderma 163: 135–140.

30. Schmidt S, Reed SC, Nemergut DR, Stuart Grandy A, Cleveland CC, et al. (2008) The earliest stages of ecosystem succession in high-elevation (5000 metres above sea level), recently deglaciated soils. Proceedings of the Royal Society B: Biological Sciences 275: 2793–2802.

31. Matejovic I (1997) Determination of carbon and nitrogen in samples of various soils by the dry combustion. Communications in Soil Science and Plant Analysis 28: 1499–1511.

32. Jeannotte R, Sommerville DW, Hamel C, Whalen JK (2004) A microplate assay to measure soil microbial biomass phosphorus. Biology and Fertility of Soils 40: 201–205.

33. Lajtha K, Driscoll C, Jarrell W, Elliot ET (1999) Standard soil methods for long-term ecological research. New York: Oxford University Press. 486 p.

34. Weaver RW, Angler S, Bottomly P, Bezdicek D, Smith S, et al. (1994) Methods of soil analysis, part 2. Microbiological and biochemical properties. Soil Society of America, Inc.

35. Schmidt SK, Nemergut DR, Miller AE, Freeman KR, King AJ, et al. (2009) Microbial activity and diversity during extreme freeze–thaw cycles in periglacial soils, 5400 m elevation, Cordillera Vilcanota, Perú. Extremophiles 13: 807–816.

36. Hamady M, Walker JJ, Harris JK, Gold NJ, Knight R (2008) Error-correcting barcoded primers for pyrosequencing hundreds of samples in multiplex. Nature Methods 5: 235–237.

37. Fierer N, Hamady M, Lauber CL, Knight R (2008) The influence of sex, handedness, and washing on the diversity of hand surface bacteria. Proceedings of the National Academy of Sciences 105: 17994–17999.

38. Reeder J, Knight R (2010) Rapid denoising of pyrosequencing amplicon data: exploiting the rank-abundance distribution. Nature Methods 7: 668–669.

39. Edgar RC (2010) Search and clustering orders of magnitude faster than BLAST. Bioinformatics 26: 2460–2461.

40. Wang Q, Garrity GM, Tiedje JM, Cole JR (2007) Naïve Bayesian classifier for rapid assignment of rRNA sequences into the new bacterial taxonomy. Applied Environmental Microbiology 73: 5261–5267.

41. DeSantis TZ, Hugenholtz P, Keller K, Brodie EL, Larsen N, et al. (2006) NAST: a multiple sequence alignment server for comparative analysis of 16S rRNA genes. Nucleic Acids Research 34: 394–399.

42. Price MN, Dehal PS, Arkin AP (2009) FastTree: Computing large minimum evolution trees with profiles instead of a distance matrix. Molecular Biology and Evolution 26: 1641–1650.

43. Lozupone C, Hamady M, Knight R (2006) UniFrac - An online tool for comparing microbial community diversity in a phylogenetic context. BMC Bioinformatics 7: 371.

44. Lozupone CA, Hamady M, Kelley ST, Knight R (2007) Quantitative and qualitative β-diversity measures lead to different insights into factors that structure microbial communities. Applied Environmental Microbiology 73: 1576–1585.

45. Clarke, Gorley (2006) PRIMER v6: User Manual/Tutorial, PRIMER–E. Plymouth, UK.

46. R Development Core Team (2009) R: A language and environment for statistical computing. R Foundation for Statistical Computing Vienna, Austria. Available: http://www.R-project.org.

47. Oksanen J, Blanchet FG, Kindt R, Legendre P, Minchin P, et al. (2013) vegan: community ecology package. Available: http://CRAN.R-project.org/package=vegan.

48. Caporaso JG, Kuczynski J, Stombaugh J, Bittinger K, Bushman FD, et al. (2010) QIIME allows analysis of high-throughput community sequencing data. Nature Methods 7: 335–336.

49. Harpole WS, Ngai JT, Cleland EE, Seabloom EW, Borer ET, et al. (2011) Nutrient co-limitation of primary producer communities. Ecology Letters 14: 852–862.

50. Vitousek PM (1999) Nutrient limitation to nitrogen fixation in young volcanic sites. Ecosystems 2: 505–510.

51. Vitousek PM, Porder S, Houlton BZ, Chadwick OA (2010) Terrestrial phosphorus limitation: mechanisms, implications, and nitrogen–phosphorus interactions. Ecological Applications 20: 5–15.

52. Olander LP, Vitousek PM (2000) Regulation of soil phosphatase and chitinase activity by N and P availability. Biogeochemistry 49: 175–191.

53. Wang Y-P, Houlton BZ, Field CB (2007) A model of biogeochemical cycles of carbon, nitrogen, and phosphorus including symbiotic nitrogen fixation and phosphatase production. Global Biogeochemical Cycles 21: GB1018.

54. Vitousek PM, Farrington H (1997) Nutrient limitation and soil development: Experimental test of a biogeochemical theory. Biogeochemistry 37: 63–75.

55. Griffiths BS, Philippot L (2013) Insights into the resistance and resilience of the soil microbial community. FEMS Microbiology Reviews 37: 112–129.

56. Van Wijnen HJ, Bakker JP (1999) Nitrogen and phosphorus limitation in a coastal barrier salt marsh: the implications for vegetation succession. Journal of Ecology 87: 265–272.

57. Drakare S (2002) Competition between picoplanktonic Cyanobacteria and heterotrophic bacteria along crossed gradients of glucose and phosphate. Microbial Ecology 44: 327–335.

58. Yoshitake S, Uchida M, Koizumi H, Nakatsubo T (2007) Carbon and nitrogen limitation of soil microbial respiration in a High Arctic successional glacier foreland near Ny-Ålesund, Svalbard. Polar Research 26: 22–30.

59. Göransson H, Olde Venterink H, Bååth E (2011) Soil bacterial growth and nutrient limitation along a chronosequence from a glacier forefield. Soil Biology of Biochemistry 43: 1333–1340.

60. Borin S, Ventura S, Tambone F, Mapelli F, Schubotz F, et al. (2010) Rock weathering creates oases of life in a High Arctic desert. Environmental Microbiology 12: 293–303.

61. Van Der Heijden MG, Bardgett RD, Van Straalen NM (2008) The unseen majority: soil microbes as drivers of plant diversity and productivity in terrestrial ecosystems. Ecology Letter 11: 296–310.

14

Species-Specific Responses to Community Density in an Unproductive Perennial Plant Community

Michael A. Treberg¤, Roy Turkington*

Department of Botany, and Biodiversity Research Center, University of British Columbia, Vancouver, BC, Canada

Abstract

Most studies of density dependent regulation in plants consider a single target species, but regulation may also occur at the level of the entire community. Knowing whether a community is at carrying capacity is essential for understanding its behaviour because low density plant communities may behave quite differently than their high density counterparts. Also, because the intensity of density dependence may differ considerably between species and physical environments, generalizations about its effects on community structure requires comparisons under a range of conditions. We tested if: (1) density dependent regulation occurs at the level of an entire plant community as well as within individual species; (2) the intensity (effect of increasing community density on mean plant mass) and importance (the effect of increasing density, relative to other factors, on mean plant mass) of competition increases, decreases or remains unchanged with increasing fertilization; (3) there are species-specific responses to changes in community density and productivity. In 63 1 m^2 plots, we manipulated the abundance of the nine most common species by transplanting or removing them to create a series of Initial Community Densities above and below the average natural field density, such that the relative proportion of species was consistent for all densities. Plots were randomly assigned to one of three fertilizer levels. At the community level, negative density dependence of mean plant size was observed for each of the 4 years of the study and both the intensity and importance of competition increased each year. At the species level, most species' mean plant mass were negatively density dependent. Fertilizer had a significant effect only in the final year when it had a negative effect on mean plant mass. Our data demonstrate a yield-density response at the entire community-level using perennial plant species in a multi-year experiment.

Editor: Rick Edward Paul, Institut Pasteur, France

Funding: Funding for this research was provided by a Natural Sciences and Engineering Research Council (NSERC) Discovery grant to RT, an NSERC Graduate Scholarship, a UBC Graduate Scholarship, and a Northern Sciences Training Program grant to MAT. The funders had no role in study design, data collection and analysis, decision to publish, or preparation of the manuscript.

Competing Interests: The authors have declared that no competing interests exist.

* Email: royt@mail.ubc.ca

¤ Current address: Department of Geography and Environmental Studies, Carleton University, Ottawa, ON, Canada

Introduction

Most experimental studies of density dependence in plant populations are problematic in that they focus on single species within a community and seldom consider regulation at the total community level. This raises the question of whether a community-level carrying capacity can be defined. Within multi-species plant communities, if the community is at carrying capacity then any reduction in density of one species is likely to be associated with increases in density (or biomass) of other neighboring species that are likely to be potential competitors. Knowing whether a plant community is at, or close to, carrying capacity is essential for understanding its behaviour [1] and there is reason to believe that low-density plant communities will behave quite differently, and less predictably, than plant communities close to carrying capacity [2]. It is argued that most natural plant communities are at, or near, carrying capacity, and their dynamics are therefore quite predictable because the community is using all available resources that act as a constraint on plant community dynamics [2–3]. These conclusions thus far do not have broad empirical support and to test them requires manipulation of whole community densities, both below and above the natural field densities, and the

monitoring of individual species populations within these communities in response to altered densities [4–7].

Competition is important in structuring plant communities [8–9], but the intensity at which it occurs is dependent upon local conditions and has been the subject of much debate [9–13]. Much of the debate focuses on competition in low productivity communities [4] [14–15] and although many studies demonstrate that competition, as well as facilitation, is prevalent in a wide range of natural communities [16–18], they tell us little about their overall effects on the community (but see [19]) and mostly focus on the effect or response of certain species [8]. Theoretical studies suggest that individual-level data will often not predict community patterns, even on a local spatial scale [20–22] and empirical evidence likewise has shown that individual-level effects of competition could not predict community-level effects [23]. Rees [12] recently presented a simple framework for interpreting the results of short-term competition experiments along natural productivity gradients.

A method for directly examining the effects of competition on community structure using multi-species mixtures is the Community Density Series (CDS), first described in [24]. It is a multi-species (community) version of the traditional single-species (population) yield-density experiments [25] in which the yield of a single species is measured when the species is grown at a range of densities ranging from very low to very high. In this method, the density of an entire community is manipulated in the same way as a single species to obtain densities below and above the natural condition of the community. The lowest density plots, where density is low enough to preclude plant-plant interactions, characterize the "null" community and this can be compared to higher density plots where biotic interactions are affecting the plant community as a whole [24]. In addition, each plant species in the CDS can be considered separately to determine if they respond similarly to changing density. Another advantage of the CDS is that both negative and positive density dependent processes are detectable. The influence of resource level on these responses can also be investigated by examining changes in the yield-density relationship. For example, if limiting resources are increased, we may expect an increase in the constant final yield or an increase in the intensity of competition. An additional advantage of the CDS is that the slope and the R^2 from the regression of the yield-density relationship represent both the intensity and importance of competition respectively [26]. In traditional yield-density studies, the coefficient of determination, or R^2, from simple linear regression [26–27] or it's multivariate equivalent [28–29] can be interpreted as being the importance of competition. R^2 represents the importance of competition, compared with other possible factors affecting yield, because it is the proportion of variation in yield that is directly due to the density. The ability to quantify both the intensity and importance of competition can help untangle some of the debates surrounding the role of competition in structuring plant communities [13] [16]. More recently, Bennett and Cahill [13] used a novel method (based on Lamb and Cahill [19] to estimate the intensity and importance of competition in a grassland with a limited range of productivity levels. They measured the performance of 22 different species in the field, with and without neighbours, and averaged the responses. The Bennett and Cahill [13] approach focuses on the effects of neighbours on seedling survival and growth but does not address the question of density-dependent regulation at a community level.

The CDS has been successfully applied in both experimental and natural communities of annual plants in the Negev Desert [4–7] [30] in an experimental bryophyte community [31] or a single season in an old-field community [23] and in an experimental boreal understory community [32]. However, it has never been applied in a perennial system in a multi-year study. Using the CDS, we investigated the influence of competition in structuring an unproductive boreal understory plant community. Specifically, we tested if density dependent regulation occurs at the level of an entire plant community i.e. if the community is at carrying capacity, or constant final yield [2] as well as among individual species; if the intensity and importance of competition changes with increasing fertilizer addition; and if there are species-specific responses to both changes in community density and community productivity. In addition, we asked at what community density does competition begin to have an effect, and, at what community density is maximum constant final yield achieved.

Methods

Study site

The study site is located within the boreal forest close to Kluane Lake in the southwestern Yukon Territory (138° 16′ W; 61° 00′N) at approximately 1000 m above sea level. This research was done in an area that consists of Crown Land (Scientist & Explorer

permit, Heritage Branch, Yukon Government) and we have oral permission from both the Champagne and Aishihik, and the Kluane First Nations. The research did not involve endangered or protected species. This ecosystem is extensively studied for both its animal and plant components and was used for the Kluane Boreal Ecosystem Project [33–34]. Previous research has shown that the vegetation in this system is nutrient limited [35] and competition affects some of the understory plant species [36]. Beginning in 1995, an outbreak of spruce bark beetle caused the death of many of the overstory trees (White spruce; *Picea glauca* (Moench) Voss s.l.)) resulting in a rather open canopy. Although there are herbivores such as snowshoe hares, red squirrels and microtine rodents at this site, understory composition is more affected by the limited soil nutrients than by herbivores [34].

Experimental design

We used a full-factorial block design with six levels of density plus control plots, 3 levels of fertilizer and three replicates per treatment, for a total of 63 plots, each 1 m×1 m. In late May 1999, the 63 plots were marked in an area of approximately 25 m×75 m. These plots were located in patches with representative samples of the vegetation common to the understory community in this forest. Plots were in small groups of 2 to 5, with a minimum of 1 m between adjacent plots. Each group of plots was surrounded by a 1 m high 2.5 cm mesh galvanized chicken wire fence.

Percent cover of all species in the 63 plots was estimated using a point quadrat frame with 100 points per plot (Table S1). The nine most abundant species that represented 94.4% of the total cover, (97.5% of biomass in control plots in 2002) were chosen to be included in the community density series (CDS). Seven of the species are herbaceous perennials: *Achillea millefolium* L. ssp. *borealis* (Bong.) Breitung (yarrow), *Epilobium angustifolium* L. s.l. (fireweed), *Festuca altaica* Trin. (northern rough fescue), *Lupinus arcticus* Wats. (arctic lupine), *Mertensia paniculata* (Ait.) G. Don var. *paniculata* (bluebells), *Senecio lugens* Richards. (black-tipped groundsel), *Solidago multiradiata* Ait. (goldenrod). The remaining two species are prostrate woody perennials: *Arctostaphylos uva-ursi* (L.) Spreng. s.l. (bearberry) and *Linnaea borealis* L. ssp. *americana* (twinflower). Hereafter we will refer to species using their generic name following the nomenclature of Cody [37].

We estimated density of six of the species in all plots by counting stems or ramets; it was not possible to estimate density for *Arctostaphylos, Linnaea* or *Festuca* so we used percentage cover instead. Using these abundance estimates we constructed a geometric series of six Initial Community Densities consisting of 1/16, 1/8, 1/4, 1/2, 1 and x2 the average natural field density. The x1 density closely approximated the density of the natural vegetation estimated from the initial survey of the community. All plots in the CDS were manipulated by transplanting and removing plants such that the relative proportion of the nine most common species was consistent for all densities. Without exception, every plot in the CDS had some plants added and some removed to obtain the proper proportions of the nine study species. To increase the density in plots, transplants were taken from the surrounding vegetation either as large sods containing many individuals (and sometimes many species) or as single shoots. Removal was accomplished by cutting the unwanted shoots off at ground level. In the lower density plots, an attempt was made to keep the remaining plants approximately equidistant from each other. Density manipulations began in mid-June 1999, and were completed by mid-July. Some regrowth of removed plants occurred but this was removed before the end of the season survey done in the final 2 weeks of August. No plants were added at this time. In 2000, minor weeding was completed in June to adjust to the desired densities. No other density manipulation was required and plants were allowed to grow for three more growing seasons.

Fifty-four of the 63 1 m^2 plots were randomly assigned to the CDS; the remaining 9 plots were used as controls and did not have any density manipulation. All 63 plots were randomly assigned to one of the three fertilizer levels. At the beginning of each growing season, the soil surrounding all plots was cut to a depth of approximately 25 cm just outside of the 1×1 m perimeter to sever any belowground connections between plant inside and outside the plots.

Three levels of fertilization were used - the low fertilizer treatment (control) had no fertilizer added. Granular fertilizer (N-P-K; 21-7-7) was added after snowmelt, at the end of May or early June for each of the 4 years of the study. Fertilizer was added at a rate of 13 g N m^{-2} y^{-1}, 4.4 g P m^{-2} y^{-1} and 4.4 g K m^{-2} y^{-1} for the medium fertilizer treatment and at double this rate for the high treatment. These are within the range of application rates from other studies in this area that demonstrated a significant effect of fertilizer addition [34]. These nutrients and the medium rate of application were used following the protocols established by the 10-yr Kluane Boreal Forest Project [33–34]; preliminary tests had shown N and P to be limiting and the rate of application was the lowest of the range of application rates used by foresters when fertilizing forests.

As a response variable, in each year the average plant biomass was estimated or measured. However, because we needed to estimate biomass within the plots, yet could not destructively harvest, surrogate measures of biomass were used to approximate the biomass for each species. In July 1999, 20 plots were randomly chosen nearby the experimental plots and were sampled for percent cover of each species. The plots were clipped to ground level, sorted to species, dried, and the relationship between each species cover and biomass was determined. The biomass of *Arctostaphylos*, *Festuca*, and *Linnaea* was accurately estimated by percent cover. The width of the widest leaf, length of the longest leaf, the number of leaves and maximum height were measured for random individuals of the remaining herbaceous species. These were then cut at ground level and dried. These measurements were then related to shoot mass and the best fitting relationships were determined (Table S2). All relationships were statistically significant and R^2 values ranged from 0.62 to 0.97. For each year of the study, during peak biomass, which occurs approximately at the end of July, all individuals in each CDS plot and control plot were measured and the plot biomass estimated (Table S3).

At the end of August 2002, all plants were counted in all plots. All aboveground biomass was removed and each individual plant was bagged separately before being air dried for transport to the University of British Columbia. All samples were oven dried at 60°C for 48 hours and weighed to the nearest mg.

Analysis

The effects of density and fertilizer addition on the average plant mass were examined using analysis of covariance (AN-COVA) with density as the covariate and fertilizer as a categorical variable with 3 levels. All analyses were done using JMP 4 [38]. The effect of density on the understory community was analyzed using an individual performance approach [4]. By examining the average performance of the individual in relation to density, any non-zero slope would indicate density dependence. For example, an increase in performance with density, whether linear or nonlinear, would indicate positive density dependence or facilitation. A negative slope would indicate negative density dependence

or competition. Because the ANCOVA requires a linear covariate, four transformations were used to linearize the data: linear, power, semilog and reciprocal. The best fitting model, with the highest R^2, is reported. These transformations also assisted in making the ANCOVAs better meet the usual statistical assumptions, namely, normality of residuals, homogeneity of variances, homogeneity of regression slopes, linearity of regression, and independence of error terms. If density were not significantly related to the response variable, standard ANOVA was used to determine the effect of fertilizer addition.

At the community level, the effect of density was examined on the mean plant mass for each of the 4 years of the experiment. We used a mean plant size index calculated as the total biomass of the entire plot divided by the initial density, and the relative densities from the Initial Community Densities were used as the covariate for all ANCOVAs.

The density at which competition began to reduce the mean plant mass was determined by stepwise regression of mean plant mass against density, beginning with the low density plots. The density at which a final constant yield was reached was determined by regressing the final plot yield (total biomass) of the two highest densities against density. Lower density plots were added until the slope became significant. The last density that had a non-significant slope is the density where a final constant yield was reached.

The effect of density on the diversity in the CDS plots was also examined. The species richness of each plot did not change in the 4 years of the study; however, if density affects species differently, we may expect to see changes in their relative abundance or evenness. In the final year of the study, we used each species' mass in the plots to calculate an evenness index (E_{var}) [39]. Evenness values close to 1 indicate that species are nearly uniformly abundant (i.e. are of similar mass) and values close to 0 indicate that one or a few species are much more abundant than the others are. Therefore, if we detect any change in evenness with changing density, there would have to be species-specific changes in abundance.

The effect of fertilizer addition on the plot biomass of the unmanipulated controls, those plots not part of the CDS, was also examined using ANOVA. The controls were also compared to the x1 CDS plots to determine if the final plot mass was similar between the manipulated plots and the unmanipulated plots and whether there was any difference in their response to fertilizer addition.

Species-specific effects of density and fertilizer addition were examined on the final mean biomass for each plant species using ANCOVA with the relative planting density being used as the covariate. If density was not related to the final mean plant mass, ANOVA was used to determine the effect of fertilizer addition.

Results

Community-level responses

Negative density dependence of the mean plant size index was observed each year of the CDS experiment with the negative slope becoming increasingly steeper each subsequent year (Table 1, Fig. 1). Generally, mean plant size increases each year at low densities, but remains constant or decreases at higher densities. Therefore, the intensity of competition (the slope) increased each year. Similarly, the importance of competition (R^2) also increased with time. In all years, the relationship between mean plant size and density was nonlinear (Table 1). Mean plant mass was significantly higher below the x1 density (the natural density of the community) (Fig. 1). The density at which competition began to

Table 1. Regression coefficients for the mean plant size index (total plot mass divided by the density) and density relationships for the years 1999 through 2002 and the evenness and density relationship in 2002.

Variable	Model	Intercept	Slope	R^2	P
Mean plant size index 1999	power	4.658	−0.113	0.495	**<0.001**
Mean plant size index 2000	power	4.488	−0.306	0.617	**<0.001**
Mean plant size index 2001	power	4.427	−0.407	0.613	**<0.001**
Mean plant size index 2002	power	5.096	−0.663	0.742	**<0.001**
Evenness (E_{var})	semilog	0.338	5.97×10^{-2}	0.309	**<0.001**

Significant values (P<0.05) are in **bold**. These data are plotted in Figures 1 and 3. The negative slopes indicate negative density dependence (or competition) for the plant size index. Model type refers to the data transformation that best linearizes the data. The degrees of freedom (df = 53) are for the model and error combined.

reduce mean plant mass was at x1/8 for years 3 and 4. Final constant yield was only reached in 2001 and 2002 at x1, the average natural density observed in the field.

The mean plant size index was affected each year by plant density; however, fertilizer only had a significant effect in the final year (Fig. 2A), although there was a significant fertilizer and density interaction in the first year of the study (Table 2). In the final year, fertilizer surprisingly had a negative effect on mean plant size with the highest growth in the unfertilized plots (Fig. 2A).

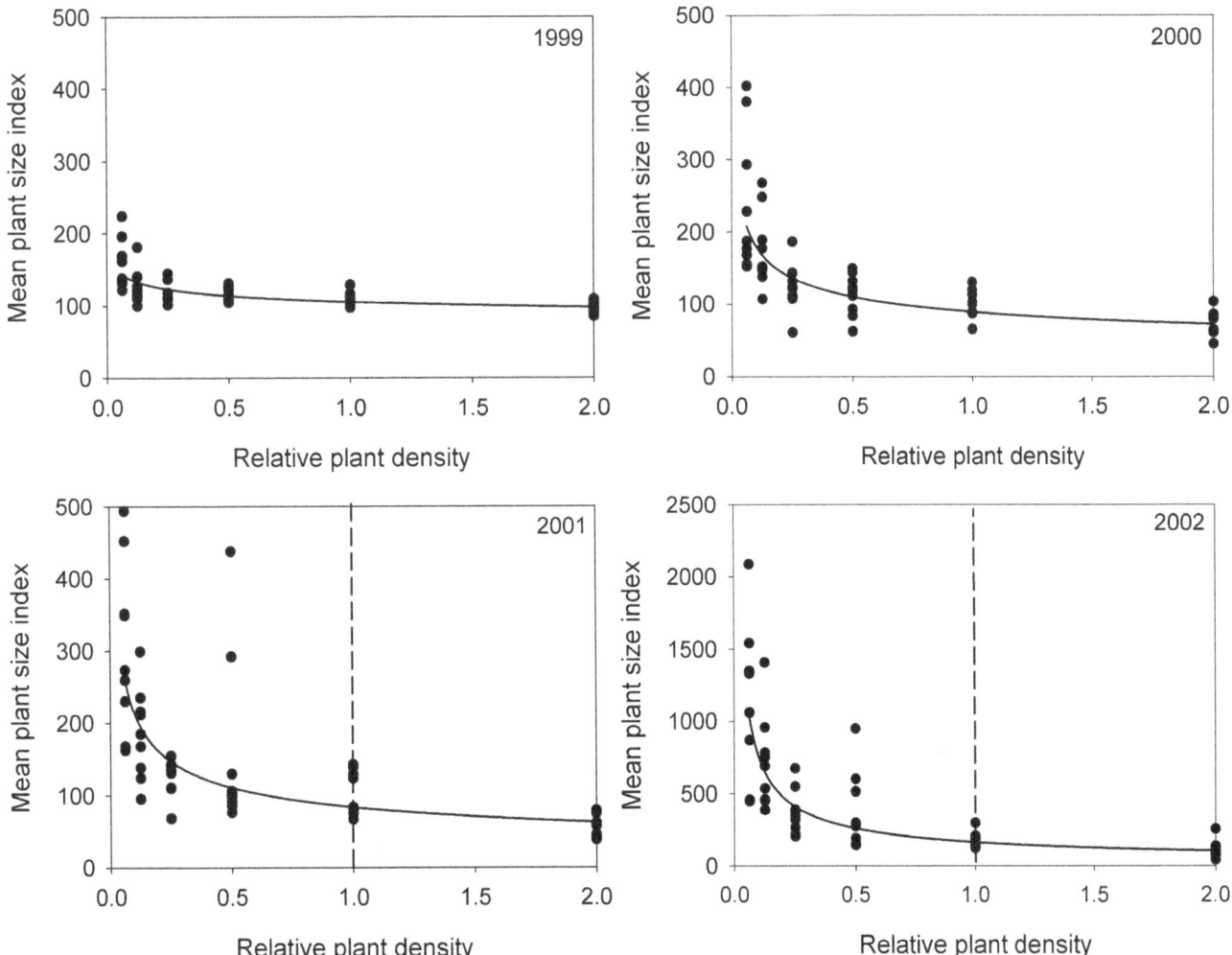

Figure 1. The effect of density on the mean plant size index (the total plot mass divided by density) for the years 1999 through 2002. All curves are statistically significant (P<0.001) and the coefficients for the best fit curve are given in Table 1. The y-axis for 2002 has a different scale than the graphs for other years. The natural field density is x1. The density that competition began to reduce mean plant size was at x1/8 (i.e. 0.125) for all graphs. The vertical dashed line represents the density that constant final yield is reached.

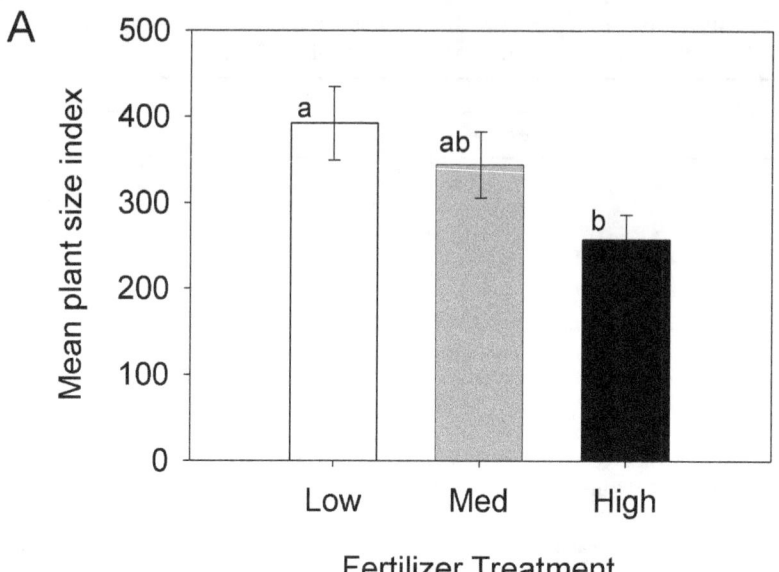

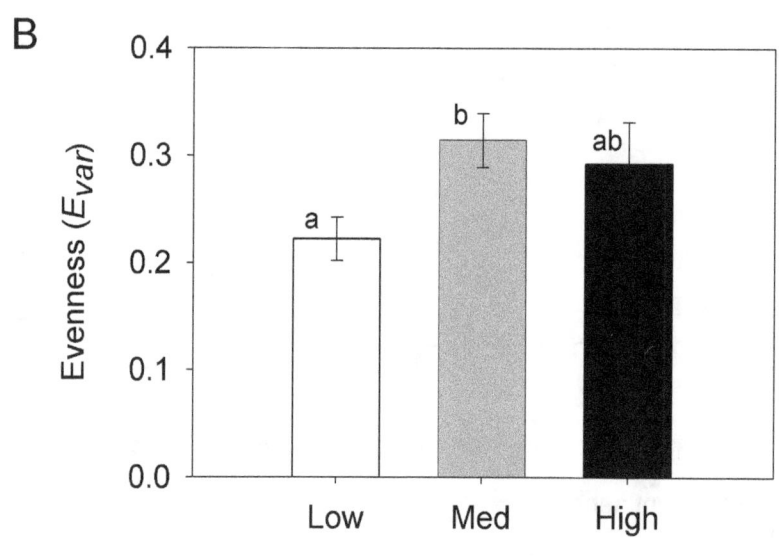

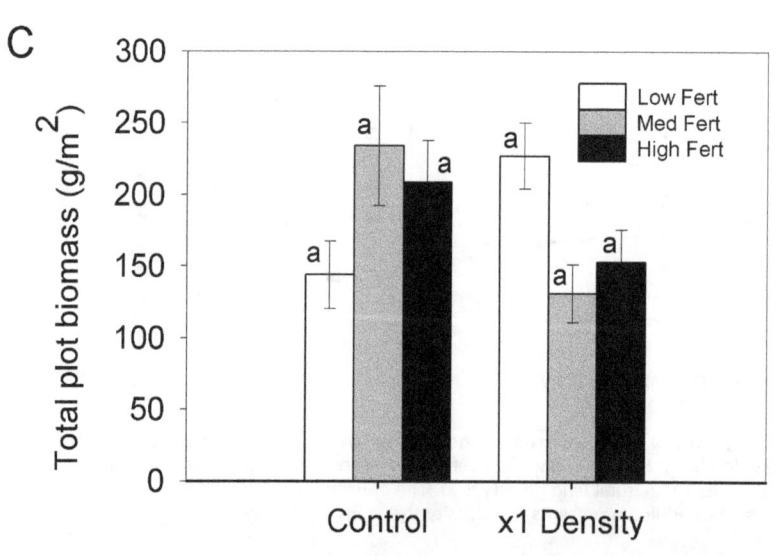

Figure 2. The effect of fertilizer level on (A) mean plant size index (total plot mass divided by density) in the CDS, (B) evenness in the CDS and (C) the total plot biomass for the controls (density not manipulated) and the x1 density plots in the CDS experiment. Error bars are ± 1 S.E. All data are for 2002. Columns that share the same letter are not statistically different (Tukey's HSD, $P>0.05$).

Species evenness in the community in the final year was also affected by density (Table 1, Fig. 3) with the highest evenness in the higher density plots. The significant nonlinear relationship between evenness and density means that the relative proportion of each species' biomass was not constant, i.e. the species' responses were not consistent along a density gradient. Evenness was also significantly different between fertilizer levels, regardless of density, with the lowest evenness in the unfertilized plots (Table 2, Fig. 2B).

There was no difference in the overall plot biomass between the unmanipulated controls and the x1 density plots (Table 2, Fig. 2C). There was also no effect of fertilizer addition on the plot biomass for either the control or the x1 density plots (Table 2, Fig. 2C).

Species-level responses

Most species' mean plant mass were negatively and nonlinearly density dependent (Tables 3, 4, Fig. 4). Only two species' mean mass, *Epilobium* and *Senecio*, were not related to density, although *Mertensia* and *Solidago* were only related to density at $P<0.10$ (Table 3). The intensity of competition was highest on *Linnaea*

and *Arctostaphylos* (slopes of -0.897 and -0.851, respectively) and was also high on *Festuca* (-0.687). The importance of competition was highest on *Festuca* (R^2 of 0.719) with high values also on *Arctostaphylos* (0.402) and *Linnaea* (0.477). No species displayed positive density dependence (facilitation).

Species-specific responses to fertilizer addition were more varied than the response to density. The prostrate woody shrubs, *Arctostaphylos* and *Linnaea*, were negatively affected by fertilizer addition while *Epilobium* and *Mertensia* responded favorably to fertilizer addition (Table 4, Fig. 5). The remaining 5 species had no response to fertilizer addition, although *Achillea* and *Senecio* had marginally significant responses ($0.05<P<0.10$).

A switch between no density dependence to density dependence at low densities was observed for three species, *Achillea*, *Arctostaphylos*, and *Festuca* (Fig. 4). The density that competition began to be important could not be determined for the other species because all densities needed to be included before the regression had a significant slope. Constant yield was reached for three species - for *Achillea*, and *Festuca* at natural field density and for *Lupinus* at 0.5 density.

Table 2. Summary of ANCOVAs for the mean plant size index in the CDS for 1999 to 2002 and ANOVA for total plot biomass in the control plots in 2002.

Variable	Source	df	SS	F-ratio	P
Mean plant size index 1999	Density	1	0.961	57.362	**<0.001**
	Fertilizer	2	0.018	0.539	0.587
	Density x Fertilizer	2	0.158	4.712	**0.014**
	Error	48	0.804		
Mean plant size index 2000	Density	1	7.077	91.225	**<0.001**
	Fertilizer	2	0.421	2.715	0.076
	Density x Fertilizer	2	0.250	1.612	0.210
	Error	48	3.723		
Mean plant size index 2001	Density	1	12.546	78.564	**<0.001**
	Fertilizer	2	0.080	0.251	0.779
	Density x Fertilizer	2	0.172	0.538	0.588
	Error	48	7.665		
Mean plant size index 2002	Density	1	33.254	162.465	**<0.001**
	Fertilizer	2	1.603	3.916	**0.027**
	Density x Fertilizer	2	0.124	0.303	0.740
	Error	48	9.825		
Evenness (E_{var})	Density	1	0.269	25.455	**<0.001**
	Fertilizer	2	0.084	3.977	**0.025**
	Density x Fertilizer	2	0.011	0.526	0.595
	Error	48	0.508		
Controls (Total Plot Biomass)	Control	1	3197	0.671	0.430
	Fertilizer	2	104.858	0.011	0.989
	Control x Fertilizer	2	28025	2.941	0.095
	Error	11	52406		

The control treatment for the total plot biomass ANOVA compares the mean of the unmanipulated control plots to the x1 density in the CDS plots. Significant values ($P<0.05$) are in **bold**.

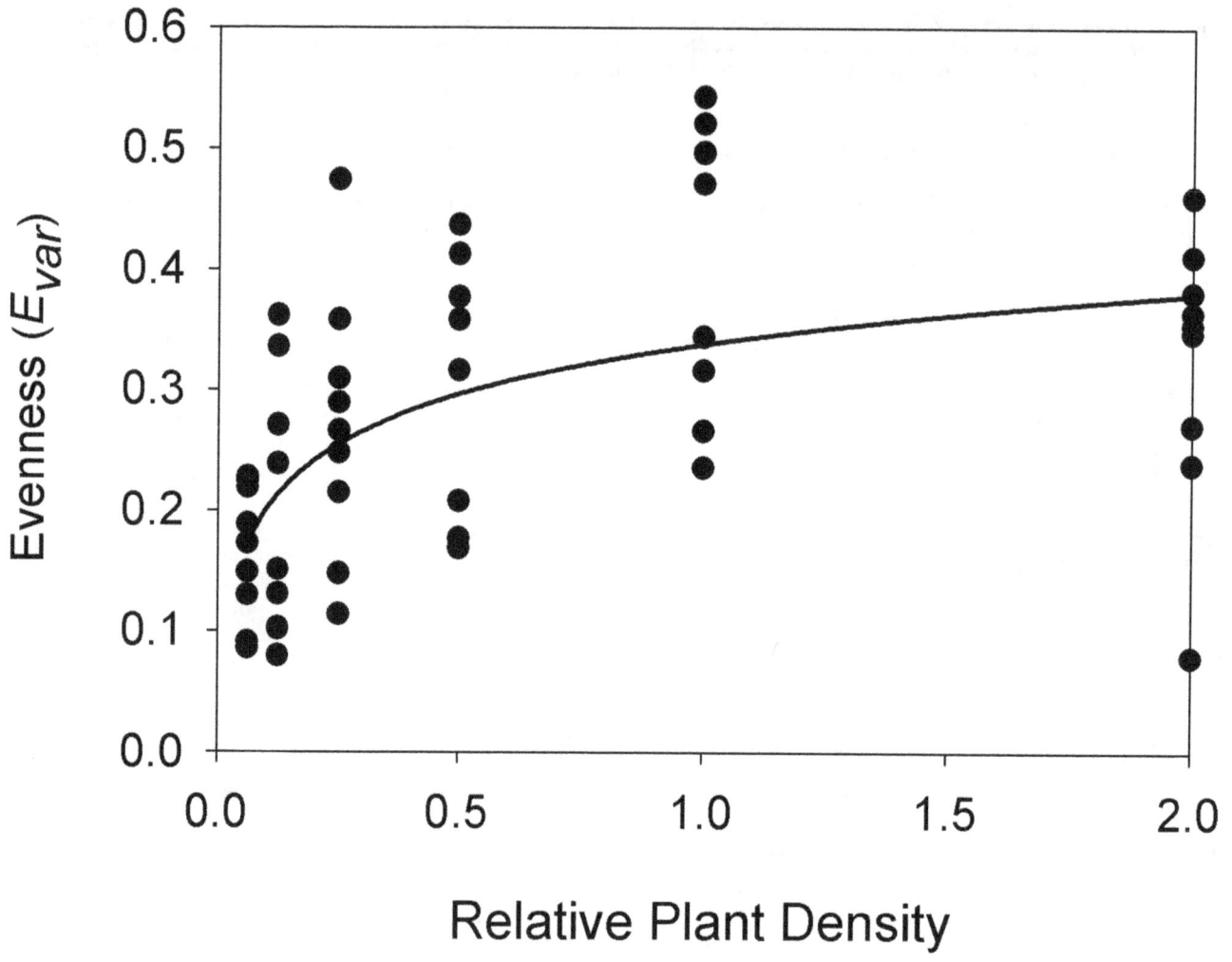

Figure 3. The effect of plant density on evenness (E_{var}). The natural field density is x1. The best fit curve is statistically significant ($P<0.001$) and the coefficients for the curve are given in Table 1.

Discussion

There are now a number of examples of entire plant communities following the same pattern as observed in single-species experiments that show increasing yield with increasing plant density until an asymptote is reached i.e. the constant final yield [4–7] [23] [30–31]. Most of these studies were done using annual species in the Negev desert. In common, the Negev studies and the current study show that the initial density had significant effects on final species composition but there was no convergence of species composition or even in functional groups. In contrast, the Negev communities did not converge to a common biomass (after 3 years) whereas our communities did converge to a constant biomass at the natural (1-X) density (after 4 years). Both sets of studies demonstrated that density dependent regulation occurs at the level of the entire community; the current study shows that the intensity and importance of competition increased in each subsequent year but the Negev studies did not make this distinction. Therefore, while both systems showed some degree of convergence of abundance, neither of them showed convergence of species composition.

Knowing whether a plant community is at, or close to, carrying capacity is essential for understanding its behaviour and there is reason to believe that low-density plant communities will behave

quite differently, and less predictably, than plant communities close to carrying capacity [2]. At carrying capacity, a community will be using all available resources that act as a limitation on plant community dynamics, increasing predictability [3]. At lower community densities, the plant dynamics are transient and initial conditions and external factors play a crucial role. In the desert, this can be easily understood due to the wide variations of seed-bank density across the landscape, and the variable nature of the species composition of the seed-bank. The spatial variation is influenced by the location and size of adult plants that are dispersing seeds, the nature of the terrain, wind speed and direction when seeds are being scattered, the abundance of seed predators, and other effects. Therefore, the species composition and structure of these annual desert communities of annual plants, although strongly influenced by competition and herbivory, are seemingly indeterminate at the local scale and may be substantially due to chance and unpredictable events. A similar proposal may be made for the boreal perennial community although it must be seen in the framework of a much longer time-scale. Our study area was most recently burned about 80 years before this research. The initial recolonization after the fire would have depended on time of year, location and abundance of species able to disperse into the site, and on the composition of the seed bank. Once established,

Table 3. Regression coefficients for the relationship between each species mean plant mass and density.

Variable	Model	df	Intercept	Slope	R^2	P
Achillea millefolium ssp. *borealis*	semilog	53	0.226	−0.119	0.235	**<0.001**
Arctostaphylos uva-ursi	power	51	1.965	−0.851	0.402	**<0.001**
Epilobium angustifolium	linear	53	2.016	−0.355	0.026	0.247
Festuca altaica	power	53	3.892	−0.687	0.719	**<0.001**
Linnaea borealis	power	53	3.492	−0.897	0.477	**<0.001**
Lupinus arcticus	power	51	−0.762	−0.211	0.161	**0.003**
Mertensia paniculata	semilog	53	1.215	−0.263	0.068	*0.056*
Senecio lugens	linear	53	0.354	−0.114	0.045	0.124
Solidago multiradiata	linear	53	0.424	−0.107	0.063	*0.068*

Significant values (P<0.05) are in **bold** and values where P<0.10 are in *italics*. The model type is the transformation that best linearized the data. The degrees of freedom (df) are for the model and error combined. These data are plotted in Figure 4.

Table 4. Summary of ANCOVAs and ANOVAs on each species' mean plant mass in the CDS in 2002, in response to manipulations of density and fertilizer.

Species	Source	df	SS	F-ratio	P
Achillea millefolium ssp. *borealis*	Density	1	1.231	22.178	**<0.001**
	Fertilizer	2	0.313	2.819	0.071
	Density x Fertilizer	2	0.265	2.389	0.104
	Error	44	2.441		
Arctostaphylos uva-ursi	Density	1	53.322	42.204	**<0.001**
	Fertilizer	2	22.371	8.853	**<0.001**
	Density x Fertilizer	2	0.154	0.061	0.941
	Error	44	55.591		
Epilobium angustifolium	Fertilizer	2	36.434	10.941	**<0.001**
	Error	51	84.914		
Festuca altaica	Density	1	779900	35.754	**<0.001**
	Fertilizer	2	22318	0.512	0.603
	Density x Fertilizer	2	12363	0.283	0.755
	Error	44	959761		
Linnaea borealis	Density	1	63.621	57.632	**<0.001**
	Fertilizer	2	9.488	4.297	**0.020**
	Density x Fertilizer	2	3.585	1.624	0.209
	Error	44	48.573		
Lupinus arcticus	Density	1	3.261	9.404	**0.004**
	Fertilizer	2	0.462	0.666	0.519
	Density x Fertilizer	2	0.572	0.825	0.449
	Error	44	15.259		
Mertensia paniculata	Fertilizer	2	26.201	13.281	**<0.001**
	Error	51	50.305		
Senecio lugens	Fertilizer	2	0.620	2.422	0.099
	Error	51	6.532		
Solidago multiradiata	Fertilizer	2	0.018	0.101	0.904
	Error	51	4.489		

Significant values (P<0.05) are in **bold**. An ANOVA with just fertilizer as an effect was done when the effect of density was not significant (P<0.05, Table 3) on the mean plant mass.

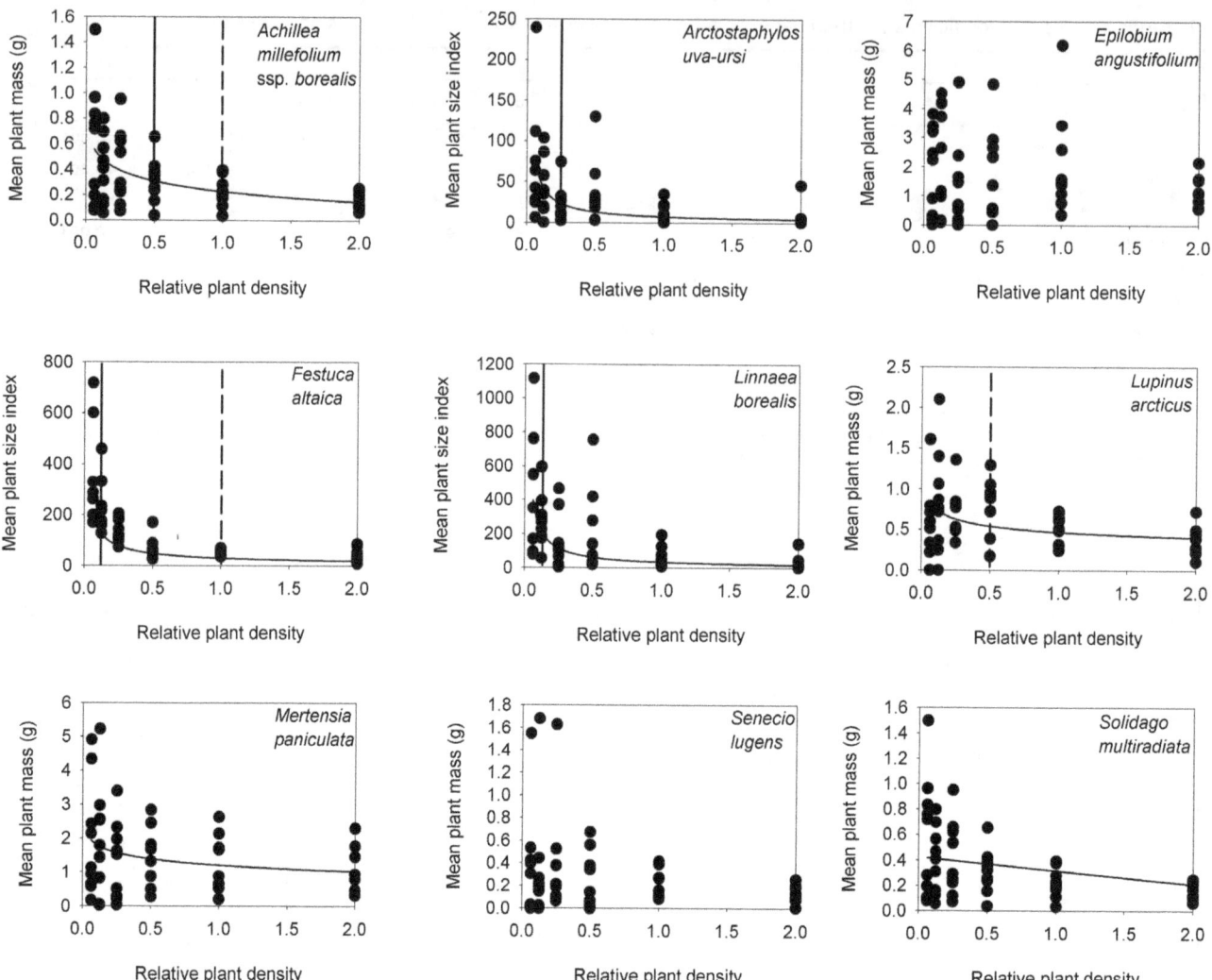

Figure 4. The effect of plant density on the mean plant mass or mean plant size index (total plot mass divided by density) for species in the CDS plots in 2002. All curves are statistically significant at *P*<0.05, except for *Mertensia* and *Solidago*, which are significant at *P*< 0.10. The coefficients for the best fit curve are given in Table 3. The natural field density is x1. Solid vertical lines indicate the density that competition begins to reduce the mean plant mass. Dashed vertical lines indicate the density where the final constant yield for that species was reached.

most of these species rely much more on clonal growth than on seed dispersal and further seedling establishment. After 10 or 20 years, the composition of this community was quite likely still influenced to some extent by the nature of the founder populations. Likewise, after our treatment perturbations, subsequent recolonization may have been strongly influenced by various uncontrolled factors, and even though competition is clearly an important factor structuring this community, chance may also play a role in the structure of the boreal community. Of theoretical significance is that these results indicate that the community response is not simply an additive effect of multiple species responses, but likely also due to history and other interactions that are more complex. When carrying capacity is finally reached, internal processes become more dominant [2]. However, carrying capacity is clearly a function of the plant species composition as well as resource levels. It represents the maximum biomass for a particular assemblage of species in an environment after a period of growth and therefore serves as a baseline for the measurement of disturbance in the community.

Competition began to affect the structure of this experimental community at density levels much lower (x1/8 density) than the natural density and this was apparent in all years of the study. The density at which the community reached constant final yield occurred at the x1 density or natural density in the final two years of the study. In the first two years, constant final yield was not reached likely because insufficient time had elapsed since the densities were manipulated for the mean plant mass to show the effect of competition.

Other than Zamfir and Goldberg [31], this is the only other study to present both species-specific responses and community-level responses using this technique, though many others have reported species diversity changes within a Community Density Series [5] [23] [40]. For competition to significantly affect community structure, and therefore diversity, it must affect species differently [23]. In this experiment, the community as a whole was negatively affected by increasing density with most species showing a decrease in mean plant mass, although two species, *Epilobium* and *Senecio*, were not affected. Similarly, the intensity and importance of competition varied for different species. The effects

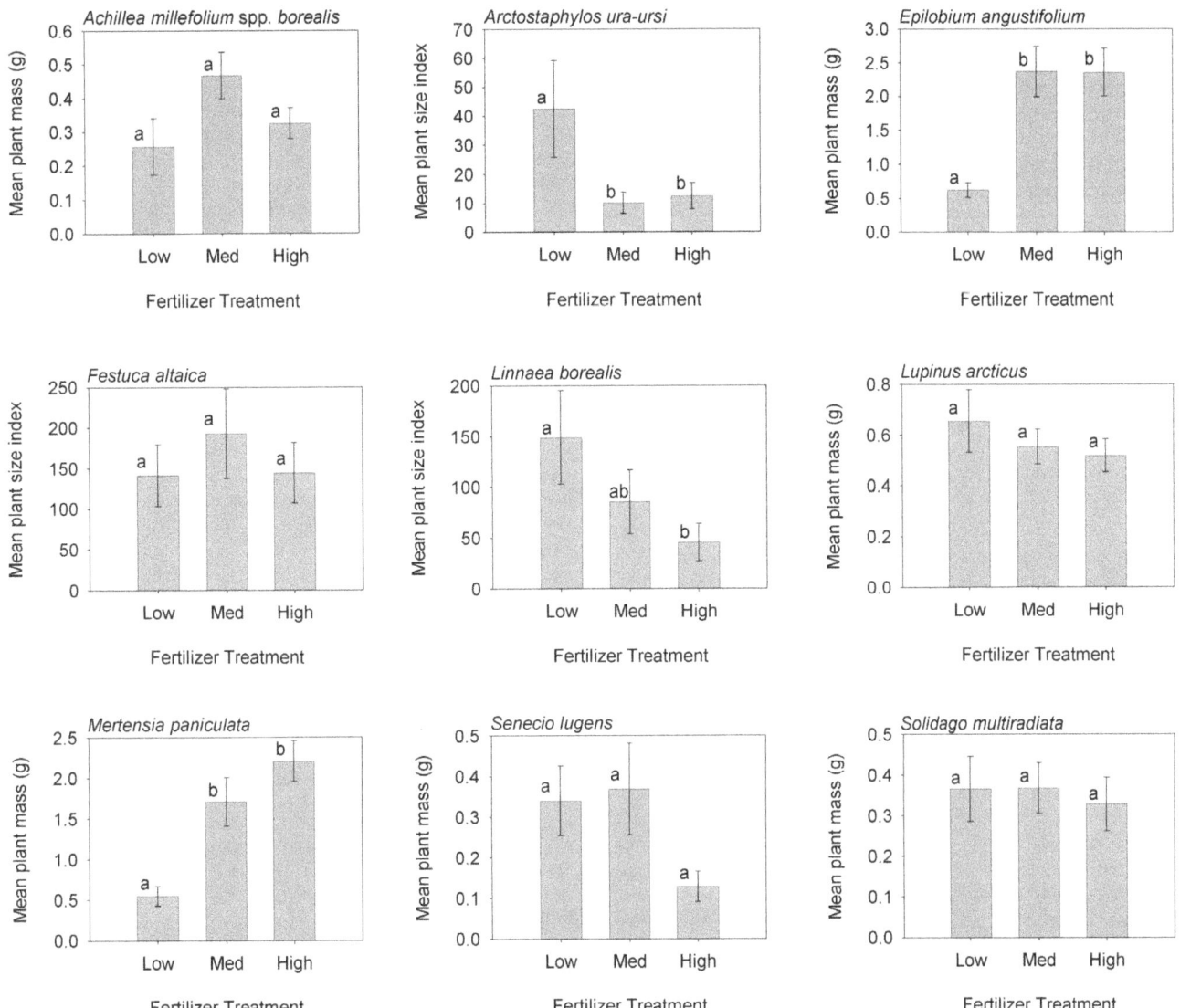

Figure 5. The effect of fertilizer level on the mean plant mass (±1 S.E.) or mean plant size index (±1 S.E.) for each species in the CDS in 2002. Columns sharing the same letter are not statistically different (Tukey's HSD, $P>0.05$).

of competition on the community began with *Festuca* and *Linnaea* at low densities (x1/8), while *Achillea*, *Arctostaphylos* and *Lupinus* did not respond until x1/2. The community reached constant final yield at x1, the natural field density, but only *Achillea* and *Festuca* reached constant final yield (also at x1). No changes in species richness occurred in this experiment, but there were changes in evenness that increased with increasing density. Because evenness expresses how equally abundant species are in a sample, the lowest density plots, or null community without competitive interactions, had some species become much more abundant relative to others, and as density and competition increased, these species were more affected than other species. Evenness was also affected by fertilizer rates with the higher evenness in the fertilized plots. This result is opposite to what has been observed in other research done in this plant community [41] and is contrary to the usual observations [42–44]. In addition, competition was neither intense nor relatively important at high productivity for either plant survival or size, suggesting that our results are also not consistent with an

increase in the likelihood of competitive exclusion as predicted by Newman, at least not over this range of productivity

Bennett and Cahill's [13] approach focuses on the effects of neighbours on seedling survival and growth. For both seedling survival and growth, relative competitive importance and competitive intensity declined with some measure of productivity; neighbour effects on survival declined with standing crop, while effects on growth declined with gross water supply. These results add to the growing evidence that plant-plant interactions vary among life history components with different life history components contingent upon separate environmental factors [13].

In this boreal understory experiment, the addition of fertilizer had a negative effect on the mean plant mass in the Community Density Series. Another study [45] demonstrated increased productivity in response to fertilizing a low-productivity sand system, but without an effect on diversity. Although the boreal system is generally considered to be nutrient limited, these results are not contradictory to other long-term studies done in this system that have observed both positive and negative effects of increased fertility [34] [41]. Some short-term studies have reported decreases in survival with increased fertilizer [46] while others indicate either no effect of fertilization [47] or positive effects [36]. The species-specific responses to fertilizer addition here correspond well with responses observed by Turkington et al. [41]. Species that decreased with increased fertilizer in both studies include *Arctostaphylos* and *Linnaea*, which are both low-growing prostrate shrubs, while *Epilobium* and *Mertensia*, taller erect herbaceous species, increased in both studies. The biggest difference here is the lack of response of some species that usually increase with added fertilizer such as *Festuca* and *Achillea* [36] [41]. This lack of a positive response, especially for the graminoid, *Festuca*, may in part be due to an unusually high abundance of microtine rodents in 2002 and it is well known that many plants experience increased herbivory when fertilized, especially species growing in low-nutrient environments. There is evidence to suggest that these rodents (voles) may be specifically attracted to the fertilizer added to our experimental plots [48]. Fertilizer level was positively related to the number of over-winter vole nests found in the experimental CDS and unmanipulated plots. There was no evidence to support the idea that fertilizer addition affects the role of competition in structuring this community at either the species or community level, nor was their evidence to support the idea that facilitative interactions may be important in this community.

We have demonstrated that density dependence is important in structuring this boreal understory community utilizing the community density series. This CDS approach allows us to quantify both the intensity and importance of plant competition at the community and species levels and to determine whether the importance of these biotic interactions depend on abiotic factors. While fertilizer addition did have minor effects on the community, it did not change the intensity of competition. The results presented here clearly show that species-specific responses to biotic interactions are not necessarily the same as community level ones and if we are to understand what occurs at the community level, it is necessary to use appropriate methodological approaches such as the CDS.

Supporting Information

Table S1 Initial abundance of species. The abundance of all species found in the 63 1 m² Community Density Series plots during the initial survey in 1999. Frequency is the percent occurrence in the 63 plots. Percent cover was estimated using a point frame with 100 pin drops per m². Density was assessed by counting all individuals in the 1 m² plot. The densities of *Arctostaphylos uva-ursi*, *Festuca altaica* and *Linnaea borealis* were not estimated (*n/a*) due to the impossibility of identifying distinct individuals.

Table S2 Equations used to estimate plant biomass for all species. The equations used to estimate biomass for the years 1999 through 2001 for all species growing in the Community Density Series plots and control plots. These equations were the best fitting curves between biomass and various surrogates for biomass and were based on destructive sampling done in 1999. In the equations: C = cover, H = height, L = length of the longest leaf, N = number of leaves, and W = width of longest leaf. All equation components are measured in mm and all estimated masses are in grams.

Table S3 Density and biomass estimates for all species in all plots for all years.

Acknowledgments

We thank our many field assistants - Saskia Arnesen, Andrew Bachmann, Rebecca Best, Maureen Bezanson, Lesley Dampier, Kate Edwards, Sarah Lord, Marilyn Makortoff, Jennie McLaren, Erika Olson, and Alda Ngo, Pippa Seccombe-Hett, and the staff of The Arctic Institute of North America base camp at Kluane Lake – Andy and Sian Williams, Lance Goodwin, Jessica Logher and Dan Gillis. We are grateful to the Champagne and Aishihik First Nations for allowing us to do research on their traditional lands.

Author Contributions

Conceived and designed the experiments: MAT RT. Performed the experiments: MAT. Analyzed the data: MAT. Contributed reagents/materials/analysis tools: MAT RT. Wrote the paper: MAT RT.

References

1. He J-S, Wolfe-Bellin KS, Schmid B, Bazzaz FA (2005) Density may alter diversity-productivity relationships in experimental plant communities. Basic Appl Ecol 6: 505–517.

2. Weiner J, Freckleton RP (2010) Constant Final Yield. Ann Rev Ecol Evol Syst 41: 173–192.

3. Kerkhoff AJ, Enquist BJ (2007) The implications of scaling approaches for understanding resilience and reorganization in ecosystems. BioSci 57, 489–99.

4. Goldberg DE, Turkington R, Olsvig-Whittaker L, Dyer AR (2001) Density-dependence in an annual plant community: variation among life history stages. Ecol Monographs 71: 423–446.

5. Lortie CJ, Turkington R (2002) The effect of initial seed density on the structure of a desert annual plant community. J Ecol 90: 435–445.

6. Rajaniemi TK, Turkington R, Goldberg DE (2009) Community-level consequences of species interactions in an annual plant community. J Veg Sci 20: 836–846.

7. Shilo-Volin H, Novoplansky A, Goldberg DE, Turkington R (2005) Density regulation in annual plant communities under variable resource levels. Oikos 108: 241–252.

8. Goldberg DE, Barton AM (1992) Patterns and consequences of interspecific competition in natural communities: a review of field experiments with plants. Am Nat 139: 771–801.

9. Goldberg DE, Rajaniemi TK, Gurevitch J, Stewart-Oaten A (1999) Empirical approaches to quantifying interaction intensity: competition and facilitation along productivity gradients. Ecology 80: 1118–1131.

10. Grime JP (1977) Evidence for the existence of three primary strategies in plants and its relevance to ecological and evolutionary theory. Am Nat 111: 1169–1194.

11. Tilman D (1988) Plant strategies and the dynamics and structure of plant communities., Princeton NJ: Princeton University Press.

12. Rees M (2013) Competition on productivity gradients – what do we expect? Ecology Letters (2013) 16: 291–298

13. Bennett JA, Cahill JF Jr (2012) Evaluating the Relationship between Competition and Productivity within a native Grassland. PLoS ONE 7(8): e43703. doi:10.1371/journal.pone.0043703

14. Goldberg DE, Novoplansky A (1997) On the relative importance of competition in unproductive environments. J Ecol 85: 409–418.

15. Kadmon R (1995) Plant competition along soil moisture gradients: a field experiment with the desert annual *Stipa capensis*. J Ecol 83: 253–262.

16. Brooker RW, Maestre FT, Callaway RM, Lortie CL, Cavieres LA, et al. (2008) Facilitation in plant communities: the past, the present, and the future. J Ecol 96: 18–34.

17. Callaway RM, Brooker RW, Choler P, Kikvidze Z, Lortie CJ, et al. (2002) Positive interactions among alpine plants increase with stress. Nature 417: 844–848.

18. Holmgren M, Scheffer M (2010) Strong facilitation in mild environments: the stress gradient hypothesis revisited. J Ecol 98: 1269–1275. doi: 10.1111/j.1365-2745.2010.01709.x

19. Lamb EG, Cahill JF (2008) When competition does not matter: Grassland diversity and community composition. Am Nat 171: 777–787. doi: 10.1086/587528

20. Goldberg DE (1994) Influence of competition at the community level: an experimental version of the null models approach. Ecology 75: 1503–1506.

21. Kareiva P (1994) Higher order interactions as a foil to reductionist ecology. Ecology 75: 1527–1528.

22. Sandel B, Smith AB (2009) Scale as a lurking factor: incorporating scale-dependence in experimental ecology. Oikos 118: 1284–1291.

23. Rajaniemi TK, Goldberg DE (2000) Quantifying individual- and community-level effects of competition using experimentally determined null species pools. J Veg Sci 11: 433–442.

24. Goldberg DE, Turkington R, Olsvig-Whittaker L (1995) Quantifying the community-level consequences of competition. Folia Geobot Phytotax 30: 231–242.

25. Harper JL (1977) Population biology of plants. London: Chapman and Hall.

26. Welden CW, Slauson WL (1986) The intensity of competition versus its importance: an overlooked distinction and some implications. Quart Rev Biol 61: 23–44.

27. Weigelt A, Jolliffe P (2003) Indices of plant competition. J Ecol 91,707–720.

28. McLellan AJ, Law R, Fitter AH (1997) Response of calcareous grassland plant species to diffuse competition: results from a removal experiment. J Ecol 85: 479–490.

29. Sammul M, Kull K, Oksanen L, Veromann P (2000) Competition intensity and its importance: results of field experiments with *Anthoxanthum odoratum*. Oecologia 125:18–25.

30. Turkington R, Goldberg DE, Olsvig-Whittaker L, Dyer AE (2005) Effects of density on timing of emergence and its consequences for survival and growth in two communities of annual plants. J Arid Env 61: 377–396.

31. Zamfir M, Goldberg DE (2000) The effect of initial density on interactions between bryophytes at individual and community levels. J Ecol 88: 243–255.

32. Treberg MA, Turkington R (2010) Density dependence in an experimental boreal forest understory community. Botany 88: 753–764.

33. Krebs CJ, Boutin S, Boonstra R (2001) Ecosystem Dynamics of the Boreal forest: The Kluane Project. New York: Oxford University Press.

34. Krebs CJ, Boonstra R, Boutin S, Sinclair ARE, Smith JNM, et al. (2014) Trophic dynamics of the boreal forests of the Kluane region. Arctic http://dx.doi.org/10.14430/arctic.2012.12-114.

35. Turkington R, McLaren JR, Dale MRT (2014) Determinants of herbaceous community structure and function in the Kluane region. Arctic http://dx.doi.org/10.14430/arctic4351

36. Arii K, Turkington R (2002) Do nutrient availability and competition limit plant growth of herbaceous species in the boreal forest understorey? Arct Antarct Alp Res 34: 251–261.

37. Cody WJ (2000) Flora of the Yukon Territory, 2nd edition. Ottawa, ON: National Research Council Press.

38. SAS (1995) JMP. SAS Institute Inc., Cary, North Carolina.

39. Smith B, Wilson JB (1996) A consumer's guide to evenness indices. Oikos 76: 70–82.

40. Goldberg DE, Estabrook GF (1998) Separating the effects of number of individuals sampled and competition on species diversity: an experimental and analytic approach. J Ecol 86: 983–988.

41. Turkington R, John E, Watson S, Seccombe-Hett P (2002) The effects of fertilization and herbivory on the herbaceous vegetation of the boreal forest in north-western Canada: a 10-year study. J Ecol 90: 325–337.

42. DiTomamaso A, Aarssen LW (1989) Resource manipulations in natural vegetation: a review. Vegetatio 84: 9–29.

43. Gough L, Shaver GR, Carroll J, Royer DL, Laundre JA (2000) Vascular plant species richness in Alaskan arctic tundra: the importance of soil pH. J Ecol 88: 54–67.

44. Rajaniemi TK (2003) Explaining productivity-diversity relationships in plants. Oikos 101: 449–457.

45. Storm C, Suss K (2008) Are low-productivity plant communities response to nutrient addition? Evidence from sand pioneer grassland. J Veg Sci 19: 343–354.

46. Dlott F, Turkington R (2000) Regulation of boreal forest understorey vegetation: the roles of resources and herbivores. Pl Ecol 151: 239–251.

47. Graham SA, Turkington R (2000) Population dynamics response of Lupinus arcticus to fertilization, clipping, and neighbour removal in the understorey of the boreal forest. Can J Bot 78: 753–758.

48. Treberg MA, Edwards K, Turkington R (2010) Voles are attracted to fertilizer in field experiments. Arct Antarct Alp Res 42: 113–116.

15

Effects of Slag-Based Silicon Fertilizer on Rice Growth and Brown-Spot Resistance

Dongfeng Ning[1], Alin Song[1], Fenliang Fan[1], Zhaojun Li[1], Yongchao Liang[1,2]*

1 Ministry of Agriculture Key Laboratory of Crop Nutrition and Fertilization, Institute of Agricultural Resources and Regional Planning, Chinese Academy of Agricultural Sciences, Beijing, China, **2** Ministry of Education Key Laboratory of Environment Remediation and Ecological Health, College of Environmental and Resource Sciences, Zhejiang University, Hangzhou, China

Abstract

It is well documented that slag-based silicon fertilizers have beneficial effects on the growth and disease resistance of rice. However, their effects vary greatly with sources of slag and are closely related to availability of silicon (Si) in these materials. To date, few researches have been done to compare the differences in plant performance and disease resistance between different slag-based silicon fertilizers applied at the same rate of plant-available Si. In the present study both steel and iron slags were chosen to investigate their effects on rice growth and disease resistance under greenhouse conditions. Both scanning electron microscopy (SEM) and transmission electron microscopy (TEM) were used to examine the effects of slags on ultrastructural changes in leaves of rice naturally infected by *Bipolaris oryaze*, the causal agent of brown spot. The results showed that both slag-based Si fertilizers tested significantly increased rice growth and yield, but decreased brown spot incidence, with steel slag showing a stronger effect than iron slag. The results of SEM analysis showed that application of slags led to more pronounced cell silicification in rice leaves, more silica cells, and more pronounced and larger papilla as well. The results of TEM analysis showed that mesophyll cells of slag-untreated rice leaf were disorganized, with colonization of the fungus (*Bipolaris oryzae*), including chloroplast degradation and cell wall alterations. The application of slag maintained mesophyll cells relatively intact and increased the thickness of silicon layer. It can be concluded that applying slag-based fertilizer to Si-deficient paddy soil is necessary for improving both rice productivity and brown spot resistance. The immobile silicon deposited in host cell walls and papillae sites is the first physical barrier for fungal penetration, while the soluble Si in the cytoplasm enhances physiological or induced resistance to fungal colonization.

Editor: Guoping Zhang, Zhejiang University, China

Funding: This work was jointly supported by The 12th Five-year Key Programs entitled "Techniques for Agricultural Use of Steel and Iron Slag: Research and Demonstration" and "Study of Key Technologies for Alleviating Obstacle Factors and Improving Productivity of Low-Yield Cropland" supported by Ministry of Science and Technology, China and The National Natural Science Foundation of China entitled "The quantitative study of influence of straw returning on silicon releasing in typical rice soil of Southern China (41301310)". The funders had no role in study design, data collection and analysis, decision to publish, or preparation of the manuscript.

Competing Interests: The authors have declared that no competing interests exist.

* Email: ycliang@zju.edu.cn

Introduction

Silicon (Si) is the second most abundant element in soils [1,2]. Although Si has not been proven to be an essential element for plant growth and development, its beneficial roles in stimulating plant growth, grain yield and resistance to abiotic (metal toxicity, salt and drought stress, nutrient imbalance, extreme temperature) and biotic stress (plant diseases and insect pests) have been well documented [3–7].

Rice (*Oryza sativa* L.) is the second most widely grown crop in the world, and the major staple food for more than half of the world's population [8,9]. Rice is also a typical Si hyper-accumulating plant species, containing Si up to 10% in shoots on a dry weight basis [2]. Rice roots take up Si in the form of silicic acid (H_4SiO_4) from the soil solution [10]. In tropical and subtropical areas, because of heavy desilication-aluminization arising from high temperature and rainfall, plant- available Si is low in these highly-weathered soils [11]. In addition, repeated mono-cropping with rice may greatly decrease plant-available Si in soil. It is estimated that producing a total rice grain yield of 5000 kg ha^{-1} will remove Si at 230–470 kg ha^{-1} from the soil [5],

and Si may then become a yield-limiting element for rice production [12–14]. Therefore, it maybe is necessary to provide exogenous Si fertilizer for an economic and sustainable rice production system [15–18].

Brown spot caused by the fungus (*Bipolaris oryzae*) is one of the most devastating and prevalent diseases of rice. Brown spot may cause significant yield losses [19,20]. The major method to control brown spot in agriculture is through application of fungicides [19]. However, there is a need to explore more eco-friendly management practices in consideration of the public's concerns with health and environmental issues. The physiological condition of rice plant, which is strongly influenced by soil conditions, particularly soil nutrient status (e.g. potassium, calcium, magnesium, manganese, iron, and silicon etc.), is one of the main factors governing brown spot severity [19,21]. Some authors suggest that application of Si fertilizer to rice fields is an alternative approach to control brown spot, especially in soils where plant-available Si is very low [22–24].

Steel slags or iron slags are byproducts of steel or iron industries, which account for 15–20% of total steel production. Large

Table 1. The main chemical characteristics of two slag-based silicon fertilizers tested in the present study (%).

Si fertilizer	CaO	SiO$_2$	MgO	Al$_2$O$_3$	Fe$_2$O$_3$	MnO	TiO$_2$
Q	43.6	26.9	8.1	10.9	3.1	0.9	1.2
H	50.9	21.0	7.7	6.0	5.0	1.5	0.6

amounts of slag are produced in China annually [25]. Slags are not merely metallurgical wastes, but they have been successfully used in agriculture in many developed counties [26,27]. In contrast, only 10% of the total slag is recycled in China [28]. Slags contain sufficient amounts of Si (10–28%); therefore, they may potentially be used as a Si fertilizer source. Application of such kind of Si fertilizer has been shown to improve degraded paddy soils, as well as rice growth and disease resistance [2,29–34]. So, slag applied to paddy rice fields as Si fertilizer is beneficial not only for rice health and growth, but also from economic and environmental perspectives. However, variation exists in the ore and coke, as well as in the cooling process; consequently, the composition and property of slags may vary widely [35]. Therefore, plant-available Si content and Si availability in slags vary widely too [36]. Previous studies have demonstrated positive effects of wollastonite or calcium silicate as Si resource on rice growth and disease resistance [30,32]. However, there are only a few reports to compare the agronomic benefits of different sources of slag used in rice.

The objective of this study was to assess the effects of steel slag and iron slag applied at the same rates of plant-available Si on rice growth and brown spot development in rice and to investigate the relationship between Si-mediated ultrastructural changes and brown spot disease infection in rice.

Materials and Methods

Soil and plant material preparation

The soil used was sampled from Qionghai, Hainan province of South China (N 19°09′16.2″, E 110°17′35.3″) (no specific permissions were required for soil sampling in this location and the field in this study did not involve endangered or protected species). It was a latosol derived from basalt with a plant-available Si concentration of 41.8 mg kg^{-1} (extracted by 0.025 M citric acid) and a pH value of 5.16. The soil was air-dried and sieved (2.0 mm). The rice variety tested is a hybrid (*Oryza sativa* L. cv. Fengyuanyou 299), characterized by its mid-late maturity. Seeds

were sterilized with 10% (v/v) H$_2$O$_2$ for 15 min, rinsed with distilled water, soaked in water for 24 hours, and then transferred into culture dishes for germination at 25°C in the dark. Two days later, the germinated seeds were placed on a float tray (10×15 cm) in a controlled environment with a day/night temperature of 25°C (12 h): 25°C (12 h).

Experimental design

A pot experiment factorially arranged in a 2×4 randomized, complete block design was conducted with three replicates per treatment, giving a total of 24 pots. The entire experiment was duplicated. Two different Si fertilizers were chosen for the pot experiment. One was derived from air-cooling steel slag, with HCl-soluble Si content of 7.61%, referred to as H, and the other was based on water-cooling iron slag, with HCl-soluble Si content of 9.35%, referred to as Q. The main chemical properties of the two slags are presented in Table 1. Four Si treatments with three replicates each were established. The rate of Si applied, equivalent to 0.5 M HCl-soluble Si, was 0 (Si$_0$), 187 (Si$_1$), 560 (Si$_2$) and 935 (Si$_3$) mg Si kg^{-1}. The Si fertilizer was thoroughly mixed with soil prior to potting. Basal fertilizers supplied were 0.2 g N kg^{-1} as urea, 52 mg P kg^{-1} as potassium dihydrogen phosphate, and 84 mg K kg^{-1} as potassium sulfate. Each plastic pot was filled with 5 kg of air-dried and sieved (2.0 mm) soil. Uniform seedlings with three leaves fully expanded were transplanted at two seedlings per pot. During the rice growing period, distilled water was applied to maintain a 2-cm water layer but no pesticides were applied.

Plant sampling

Rice plants were harvested at maturity, and separated into stem, leaf, and grain, and then washed thoroughly with distilled water. The dry weight of these tissues was recorded after being oven-dried at 75°C till a constant weight. These tissues were then ground to pass through a 0.5-mm sieve for Si analysis.

Table 2. Effects of different silicon treatments on dry weight of rice organs (%).

Fertilizer	Rate	Leaf	Stem	Grain
Control	Si$_0$	7.72±1.60 b	13.6±4.73 b	4.71±1.52 c
Q	Si$_1$	13.1±2.02 a	20.8±1.11 a	13.7±0.69 b
	Si$_2$	13.7±2.02 a	20.8±1.07 a	13.4±2.37 b
	Si$_3$	13.0±2.27 a	20.2±1.46 a	11.8±1.99 b
H	Si$_1$	12.8±2.71 a	19.4±1.30 b	15.1±2.33 ab
	Si$_2$	11.5±1.08 a	20.2±1.24 a	15.0±1.53 ab
	Si$_3$	12.5±1.15 a	23.7±2.40 a	16.9±2.08 a

Si$_0$: no Si fertilizer; Si$_1$: slag fertilizer applied at a rate of 187 mg plant-available Si per kg soil; Si$_2$: slag fertilizer applied at a rate of 560 mg plant-available Si per kg soil; Si$_3$: slag fertilizer applied at a rate of 935 mg plant-available Si per kg soil; H: slag fertilizer H, Q: slag fertilizer Q; Data are means ± SD of three replicates; mean values followed by different letters (a, b, c) are significantly different (P≤0.05).

Table 3. Analysis of variance of the effects slag-based silicon fertilizer (slag) and application rate of Si (Si-R) on dry weight of rice organs (%).

| Sources of variation | Df | F values | | |
		Leaf	Stem	Grain
Slag	1	0.816 ns	0.142 ns	6.41*
Si-R	3	9.93*	11.61*	36.30**
Slag×Si-R	3	0.388 ns	1.03 ns	1.90 ns

Levels of probability: ns = non significant, significantly different *p≤0.05 and **p≤0.01.
Levels of probability: ns = non significant and *p≤0.05, **p≤0.01.

Disease index survey

Rice leaves were naturally infected by *Bipolaris oryaze*, the causal agent of brown spot at the joining stage. Disease severity, based on the percentage of infected leaf surface area and the percentage of infected leaves per pot, was determined two weeks after infection. In this study, disease severity (DS) was classified into nine grades based on the following: DS0 = healthy plants, DS1≤1%, DS3 = 2–5%, DS5 = 6–15%, DS7 = 16–25% and DS9≥25%. Disease index (%) = $[\sum(S*n_s)/(9*Ns)]*100$. Where S is the severity value, n_s is the number of infected leaves with a severity of S and Ns is the number of leaves evaluated [37].

Scanning electron microscopy

The deposition of Si in the leaf was observed using scanning electron microscopy (SEM). Since similar slag effects on plant growth and brown spot resistance were observed for both Si sources, only leaf samples of rice plants treated with slag H were collected for microscopic examination. At the anthesis stage, fresh specimens of the top-second leaf of rice plants grown without slag (control) or with slag (H) applied at a rate of 935 mg plant-available Si per kg soil were randomly sampled from two plants per pot. They were first fixed with 2.5% (v/v) glutaraldehyde in 0.1 M phosphate buffer solution (pH 7.4) under vacuum for 2–3 h at 20°C, and then post-fixed with 1% (w/v) osmium tetroxide in the phosphate buffer solution for 30 min [38]. Afterwards, they were dehydrated through a graded series of ethanol [50, 70, 80, 90 and 100% (v/v)], dried by a critical-point drying method with liquid CO_2 and coated with metal and then loaded onto the instrument [38]. The surface scan was performed using a scanning electron microscope (FEI QUANTA200, Japan).

Transmission electron microscopy

Squares were excised with scissors from the top-second leaf at the anthesis stage. The leaf samples of rice plants grown without slag (control) or with slag (H) applied at a rate of 935 mg plant-available Si per kg soil were collected and fixed immediately with 2% (v/v) glutaraldehyde and 2% (v/v) paraformaldehyde in 0.05 M sodium cacodylate buffer (pH 7.2) at room temperature overnight and then washed with the same buffer three times for 10 min each [38]. Afterwards, samples were postfixed with 1% (w/v) osmium tetroxide in the same buffer at room temperature for 2 h and washed twice with distilled water. The post-fixed samples were stained with 0.5% (w/v) uranyl acetate at 4°C overnight. They were then dehydrated in a graded series of ethanol [30, 50, 70, 80, 95, and 100% (v/v)] and three times in 100% ethanol for 10 min each [38]. Ultrathin sections (approximately 50 nm in thickness) were made with a diamond knife by an ultramicrotome (LKBVI). The sections were mounted on copper grids and stained for 7 min each with 2% (w/v) uranyl acetate and Reynolds' lead citrate [38]. The sections were examined by transmission electron microscopy (Phillips EM 400 ST, the Netherlands).

Chemical analysis

The main chemical components of slag fertilizers were measured by SEM. Scanning electron microscopy was performed in a JSM-6510 SEM at accelerating voltage of 20 kV attached with an X-ray energy-dispersive spectrometer, EDS (Genesis XM2). Before the scanning process, all samples were dried and coated with gold to enhance the electron conductivity.

The available Si content in slag was determined following extraction with 0.5 M HCl [slag/(HCl) ratio of 1:50, shaking at

Table 4. Effects of different silicon (Si) treatments on Si concentrations in rice organs (SiO$_2$%).

Fertilizer	Rate	Leaf	Stem	Grain
Control	Si$_0$	11.1±0.31 c	5.17±0.42 d	0.19±0.031 c
Q	Si$_1$	12.3±0.36 b	5.57±0.47 cd	0.22±0.04 bc
	Si$_2$	12.3±0.35 b	5.34±0.27 cd	0.23±0.04 bc
	Si$_3$	12.1±0.31 b	5.76±0.14 c	0.27±0.02 ab
H	Si$_1$	11.9±0.22 b	5.54±0.30 cd	0.21±0.03 c
	Si$_2$	12.6±0.13 ab	6.31±0.39 b	0.26±0.02 ab
	Si$_3$	12.9±0.27 a	6.87±0.37 a	0.28±0.04 a

Si$_0$: no Si fertilizer; Si$_1$: slag fertilizer applied at a rate of 187 mg plant-available Si per kg soil; Si$_2$: slag fertilizer applied at a rate of 560 mg plant-available Si per kg soil; Si$_3$: slag fertilizer applied at a rate of 935 mg plant-available Si per kg soil; H: slag fertilizer H, Q: slag fertilizer Q; Data are means ± SD of three replicates; mean values followed by different letters (a, b, c) are significantly different (P≤0.05).

Table 5. Analysis of variance of the effects of slag-based silicon fertilizer (slag) and application rate of Si (Si-R) on Si concentrations in rice organs (SiO_2%).

Sources of variation	Df	F values		
		Leaf	Stem	Grain
Slag	1	2.09 ns	21.05**	0.602 ns
Si-R	3	30.93**	18.51**	13.74*
Slag×Si-R	3	1.80 ns	7.56*	0.786 ns

Levels of probability: ns = non significant, significantly different *p≤0.05 and **p≤0.01

300 rpm for 1 h] and analyzed by the colorimetric silicon molybdenum blue method [39]. Slag pH and EC were measured at a water/soil ratio of 2.5.

Plant-available Si content in soil was extracted by 0.25 M citric acid [soil/(citric acid) ratio of 1:5] for 5 hrs, and analyzed by the colorimetric silicon molybdenum blue method [40]. The soil pH was measured at a water/soil ratio of 2.5.

The silicon content in rice plants was determined by the colorimetric silicon molybdenum blue method [41–42]. Briefly, 100 mg of plant tissue was mixed with 3 mL of 50% (w/v) NaOH in a polyethylene tube. These tubes were covered with loose-fitting plastic caps and autoclaved at 125°C for 1 h and analyzed by the colorimetric silicon molybdenum blue method.

Statistical analysis

All data in figures and tables are shown as means ± SD of three replicates. Two -way ANOVA was used for statistical analysis and Fisher's L.S.D. test was adopted to detect the significant difference ($p≤0.05$) between the means of different treatments. All statistical analyses were done using the Excel 2007 and SPSS (PASW Statistics 18.0).

Results

Dry weight and silicon concentration of different rice tissues

Table 2–3 show that application of both iron slag (Q) and steel slag (H) fertilizers significantly increased dry weight of leaf and stem, and grain yield compared with the control (Si$_0$) treatment. However, there was no significant difference among different application rates of silicon (except Si$_0$). Dry weight of leaf and stem showed no significant difference between the two Si fertilizers

tested, but Si fertilizer H produced significantly more grain weight than Si fertilizer Q.

Table 4–5 show that the Si concentration was significantly different among different organs, with the order of leaf > stem > grain. Application of both Si fertilizers significantly increased the Si concentration in leaf, stem and grain compared with the Si$_0$ treatment. The Si concentration in rice organs tended to increase with increasing application rate of Si, and there was a significant difference between the Si$_3$ treatment and Si$_1$ treatment. The Si concentration of stem was significantly higher in Si fertilizer H than in Si fertilizer Q. However, no significant difference in leaf or grain Si concentration was noted between the two Si fertilizers used.

Disease severity

Under greenhouse conditions, rice leaves were naturally infected with brown spot caused by *Bipolaris oryzae*. At the anthesis stage, disease severity showed visible differences among treatments. The data in Table 6–7 demonstrate that rice leaf lesion of the control treatment (Si$_0$) was most severe with an incidence of 39.6%, and a disease index of 56.0%. Application of both Si fertilizers significantly decreased brown spot development. Meanwhile, disease severity of fertilizer H treatments was lower than that of fertilizer Q treatments.

Transmission electron microscopic analysis of silica cell

Ultrathin sections of leaf samples were observed by transmission electron microscope (TEM). The ultrastructural details demonstrated that the numbers of fungal cells and fungal colonization in the leaf epidermis were different between Si-untreated and Si-treated rice plants. The leaf mesophyll cells of silicon-untreated

Table 6. Effects of different silicon (Si) treatments on rice brown spot development at the anthesis stage (%).

Fertilizer	Rate	Incidence of disease	Disease index
Control	Si$_0$	39.7±2.11a	56.1±2.60 a
Q	Si$_1$	4.67±0.60 b	22.0±3.50 b
	Si$_2$	1.33±1.53 b	8.64±2.71bc
	Si$_3$	0.33±0.58 b	2.22±3.85 c
H	Si$_1$	3.67±2.08 b	18.0±2.26 bc
	Si$_2$	1.33± 0.58 b	8.89±2.67bc
	Si$_3$	0.00±0.0 b	0.00±0.00 c

Si$_0$: no Si fertilizer; Si$_1$: slag fertilizer applied at a rate of 187 mg plant-available Si per kg soil; Si$_2$: slag fertilizer applied at a rate of 560 mg plant-available Si per kg soil; Si$_3$: slag fertilizer applied at a rate of 935 mg plant-available Si per kg soil; H: slag fertilizer H, Q: slag fertilizer Q; Data are means ± SD of three replicates; mean values followed by different letters (a, b, c) are significantly different (P≤0.05).

Table 7. Analysis of variance of the effects of slag-based silicon fertilizer (slag) and application rate of Si (Si-R) on rice brown spot development at the anthesis stage (%).

Sources of variation	Df	F values	
		Incidence of disease	Disease index
Slag	1	0.0145 ns	13.20*
Si-R	3	47.05**	3545**
Slag×Si-R	3	0.00727 ns	5.85 ns

Levels of probability: ns = non significant, significantly different *p≤0.05 and **p≤0.01.

plants were disorganized at the stage of the fungal (*Bipolaris oryzae*) colonization. The cytoplasm was disintegrated with a consequence of chloroplast degradation and cell-wall alterations. Abundant amorphous materials were noticed in a mesophyll cell colonized by the fungus (Figure 1). The Si layers were observed in Si-treated epidermal cell walls, and the thickness of the silicon layer was seen to be increased by Si application (Figure 2). The chloroplast thylakoid lamella of mesophyll cells of Si-untreated leaves became swollen, and stroma lamellae and grana lamellae of chloroplasts were distorted. In contrast, the chloroplast structure of mesophyll cells of Si-treated rice leaves was relatively intact, with thylakoid lamellae stacked in order, grana lamellae accumulated compactly and some starch grains visible (Figure 3).

Scanning electron microscopic analysis of silica cells

Morphology of silica cells on the surface of the top-second leaf at the anthesis stage differed among treatments (Figure 4, 5). There were many silica cells, wart-like protuberances (papillae) and stomata on the leaf surface. The silica cells had a dumbbell shape and were distributed in rows along the leaf veins. However, the morphology and number of these silica cells varied among treatments. Silicon application led to more pronounced cell silicification in rice leaves, more silica cells and larger papillae.

Discussion

In this study, the concentration of plant-available Si in the soil tested was 41.8 mg (Si) kg^{-1}. In China, the critical value for plant-

available Si concentration in acid paddy soil is 44.4–51.4 mg kg^{-1} (Si), below which positive rice responses to silicate fertilizer can be expected [29]. Our results show that silicon fertilizers from steel slag and iron slag both significantly promoted rice growth and rice yield. Silicon fertilizer H produced significantly higher grain weight than silicon fertilizer Q at the same plant-available Si application rate. Two factors may account for this observation. First, the composition and cooling process of slags influence Si-dissolution from slags. Slag H, which was cooled slowly, had higher Si-availability to plants compared with slag Q, which was more rapidly cooled in water. This result was consistent with a report by Takahashi (1981) [43], suggesting that Si availability of slag to plants cannot be precisely determined only by the extraction method using 0.5 M HCl. It is necessary to estimate the Si-releasing process from slags in paddy soils and to analyze the factors affecting the solubility of the slags in future studies. Second, other nutrients provided by slag might be also beneficial for rice growth, such as Ca, Mg, Fe and Mn etc. In this study, the plant-available Si concentration was lower in slag H than in slag Q, thus, at the same available-Si application rate, the real application rate of the slag H was higher than that of slag Q. In this case, the amount of other nutrients such as Ca, Fe and Mn provided by slag H might be higher than that by slag Q because not only the real application rate of slag H was higher than that of slag Q but also the content of Ca, Fe and Mn was higher in slag H than in slag Q (Table 1). It could be supposed that other nutrients provided by slags also contributed to the final rice performance,

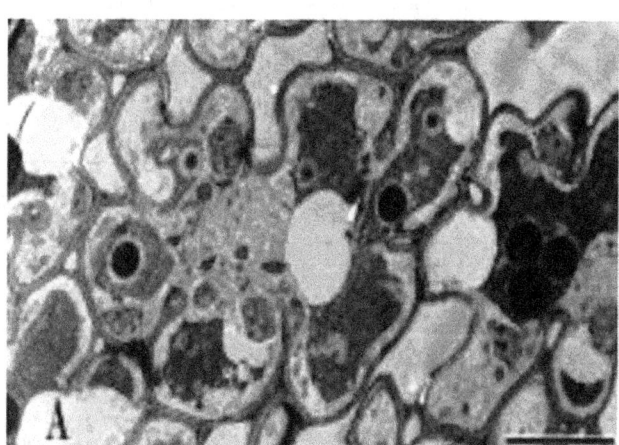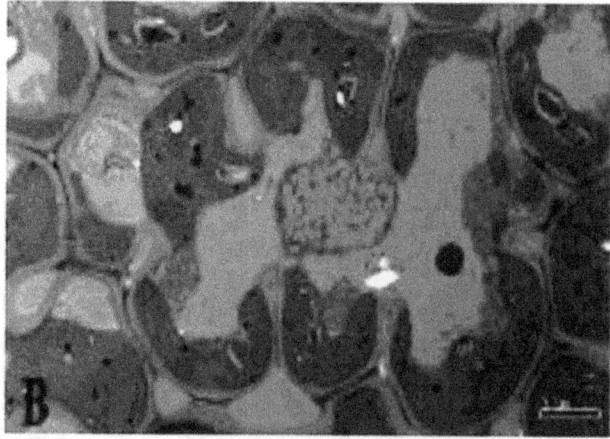

Figure 1. Transmission electron micrographs of mesophyll cells of rice leaves. Scale bars = 5 μm. **A**: Mesophyll cells of a control plant grown without silicon fertilizer at the anthesis stage; **B**: Mesophyll cells of a silicon-treated plant grown with slag (H) applied at a rate of 935 mg plant-available Si per kg soil at the anthesis stage.

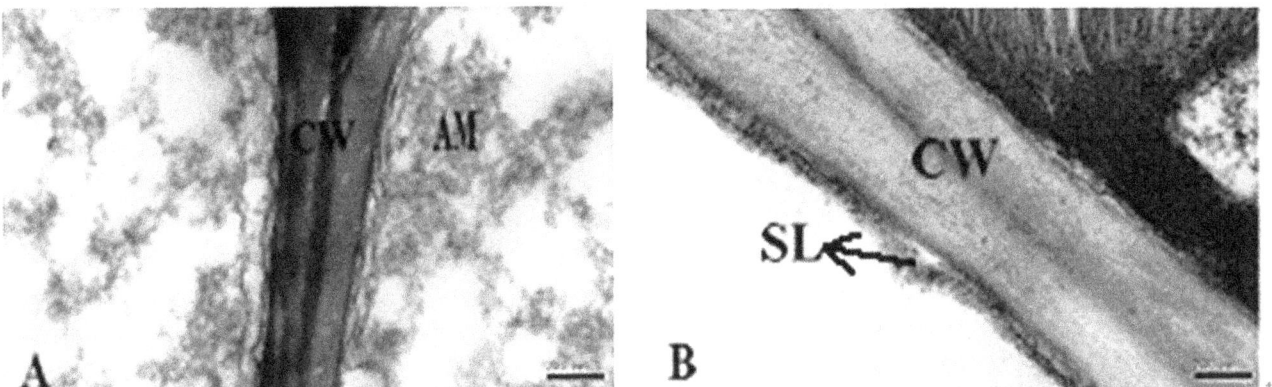

Figure 2. Transmission electron micrographs of cell wall from leaves of rice. CW, cell wall; AM, amorphous material; SL, silicon layer. A: Leaf epidermis of a control plant grown without silicon fertilizer at the anthesis stage; B: Leaf epidermis of a silicon treated plant with slag (H) applied at a rate of 935 mg plant-available Si per kg soil at the anthesis stage.

dry weight and rice yield, which, however, needs further validation.

Silicon fertilizer could be an environmentally-friendly alternative to control rice diseases [14,44–46]. In this study, rice leaves were naturally infected with brown spot disease caused by *Bipolaris oryzae* at the jointing stage. The leaves of rice plants that were not treated with slag showed disease symptoms 5 days earlier than those treated with slag. At anthesis, visible differences in disease severity appeared among treatments. We found that application of both steel slag and iron slag fertilizers showed significantly lower brown spot incidence and severity (Table 6–7). Lesion areas of leaves showed a decreasing, but non-significant trend with increasing Si application rates (Table 4–5). This result was consistent with rice yield (Table 2). The ultrastructural characteristics showed that the chloroplast thylakoid lamellae of mesophyll cells of untreated rice leaves became swollen, and stroma and grana lamellae of chloroplast were distorted at anthesis. However, the chloroplast structure of mesophyll cells of Si-treated leaves was relatively intact (Figure 3).

Brown spot severity has been reported to be negatively correlated with Si concentration in rice tissue [15,24,47]. An active Si uptake by lateral roots of rice plants plays a key role in

rice resistance to brown spot [24]. Application of the two Si fertilizers significantly increased the Si concentration in leaves (Table 2). There have been debates of the mechanisms involved in Si-mediated plant disease resistance. Some authors suggest that a mechanical or physical barrier provided by Si deposition in cell walls contributes to enhanced resistance [38,48–49], while more recent studies suggest that Si plays a biochemical role in mediating plant resistance to pathogens [33,46,50]. Our results show that Si application led to more pronounced cell silicification in rice leaves and more elaborate and larger papillae (Figure 4, 5). The elaborate papillae formed in Si-treated leaf epidermal surface might increase the resistance to fungal penetration [37,51]. The Si layers were observed in Si-treated epidermal cell walls, and their thickness was increased by Si treatment (Figure 2). The Si layers in epidermal cell walls supposedly confer enhanced host resistance to brown spot, which is in line with the previous reports that the cuticular Si double layer developed on rice leaf cells constituted a physical barrier to impede fungal penetration and colonization [38,48–49].

In this study, we also found apparent differences in the number of fungal cells and fungal colonization in the leaf epidermis between Si-untreated and Si-treated plants (Figure 1). We surmise that soluble Si may induce physiological resistance to restrain the

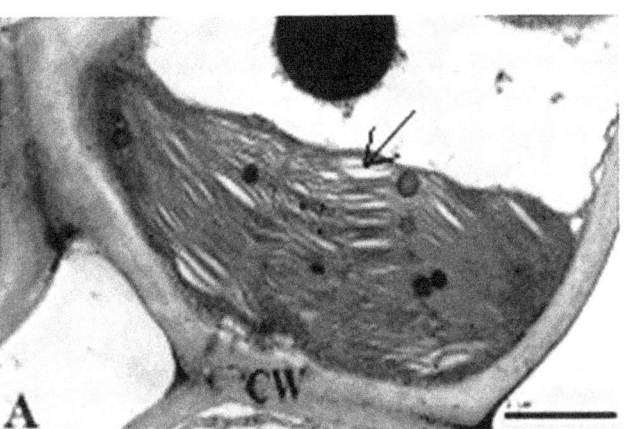
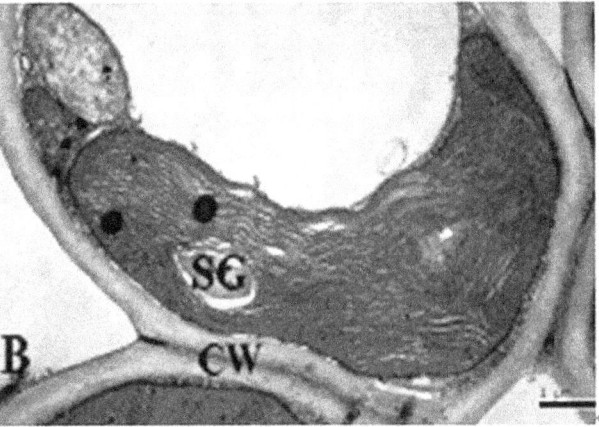

Figure 3. Transmission electron micrographs of chloroplasts from leaves of rice. Scale bars =1 μm. CW, cell wall; SG, starch grain. **A:** Chloroplast of a control plant grown without silicon fertilizer at the anthesis stage; **B:** Chloroplast of a silicon treated plant grown with slag (H) applied at a rate of 935 mg plant-available Si per kg soil at the anthesis stage.

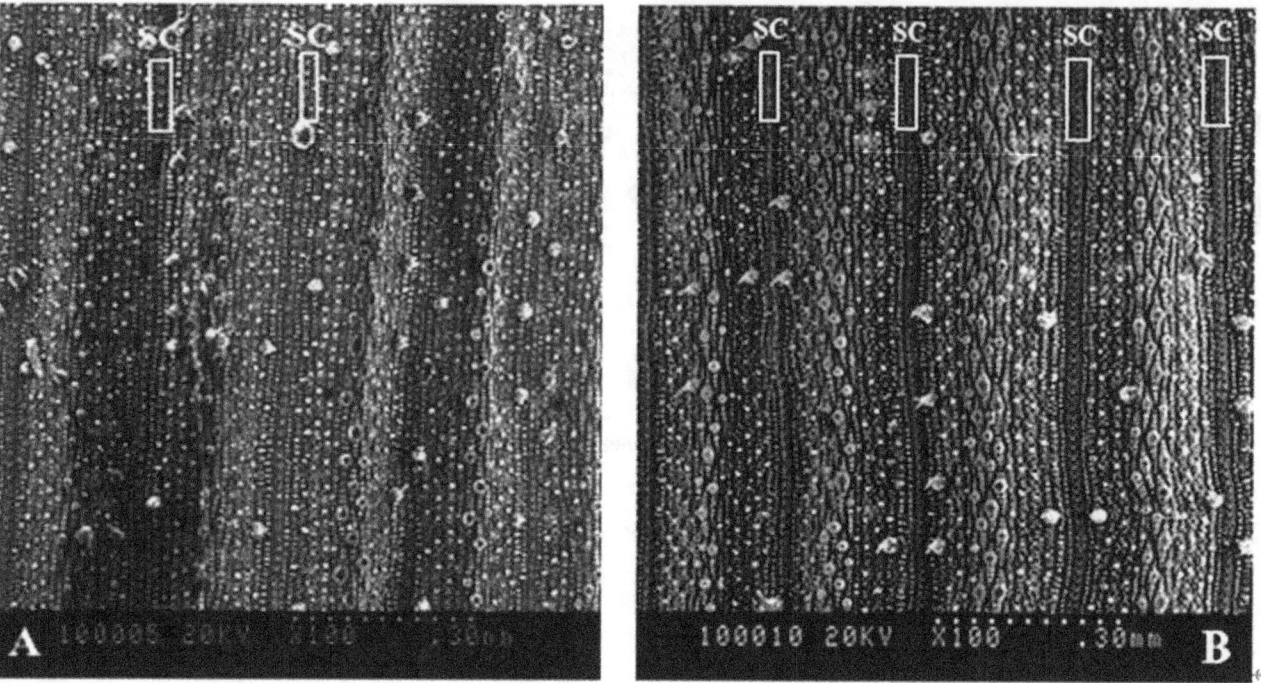

Figure 4. Scanning electron micrographs with 100 magnification of rice leaves. Scale bars = 30 mm. SC, silica cell. **A:** the top second rice leaf epidermis of a control plant without silicon fertilizer at the anthesis stage; **B:** the top second rice leaf epidermis of a silicon-treated rice plant grown with slag (H) applied at a rate of 935 mg plant-available Si per kg soil at the anthesis stage.

growth of *Bipolaris oryzae* and keep host cells relatively intact. Rodrigues et al. (2003a, 2005b) suggested that Si induced accumulation of phenolic compounds or phytoalexins, which played a primary role in rice defense against infection by

Magnaporthe grisea [31,52]. Dallagnol et al. (2011) found that the concentrations of soluble phenolics and lignin and activities of peroxidase and chitinase were higher in Si-treated rice leaves infected by *Bipolaris oryzae*, which contributed to rice resistance to

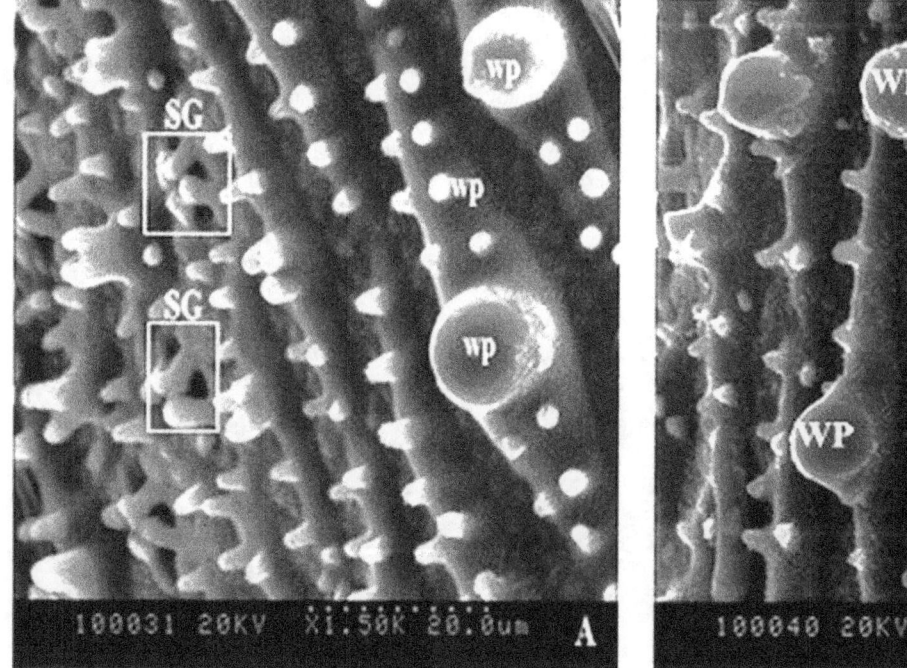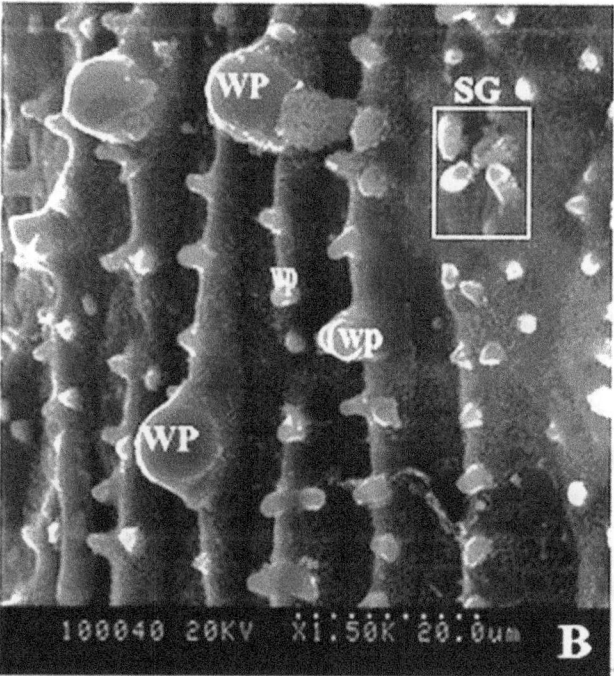

Figure 5. Scanning electron micrographs with 150 K magnification of rice leaves. Scale bars = 20 μm. SC, silica cell; WP, wart-like protuberance; SG, stomatal guard cell. **A:** the top second leaf epidermis of a control plant without silicon fertilizer at the anthesis stage; **B:** The top second leaf epidermis of a silicon-treated rice plant grown with slag (H) applied at a rate of 935 mg plant-available Si per kg soil at the anthesis stage.

brown spot [47]. Other reports suggest that after inoculation with *M. grisea*, Si-treated rice plants significantly increased the activities of pathogenesis-related proteins (PRs) in leaves, such as peroxidase (POD), polyphenol oxidase (PPO), phenylalanine ammonia lyase (PAL), and catalase (CAT) [37,46]. Therefore, we believe that Si-enhanced plant disease resistance plus the role of Si as physical barrier as suggested by Sun et al. (2010) in rice blast resistance [46] also contributed to the Si-enhanced resistance to rice brown spot observed in the present study.

Conclusions

Applying Si fertilizer to Si-deficient paddy soil is necessary for both high rice yield and brown spot resistance. Both steel slag and iron slag are effective in this regard. In this experiment, silicon fertilizer H produced significantly higher grain weight than silicon fertilizer Q at the same plant-available Si application rate. Composition and cooling process of slags influence Si-dissolution from slags. Si availability of slag to plants cannot be precisely determined only by the extraction method using 0.5 M HCl. The immobile silicon deposited in host cell walls and papillae sites is the first physical barrier for fungal (*Bipolaris oryzae*) penetration and soluble Si in the cytoplasm enhances physiological or induced resistance to restrain fungal colonization.

Author Contributions

Conceived and designed the experiments: DFN ALS FLF ZJL YCL. Performed the experiments: DFN. Analyzed the data: DFN YCL. Contributed reagents/materials/analysis tools: YCL. Wrote the paper: DFN YCL.

References

1. Epstein E (1994) The anomaly of silicon in plant biology. Proc Natl Acad Sci USA 91: 11–17.
2. Ma JF, Takahashi E (2002) Soil, Fertilizer, and Plant Silicon Research in Japan. Elsevier, Amsterdam, pp. 1–2.
3. Epstein E (1999) Silicon. Ann Rev Plant Physiol Plant Mol Biol 50: 641–664.
4. Ma JF (2004) Role of silicon in enhancing the resistance of plants to biotic and abiotic stresses. Soil Sci Plant Nutr 50: 11–18.
5. Rodrigues FÁ, Datnoff LE (2005a) Silicon and rice disease management. Fitopatol Bras 30: 457–469.
6. Liang YC, Sun WC, Zhu YG, Christie P (2007) Mechanisms of silicon-mediated alleviation of abiotic stresses in higher plants: A review. Environ Pollut 147: 422–428.
7. Catherine Keller FG, Meunier JD (2012) Benefits of plant silicon for crops: a review. Agron Sust Develop 32: 201–213.
8. Wailes EJ, Cramer GL, Chavez EC, Hansen JM (1997) Arkansas global rice model: international baseline projections for 1997–2010. Arkansas Agric Exp Stat, Arkansas. pp. 1–46.
9. van Nguyen N, Ferrero A (2006) Meeting the challenges of global rice production. Paddy Water Environ 4: 1–9.
10. Ma JF, Yamaji N (2006) Silicon uptake and accumulation in higher plants. Trends Plant Sci 11: 392–397.
11. Raven JA (2003) Cycling silicon–the role of accumulation in plants. New Phytol 158: 419–430.
12. Foy CD (1992) Soil chemical factors limiting plant root growth. Adv Soil Sci 19: 97–149.
13. Winslow MD, Okada K, Correa-Victoria F (1997) Silicon deficiency and the adaptation of tropical rice ecotypes. Plant Soil 188: 239–248.
14. Datnoff LE, Deren CW, Snyder GH (1997) Silicon fertilization for disease management of rice in Florida. Crop Prot 16: 525–531.
15. Deren CW, Datnoff LE, Snyder GH, Martin FG (1994) Silicon concentration, disease response, and yield components of rice genotypes grown on flooded organic histosols. Crop Sci 34: 733–737.
16. Savant NK, Snyder GH, Datnoff LE (1996) Silicon management and Sustainable rice production. Adv Agron 58: 151–199.
17. Alvarez J, Datnoff LE (2001) The economic potential of silicon for integrated management and sustainable rice production. Crop Prot 20: 43–48.
18. Bocharnikova EA, Loginov SV, Matychenkov VV, Storozhenko PA (2010) Silicon fertilizer efficiency. Russ Agric Sci 36: 446–448.
19. Ou SH (1985) Rice diseases, 2nd ed. Kew, Surrey, UK, Commonwealth Mycological Institute.
20. Motlagh MR, Kaviani B (2008) Characterization of new bipolaris spp.: the causal agent of rice brown spot disease in the North of Iran. Int J Agric Biol 10: 638–642.
21. Marchetti MA, Peterson HD (1984) The role of *Bipolaris oryzae* in floral abortion and kernel discoloration in rice. Plant Dis 68: 288–291.
22. Lee TS, Hsu LS, Wang CC, Jeng YH (1981) Amelioration of soil fertility for reducing brown spot incidence in the patty field of Taiwan. J Agric Res Chin 30: 35–49.
23. Datnoff LE, Snyder GH, Raid RN, Jones DB (1991) Effect of calcium silicate on blast and brown spot intensities and yields of rice. Plant Dis 75: 729–732.
24. Dallagnol LJ, Rodrigues FÁ, Mielli MVB, Ma JF, Datnoff LE (2009) Defective active silicon uptake affects some components of rice resistance to brown spot. Phytopathol 99: 116–121.
25. Wu SP, Xue YJ, Ye QS, Chen YC (2007) Utilization of steel slag as aggregates for stone mastic asphalt (SMA) mixtures. Build Environ 42: 2580–2585.
26. Motz H, Geiseler J (2001) Products of steel slags an opportunity to save natural resources. Waste Manage 21: 285–293.
27. Shen HT, Forssberg E (2003) An overview of recovery of metals from slags. Waste Manage 23: 933–949.
28. Zhu GL (2010) The current state and developing of comprehensive disposal of steel and iron slag. Iron Steel Scrap 1: 12–16 (in Chinese).
29. Wang HL, Li CH, Liang YC (2001) Agricultural utilization of silicon in China. In: Silicon in Agriculture. Datnoff LE, Snyder GH, Korndorfer GH. ed. Elsevier, Amsterdam, pp. 343–358.
30. Seebold KW, Kucharek TA, Datnoff LE, Correa Victoria FJ, Marchetti MA (2001) The influence of silicon on components of resistance to blast in susceptible, partially resistant, and resistant cultivars of rice. Phytopathol 91: 63–69.
31. Rodrigues FÁ, Benhamou N, Datnoff LE, Jones JB, Bélanger RR (2003a) Ultrastructural and cytochemical aspects of silicon-mediated rice blast resistance. Phytopathol 93: 535–546.
32. Rodrigues FÁ, Valeb FXR, Korndörfer GH, Prabhud AS, Datnoff LE, et al. (2003b) Influence of silicon on sheath blight of rice in Brazil. Crop Prot 22: 23–29.
33. Rodrigues FÁ, Vale FXR, Datnoff LE, Prabhu AS, Korndörfer GH (2003c) Effect of rice growth stages and silicon on sheath blight development. Phytopathol 93: 256–261.
34. Rodrigues FÁ, McNally DJ, Datnoff LE, Jones JB, Labbé C, et al. (2004) Silicon enhances the accumulation of diterpenoid phytoalexins in rice: A potential mechanism for blast resistance. Phytopathol 94: 177–183.
35. Cha W, Kim J, Choi H (2006) Evaluation of steel slag for organic and inorganic removals in soil aquifer treatment. Water Res 40: 1034–1042.
36. Naoto K, Naoto O (1997) Dissolution of Slag Fertilizers in a Paddy Soil and Si Uptake by Rice Plant. Soil Sci Plant Nutr 43: 329–341.
37. Cai KZ, Gao D, Luo SM, Zeng RS, Yang JY, et al. (2008) Physiological and cytological mechanisms of silicon-induced resistance in rice against blast disease. Physiol Plant 134: 324–333.
38. Kim SG, Kim KW, Park EW, Choi D (2002) Silicon-induced cell wall fortification of rice leaves: A possible cellular mechanism of enhanced host resistance to blast. Phytopathol 92: 1095–1103.
39. Buck GB, Korndörfer GH, Datnoff LE (2011) Extractors for estimating plant available silicon from potential silicon fertilizer sources. J Plant Nutr 34: 272–282.
40. Lu RK (2000) Analytical methods for soil and agro-chemistry. Chinese Agricultural Technology, Beijing, pp. 201–203 (in Chinese).
41. Nanayakkara UN, Uddin W, Datnoff LE (2008) Effects of soil type, source of silicon, and rate of silicon source on development of gray leaf spot of perennial ryegrass turf. Plant Dis 92: 870–877.
42. Dai WM, Zhang KQ, Duan BW, Sun CX, Zheng KL, et al. (2005) Rapid determination of silicon content in rice (Oryza sativa). Chinese J Rice Sci 19: 460–462 (in Chinese).
43. Takahashi K (1981). Effects of slags on the growth and the silicon uptake by rice plants and the available silicates in paddy soils. Bulletin Shikoku Agric Exp Stat 38: 75–114.
44. Liang YC, Ma TS, Li FJ, Feng YJ (1994) Silicon availability and response of rice and wheat to silicon in calcareous soils. Commun Soil Sci Plant Anal 25(13&14): 2285–2297.
45. Seebold KW, Datnoff LE, Correa-Victoria FJ, Kucharek TA, Snyder GH (2004) Effects of silicon and fungicides on the control of leaf and neck blast in upland rice. Plant Dis 88: 253–258.
46. Sun WC, Zhang J, Fan QH, Xue GF, Li ZL, et al. (2010) Silicon-enhanced resistance to rice blast is attributed to silicon-mediated defence resistance and its role as physical barrier. Eur J Plant Pathol 128: 39–49.
47. Dallagnol LJ, Rodrigues FÁ, DaMatta FM, Mielli M B, Pereira SC (2011) Deficiency in silicon uptake affects cytological, physiological, and biochemical events in the rice–*Bipolaris oryzae* interaction. Phytopathol 101:92–104.
48. Yoshida S (1965) Chemical aspects of the role of silicon in physiology of the rice plant. Bull Shikoku Agr Exp Stat 15: 1–58.
49. Hayasaka T, Fujii H, Ishiguro K (2008) The role of silicon in preventing appressorial penetration by the rice blast fungus. Phytopathol 98: 1038–1044.

50. Liang YC, Sun WC, Si J, Römheld V (2005) Effect of foliar- and root-applied silicon on the enhancement of induced resistance in *Cucumis sativus* to powdery mildew. Plant Pathol 54: 678–685.

51. Zhang GL, Gen DQ, Zhang HC (2006) Silicon Application enhances resistance to sheath blight (*rhizoctoniasolani*) in rice. J Plant Physiol Mol 32: 600–606 (in Chinese).

52. Rodrigues FÁ, Jurick WM, Datnoff LE, Jones JB, Rollins JA (2005b) Silicon influences cytological and molecular events in compatible rice-*Magnaporthe grisea* interactions. Physiol Mol Plant Pathol 66: 144–159.

16

Remotely Sensed Rice Yield Prediction Using Multi-Temporal NDVI Data Derived from NOAA's-AVHRR

Jingfeng Huang[1,2,3]**, Xiuzhen Wang**[4]*****, Xinxing Li**[1,2,3]**, Hanqin Tian**[5,6]**, Zhuokun Pan**[1,3]

1 Institute of Agricultural Remote Sensing & Information Application, Zijingang Campus, Zhejiang University, Hangzhou, China, **2** China Ministry of Education Key Laboratory of Environmental Remediation and Ecological Health, Zhejiang University, Hangzhou, China, **3** Key Laboratory of Agricultural Remote Sensing and Information System, Zhejiang Province, China, **4** Institute of Remote Sensing and Earth Sciences, Hangzhou Normal University, Hangzhou, China, **5** International Center for Climate and Global Change Research, Auburn University, Auburn, Alabama, United States of America, **6** Ecosystem Dynamics and Global Ecology (EDGE) Laboratory, School of Forestry and Wildlife Sciences, Auburn University, Auburn, Alabama, United States of America

Abstract

Grain-yield prediction using remotely sensed data have been intensively studied in wheat and maize, but such information is limited in rice, barley, oats and soybeans. The present study proposes a new framework for rice-yield prediction, which eliminates the influence of the technology development, fertilizer application, and management improvement and can be used for the development and implementation of provincial rice-yield predictions. The technique requires the collection of remotely sensed data over an adequate time frame and a corresponding record of the region's crop yields. Longer normalized-difference-vegetation-index (NDVI) time series are preferable to shorter ones for the purposes of rice-yield prediction because the well-contrasted seasons in a longer time series provide the opportunity to build regression models with a wide application range. A regression analysis of the yield versus the year indicated an annual gain in the rice yield of 50 to 128 kg ha^{-1}. Stepwise regression models for the remotely sensed rice-yield predictions have been developed for five typical rice-growing provinces in China. The prediction models for the remotely sensed rice yield indicated that the influences of the NDVIs on the rice yield were always positive. The association between the predicted and observed rice yields was highly significant without obvious outliers from 1982 to 2004. Independent validation found that the overall relative error is approximately 5.82%, and a majority of the relative errors were less than 5% in 2005 and 2006, depending on the study area. The proposed models can be used in an operational context to predict rice yields at the provincial level in China. The methodologies described in the present paper can be applied to any crop for which a sufficient time series of NDVI data and the corresponding historical yield information are available, as long as the historical yield increases significantly.

Editor: Wengui Yan, National Rice Research Center, United States of America

Funding: The authors' work was supported from National Key Technology R&D Program of China Grant(2011BAD32B01), the National Natural Science Foundation of China (NSFC) grant (40875070 and 40871158) and Zhejiang Provincial Natural Science Foundation of China grant (Y5100021). The funders had no role in study design, data collection and analysis, decision to publish, or preparation of the manuscript.

Competing Interests: The authors have declared that no competing interests exist.

* E-mail: wxz05160516@126.com

Introduction

Paddy rice is one of the most important and widely grown crops in China. The total paddy-rice production in 2009 reached 195.1 million tons, and it accounted for 40.5% of the total grain production in China (481.563 million tons) [1]. Timely, objective and quantitative information regarding to paddy-rice yield can provide important information for government agencies and producers that can be used for planning harvest, storage and marketing activities. Therefore, paddy-rice-yield prediction is important for the food security of China and is considered to be one of the most challenging tasks in agricultural research [2]. The traditional approach of crop-yield forecasting, the use of ground-based data collection is expensive, time-consuming, labor-intensive, and often difficult [3]. Crop-yield prediction using remotely sensed data has already represented a very active field of research and application [3–5]. Notable advances in remote-sensing technology over the last several decades are now providing scientists with valuable information for yield and production

forecast. Time series of normalized-difference-vegetation-index (NDVI), derived from the satellite data, have been used for crop-yield predictions since the 1980's. Most of the studies that related NDVI measurements to crop yield have been concentrated on staple crops such as wheat [4,6–38] and maize [3,13,18,20,21,24,29,39–49] and rice [2,15,37,44,50–52]. Many researchers have also found that NDVI variables are very good at grain yield predictors of millet [53–57], sorghum [24,56,58,59], barley [19,24,29,60,61], soybean [3,24,62,63], ground nut [54,59], sugar beet [29], alfalfa [29], rye [29], pea [19,29], and canola [19] (Literature review was summarized in Table 1). However, remotely sensed yield prediction appears limited in rice.

Different methods have been developed to predict crop yields using remotely sensed data, and the most common approach is, by generating regression model, to develop direct empirical relationships between the NDVI measurements and the crop yield [15,19,45,57]. These approaches assume that measures of the photosynthetic capacity from spectral-vegetation indices are directly related to crop yield. This assumption is used because many of the conditions that affect crop growth, development and ultimately yield could be captured through spectra measurements such as the NDVI [64]. By using long-term historical-yield data as a dependent variable and remotely sensed data as an independent variable, a statistical regression function was generated to perform crop-yield predictions, whereas the actual crop yields depend on many more factors than the presence of spectral-vegetation indices [37]. Tilman et al. [65] noted that increased yields in cereal are mainly the result of greater inputs of fertilizer, water and pesticides, new crop species, and the improvement of management over the last decades. For all developing countries, modern varieties accounted for 21% of the growth in crop yields during the early Green Revolution period [66]. In Asia, rice production has more than doubled as a result of the expansion of cultivated area, the adoption of modern cultivars, increased investments in irrigation, and an increased use of fertilizer over the past 4 decades [67]. Hafner [68] found that linear growth has been the most common trend in maize, rice, and wheat yields for 188 nations over the past 40 years. This scenario also occurs in China. Although the inter-annual variability of NDVI (probably due to unexpected weather conditions or disasters) can reveal crop yield fluctuations [19,59]; however, remotely sensed-NDVI cannot detect those human-induced factors that resulted in increase of rice yield. Therefore, to monitor and predict crop-yield cannot use NDVI measurements solely.

For unit-yield estimation, using one simple regression function (usually known as: Y = a+b * NDVI) would be incompatible as the advance of years, because simple regression would be likely neglect those man-induced factors in yield increase. However, few studies have analyzed the time trends of crop yields, which reflect the influence of technology development, fertilizer application, and management improvement. Moreover, the regression model between statistical data and NDVI cannot be extendable [19,45]

because cropping system and rice yield level is natural condition-dependent in China.

In consideration of social factors and regional differences for remotely sensed crop yield estimation in China, the objective of the present paper was to develop a methodological framework that may be adopted for the regional-, national- and international-scale prediction of crop yields. This methodology was based on a time series analysis of historical-yield information. Paddy rice was chosen to test the proposed methodology. To accomplish this objective, we needed to: (1) geographically regionalize rice cultivation area for remotely sensed monitoring; (2) analyze the historical trends in the grain yield of rice; (3) decompose the remotely sensed yield of rice from the long-term historical data; (4) select the optimal predictors, based on a correlation analysis between the remotely sensed yield and the AVHRR-derived NDVIs; (5) construct prediction models for rice yield; and (6) evaluate the potential for rice-grain-yield prediction in China using AVHRR NDVI data as predictors.

Materials and Methodology

2.1. The Remote-Sensing dataset

The research presented in this paper relies on a time series of AVHRR NDVI composite imagery from July 1981 to December 2006, derived from the National Oceanic and Atmospheric Administration's (NOAA) series of Advanced Very High Resolution Radiometer (AVHRR) instruments, with a spatial resolution of 8 km, by the NASA Global Inventory Monitoring and Modeling Systems (GIMMS) group at the Laboratory for Terrestrial Physics. There are two 15-day composites per month: the first (15a) is a maximum value composite from the first day to 15th of the month; and the 15b composite is from days 16 till the end of the month. All data are available from the University of Maryland Global Land Cover Facility (http://glcf.umiacs.umd.edu/data/gimms/).

Table 1. Relevant literatures that linked with crop yield forecast using remotely sensed data literatures are sorted according to the crop types.

Crop	reference
wheat	MacDonald et al., 1980; Rudorff et al., 1991; , Bullock, 1992; Benedetti et al., 1993; Gupta et al., 1993; Benedetti et al., 1993; Cheng, 1994; Dubey et al., 1994; Sridhar et al., 1994; Doraiswamy et al., 1995, 2003; Smith et al., 1995; Hochheim et al., 1998; Huang et al., 1999; Maselli et al., 2001; Boken et al., 2002; Labus et al., 2002; Manjunath et al., 2002; Mika et al., 2002; Bastiaanssen, et al., 2003; Kalubarme et al., 2003; Ferencz et al., 2004; Zhang et al., 2004; Kastensa et al., 2005; Mo et al., 2005; Wang et al., 2005; Patel et al., 2006; Ren et al., 2006; Moriondo et al., 2007; Prasad et al., 2007; Balaghi et al., 2008; Ren et al., 2008; Wall et al., 2008; Schut et al., 2009; Becker-Reshef et al., 2010; Mkhabela et al., 2011
maize	Quarmby et al., 1993; Hayes et al., 1996; Unganai et al., 1998; Lewis et al., 1998; Lee et al., 1999; Reynolds et al., 2000; Seiler et al., 2000; Maselli et al.,2001; Mika et al., 2002; Wannebo et al., 2003; Ferencz et al., 2004; Kastensa et al., 2005; Mkhabela et al., 2005; Mo et al., 2005; Prasad et al., 2006; Rojas, 2007; Ren, et al., 2008; Funk et al., 2009
millet	Rasmussen, 1992, 1997, 1998; Groten, 1993; Maselli et al.,2000
sorghum	Potdar, 1993; Fuller, 1998; Maselli et al., 2000; Kastensa et al., 2005
barley	Wendroth et al., 2003; Ferencz et al., 2004; Kastensa et al., 2005; Weissteiner et al., 2005; Mkhabela et al., 2011
soybean	Liu et al., 2002; Kastensa et al., 2005; Prasad et al., 2006; Esquerdo et al., 2011
ground nut	Rasmussen, 1997; Fuller, 1998
sugar beet	Ferencz et al., 2004
alfalfa	Ferencz et al., 2004
rye	Ferencz et al., 2004
pea	Ferencz et al., 2004; Mkhabela et al., 2011
canola	Mkhabela et al., 2011
rice	Tennakoon et al., 1992; Quarmby et al., 1993; Huang et al., 2002; Wang et al., 2002; Bastiaanssen, et al., 2003; Prasad et al., 2007; Huang et al., 2010

Table 2. NVDI variables and their calculation formulas.

	NDVIs	Description of formulas
1	$NDVI_{maxb1}$	the first biweekly NDVI before $NDVI_{max}$
2	$NDVI_{maxb2}$	the second biweekly NDVI before $NDVI_{max}$
3	$NDVI_{maxb3}$	the third biweekly NDVI before $NDVI_{max}$
4	$NDVI_{maxb4}$	the fourth biweekly NDVI before $NDVI_{max}$
5	$NDVI_{max}$	the maximum NDVI during the growth period
6	$NDVI_{maxa1}$	the first biweekly NDVI after $NDVI_{max}$
7	$NDVI_{maxa2}$	the second biweekly NDVI after $NDVI_{max}$
8	$mNDVI_{maxb4-b3}$	$(NDVI_{maxb4}+ NDVI_{maxb3})/2$
9	$mNDVI_{maxb4-b2}$	$(NDVI_{maxb4}+ NDVI_{maxb3}+ NDVI_{maxb2})/3$
10	$MNDVI_{maxb4-b1}$	$(NDVI_{maxb4}+ NDVI_{maxb3}+ NDVI_{maxb2}+ NDVI_{maxb1})/4$
11	$mNDVI_{maxb4-max}$	$(NDVI_{maxb4}+ NDVI_{maxb3}+ NDVI_{maxb2}+ NDVI_{maxb1}+ NDVI_{max})/5$
12	$mNDVI_{maxb4-a1}$	$(NDVI_{maxb4}+ NDVI_{maxb3}+ NDVI_{maxb2}+ NDVI_{maxb1}+ NDVI_{max}+ NDVI_{maxa1})/6$
13	$mNDVI_{maxb4-a2}$	$(NDVI_{maxb4}+ NDVI_{maxb3}+ NDVI_{maxb2}+ NDVI_{maxb1}+ NDVI_{max}+ NDVI_{maxa1}+ NDVI_{maxa2})/7$
14	$mNDVI_{maxb3-b2}$	$(NDVI_{maxb3}+ NDVI_{maxb2})/2$
15	$mNDVI_{maxb3-b1}$	$(NDVI_{maxb3}+ NDVI_{maxb2}+ NDVI_{maxb1})/3$
16	$mNDVI_{maxb3-max}$	$(NDVI_{maxb3}+ NDVI_{maxb2}+ NDVI_{maxb1}+ NDVI_{max})/4$
17	$mNDVI_{maxb3-a1}$	$(NDVI_{maxb3}+ NDVI_{maxb2}+ NDVI_{maxb1}+ NDVI_{max}+ NDVI_{maxa1})/5$
18	$mNDVI_{maxb3-a2}$	$(NDVI_{maxb3}+ NDVI_{maxb2}+ NDVI_{maxb1}+ NDVI_{max}+ NDVI_{maxa1}+ NDVI_{maxa2})/6$
19	$mNDVI_{maxb2-b1}$	$(NDVI_{maxb2}+ NDVI_{maxb1})/2$
20	$mNDVI_{maxb2-max}$	$(NDVI_{maxb2}+ NDVI_{maxb1}+ NDVI_{max})/3$
21	$mNDVI_{maxb2-a1}$	$(NDVI_{maxb2}+ NDVI_{maxb1}+ NDVI_{max}+ NDVI_{maxa1})/4$
22	$mNDVI_{maxb2-a2}$	$(NDVI_{maxb2}+ NDVI_{maxb1}+ NDVI_{max}+ NDVI_{maxa1}+ NDVI_{maxa2})/5$
23	$mNDVI_{maxb1-max}$	$(NDVI_{maxb1}+ NDVI_{max})/2$
24	$mNDVI_{maxb1-a1}$	$(NDVI_{maxb1}+ NDVI_{max}+ NDVI_{maxa1})/3$
25	$mNDVI_{maxb1-a2}$	$(NDVI_{maxb1}+ NDVI_{max}+ NDVI_{maxa1}+ NDVI_{maxa2})/4$
26	$mNDVI_{max-a1}$	$(NDVI_{max}+ NDVI_{maxa1})/2$
27	$mNDVI_{max-a2}$	$(NDVI_{max}+ NDVI_{maxa1}+ NDVI_{maxa2})/3$
28	$mNDVI_{maxa1-a2}$	$(NDVI_{maxa1}+ NDVI_{maxa2})/2$

Pinzon et al. [69] and Tucker et al. [70] described in detail how the GIMMS data set was developed. A number of improvements have been made on the GIMMS NDVI database, with respect to previous NDVI data sets, including corrections for: (1) sensor degradation; (2) inter-sensor differences; (3) solar-illumination angle and sensor-view angle effects due to satellite drift; (4) volcanic stratospheric aerosol corrections for 1982–1984 and 1991–1994; (5) missing data in the Northern Hemisphere during winter, using interpolation; and (6) short-term atmospheric aerosol effects, atmospheric water-vapor effects, and cloud-cover physics [69,70]. This data set is considered to be the most accurate, long-term AVHRR data record [71]. By comparing these data to new, improved coarse-resolution remotely sensed data from SPOT Vegetation instrument and MODIS instruments, recent study confirmed its suitability for long-term vegetation studies [72].

2.2. NDVI Variables

A large number of studies found a close relationship between crop yields and NDVI variables. The theory is: the NDVI value presents the yield level corresponding to every single pixel. Therefore, a simple regression function can be explained the yield: yield = a*NDVI + b; then the total yield can be obtained by multiplying planting area. By literature review, previous studies suggest three types of NDVI variables: original NDVI [13,23,42,63], cumulative NDVI [8,23,38,42,45,63,73,74], and average NDVI [34,45,63]. The cumulative NDVI and the corresponding average NDVI for the same period were highly correlated because of the linear nature of the operations involved. Only the original NDVIs and the average NDVIs were selected as input data for the prediction models in the present paper.

NDVI variables around the time of the maximum are strongly correlated with final yields [31,35,75]. Specifically, the rice yield is most determined by crop conditions during the heading (i.e. peak phenological phase of growth); and yield-reflectance relationships are typically the strongest after mid-season. In contrast, NDVI value changes that occur outside of the rice-growing period maybe not positively related to yield [52]. These relationships within changes of NDVI value suggests that the NDVIs during the mid-to-late growing period should be a good indicator of rice yield; meanwhile this phenomenon provides an approach to discriminate rice planting area from remote sensing image. Therefore, the first step of this study was to extract the maximum NDVI during the rice-growth period ($NDVI_{max}$) for each studied province from the remote sensing dataset from the year 1982 to 2006. The maximum NDVI is equal to the peak value of the seasonal NDVI profile. Then, six other original NDVIs were calculated: the first, second,

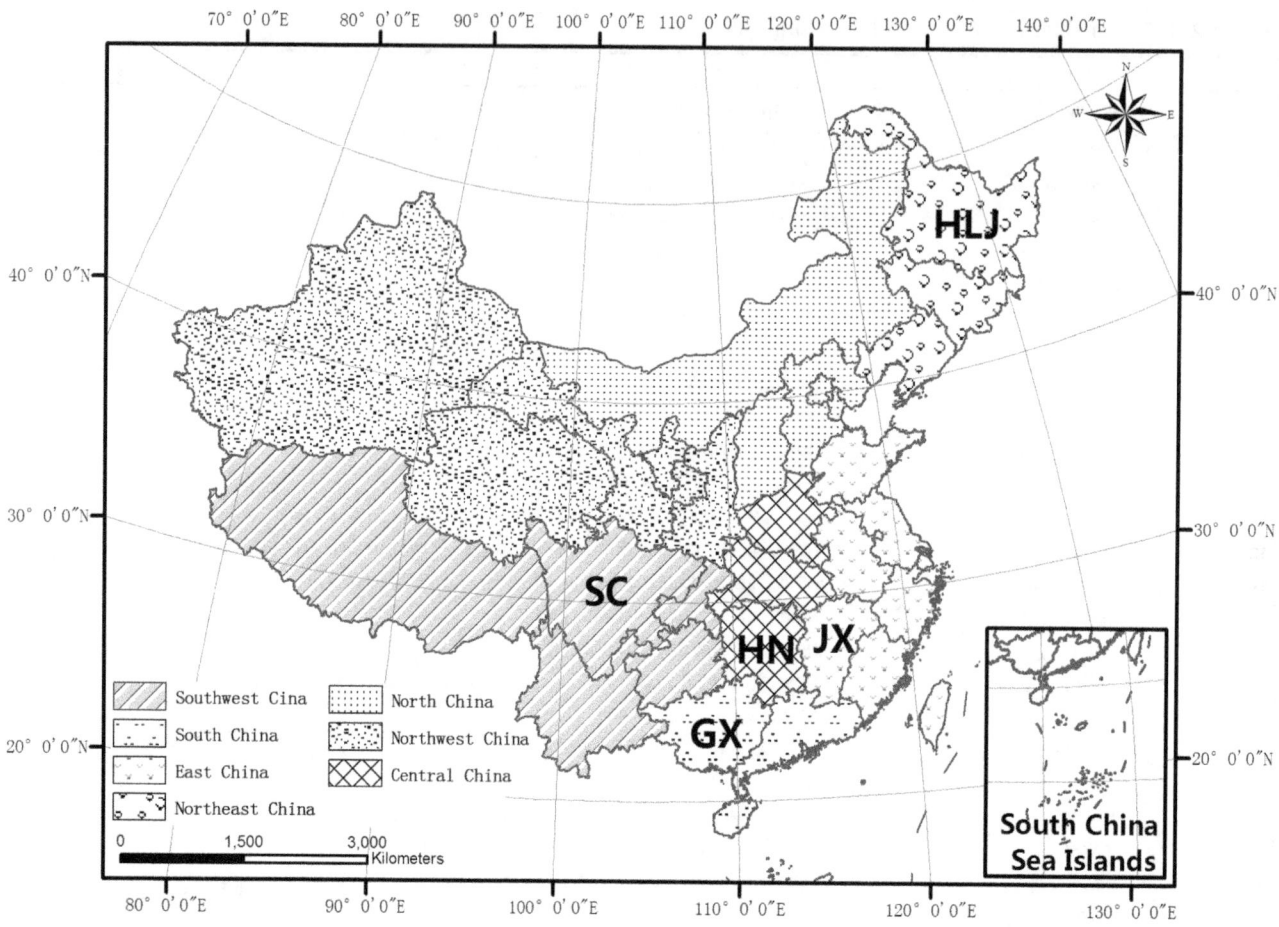

Figure 1. The locations of the study areas within Mainland China. Heilongjiang is designated by HLJ, Jiangxi by JX, Guangxi by GX, Sichuan by SC, and Hunan by HN.

third and fourth biweekly NDVIs prior to the $NDVI_{max}$ ($NDVI_{maxb4}$) and the first and second biweekly NDVIs after the $NDVI_{max}$ ($NDVI_{maxa2}$). These seven biweekly composites span 3 months of raw AVHRR imagery, corresponding to the rice-growth period. Focusing on the NDVI response during the rice-growth period helps to identify rice-specific vegetation changes.

Hochheim and Barber [27] also found that NDVI estimators with longer integration periods minimized variability in yield prediction. Therefore, based on the seven original NDVIs, twenty-one average NDVIs, clustered around the time of the peak NDVI, were calculated using a rigorous arithmetic mean framework (Table 2). In total, 28 NDVI variables were generated. They include all of the possible combinations of the original seven NDVIs.

2.3. Official Statistical Data of Rice Yield

Historical rice-yield data were acquired from the China Statistical Year Book by the National Bureau of Statistics of China (NBSC) from the years 1979 to 2009 [1]. The NBSC is the agency responsible for collecting and publishing agricultural statistics at the national and provincial levels. The NBSC crop statistics are based on data obtained from sub-province sample surveys and released in official documents. Customarily, Chinese provinces have been geographically grouped into 7 regions to present a spatial pattern for paddy rice planting area: Northeastern China (Heilongjiang, Jilin, and Liaoning), Northern China (Inner

Mongolia, Hebei, Shanxi, Beijing, and Tianjin), Northwestern China (Ningxia, Shaanxi, Gansu, Qinghai, and Xinjiang), Central China (Henan, Hunan, and Hubei), Eastern China (Shandong, Jiangsu, Shanghai, Zhejiang, Anhui, and Jiangxi), Southwestern China (Chongqing, Sichuan, Guizhou, Yunnan, and Xizang), Southern China (Guangdong, Guangxi, Hainan) (see Figure 1). Unfortunately, rice planting area and yield information for Hong Kong- Macao-Taiwan areas was not available. According to NBSC crop statistical data (see Table 3), Eastern China was the region with the highest rice acreage and production levels (9808.60 kha and 64984.00 kt, respectively) in 2009. Central China ranked second in both rice acreage and production (6703.60 kha and 46215.00 kt, respectively). The third-largest rice cultivation and production area was Southwestern China (4448.10 kha and 31214.00 kt, respectively). Southern China and Northeastern China ranked fourth and fifth, respectively, in both rice acreage and production (4402.40 kha and 23499.00 kt; 3777.90 kha and 25855.00 kt, respectively). The total rice cultivation area in Eastern China, Central China, Southwestern China, Southern China, and Northeastern China is 29140.60 kha and accounts for 98.36% of the total rice cultivation area in the conterminous China. The total rice production in Eastern China, Central China, Southwestern China, Southern China, and Northeastern China was 191767.00 kt and accounted for 98.29% of the total rice production in the conterminous China in 2009. Northern China and Northwestern China constitute less

Table 3. Planted area and production changes for rice between 1979 and 2009 for different regions in the conterminous China.

Regions	Area (Kha)				Production (Kt)			
	1979	% of China	2009	% of China	1979	% of China	2009	% of China
Northeastern China	841.73	2.49	3777.90	12.75	3860.00	2.69	25855.00	13.25
Northern China	264.07	0.78	204.40	0.69	1165.00	0.81	1343.00	0.69
Northwestern China	315.27	0.93	281.70	0.95	1305.00	0.91	1993.00	1.02
Central China	7639.13	22.55	6703.60	22.63	34260.00	23.83	46215.00	23.69
Eastern China	12926.33	38.16	9808.60	33.11	56230.00	39.12	64984.00	33.31
Southwestern China	4803.73	14.18	4448.10	15.01	21440.00	14.91	31214.00	16.00
Southern China	7082.40	20.91	4402.40	14.86	25490.00	17.73	23499.00	12.04
Total	33872.67	100.00	29626.70	100.00	143750.00	100.00	195103.00	100.00

than 2% of the national rice harvested area and production and were less important on a national scale in 2009.

2.4. Description of Study Area

We divided China into 7 regions together with 5 representative provinces selected to convey the information of paddy rice planting area: Heilongjiang (HLJ) in Northeastern China, Hunan (HN) in Central China, Jiangxi (JX) in Eastern China, Sichuan (SC) in Southwestern China, and Guangxi (GX) in Southern China. These provinces were selected as the study areas for the present research because these locations: (1) represented the typical cropping system in China, (2) are located in primary rice-production regions, and (3) are geographically and climatologically different (see Figure 1 and Table 4). The life span, cropping system, and planting schedule are all depend on regional hydro-thermal condition. The general information on life span, cropping system, total annual rainfall (mm), annual accumulated temperature (°C), area (kha), and production levels (kt) for the selected provinces is shown in Table 4. The total combined rice-cultivation area in Heilongjiang (HLJ), Hunan (HN), Jiangxi (JX), Sichuan (SC), and Guangxi (GX) is 13942.2 kha, and these regions accounted for 47.06% of the total rice-cultivation area in China in 2009. The total combined rice production in Heilongjiang (HLJ), Hunan (HN), Jiangxi (JX), Sichuan (SC), and Guangxi (GX) was 87251 kt and accounted for 44.72% of the total rice production in China in 2009. The time series of the NBSC province-level rice yields were used to train and develop the prediction models for these five provinces.

2.5. Calibration of Rice-Yield Prediction Models

The gradual trend in yields is due to the influence of technological development, fertilizer application, and improved management on the rice cultivation. The results of this analysis suggest that the most common trend of rice yield is a linear growth. The province-specific intercepts account for spatial variations in rice management and soil quality; province-specific time trends account for yield growth due to technology gains. This indicates us the yield is composed from the intrinsic and extrinsic factors. Therefore, we decomposed the historical rice yield Y into the trend yield Y_t and the remotely sensed yield Y_{RS}, using the following equation:

$$Y = Y_t + Y_{RS} \qquad (1)$$

Y_t, represents the component that is regulated by agricultural technology, including (1) the usual biological-chemical technologies (new varieties, fertilizers, herbicides, insecticides, etc.) and the mechanical technologies (machinery, equipment, etc.); (2) the management practices, which involve changes such as the timing of field operations and other practices which may or may not be involved in the purchase of new inputs. Y_{RS} is defined as the component regulated by natural environmental conditions, such as temperature, precipitation, pests and disease; these environment factors can be detected by a remote sensor.

To quantify past trends in yields, many different yield de-trend methods have been reported, including: least-squares regressions [76,77], moving averages [78,79], exponential algorithms [80], and polynomial regressions [81]. For rice-yield predictions in the present investigation, a linear regression model and a moving average are both generated to fit each separated provincial rice dataset (also see in Figure 2):

$$Y_t = \alpha + \beta t \qquad (2)$$

where Y_t is the trend yield in a given province during a given year (kg ha^{-1}), t represents the year of harvest (the year 1979 was numeral 1979, 1980 was numeral 1980, etc., until 2009 was numeral 2009), α and β are the province-specific linear regression coefficients.

In our study, a moving average is used with historical crop-yield data to smooth out short-term fluctuations and highlight longer-term trends. Rice yields were de-trended using their deviations from the 5-year moving average. The mean changes in provincial historical rice yield (Y), the trend yield (Y_t) and the remotely sensed yield (Y_{RS}) were calculated for each period as an average of the changes from each single preceding year to the next by using a moving average method. Generally, the moving average method is used to calculate arithmetic mean of each five of the entire dataset: $y_{i-n}, y_{i-n+1}, \ldots, y_i, \ldots, y_{i+n}$. Such method has been usually employed in meteorological data analysis to remove the stochastic errors from long-time series of data. Hence, an algorithm for a 5-year moving average is as follows:

Table 4. General information on Rice cropping system, Life span, Total annual rainfall (mm), Annual accumulated temperature (≥10°C), Area (kha) and Production (kt) for the study areas.

Provinces	Climate region	Rice cropping system	Life span	Total annual rainfall (mm)	Annual accumulated temperature (≥10 °C)	Planting Area in 2009(kha)	Percent age of China (%)	Production in 2009 (kt)	Percenta ge of China (%)
Heilongjiang (HLJ)	Temperate continental monsoon climate	Single cropping	May – Oct	450–650	2000–3700	2460.80	8.31	15745.00	8.07
Hunan (HN)	Subtropical monsoon climate	Double cropping	Mar – Aug, Jun – Nov	1200–1700	4500–6500	4047.20	13.66	25786.00	13.22
Jiangxi (JX)	Subtropical monsoon climate	Double cropping	Mar – Aug, Jun - Nov	1300–2000	4500–6500	3282.10	11.08	19059.00	9.77
Sichuan (SC)	Subtropical humid climate	Single cropping	Mar – Aug	950–1200 (Sichuan Basin)	4000–6000 (Sichuan Basin)	2027.10	6.84	15202.00	7.79
Guangxi (GX)	Subtropical monsoon climate	Double cropping	Mar – Aug, Jun - Nov	1300–2000	5800–9300	2125.00	7.17	11459.00	5.87

$$\begin{cases} Y_{t,i} = \frac{Y_{t,i-2} + Y_{t,i-1} + Y_{t,i} + Y_{t,i+1} + Y_{t,i+2}}{5} \ (3 \leq i \leq n-2) \\ Y_{t,1} = \frac{3Y_{t,1} + 2Y_{t,2} + Y_{t,3} - Y_{t,5}}{5} \\ Y_{t,2} = \frac{4Y_{t,1} + 3Y_{t,2} + 2Y_{t,3} + Y_{t,4}}{5} \\ Y_{t,n-1} = \frac{Y_{t,n-3} + 2Y_{t,n-2} + 3Y_{t,n-1} + 4Y_{t,n}}{5} \\ Y_{t,n} = \frac{-Y_{t,n-4} + Y_{t,n-2} + 2Y_{t,n-1} + 3Y_{t,n}}{5} \end{cases} \tag{3}$$

Where $Y_{t,i}$ is the trend yield in a given province during a given year (kg ha^{-1}); n represents the number of data points; i represents the year of the harvest (e.g. the year 1979 was numeral 1, 1980 was numeral 2, etc., until 2006 it should be numeral 31); $Y_{t,1}$ and $Y_{t,2}$ are the trend yields for the first two harvested years; then $Y_{t,n-1}$ and $Y_{t,n}$ are the trend yields for the last two harvested years within the 5 years.

Consequently, the trend yield Y_t was obtained. To remove the technological influences, it is necessary to remove the yield trend to produce a new time series that is directly related to the NDVIs. We defined this new time series as the remotely sensed yield. According to Eq. (1), the remotely sensed yield can be calculated by the following equation:

$$Y_{RS} = Y - Y_t \tag{4}$$

Next, correlation analysis was performed between the remotely sensed yield and the NDVI variables. The correlation coefficient is a measure of the strength and the direction of a linear relationship between two variables. The symbol r in Eq. (5) represents the samples' correlation coefficient; x and y represent the remotely sensed yield and the NDVI variables respectively; n is the number of data pairs.

$$r = \frac{n\sum xy - (\sum x \sum y)}{\sqrt{n\sum x^2 - (\sum x)^2}\sqrt{n\sum y^2 - (\sum y)^2}} \tag{5}$$

Statistical regression models are the most commonly used method for crop-yield prediction based on remotely sensed data [8,36]. They do not require numerous inputs and can be performed directly; also because it requires little computing power and the selected variables are distinctive and non-overlapping. Therefore, each of the provincial Y_{RS} and NDVI dataset was analyzed separately by means of stepwise regression techniques. These models were constructed via the 'STEPWISE' regression process which was available in software Statistical Product and Service Solutions (SPSS) 17.0 [82]. The probability significance thresholds for the entry and retention of candidate independent variables in the model were both set to $\alpha = 5\%$.

2.6. Evaluation of Rice-Yield Prediction Models

The rice-yield prediction models were evaluated using the following indicators:

Root mean square error (RMSE):

$$RMSE = \sqrt{\frac{\sum_{i=1}^{n}(Y_i - Y_i')^2}{n}} \tag{6}$$

Coefficient of determination (R^2):

$$R^2 = \frac{\sum_{i=1}^{n}(Y_i' - \bar{Y})^2}{\sum_{i=1}^{n}(Y_i - \bar{Y})^2} \tag{7}$$

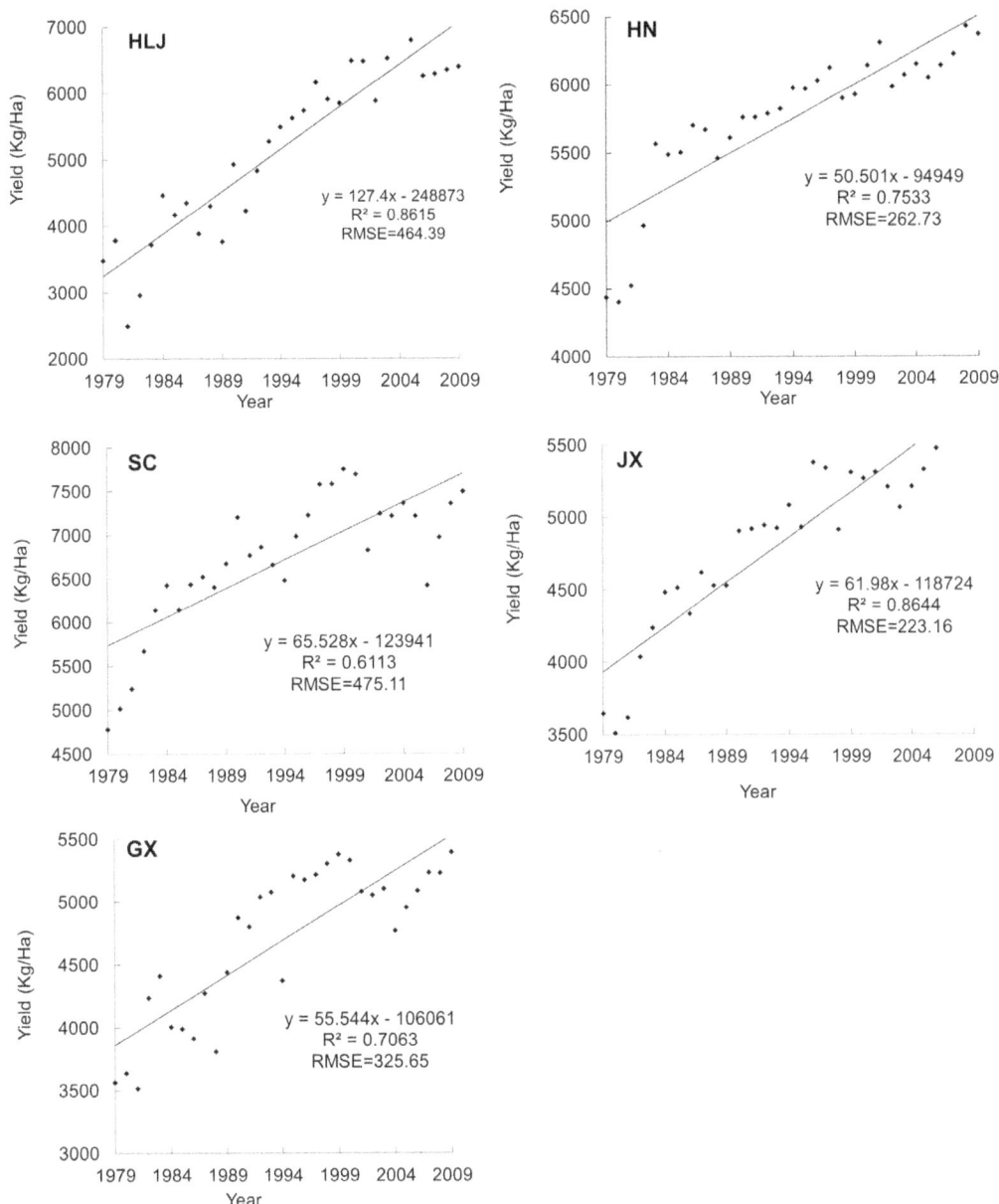

Figure 2. Rice yield trends for the provinces' of Heilongjiang (HLJ), Hunan (HN), Jiangxi (JX), Sichuan (SC) and Guangxi (GX) from 1979 to 2006.

F-value (F):

$$F = \frac{\sum\limits_{i=1}^{n} (Y_i' - \bar{Y})^2 / k}{\sum\limits_{i=1}^{n} (Y_i - Y_i')^2 / (n - k - 1)} \quad (8)$$

and relative error (RE):

$$RE = \frac{Y_i - Y_i'}{Y_i} \quad (9)$$

Together with the above, where n is the number of comparisons; k is the number of predictors; Y_i is the statistical rice yield; $\bar{Y}$ is the average rice yield, and Y_i' is the predicted yield.

Results and Discussion

3.1. Rice Yield Trend Analysis

Figure 2 presents the evolution of the average rice-grain yield in Heilongjiang (HLJ), Jiangxi (JX), Guangxi (GX), Sichuan (SC), and Hunan (HN) from 1979 to 2009; according to their R-square and RMSE, all rice yields showed a visible and significant growth trend over time. Understanding the past rice-yield trends can help us to gauge the importance of the preprocessing procedure for rice-yield prediction using remotely sensed data. The statistical data of rice yield together with average yield growth trend from 1979 to 2009 in five provinces of China is summarized in Table 5.

As analysis above (see Figure 2), the social input and advance of technology account for the linear trend of the rice-yield growth, whereas such human-induced factors could not be detected using remotely sensed data. To overcome this problem and make rice-yield prediction methods more robust and easily exportable, one possible strategy is to integrate remote-sensing data with the rice yield time series analysis. De-trending is necessary to properly identify the remote-sensible effects in these panel datasets. Therefore, before the rice-yield predicting models are established using remotely sensed variables as predictors, we suggest that the statistical yield should be decomposed into the trend yield and the remotely sensed yield, methodology was described in Section 2.5.

3.2. Correlation Coefficients between the Remotely Sensed Yield and NDVI Variables

The correlation coefficients between Y_{RS} and the NDVI variables for the rice-growth period from the fourth 15-day period before $NDVI_{max}$ ($NDVI_{maxb4}$) to the second 15-day period after $NDVI_{max}$ ($NDVI_{maxa2}$) for each of the studied provinces are summarized in Table 6. By comparing the correlation coefficients (Column 2 and 3 in Table 6), the Y_{RS} that was de-trended by linear regression performed better than the Y_{RS} that was de-trended by a 5-year moving average against the NDVI variables.

The correlation coefficients between the Y_{RS} that were de-trended by linear regression and the NDVI variables were generally high in HN and SC. According to Table 6, for HN, the correlation coefficients were significant at the 0.01 level between the Y_{RS} that was de-trended by linear regression and $NDVI_{maxb1}$, $NDVI_{max}$, $NDVI_{maxa1}$, $mNDVI_{maxb4-a2}$, $mNDVI_{maxb3-a1}$, $mNDVI_{maxb3-a2}$, $mNDVI_{maxb2-b1}$, $mNDVI_{maxb2-max}$, $mNDVI_{maxb2-a1}$, $mNDVI_{maxb2-a2}$, $mNDVI_{maxb1-max}$, $mNDVI_{maxb1-a1}$, $mNDVI_{maxb1-a2}$, $mNDVI_{max-a1}$, $mNDVI_{max-a2}$, and $mNDVI_{maxa1-a2}$; the correlation coefficients were significant at the 0.05 level between the Y_{RS} that was de-trended by linear regression and $NDVI_{maxa2}$, $mNDVI_{maxb4-a1}$, and $mNDVI_{maxb3-max}$. For SC, the correlation coefficients were significant at the 0.01 level between the Y_{RS} that was de-trended by linear regression and $NDVI_{maxb4}$, $NDVI_{maxb3}$, $NDVI_{maxb2}$, $mNDVI_{maxb4-b3}$, $mNDVI_{maxb4-b2}$, $mNDVI_{maxb4-b1}$, $mNDVI_{maxb4-max}$, $mNDVI_{maxb4-a1}$, $mNDVI_{maxb4-a2}$, $mNDVI_{maxb3-b2}$, $mNDVI_{maxb3-b1}$, and $mNDVI_{maxb3-max}$, $mNDVI_{maxb3-a1}$; the correlation coefficients were significant at the 0.05 level between the Y_{RS} that was de-trended by linear regression and $mNDVI_{maxb3-a2}$, $mNDVI_{maxb2-b1}$, $mNDVI_{maxb2-max}$, and $mNDVI_{maxb2-a1}$. The highest correlation coefficient between the Y_{RS} that was de-trended by linear regression and the NDVI variables occurred in the second 15-day period after $NDVI_{max}$ ($NDVI_{maxa2}$) and was significant at the 0.05 level for HLJ. The correlation coefficients between the Y_{RS} that was de-trended by linear regression and $NDVI_{maxb1}$,

$mNDVI_{maxb1-max}$, $mNDVI_{maxb1-a1}$, and $mNDVI_{maxb1-a2}$ were significant at the 0.05 level in JX. The correlation coefficients between the Y_{RS} that was de-trended by linear regression and the NDVI variables ranged from − 0.14 to 0.38 in GX.

The correlation coefficients between the Y_{RS} that were de-trended by a 5-year moving average and the NDVI variables were generally low in HLJ, HN, and JX. For SC, the correlation coefficients were significant at the 0.01 level between the Y_{RS} that was de-trended by a 5-year moving average and $NDVI_{maxb4}$, and the correlation coefficients were significant at the 0.05 level between the Y_{RS} that was de-trended by a 5-year moving average and $NDVI_{maxb3}$, $mNDVI_{maxb4-b3}$, $mNDVI_{maxb4-b2}$, $mNDVI_{maxb4-b1}$, $mNDVI_{maxb4-max}$, and $mNDVI_{maxb3-b2}$. The correlation coefficients were significant at the 0.01 level between the Y_{RS} that was de-trended by a 5-year moving average and $NDVI_{maxb4}$, $NDVI_{maxb3}$, and $NDVI_{maxb4-b3}$.

3.3. Remotely Sensed Yield-Prediction Models

Conclusions drawn in the yield-trend analysis and the correlation analysis between Y_{RS} and the NDVI variables encouraged us to attempt to build a simple remotely sensed yield-prediction model for rice based on the NDVI variables. According to the correlation coefficient result summarized in Table 6, the Y_{RS} values that were de-trended by linear regression were used as dependent variables in HLJ, HN, JX, and SC. The Y_{RS} values that were de-trended by a 5-year moving average were used as dependent variables in GX. The NDVIs were used as independent variables. These models were constructed through the 'STEP-WISE' regression process in SPSS software. Each model contains variables using the data period from 1982 to 2004. The correlation coefficients of the selected models ranged from 0.42 to 0.92, and all models were significant at the 0.01 level, except for HLJ which is significant at the 0.05 level (see Table 7). This means that increases in NDVI during the rice-growth period are generally related to the final rice-grain yield. The influence of NDVI always had a positive impact on yield. These results are consistent with numerous previous studies [34,36,42,75]. Data from 2005 to 2006 were used for model validation.

3.4. Validation of Rice-Yield Prediction Models

The remotely sensed yield (Y_{RS}) of rice was calculated using the NDVI variables required by each model described in Table 7. The final rice yield (Y) was the sum of the trend yield (Y_t) and the remotely sensed yield (Y_{RS}). Figure 3 shows a scatter plot of the predicted and observed final rice yields for HLJ, HN, JX, SC, and GX from 1982 to 2004, expressed in units of kilogram per hectare. The models performed well, showing a good similarity between the predicted values and the official statistical values in HLJ, HN, JX, SC, and GX from 1982 to 2004 and capturing the fluctuations of rice yields over time. The regression line between the predicted

Table 5. Trends in rice yield for five selected-provinces in China from 1979 to 2009.

Province	Yield in 1979 (kgha^{-1})	Yield in 2009(kgha^{-1})	Annual increase, 1979–2009 (kgha^{-1}yr^{-1})
Heilongjiang (HLJ)	3480	6398.3	94.14
Hunan (HN)	4440	6371.3	62.30
Jiangxi (JX)	3645	5807	69.74
Sichuan (SC)	4777.5	7499.4	87.80
Guangxi (GX)	3562.5	5392.5	59.03

Table 6. Correlation coefficient (R) between the remotely sensed yields and NDVI variables during the rice growth period.

Variables	the remotely sensed yields de-trended by linear models					the remotely sensed yields de-trended by 5-year moving average				
	HLJ	HN	JX	SC	GX	HLJ	HN	JX	SC	GX
$NDVI_{maxb4}$	−0.02	−0.08	0.05	0.68**	0.24	−0.12	0.14	0.04	0.51**	0.54**
$NDVI_{maxb3}$	−0.16	−0.02	0.14	0.73**	0.36	−0.21	0.13	0.10	0.46*	0.52**
$NDVI_{maxb2}$	−0.08	0.38	0.21	0.57**	−0.14	−0.06	0.34	0.14	0.39	−0.30
$NDVI_{maxb1}$	−0.06	0.56**	0.42*	0.32	0.19	−0.03	0.22	−0.04	0.16	0.09
$NDVI_{max}$	0.13	0.60**	0.39	−0.06	−0.04	0.20	0.20	0.10	0.10	−0.26
$NDVI_{maxa1}$	0.42*	0.62**	0.28	0.29	−0.01	0.35	0.27	−0.05	0.08	−0.28
$NDVI_{maxa2}$	0.20	0.49*	0.32	−0.11	0.38	0.28	0.18	0.01	−0.22	0.39
$mNDVI_{maxb4-b3}$	−0.08	−0.05	0.10	0.73**	0.31	−0.16	0.14	0.08	0.50*	0.57**
$mNDVI_{maxb4-b2}$	−0.09	0.12	0.16	0.73**	0.19	−0.14	0.25	0.11	0.50*	0.32
$mNDVI_{maxb4-b1}$	−0.08	0.25	0.26	0.66**	0.22	−0.13	0.28	0.08	0.43*	0.30
$mNDVI_{maxb4-max}$	−0.07	0.33	0.29	0.64**	0.22	−0.11	0.30	0.09	0.44*	0.26
$mNDVI_{maxb4-a1}$	0.09	0.47*	0.31	0.61**	0.18	0.03	0.33	0.07	0.39	0.12
$mNDVI_{maxb4-a2}$	0.15	0.56**	0.33	0.54**	0.25	0.12	0.35	0.06	0.31	0.20
$mNDVI_{maxb3-b2}$	−0.14	0.23	0.20	0.70**	0.10	−0.15	0.28	0.14	0.46*	0.06
$mNDVI_{maxb3-b1}$	−0.12	0.37	0.32	0.60**	0.15	−0.13	0.30	0.09	0.38	0.08
$mNDVI_{maxb3-max}$	−0.10	0.45*	0.35	0.58**	0.14	−0.09	0.31	0.10	0.38	0.02
$mNDVI_{maxb3-a1}$	0.13	0.57**	0.35	0.55**	0.09	0.10	0.34	0.07	0.33	−0.10
$mNDVI_{maxb3-a2}$	0.19	0.64**	0.37	0.46*	0.18	0.19	0.35	0.06	0.24	0.03
$mNDVI_{maxb2-b1}$	−0.08	0.51**	0.35	0.47*	0.00	−0.05	0.33	0.07	0.29	−0.15
$mNDVI_{maxb2-max}$	−0.03	0.59**	0.38	0.44*	−0.01	0.01	0.34	0.08	0.29	−0.21
$mNDVI_{maxb2-a1}$	0.25	0.66**	0.37	0.43*	−0.01	0.23	0.34	0.04	0.25	−0.25
$mNDVI_{maxb2-a2}$	0.27	0.69**	0.38	0.33	0.10	0.29	0.33	0.04	0.15	−0.09
$mNDVI_{maxb1-max}$	0.02	0.64**	0.46*	0.25	0.13	0.07	0.24	0.01	0.17	−0.05
$mNDVI_{maxb1-a1}$	0.34	0.69**	0.42*	0.30	0.06	0.31	0.28	−0.01	0.15	−0.19
$mNDVI_{maxb1-a2}$	0.30	0.69**	0.41*	0.18	0.19	0.32	0.27	−0.01	0.02	0.01
$mNDVI_{max-a1}$	0.40*	0.66**	0.35	0.23	−0.02	0.36	0.27	0.00	0.12	−0.31
$mNDVI_{max-a2}$	0.33	0.66**	0.37	0.07	0.16	0.34	0.26	0.01	−0.06	−0.02
$mNDVI_{maxa1-a2}$	0.32	0.62**	0.32	0.09	0.19	0.33	0.25	−0.02	−0.09	0.04

*significant at 0.05 level; ** significant at 0.01 level, n = 23.

values and the observed values was close to the diagonal (intercept = 0, slope = 1), and the coefficients of determination for the five study areas ranged from 0.84 to 0.98, indicating that the reliability of the forecasts are very high.

The yield data for 2005 and 2006 were not included in the model construction and instead were used to evaluate the prediction models independently. These data provide independent estimates of the predictive power of the selected models (Table 8).

Table 7. Results of the stepwise regression models for remotely sensed rice yield using AVHRR-derived NDVI measures as independent variables.

Study areas	Model	R	F-test value	RMSE
HLJ	$Y_{RS} = -849.158 + 0.137 NDVI_{maxa1}$	0.42*	4.508	361.99
HN	$Y_{RS} = -1240.690 + 0.229\, mNDVI_{maxb1-a2}$	0.69**	19.342	114.57
JX	$Y_{RS} = -1553.145 + 0.261\, mNDVI_{maxb1-max}$	0.46**	5.689	166.38
SC	$Y_{RS} = -1495.515 + 0.403\, mNDVI_{maxb4-b3}$	0.73**	24.238	207.07
GX	$Y_{RS} = -1832.285 + 1.138\, mNDVI_{maxb4-b3} + 0.214 NDVI_{maxa2} - 1.315\, mNDVI_{maxb4-b2} + 0.307\, mNDVI_{maxb2-b1}$	0.92**	25.103	87.70

R: multiple correlation coefficient.
*significant at 0.05 level; ** significant at 0.01 level.

The differences between the predicted values and the official statistical values were 5% or less in seven out of ten years. These results demonstrate the potential of a NDVI rice-yield estimate that is based on model calibration with historical data at the provincial level. However, it is noticeable that the predicted relative errors were greater than 10%, but less than 19% in both 2005 and 2006 for SC and in 2006 in HLJ when compared with the official statistical data. These error rates are likely due to a number of contamination sources that can confound the potential relationship between NDVIs and rice yield. For instance, cloud and atmospheric-moisture contamination can influence the NDVI signal. Vegetation signals from before or after the selected NDVIs can impact the final yield of rice.

Conclusion

This study focused on the obvious and important role that advance of technology plays in rice yields increase. The results of this analysis suggest that the most common trend of rice yields in China during the years 1979–2009 is a linear growth. In the light of rice-yield trend could not be detected directly by a satellite remote sensor therefore, yield de-trended analysis was necessary to properly identify the remote-sensible effects and obtain an accurate prediction for rice yield. Only with de-trending analysis could we interpret the NDVI's evolution as being mainly due to variations in the photosynthetic activity and growth conditions of rice and then predict the rice yield using NDVI variables.

The AVHRR-based indices explored in the present research were useful for the remotely sensed rice yield-prediction in major

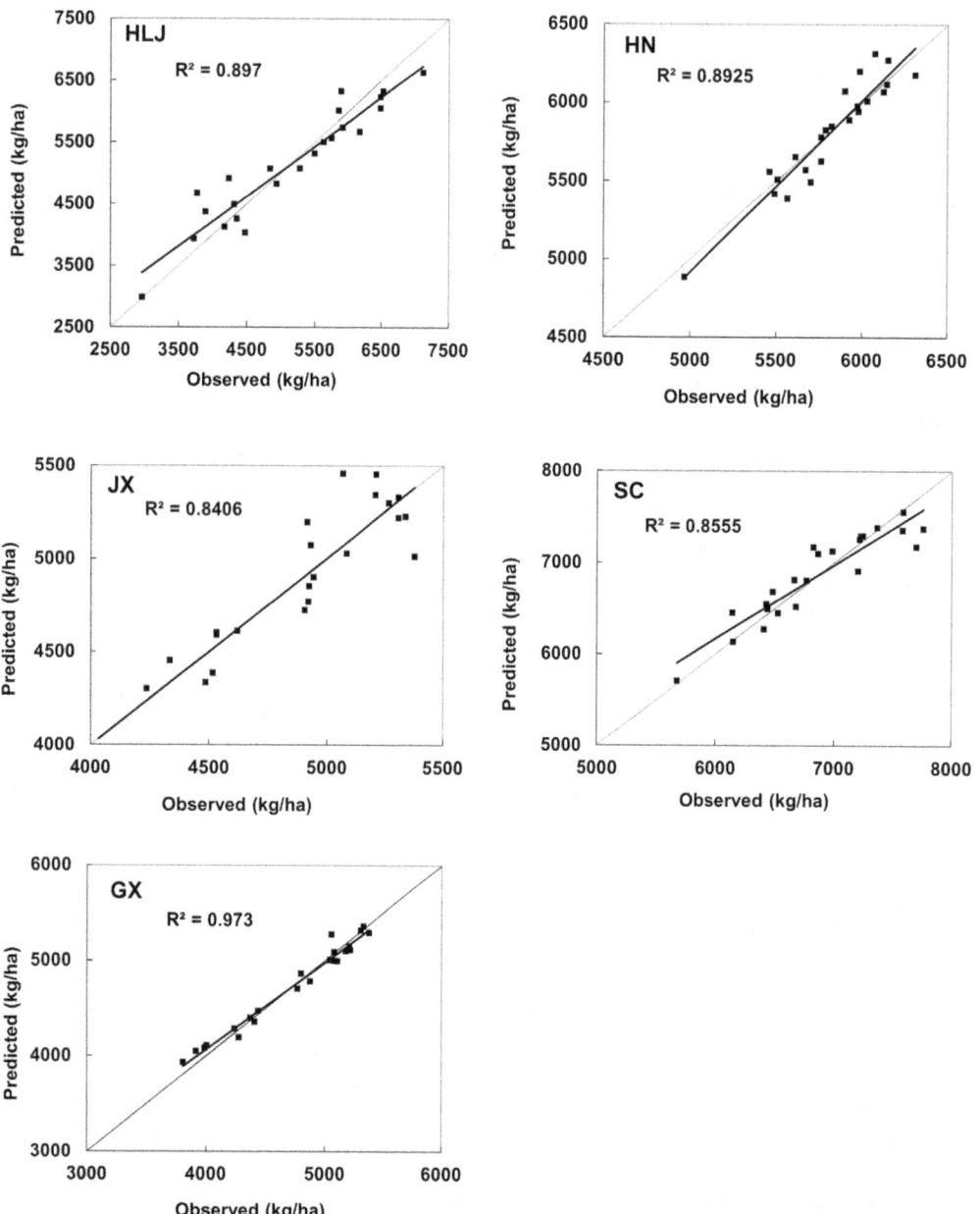

Figure 3. Observed versus predicted yields of rice (kg/ha) for the provinces of Heilongjiang (HLJ), Hunan (HN), Jiangxi (JX), Sichuan (SC) and Guangxi (GX) over the period 1982–2004.

Table 8. Observed and predicted rice yields (independent test).

Provinces	Year	Observed(kg/ha)	Predicted(kg/ha)	Relative Error (%)
Heilongjiang (HLJ)	2005	6795.7	6780.7	−0.22
	2006	6261.3	6897.8	10.17
Hunan (HN)	2005	6050.3	6337.5	4.75
	2006	6141.3	6441.2	4.88
Jiangxi (JX)	2005	5328.2	5545.9	4.09
	2006	5475.1	5634.9	2.92
Sichuan (SC)	2005	7213.0	8018.4	11.17
	2006	6420.7	7680.3	19.62
Guangxi (GX)	2005	4953.0	5028.98	1.53
	2006	5088.0	5053.44	−0.68

rice cultivation areas of China. This method allowed us to have a fine provincial estimate which satellite image could be difficult to obtain, or else a similar cost and a similar time frame data is easily available. However, it is cautious to restrict these analysis to those areas where the common trend of the crop yield is linear growth for the period considered.

The two steps for de-trending the statistical yield to obtain new time series, that are the trend yield (Y_t) and the remotely sensed yield (Y_{RS}); And by constructing the prediction models of Y_{RS} using NDVI variables enabled the development of a robust, simple, remotely sensed data-based model that was applicable at the provincial level in China. We believe the approach introduced here has a wide applicability to other rice-producing countries as well as other crops, such as wheat and corn.

More empirical studies should be performed on the use of AVHRR-derived NDVI time series as predictors for crop yield to enhance the understanding its forecasting capacity and limitations, and to validate the methods of remotely sensed yield estimation further. A future study should also include the application of a longer AVHRR NDVI time series in combination with other data sets such as SPOT-VEG, MODIS and SeaWiFS, especially in the event of one of these dataset's unexpected absence.

Acknowledgments

We thank the NASA/GSFC Global Inventory Modeling and Mapping Studies (GIMMS) group for providing access to the AVHRR NDVI data and the National Bureau of Statistics of China for providing rice yield data. We also appreciate professor Jiwei (jiwei@umkc.edu) and anonymous reviewers who provided very valuable comments.

Author Contributions

Conceived and designed the experiments: JH XW. Performed the experiments: JH XL. Analyzed the data: JH XW XL. Wrote the paper: JH XW XL ZP. Reviewed and revised this manuscript: XL HT ZP.

References

1. National Bureau of Statistics of China (2010) China Statistical Year Book. Beijing: China Statistical Press.
2. Wang RC, Huang JF (2002) Rice yield estimation using remote sensing data. Beijing: China Agriculture Press. 287 p. (in Chinese with English abstract).
3. Prasad AK, Chai L, Singh RP, Kafatos M (2006) Crop yield estimation model for Iowa using remote sensing and surface parameters. International Journal of Applied Earth Observation and Geoinformation 8: 26–33.
4. Manjunath KR, Potdar MB, Purohit NL (2002) Large area operational wheat yield model development and validation based on spectral and meteorological data. International Journal of Remote Sensing 23: 3023–3038.
5. Salazar L, Kogan F, Roytman L (2007) Use of remote sensing data for estimation of winter wheat yield in the United States. International Journal of Remote Sensing 28: 3795–3811.
6. Zhang F, Wu BF, Luo ZM (2004) Winter wheat yield predicting for America using remote sensing data. Journal of remote sensing 8: 611–617. (in Chinese with English abstract).
7. Wang CY, Lin WP (2005) Winter wheat yield estimation based on MODIS EVI. Transactions of the Chinese Society of Agricultural Engineering 21: 90–94. (in Chinese with English abstract).
8. Wall L, Larocque D, Leger PM (2008) The early explanatory power of NDVI in crop yield modelling. International Journal of Remote Sensing 29: 2211–2225.
9. Sridhar VN, Dadhwal VK, Chaudhari KN, Sharma R, Bairagi GD, et al. (1994) Wheat production forecasting for a predominantly unirrigated region in Madhya-Pradesh (India). International Journal of Remote Sensing 15: 1307–1316.
10. Smith RCG, Adams J, Stephens DJ, Hick PT (1995) Forecasting wheat yield in a Mediterranean-type environment from the NOAA satellite. Australian Journal of Agricultural Research 46: 113–125.
11. Schut AGT, Stephens DJ, Stovold RGH, Adams M, Craig RL (2009) Improved wheat yield and production forecasting with a moisture stress index, AVHRR and MODIS data. Crop & Pasture Science 60: 60–70.
12. Rudorff BFT, Batista GT (1991) Wheat yield estimation at the farm level using TM-Landsat and agrometeorological data. International Journal of Remote Sensing 12: 2477–2484.
13. Ren JQ, Chen ZX, Zhou QB, Tang HJ (2008) Regional yield estimation for winter wheat with MODIS-NDVI data in Shandong, China. International Journal of Applied Earth Observation and Geoinformation 10: 403–413.
14. Ren JQ, Chen ZX, Tang HJ (2006) Regional scale remote sensing-based yield estimation of winter wheat by using MODIS-NDVI data: A case study of Jining City in Shandong Province. Chinese Journal of Applied Ecology 17: 2371–2375. (in Chinese with English abstract).
15. Prasad AK, Singh RP, Tare V, Kafatos M (2007) Use of vegetation index and meteorological parameters for the prediction of crop yield in India. International Journal of Remote Sensing 28: 5207–5235.
16. Patel NR, Bhattacharjee B, Mohammed AJ, Tanupriya B, Saha SK (2006) Remote sensing of regional yield assessment of wheat in Haryana, India. International Journal of Remote Sensing 27: 4071–4090.
17. Moriondo M, Maselli F, Bindi M (2007) A simple model of regional wheat yield based on NDVI data. European Journal of Agronomy 26: 266–274.
18. Mo X, Liu S, Lin Z, Xu Y, Xiang Y, et al. (2005) Prediction of crop yield, water consumption and water use efficiency with a SVAT-crop growth model using remotely sensed data on the North China Plain. Ecological Modelling 183: 301–322.
19. Mkhabela MS, Bullock P, Raj S, Wang S, Yang Y (2011) Crop yield forecasting on the Canadian Prairies using MODIS NDVI data. Agricultural and Forest Meteorology 151: 385–393.
20. Mika J, Kerenyi J, Rimoczi-Paal A, Merza A, Szinell C, et al. (2002) On correlation of maize and wheat yield with NDVI: Example of Hungary (1985–1998). In: Fellous JL, LeMarshall JF, Choudhury BJ, Menenti M, Paxton LJ et al., editors. Earth's Atmosphere, Ocean and Surface Studies. 2399–2404.
21. Maselli F, Rembold F (2001) Analysis of GAC NDVI data for cropland identification and yield forecasting in Mediterranean African countries. Photogrammetric Engineering and Remote Sensing 67: 593–602.

22. MacDonald R, Hall F (1980) Global crop forecasting. Science 208: 670.

23. Labus MP, Nielsen GA, Lawrence RL, Engel R, Long DS (2002) Wheat yield estimates using multi-temporal NDVI satellite imagery. International Journal of Remote Sensing 23: 4169–4180.

24. Kastens JH, Kastens TL, Kastens DLA, Price KP, Martinko EA, et al. (2005) Image masking for crop yield forecasting using AVHRR NDVI time series imagery. Remote Sensing of Environment 99: 341–356.

25. Kalubarme AH, Potdar MB, Manjunath KR, Mahey RK, Siddhu SS (2003) Growth profile based crop yield models: a case study of large area wheat yield modelling and its extendibility using atmospheric corrected NOAA AVHRR data. International Journal of Remote Sensing 24: 2037–2054.

26. Huang JF, Wang RC, Wang XZ, Liu SM, Zhang JH (1999) Study on multiple yield estimation models of winter wheat using remote sensing data. Journal of Zhejiang University (Agric & Life Sci) 25: 519–523. (in Chinese with English abstract).

27. Hochheim KP, Barber DG (1998) Spring wheat yield estimation for Western Canada using NOAA NDVI data. Canadian Journal of Remote Sensing 24: 17–27.

28. Gupta R, Prasad S, Rao G, Nadham T (1993) District level wheat yield estimation using NOAA/AVHRR NDVI temporal profile. Advances in Space Research 13: 253–256.

29. Ferencz C, Bognar P, Lichtenberger J, Hamar D, Tarscai G, et al. (2004) Crop yield estimation by satellite remote sensing. International Journal of Remote Sensing 25: 4113–4149.

30. Dubey RP, Ajwani N, Kalubarme MH, Sridhar VN, Navalgund RR, et al. (1994) Preharvest wheat yield and production estimation for the Punjab, India. International Journal of Remote Sensing 15: 2137–2144.

31. Doraiswamy PC, Cook PW (1995) Spring wheat yield assessment using NOAA AVHRR data. Canadian Journal of Remote Sensing 21: 43–51.

32. Cheng Q (1994) The use of vegetation index for monitoring drought and winter wheat yield estimation. Remote sensing technology and application 9: 12–18. (in Chinese with English abstract).

33. Bullock PR (1992) Operational estimates of western Canadian grain production using NOAA AVHRR LAC data. Canadian Journal of Remote Sensing 18: 23–29.

34. Boken VK, Shaykewich CF (2002) Improving an operational wheat yield model using phenological phase-based Normalized Difference Vegetation Index. International Journal of Remote Sensing 23: 4155–4168.

35. Benedetti R, Rossini P (1993) On the use of NDVI profiles as a tool for agricultural statistics: the case study of wheat yield estimate and forecast in Emilia Romagna. Remote Sensing of Environment 45: 311–326.

36. Becker-Reshef I, Vermote E, Lindeman M, Justice C (2010) A generalized regression-based model for forecasting winter wheat yields in Kansas and Ukraine using MODIS data. Remote Sensing of Environment 114: 1312–1323.

37. Bastiaanssen WGM, Ali S (2003) A new crop yield forecasting model based on satellite measurements applied across the Indus Basin, Pakistan. Agriculture Ecosystems & Environment 94: 321–340.

38. Balaghi R, Tychon B, Eerens H, Jlibene M (2008) Empirical regression models using NDVI, rainfall and temperature data for the early prediction of wheat grain yields in Morocco. International Journal of Applied Earth Observation and Geoinformation 10: 438–452.

39. Wannebo A, Rosenzweig C (2003) Remote sensing of US cornbelt areas sensitive to the El Nino-Southern Oscillation. International Journal of Remote Sensing 24: 2055–2067.

40. Unganai LS, Kogan FN (1998) Drought monitoring and corn yield estimation in Southern Africa from AVHRR data. Remote Sensing of Environment 63: 219–232.

41. Seiler RA, Kogan F, Wei G (2000) Monitoring weather impact and crop yield from NOAA AVHRR data in Argentina. In: Gupta RK, editor. Remote Sensing for Land Surface Characterisation. 1177–1185.

42. Rojas O (2007) Operational maize yield model development and validation based on remote sensing and agro-meteorological data in Kenya. International Journal of Remote Sensing 28: 3775–3793.

43. Reynolds CA, Yitayew M, Slack DC, Hutchinson CF, Huete A, et al. (2000) Estimating crop yields and production by integrating the FAO Crop specific Water Balance model with real-time satellite data and ground-based ancillary data. International Journal of Remote Sensing 21: 3487–3508.

44. Quarmby N, Milnes M, Hindle T, Silleos N (1993) The use of multi-temporal NDVI measurements from AVHRR data for crop yield estimation and prediction. International Journal of Remote Sensing 14: 199–210.

45. Mkhabela MS, Mashinini NN (2005) Early maize yield forecasting in the four agro-ecological regions of Swaziland using NDVI data derived from NOAAs-AVHRR. Agricultural and Forest Meteorology 129: 1–9.

46. Lewis JE, Rowland J, Nadeau A (1998) Estimating maize production in Kenya using NDVI: some statistical considerations. International Journal of Remote Sensing 19: 2609–2617.

47. Lee R, Kastens D, Price K, Martinko E. Forecasting corn yield in Iowa using remotely sensed data and vegetation phenology information; 2000 January 10–12; Lake Buena Vista, Florida. 460–467.

48. Hayes M, Decker W (1996) Using NOAA AVHRR data to estimate maize production in the United States Corn Belt. International Journal of Remote Sensing 17: 3189–3200.

49. Funk C, Budde ME (2009) Phenologically-tuned MODIS NDVI-based production anomaly estimates for Zimbabwe. Remote Sensing of Environment 113: 115–125.

50. Tennakoon SB, Murty VVN, Eiumnoh A (1992) Estimation of cropped area and grain-yield of rice using remote-sensing data. International Journal of Remote Sensing 13: 427–439.

51. Huang JF, Yang ZE, Wang RC, Xu HW, Jiang HX (2002) The rice production forecasting models using NOAA/AVHRR data based on GIS. Remote sensing technology and application 17: 125–128. (in Chinese with English abstract).

52. Huang JF, Wang FM, Wang XZ (2010) Hyperspectral experiment for paddy rice remote sensing; Huang JQ, Chen JY, editors. Hangzhou: Zhejiang University Press. 315 p. (in Chinese with English abstract).

53. Rasmussen MS (1998) Developing simple, operational, consistent NDVI-vegetation models by applying environmental and climatic information: Part I. Assessment of net primary production. International Journal of Remote Sensing 19: 97–117.

54. Rasmussen MS (1997) Operational yield forecast using AVHRR NDVI data: Reduction of environmental and inter-annual variability. International Journal of Remote Sensing 18: 1059–1077.

55. Rasmussen MS (1992) Assessment of millet yields and production in northern Burkina Faso using integrated NDVI from the AVHRR. International Journal of Remote Sensing 13: 3431–3442.

56. Maselli F, Romanelli S, Bottai L, Maracchi G (2000) Processing of GAC NDVI data for yield forecasting in the Sahelian region. International Journal of Remote Sensing 21: 3509–3523.

57. Groten SME (1993) NDVI – Crop Monitoring and Early Yield Assessment of Burkina-Faso. International Journal of Remote Sensing 14: 1495–1515.

58. Potdar MB (1993) Sorghum yield modelling based on crop growth parameters determined from visible and near-IR channel NOAA AVHRR data. International Journal of Remote Sensing 14: 895–905.

59. Fuller DO (1998) Trends in NDVI time series and their relation to rangeland and crop production in Senegal, 1987–1993. International Journal of Remote Sensing 19: 2013–2018.

60. Wendroth O, Reuter HI, Kersebaum KC (2003) Predicting yield of barley across a landscape: a state-space modeling approach. Journal of Hydrology 272: 250–263.

61. Weissteiner C, Kühbauch W (2005) Regional Yield Forecasts of Malting Barley (Hordeum vulgare L.) by NOAA-AVHRR Remote Sensing Data and Ancillary Data. Journal of agronomy and crop science 191: 308–320.

62. Liu WT, Kogan F (2002) Monitoring Brazilian soybean production using NOAA/AVHRR based vegetation condition indices. International Journal of Remote Sensing 23: 1161–1179.

63. Esquerdo J, Zullo J, Antunes JFG (2011) Use of NDVI/AVHRR time-series profiles for soybean crop monitoring in Brazil. International Journal of Remote Sensing 32: 3711–3727.

64. Tucker CJ (1979) Red and photographic infrared linear combinations for monitoring vegetation. Remote Sensing of Environment 8: 127–150.

65. Tilman D, Cassman KG, Matson PA, Naylor R, Polasky S (2002) Agricultural sustainability and intensive production practices. Nature 418: 671–677.

66. Evenson RE, Gollin D (2003) Assessing the impact of the Green Revolution, 1960 to 2000. Science 300: 758.

67. Peng S, Laza R, Visperas R, Sanico A, Cassman KG, et al. (2000) Grain yield of rice cultivars and lines developed in the Philippines since 1966. Crop Science 40: 307–314.

68. Hafner S (2003) Trends in maize, rice, and wheat yields for 188 nations over the past 40 years: a prevalence of linear growth. Agriculture, ecosystems & environment 97: 275–283.

69. Pinzon J, Brown ME, Tucker CJ (2004) Satellite time series correction of orbital drift artifacts using empirical mode decomposition. Hilbert-Huang Transform: Introduction and Applications 10: 285–295.

70. Tucker CJ, Pinzon JE, Brown ME, Slayback DA, Pak EW, et al. (2005) An extended AVHRR 8-km NDVI dataset compatible with MODIS and SPOT vegetation NDVI data. International Journal of Remote Sensing 26: 4485–4498.

71. Fensholt R, Nielsen TT, Stisen S (2006) Evaluation of AVHRR PAL and GIMMS 10-day composite NDVI time series products using SPOT-4 vegetation data for the African continent. International Journal of Remote Sensing 27: 2719–2733.

72. Fensholt R, Rasmussen K, Nielsen TT, Mbow C (2009) Evaluation of earth observation based long term vegetation trends – Intercomparing NDVI time series trend analysis consistency of Sahel from AVHRR GIMMS, Terra MODIS and SPOT VGT data. Remote Sensing of Environment 113: 1886–1898.

73. Genovese G, Vignolles C, Negre T, Passera G (2001) A methodology for a combined use of normalised difference vegetation index and CORINE land cover data for crop yield monitoring and forecasting. A case study on Spain. Agronomie 21: 91–111.

74. Freund JT (2005) Estimating Crop Production in Kenya: A Multi-Temporal Remote Sensing Approach [Masters]. Santa Barbara: University of California. 59 p.

75. Tucker CJ, Holben BN, Elgin JH, McMurtrey JE (1980) Relationship of spectral data to grain yield variation. Photogrammetric Engineering and Remote Sensing 46: 657–666.

76. Dyson T (1999) World food trends and prospects to 2025. Proceedings of the National Academy of Sciences 96: 5929.

77. Clarke FR, Baker RJ, Depauw RM (1994) Moving mean and least-squares smoothing for analysis of grain-yield data. Crop Science 34: 1479–1483.

78. Peltonen-Sainio P, Jauhiainen L, Laurila IP (2009) Cereal yield trends in northern European conditions: Changes in yield potential and its realisation. Field Crops Research 110: 85–90.

79. Epplin FM, Peeper TF (1998) Influence of planting date and environment on Oklahoma wheat grain yield trend from 1963 to 1995. Canadian Journal of Plant Science 78: 71–77.

80. Stergiou KI, Christou ED (1996) Modelling and forecasting annual fisheries catches: Comparison of regression, univariate and multivariate time series methods. Fisheries Research 25: 105–138.

81. Foody GM, Boyd DS, Cutler MEJ (2003) Predictive relations of tropical forest biomass from Landsat TM data and their transferability between regions. Remote Sensing of Environment 85: 463–474.

82. Sirkin RM (2006) Statistics for the social sciences. Thousand OaksCalifornia: Sage Publications, Inc. 244 p.

17

Short-Term Effect of Nutrient Availability and Rainfall Distribution on Biomass Production and Leaf Nutrient Content of Savanna Tree Species

Eduardo R. M. Barbosa[1,2]*, Kyle W. Tomlinson[1,3], Luísa G. Carvalheiro[4,5], Kevin Kirkman[6], Steven de Bie[1], Herbert H. T. Prins[1,6], Frank van Langevelde[1]

1 Resource Ecology Group, Wageningen University, Wageningen, The Netherlands, 2 Departamento de Botânica, Laboratório de Termobiologia, Instituto de Ciências Biológicas, Universidade de Brasília, Brasília, DF, Brazil, 3 Community Ecology & Conservation Group, Xishuangbanna Tropical Botanical Garden, Chinese Academy of Sciences, Yunnan, China, 4 School of Biology, University of Leeds, Leeds, the United Kingdom, 5 Terrestrial Zoology, Naturalis Biodiversity Center, Leiden, The Netherlands, 6 School of Life Sciences, University of KwaZulu-Natal, Scottsville, South Africa

Abstract

Changes in land use may lead to increased soil nutrient levels in many ecosystems (e.g. due to intensification of agricultural fertilizer use). Plant species differ widely in their response to differences in soil nutrients, and for savannas it is uncertain how this nutrient enrichment will affect plant community dynamics. We set up a large controlled short-term experiment in a semi-arid savanna to test how water supply (even water supply vs. natural rainfall) and nutrient availability (no fertilisation vs. fertilisation) affects seedlings' above-ground biomass production and leaf-nutrient concentrations (N, P and K) of broad-leafed and fine-leafed tree species. Contrary to expectations, neither changes in water supply nor changes in soil nutrient level affected biomass production of the studied species. By contrast, leaf-nutrient concentration did change significantly. Under regular water supply, soil nutrient addition increased the leaf phosphorus concentration of both fine-leafed and broad-leafed species. However, under uneven water supply, leaf nitrogen and phosphorus concentration declined with soil nutrient supply, this effect being more accentuated in broad-leafed species. Leaf potassium concentration of broad-leafed species was lower when growing under constant water supply, especially when no NPK fertilizer was applied. We found that changes in environmental factors can affect leaf quality, indicating a potential interactive effect between land-use changes and environmental changes on savanna vegetation: under more uneven rainfall patterns within the growing season, leaf quality of tree seedlings for a number of species can change as a response to changes in nutrient levels, even if overall plant biomass does not change. Such changes might affect herbivore pressure on trees and thus savanna plant community dynamics. Although longer term experiments would be essential to test such potential effects of eutrophication via changes in leaf nutrient concentration, our findings provide important insights that can help guide management plans that aim to preserve savanna biodiversity.

Editor: Ben Bond-Lamberty, DOE Pacific Northwest National Laboratory, United States of America

Funding: Funding and research assistance for this project were provided by Shell Research Foundation (http://www.shellfoundation.org/). The funders had no role in study design, data collection and analysis, decision to publish, or preparation of the manuscript.

Competing Interests: The authors have declared that no competing interests exist.

* E-mail: eduardormbarbosa@gmail.com

Introduction

Recent studies predict an increase in nitrogen deposition over southern Africa during the next few decades [1], due to rising industrial emissions and changes in land use [2]. Soil nitrogen enrichment can lead to soil acidification, which reduces soil fertility by promoting leaching of certain nutrients (such as calcium and magnesium) [3]. Moreover, increased nitrogen availability might also affect the carbon flux from soils of natural ecosystems [4] through changes in plant and soil microbial communities [5]. Such environmental changes can have important impacts for African savannas, especially on the species composition and abundance. Furthermore, alterations in rainfall patterns are also expected in the region where savannas occur [6]. However, little information on the effects of changes in soil nutrient and water availability on the leaf nutrient concentration of savanna trees is found in the literature [7].

Plant productivity and above-ground biomass are thought to increase with higher soil resource availability (e.g. nitrogen, water, phosphorus) [8–10]. In drier regions (such as semi-arid savannas), highly variable rainfall may negatively affect plant nutrient uptake and storage [11,12], potentially limiting plant growth during the growing season [13]. Indeed, performance of savanna tree seedlings is suggested to be worse when grown in nutrient-rich soils than in nutrient-poor soils [14,15]. This effect may be caused by the intensification of herbaceous competition for water and not by direct negative effects of high nutrient availability on tree seedlings [15]. Moreover, increased amounts of nutrients in plant leaves might increase their quality as food for herbivores [16–18], whereas increased water availability may increase biomass but decrease leaf nutrient concentration [18]. Tree seedling recruitment is a critical stage in the regeneration of trees and overall plant population dynamics [19–22]. However, there is a lack of empirical studies involving multiple plant species [22]. Most experimental studies evaluating the growth of tree species in response to resource supply and disturbance, with and without grass competitors, focus on single species [23,24], and there are few comparative investigations on seedlings of savanna tree species either within or across communities. This lack of empirical knowledge critically limits our ability to understand how seedlings of different species in a community perform under different environmental condition, and consequently, how plant community dynamics might change under modifications in the land use and climate conditions [25,26].

As plant species differ widely in their response to differences in soil nutrients [27], changes in soil resource availability (water and nutrient availability) may change structural heterogeneity in tree cover [28] or leaf nutrient concentration, and thereby influence primary productivity [29,30]. Differences in functional traits can mechanistically explain why species differ in their performance across resource and disturbance gradients [31,32]. Qualitative trait differences between species which are associated with nutrient and water gradients have been recognised [33]. Notably, within African savannas, dystrophic or humid savannas are dominated by broad-leafed species that are also non-spinescent, non-N-fixing species, whereas eutrophic or arid savannas are dominated by fine-leafed species which may additionally be spinescent or N-fixing [33–36]. These two groups can also be distinguished on the basis of their leaf chemistry, physiology and morphology [32]. As these functional traits are already found in tree seedlings during their first season of growth (e.g. N-fixing associations can be established early as two/three weeks after planting) [37], there is reason to believe that seedlings of tree species representing these functional species groups respond differently to changes in supply rates of resources, and that these differences may in part explain why they dominate in different environments.

Savannas are often characterized by water-limited plant growth during the growing season [38]. However, the amount of rainfall within the wet season is highly unpredictable, especially in semi-arid savannas [39]. Dry periods during the wet season can have an important impact limiting tree seedling survival [28]. Such dry periods may become more frequent in the near future, as global climate models indicate rising temperatures and increasingly erratic rainfall patterns across Southern African regions [40,41]. Climatic changes may also lead to slightly extended later summer rainfall over eastern South Africa [42]. Here we evaluated the short-term effects of water variability (even water supply vs. natural rainfall) and soil nutrient availability (no fertilisation vs. NPK addition) on above-ground biomass production and leaf nutrient concentrations of seedlings of two important functional groups of semi-arid savanna trees: broad-leafed (4 species) and fine-leafed (4 species) species (Table 1). We focus on the leaf concentrations of nitrogen (N), phosphorus (P) and potassium (K) because these nutrients are important in many plant metabolism processes [43], and in the diets of herbivores [16,18].

As increases in soil resources and reduction of periods of soil moisture deficiency are thought to increase plant productivity [8–11], we expected that all species respond positively to even water availability (no dry periods during the wet season) and to increased nutrient supply by increasing above-ground plant biomass (Hypothesis 1). As fine-leafed species are the dominant tree species in nutrient-rich savannas [35], are N-fixing and may have greater photosynthetic rates [44] than broad-leafed species, we expected fine-leafed species to have always higher leaf nutrient concentrations than broad-leafed species (Hypothesis 2). However, during growth, most nutrients (50–75%) are thought to be located in the leaves (e.g. [45,46]), their concentration depending mostly on

soil nutrient availability [47,48] and soil moisture [49]. Longer periods of soil moisture availability during the growing season may decrease leaf nutrient concentration, due to dilution effects of increased plant growth [18,50]. Therefore, we expected that the two species groups would increase leaf nutrient concentration with increasing soil nutrient availability, and that it would decrease with constant water supply (Hypothesis 3).

Methods

To test whether the two functional species groups differed in their response to variation in the growth conditions, we set up a large controlled, short-term field experiment in the Lowveld savanna region [35]. The study was carried out on private land of the Southern African Wildlife College (SAWC), Limpopo Province, South Africa (24°15′20.23″S, 31°23′23.63″E). For future permissions for fieldwork at the SAWC please contact Mrs. Theresa Sowry (tsowry@sawc.org.za) or Mr. Francois Nel (fnel@ sawc.org.za). The experiments were run during the wet season of 2009–2010 (November–May), in a fenced area that excluded large herbivores. The mean rainfall during the growing season (from October till April) of the previous 10 years (2000–2010) is ca. 456 mm (Satara Camp, Kruger Park around 40 km northeast of the research site). The mean maximum temperature during January (hottest month) is 33.7°C and the mean minimum temperature for June (coolest month) is 9.4°C [42]. The vegetation is described as Granite Lowveld [35], and the area is classified as semi-arid under the Köppen-Geiger System [36]. Soils in the experimental site were shallow (ca. 1.5 m depth) and mainly derived from granite [34] with occasional gabbro extrusions. Soils derived from granite tend to be coarse-textured and nutrient-poor (i.e., low availability of N and P) on crests and mid-slopes [51], but nutrient availability may be elevated in bottom positions in the landscape, and very locally such as on termitaria or underneath large *Acacia* trees [51].

Species

We selected eight locally abundant tree species that make up a large proportion of vegetation cover in the Lowveld savanna region in South Africa. Although most of the selected species belong to the Fabaceae family (with the exception of *Combretum apiculatum*), these species are classified into two different subfamilies: Mimosoideae (fine-leafed species) and Caesalpinioideae (broad-leafed species). In African semi-arid savannas, broad-leafed and medium-leafed species are found on dystrophic soils, characterised by high fire frequency (annual to triennial) and MAPs from 600–1500 mm [14]. Fine-leafed species (largely Mimosoideae) are found on eutrophic soils or skeletal soils with low fire frequency (quintennial or longer) and MAPs of 300–800 mm [35]. The study species were separated in two different functional species groups: four species with characteristic small leaves, spines, and N-fixing associations (hereafter termed 'fine-leafed species'), and four species with characteristic broad leaves, no spines, and lacking N-fixing associations (hereafter termed 'broad-leafed species') (see Table 1). All the seeds used in the experiment were collected in areas surrounding the experimental site. Since these species are abundant in the savannas of Southern African region, changes in their populations due to varying environmental conditions will likely have substantial effects on the local vegetation structure.

Treatments

The study site was ploughed (about 20 cm deep) to homogenize the soil and to give all treatments the same starting conditions. Five

Table 1. Functional trait data for tree species used in the experiment.

Species	Family	Sub-family	N$_2$-fixing[‡]	Leaf type[‡‡]	Leaf size (cm^2)[††]	Spinescence
Fine-leafed species						
Acacia nigrescens Oliv.	Fabaceae	Mimosoideae	Yes	Bipinnate	16.0 ($\pm$3.6)	Yes
Acacia nilotica Willd.	Fabaceae	Mimosoideae	Yes	Bipinnate	12.0 ($\pm$2.1)	Yes
Acacia tortilis Hayne	Fabaceae	Mimosoideae	Yes	Bipinnate	12.5 ($\pm$7.0)	Yes
Dichrostachys cinerea Wight and Arn	Fabaceae	Mimosoideae	Yes	Bipinnate	31.7 ($\pm$28.4)	Yes
Broad-leafed species						
Colophospermum mopane J. Léonard	Fabaceae	Caesalpinioideae	No	Pinnate	47.2 ($\pm$21.6)	No
Combretum apiculatum Sond.	Combretaceae	–	No	Simple	25.3 ($\pm$5.3)	No
Schotia brachypetala Sond.	Fabaceae	Caesalpinioideae	No	Pinnate	42.6 ($\pm$17.6)	No
Peltophorum africanum Sond.	Fabaceae	Caesalpinioideae	No	Bipinnate	99.5 ($\pm$81.8)	No

For continuous values the standard deviation is indicated between brackets. Sources of data are indicated in postscripts.
[‡][51], [70].
[††]Obtained from the experimental seedlings from the treatment W0N0 (Natural rainfall-No nutrient addition).
[‡‡][70].

blocks were laid out in a restricted area (90×90 m) in the experimental area (maximum distance between the blocks was 40 m). Inside each block, four 4-m^2 plots were located, separated by a 2 m gap between the plots (Figure 1) Seedlings were subjected to two different water regimes: one covered with a rain-out shelter (W1 - even water supply) and another exposed to natural rainfall conditions (W0 - natural rainfall or uneven water supply). The nutrient treatment was separated in two different nutrient applications (N0 - no nutrient supply vs. N1 - nutrient supply), leading to a total of 20 experimental plots.

Three weeks before the experiment all seeds were sown in nursery bags (using the same soil of the experimental area). At four weeks after germination, 20 seedlings per species were randomly transplanted in treatment combination plots (five replicate blocks, each with four seedlings per plot). The seedling positions in the plots were randomly selected. The seedling density inside of the plots (20 seedlings per m^2) was lower than the normal early seedling density in savannas (more than 50 seedlings per m^2 in the seed/seedling bank [52]. The seedlings were then followed for six months (November 2009 to May 2010). Although our experiments were performed during a short period of time (six months), this period of time is equivalent to a growing season in the area where the study was conducted. As savanna tree species show great

differences in growth strategies, which allow them to cope with the high unpredictability of the amount of annual rainfall [39], we expected differences in the responses to the variation in resource avaliabiity between species even in short-term experiments.

Plots within the uneven water supply treatment (W0) received 623 mm of water from natural rainfall during the period of the experiment, which was higher than the mean rainfall of the previous 10 years for the area (456 mm) (Figure 2). The distribution was uneven during the experiment: 206 mm in November, 114 mm in December, 55 mm in January, 31 mm in February, 57 mm in March and 160 mm in April. For the even water treatment (W1), natural rainfall was excluded from the treatments by rain-out shelters (200 µm clear greenhouse polyethylene film, allowing around 95% of sun light irradiation) and we supplied a fixed amount of 46.3 mm (185 l per 2×2 m plots) of water to the seedlings every two weeks for the six months of the experiment, yielding a total of 556 mm water over the experiment, using sprinkler irrigation systems. Due to the lack of the rainfall data from the research site, the amount of water applied in W1 was based on the water deficit rules as defined in the Köppen-Geiger climate classification (550 mm per season), based on a recent update of these regional classifications [36].

To increase the nutrient availability for the tree seedlings in the high nutrient treatment (treatment N1), we used a granular slow-release inorganic fertilizer containing nitrogen (N), phosphorus (P) and potassium (K) in the ratio 3:1:2 (Osmocote Exact Standard 15:9:11, Scotts International, The Netherlands). The fertilizer was added once before the seedlings being transplanted in treatment seedlings at a rate of 4 g N m^{-2} (640 g per plot), following rates previously applied [15]. Normal annual amount of nitrogen mineralized in the study region was estimated at 5.8 g N m^{-2} [53], so N1 treatment increased local nitrogen availability ca.1.7 times.

Shoot Foliar Nutrient Concentration and Biomass

The shoots of seedlings were harvested six months after planting in May 2010. These were oven-dried at 70°C for at least 48 h, and their dry weights were measured. To quantify the concentration of the elements N, P and K in leaves, the leaf material was digested with a mixture of H$_2$SO$_4$, Se and salicylic acid [54]. The concentrations of N and P in the leaves were measured with a Skalar San-plus auto-analyzer, and K was measured with an

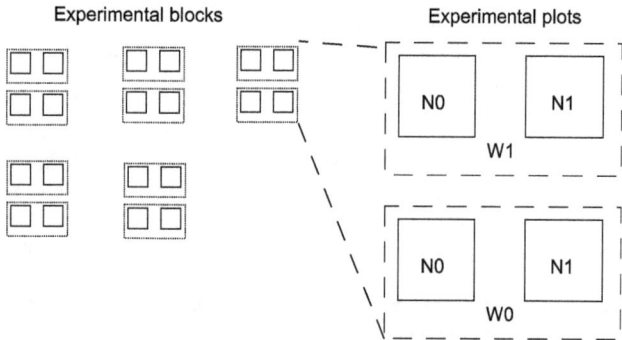

Experimental blocks   Experimental plots

Figure 1. The experimental design. W0 - natural rainfall treatment, W1– even watering treatment, N0– no addition of nitrogen-phosphorus-potassium fertilizer, N1 addition of nitrogen-phosphorus-potassium fertilizer.

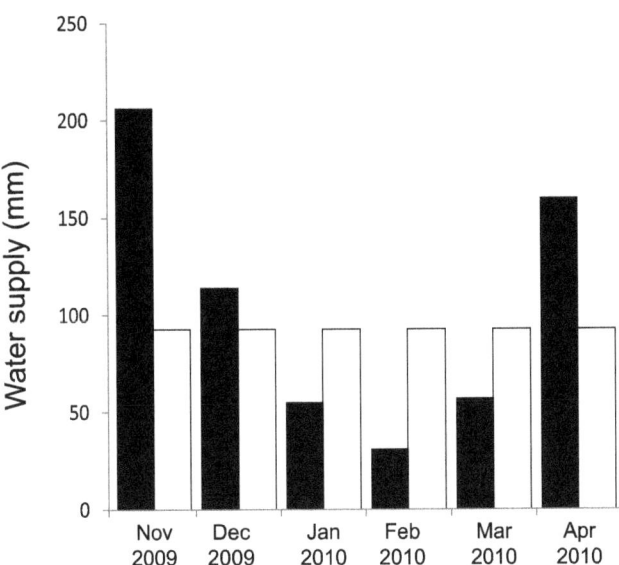

Figure 2. Monthly water availability (mm) during the experimental period. The black bars represent the monthly rainfall in the area during the wet season of 2009–2010 (natural rainfall treatment-W0). The white bars represent the monthly water supplied in the treatment W1. The rainfall data are from Satara Camp, Kruger Park around weather station (40 km northeast from the research site).

Atomic Absorption Spectrometer (AAS) from Varian (Palo Alto, CA, USA).

Data Analysis

To test how the different treatments affected leaf nutrient concentrations and biomass production, we used general linear mixed models (GLMM), using maximum likelihood [55]. Water regime (W) and nutrient addition (N) and functional species group (FG) were included as fixed variables. To account for inter-specific variability, species was treated as a random factor in the model (Species, 8 levels), and plot within experimental block. As the inclusion of block and plot position did not significantly improve the model (all plots were very close to each other), these two random factors were dropped from the final model. The individual species analyses are provided in Table S1.

Mixed model analyses were conducted in R (R Development Core Team, 2013 - version 3.0.2) using the *lmer* function of the package lme4 [55]. To test the significance of the terms in the statistical model we ran Monte Carlo Markov Chain simulations (100,000 iterations) using the LanguageR package (http://cran.r-project.org/web/packages/languageR/languageR.pdf) to analyse the seedling biomass production and leaf nutrient concentrations.

The data used for this manuscript is made available via SANParks Data Repository website (https://knb.ecoinformatics.org/knb/style/skins/sanparks/index.jsp) and can also be obtained from the corresponding author.

Results

Contrary to our expectations (Hypothesis 1), short-term changes in soil resource availability (water and nutrient availability) did not affect above-ground biomass production of any of the tree species (Table 2 and Figure 3).

In relation to leaf quality, we expected that fine-leafed species would present higher leaf nutrient concentrations than broad-leafed species (Hypothesis 2). Indeed, overall leaf N concentration

Table 2. The effect of functional species groups (FG) (fine-leafed vs. broad-leafed) of water (regular water supply vs. natural rainfall) and nutrient (no addition vs. NPK addition) treatments on leaf nutrient concentrations and above-ground biomass.

	Leaf nitrogen concentration			Leaf phosphorus concentration			Leaf potassium concentration			Total Biomass		
	Post mean	effective samples	p- MCMC	Post mean	effective samples	p- MCMC	Post mean	effective samples	p- MCMC	Post mean	effective samples	p- MCMC
FG	**-0.49**	**-0.14**	**0.011**	0.004	0.024	0.96	-0.01	0.20	0.93	-0.24	0.23	0.27
Nutrients	**-0.13**	**-0.01**	**0.029**	**-0.01**	**-0.003**	**0.015**	-0.03	0.05	0.44	-0.07	0.10	0.45
Water	**-0.24**	**-0.12**	**<0.0001**	**-0.01**	**-0.003**	**0.015**	**-0.10**	**0.00**	**0.040**	-0.07	0.10	0.46
FG × Nutrients	-0.09	0.09	0.31	-0.01	0.008	0.42	0.02	0.15	0.73	0.02	0.28	0.85
FG × Water	-0.01	0.17	0.92	0.005	0.020	0.49	-0.07	0.06	0.28	-0.07	0.18	0.58
Water × Nutrients	**0.20**	**0.37**	**0.024**	**0.03**	**0.045**	**<0.0001**	0.12	0.25	0.06	0.09	0.34	0.47
FG x Nutrients x Water	-0.07	0.17	0.60	0.01	0.026	0.56	-0.03	0.15	0.73	-0.08	0.26	0.68

Species was used as a random variable to correct for the variation among the different species. P values were obtained with Monte Carlo Markov Chain simulations (100000 iterations), using the MCMCglmm package and LanguageR package for R software (R Development Core Team, 2013, version 3.0.2). The significant values are represented in bold.

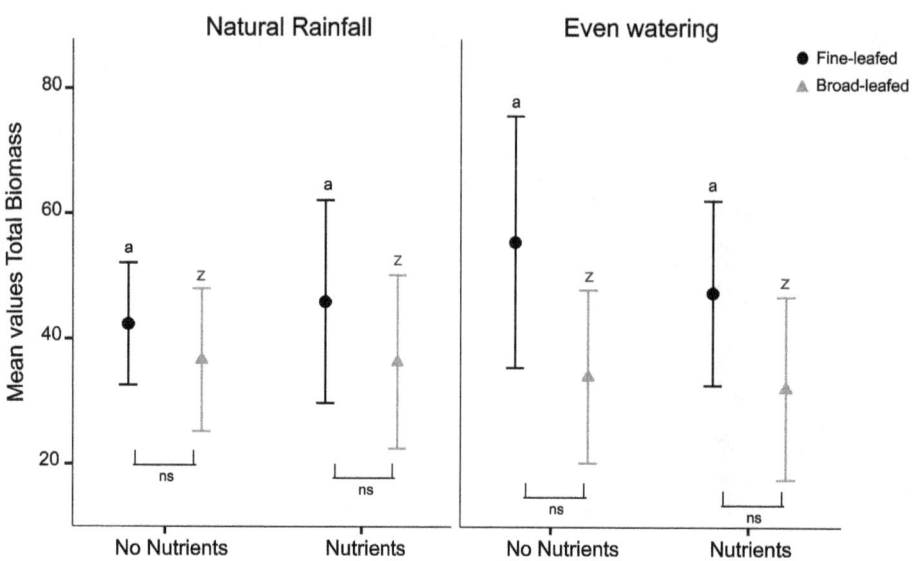

Figure 3. Effect of soil water and nutrient supply on the mean above-ground dry biomass (g). Bars represent 95% confidence intervals. The black circles represent the fine-leafed species, and the grey triangles represent the broad-leafed species. Nutrients represents the where Nitrogen-Phosphorus-Potassium (NPK) fertilizer was added, and No nutrients represents the plots where no NPK fertilazer was added. Statistical details are presented in Table 2. The letters represent differences between the treatments. Results of pairwise comparisons between the species groups within a treatment combination are indicated by the brackets (*: $p<0.05$; **: $p<0.01$; ***: $p<0.001$; ns: non-significant).

was lower in broad-leafed species, but no significant differences were found between the two groups for leaf K and P concentration. In relation to the responses of leaf quality to treatments, we expected that leaf nutrient concentration would increase with increasing soil nutrient availability, and that it would decrease with constant water supply (Hypothesis 3). However, leaf P concentration only increased with NPK fertilizer input under even water supply in both species groups (Figure 4). A non-significant positive trend in leaf K concentration was also apparent. In contrast, under uneven water supply (natural rainfall), foliar concentrations of P and N were lower under the nutrient addition treatment (Table 2 and Figure 4). This trend was more accentuated for broad-leafed species with respect to leaf N concentration. Moreover, leaf K concentration of broad-leafed species was significantly lower than fine-leafed species when grown under constant water supply.

Discussion

Plant productivity and above-ground biomass are thought to increase with higher soil resources (e.g. nitrogen, phosphorus) and water availability [8–10]. However, our results show that changes in leaf nutrient concentration varied with changes in the soil resource availability, even when biomass is not affected. Here we discuss the variability of responses to nutrient and water supply of two functional groups of tree species that are representative for the African savanna biome as a whole.

Effect of Water and Nutrient Availability on Shoot Growth and Leaf Nutrient Concentration

Contrary to our expectations (Hypothesis 1), increased short-term nutrient input and water availability did not significantly influence above-ground biomass production of the two functional species groups studied. Three plausible explanations arise. Firstly, the nutrient additions may have been insufficient to cause a difference in growth between the unfertilised and fertilised plots (raised the available N in the soil by at most 70%). Indeed, as we

used a slow release fertilizer, nutrients added may not have been immediately available. However, nutrient addition had a strong effect on leaf chemical composition, suggesting that nutrient additions did increase nutrient uptake by the tree seedlings. A second explanation is that seedlings of the considered species are not limited by soil nutrient availability in the study, savanna tree species being able to cope with low resource conditions. While our study included fine-leafed species, which can be found in regions with high nutrient soils (e.g. *Acacia* species), the seeds used in this study were collected in areas with relative low soil nutrients. It is, therefore, possible that the source populations of the seeds used in this study are adapted to grow in relatively infertile soils exhibiting lower maximum potential growth rates, and responding less to nutrient addition [56]. Moreover, under frequent water supply and high nutrient availability, it is possible that plant species of semi-arid environments allocate more resources to the root system [57] while above ground biomass remains constant.

The lack of growth response to improved fertility has been observed previously for tree seedlings growing in low nutrient environments [32,58]. While the application of nutrient fertilizers may mitigate the adverse effects of water stress on plant development [59], the potential effect of nutrient addition on plants depends on their growth potential [58]. Therefore, changes in tree species composition are gradual, potentially taking a long time to be noticeable [60]. A longer term experiment (e.g. several years) would be essential to verify if nutrient enrichment and changes in water supply have delayed effects on savanna tree growth and biomass.

In relation to leaf-nutrient concentration, as expected fine-leafed species had higher leaf nutrient concentrations than broad-leafed species (Hypothesis 2). Fine leafed species are dominant on eutrophic soils [33–36], and hence are likely to be adapted to this high nutrient availability. Indeed all fine-leafed species studied here are able to fix atmospheric nitrogen [51], leading to a higher access to nitrogen. This extra N may be stored in leaves for future use, explaining the higher nitrogen values found in this study. Such storage might be important for shoot biomass recovery after

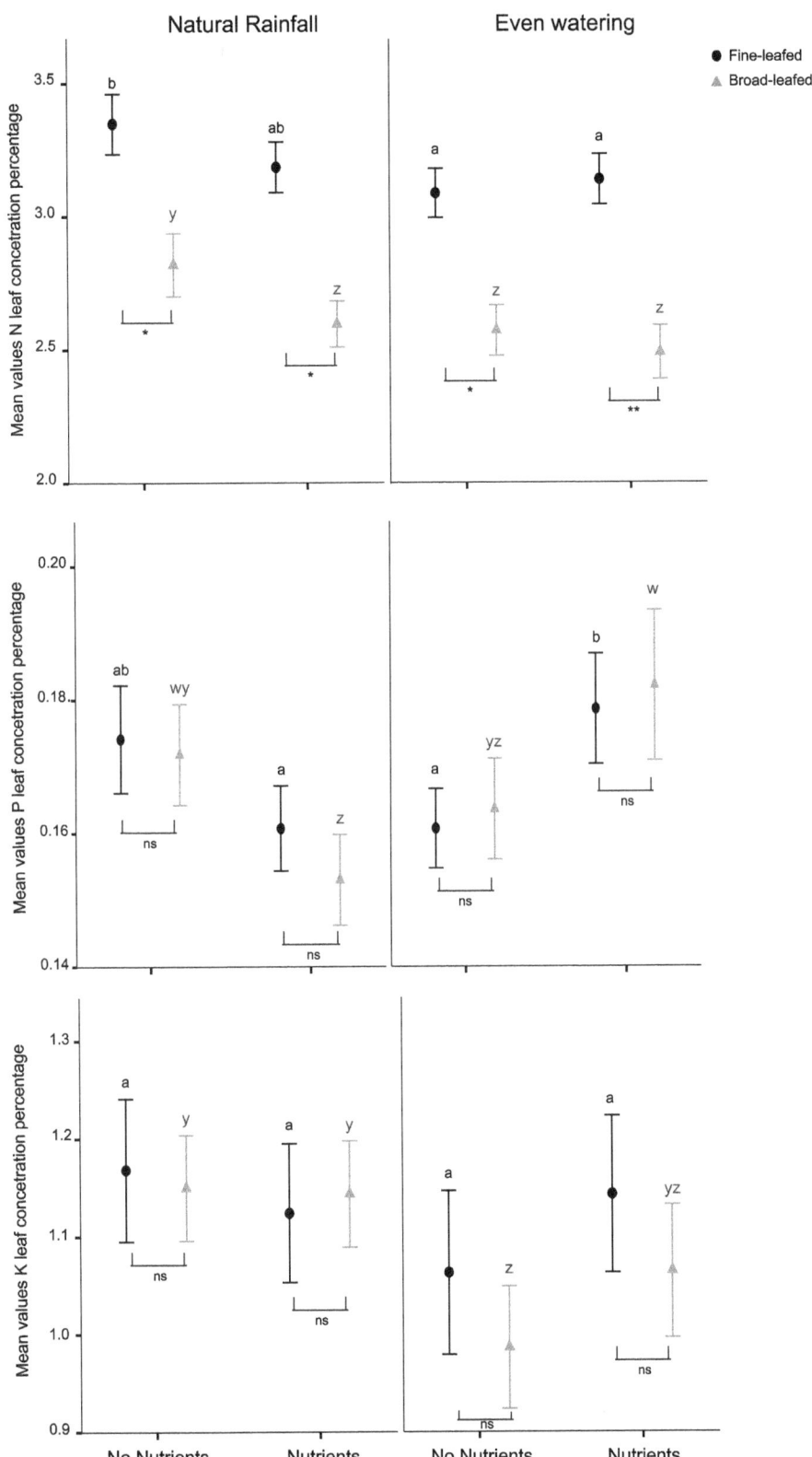

Figure 4. Effect of soil water and nutrient supply on the mean leaf nutrient concentration. Bars represent 95% confidence intervals. N = leaf nitrogen concentration, P = leaf phosphorus concentration and K = leaf potassium concentration. The black circles represent the fine-leafed species, and the grey triangles represent the broad-leafed species. "Nutrients" indicates where Nitrogen-Phosphorus-Potassium (NPK) fertilizer was added, and "No nutrients" represents the plots where no NPK fertilizer was added. Statistical details are presented in Table 2. The letters represent differences between the treatments. Results of pairwise comparisons between the species groups within a treatment combination are indicated by the brackets (*: p<0.05; **: p<0.01; ***: p<0.001; ns: non-significant).

intense fire or herbivory, as the N stored in the leaves not lost in the defoliation event can enhance shoot and leaf growth rates [27], which characterize nutrient rich savannas [35,36]. Fine leafed species could, hence, show high growth rate [61]. Indeed, in all treatments the average biomass of the fine-leafed species was higher than for broad-leafed species, which could be due to this extra nitrogen availability. However, these differences between the two groups were not significant, which possibly due to the short period of the experiment.

In contrast to plant productivity, plant quality (measured as the nutrient concentration in leaves) did significantly vary with nutrient input and water availability (Figure 4). A previous study of leaf nutrient concentration of grasses [18] suggested that plant quality (i.e. leaf nutrient concentration) increases with soil nutrient concentration and decreases with water availability (similar to Hypothesis 3). Indeed, for both functional groups, leaf nitrogen concentrations were higher under natural conditions (natural rainfall/no NPK input) than in other treatments (Fig. 4). For plants growing under high N levels, plants may invest mostly in growth, leading to a dilution effect of N content of leaves [18,50], potentially explaining the lower N content found in treatments with NPK addition. At low soil N availability, most N is stored in the leaves as amino acids, amides, or protein (enzymes such as Rubisco) [58]. The total Rubisco in the leaves increases linearly with increase of leaf N content, being essential for photosynthesis [62]. However, the activation of the Rubisco is regulated by CO_2 levels in the leaves [63]. As leaf N content and CO_2 assimilation by the leaves have a non-linear relationship, and hence most of the Rubisco in the leaves is inactive [64], not being used in photosynthesis. Such storage of N (e.g. as amino acids, or inactive Rubisco) can, however, be exported to support growth of other parts of the plant, whenever is needed [65].

In contrast with our expectations, our empirical results with savanna tree seedlings show that nutrient input increased leaf phosphorus concentration only when water input was regular, whereas decreases in leaf concentrations for this element occurred when water availability was uneven. Seedling dependence on water for a positive effect of nutrient (K and P) availability in leaves can be explained by the fact that nutrient uptake depends on water movement within plants [66]. Furthermore, the negative effect of nutrient input under the uneven water supply treatment (natural rainfall) also suggests that such irregularity in water provision stimulates allocation of resources away from leaves towards other organs, such as roots, that can support growth and survival when soil reserves are unavailable [67] for example during the dry (non-growing) season. This re-allocation is more likely to be noticeable at the end of the growing season, when our measurements were taken. Further studies on root production in tree seedlings across soil nutrient and moisture gradients would help to confirm where absorbed nutrients are allocated.

Implications for Herbivore-plant Relationships

The availability of soil nutrients is influenced by herbivore density through dung and urine [7]. The results of this short-term experiment suggests that when combined with the natural (i.e. uneven) rainfall patterns, high soil nutrient availability may lead to a decrease of the leaf quality of tree seedlings as forage for herbivores (due to lower nutrient concentration in leaves, in this study mostly nitrogen and phosphorus), even when overall biomass does not change (Figure 3). Tree seedlings are a common food source for herbivores, especially due the high nutrient levels, low levels of defensive structures, and secondary defensive compounds [68]. This decline in leaf quality might, hence, increase the need of consumption by herbivores to acquire the amounts of nutrients needed by them, magnifying the effects of high herbivore density. This increase in browsing may affect tree recruitment, potentially impacting long-term dynamics and vegetation structure in savannas [22,69]. Further changes in soil resource levels (e.g. higher N deposition, changes in wet season rainfall patterns) can lead to further accentuation of impacts of increased herbivore density for plant community dynamics in savannas. As several herbivore species are limited by the nutrient concentration of tree leaves, especially in pregnant and lactating animals [43], such changes may affect their population dynamics (e.g., reproduction, breeding times, and foraging range [14,16,18,43]. Future studies would be needed to test such potential effects of soil nutrient input on herbivores, via change in tree seedlings nutrient content.

Concluding Remarks

The results of our short-term multi-species experiment show that differences in soil resource availability lead to changes in leaf quality (leaf nutrient concentration). The effect of nutrient input on leaf quality (especially nitrogen and phosphorus concentrations) depends on water availability. Under more uneven water availability, leaf nutrient concentration decreases, while under regular rainfall it increases. While changes in the soil conditions might not directly affect plant species distribution [7], the changes in leaf quality may affect browsing pressures, and consequently affect overall vegetation structure. Our results hence suggest that, in response to the predicted changes in the rainfall distribution during the wet season in Southern African savannas (which is expected to become more erratic, with increases of the interval between each rainfall event [3,41], leaf quality of tree seedlings for a large number of species will change, potentially affecting vegetation communities and herbivore population dynamics. Long-term experiments across multiple growing seasons are essential to confirm the robustness of the results obtained in this study. Moreover, close monitoring of how vegetation and herbivore communities will change in response to climate and land-use changes is essential both to understand the full extent of the ecological and consequences and to contributing to the development of adequate policies and management plans that aim to preserve biodiversity.

Supporting Information

Table S1 Effect of water (regular water supply vs. natural rainfall) and nutrient (no addition vs. NPK addition) treatments on leaf nutrient concentrations and above-ground biomass in all study species used in this research. P values were obtained with Monte Carlo Markov Chain simulations (100000 iterations), using the MCMCglmm package and LanguageR package for R software (R Development Core Team, 2013).

Acknowledgments

We are thankful to the staff of Southern African Wildlife College, SANParks, School of Biological and Conservation Sciences (University of KwaZulu-Natal) and the Resource Ecology Group (Wageningen University) for technical support, especially A. van den Driessche, C. Y.Ravenstein, C. Mthabine, P. Mthabine, L. Mthabine, D. Mlambo and P. Ngomana.

Author Contributions

Conceived and designed the experiments: ERMB KWT FVL. Performed the experiments: ERMB. Analyzed the data: ERMB LGC KWT. Wrote the paper: ERMB. Provided comments in several stages of the project and manuscript writing: FVL KWT LGC KK SDB HHTP.

References

1. Miyazaki K, Eskes HJ, Sudo K (2012) Global NOx emission estimates derived from an assimilation of OMI tropospheric NO2 columns. Atmos Chem Phys 12: 2263–2288.
2. Adams M, Ineson P, Binkley D, Cadlsch G, Tokuchi N, et al. (2004) Soil function responses to excess nitrogen inputs at global scale. Ambio 33: 530–536.
3. Allison SD, Treseder KK (2008) Warming and drying suppress microbial activity and carbon cycling in boreal forest soils. Glob Change Biol 14: 2898–2909.
4. Neff JC, Townsend AR, Gleixner G, Lehman SJ, Turnbull J, et al. (2002) Variable effects of nitrogen additions on the stability and turnover of soil carbon. Nature 419: 915–917.
5. Xu LK, Baldocchi DD, Tang JW (2004) How soil moisture, rain pulses, and growth alter the response of ecosystem respiration to temperature. Glob Biogeochem Cycles: 18 GB4002.
6. IPCC (Intergovernmental Panel on Climate Change) (2007) Climate Change 2007: The Physical Science Basis. Cambridge: Cambridge University Press582 p.
7. Van der Waal C, Kool A, Meijer SS, Kohi E, Heitkonig IMA, et al. (2011) Large herbivores may alter vegetation structure of semi-arid savannas through soil nutrient mediation. Oecologia 165: 1095–1107.
8. Tilman D (1984) Plant dominance along an experimental nutrient gradient. Ecology 65: 1445–1453.
9. Gower ST, Vogt KA, Grier CC (1992) Carbon Dynamics of Rocky Mountain Douglas-Fir: Influence of Water and Nutrient Availability. Ecol Monogr 62: 43–65.
10. Polis GA (1999) Why Are Parts of the World Green? Multiple Factors Control Productivity and the Distribution of Biomass. Oikos 86: 3–15.
11. Parks SE, Haigh AM, Creswell GC (2000) Stem tissue phosphorus as an index of the phosphorus status of Banksia eri cifolia L. Plant Soil 227: 59–65.
12. Shane MW, Cramer MD, Funayama-Noguchi S, Cawthray GR, Millar HA, et al. (2004) Developmental physiology of cluster-root carboxylate synthesis and exudation in Hakea prostrata R. Br. (Proteaceae): expression of PEP-carboxylase and the alternative oxidase. Plant Physiol 135: 549–560.
13. Kambatuku JR, Cramer MD, Ward D (2011) Savanna tree-grass competition is modified by substrate type and herbivory. J Veg Sci 22: 225–237.
14. Kraaij T, Ward D (2006) Effects of rain, nitrogen, fire and grazing on tree recruitment and early survival in bush-encroached savanna, South Africa. Plant Ecol 186: 235–246.
15. Van der Waal C, De Kroon H, de Boer F, Heitkönig IMA, Skidmore AK, et al. (2009) Water and nutrients alter herbaceous competitive effects on tree seedlings in a semi-arid savanna. J Ecol 97: 430–439.
16. Ahrestani FS, Heitkönig IMA, van Langevelde F, Vaidyanathan S, Madhusudand MD, et al. (2011) Moisture and nutrients determine the distribution and richness of India's large herbivore species assemblage. Basic Appl Ecol 12: 634–642.
17. Du Toit JT, Bryant JP, Frisby K (1990) Regrowth and palatability of Acacia shoots following pruning by African savanna browsers. Ecology 71: 149–154.
18. Olff H, Ritchie ME, Prins HHT (2002) Global environmental controls of diversity in large herbivores. Nature 415: 901–04.
19. Higgins SI, Bond WJ, Trollope WSW (2000) Fire, resprouting and variability: a recipe for grass-tree coexistence in savanna. J Ecol 88: 213–229.
20. Kitajima K, Fenner M (2000) Ecology of seedling regeneration. In: Fenner M, editor. Seeds: the ecology of regeneration in plant communities. Wallingford: CAB, International. 331–360.
21. Wiegand K, Saltz D, Ward D (2006) A patch dynamics approach to savanna dynamics and bush encroachment – Insights from an arid savanna. Perspect Plant Ecol Evol Syst 7: 229–242.
22. van Langevelde F, Tomlinson KW, Barbosa ERM, de Bie S, Prins HHT, et al. (2011) Understanding tree-grass coexistence and impacts of disturbances and resource variability in savannas. In: Hill M, Hanan N editors. Ecosystem function in savannas. Boca Raton: CRC Press. 257–271.
23. Kambatuku JR, Cramer MD, Ward D (2011) Savanna tree–grass competition is modified by substrate type and herbivory. J Veg Sci 22: 225–237.
24. Ward D, Esler KJ (2011) What are the effects of substrate and grass removal on recruitment of Acacia mellifera seedlings in a semi-arid environment? Plant Ecol 212: 245–250.
25. Ward D (2005) Do we understand the causes of bush encroachment in African savannas? Afr J Range Forage Sci 22: 101–105.
26. Lehmann CER, Ratman J, Hetley LB (2009) Which of these continents is not like the other? Comparisons of tropical savanna systems: key questions and challenges. New Phytol 181: 508–511.
27. Lambers H, Chapin FS, Pons TL (2008) Plant physiological ecology. New York: Springer. 610 p.
28. Wijesinghe DK, John EA, Hutchings MJ (2005) Does pattern of soil resource heterogeneity determine plant community structure? An experimental investigation. J Ecol 93: 99–112.
29. Sankaran M, Ratnam J, Hanan NP (2008) Woody cover in African savannas: The role of resources, fire and herbivory. Glob Ecol Biogeogr 17: 236–245.
30. Van Langevelde F, Van De Vijver C, Kumar L, Van De Koppel J, De Ridder N, et al. (2003) Effects of fire and herbivory on the stability of savanna ecosystems. Ecology 84: 337–350.
31. Chapin FS, Autumn K, Pugnaire F (1993) Evolution of suites of traits in response to environmental stress. Am Nat 142: S78–S92.
32. Barbosa ERM, Van Langevelde F, Tomlinson KW, Carvalheiro LMGR, Kirkman K, et al. (2013) Tree species from different functional groups respond differently to environmental changes during establishment. Oecologia. "In press". DOI: 10.1007/s00442-013-2853-y.
33. Scholes RJ (1997) Savanna. In: Cowling RM, Richardson DM, Pierce SM, editors. Vegetation of southern Africa. Cambridge: Cambridge University Press. 258–277.
34. Venter FJ, Scholes RJ, Eckhardt HC (2003) The abiotic template and its associated vegetation pattern. In: du Toit JT, Rogers KH, Biggs HC, editors. The Kruger Experience. Washington DC: Island Press. 83–129.
35. Mucina L, Rutherford MC (2006) The Vegetation Map of South Africa, Lesotho and Swaziland. Pretoria: SANBI. 807 p.
36. Kottek M, Grieser J, Beck C, Rudolf B, Rubel F (2006) World map of the Köppen-Geiger climate classification updated. Meteorol Z 15: 259–263.
37. Bohrer G, Kagan-Zur V, Roth-Bejerano N, Ward D (2001) Effects of environmental variables on vesicular-arbuscular mycorrhizal abundance in wild populations of Vangueria infausta. J Veg Sci 12: 279–288.
38. Kambatuku JR, Cramer MD, Ward D (2012) Overlap in soil water sources of savanna woody seedlings and grasses. Ecohydrology 6: 464–473.
39. Ward D (2009) The biology of deserts. Oxford: Oxford University Press. 304 p.
40. IFPRI (International Food policy Research Institute) (2013) African Agriculture and Climate Change Country Summary – South Africa report. Johnston P, Hachigonta S, Sibanda L, Thomas M, Timothy S, editors. Washington DC: IFPRI. 38 p.
41. IPCC (Intergovernmental Panel on Climate Change) (2012) Managing the Risks of Extreme Events and Disasters to Advance Climate Change Adaptation. Field CB, Barros V, Stocker TF, Qin D, Dokken DJ, Ebi KL, et al., editors. Cambridge: Cambridge University Press. 582 p.
42. Marschner H (1995) Mineral Nutrition of Higher Plants. London: Academic Press. 889 p.
43. Prins HHT, Van Langevelde F (2008) Assembling a diet from different places. In: Prins HHT, van Langevelde F, editors. Resource Ecology: Spatial and Temporal Dynamics of Foraging. Berlin: Springer-Verlag. 129–155.
44. Kgope BS (2004) Differential photosynthetic responses of broad- and fine-leafed savanna trees to elevated temperatures. S Afr J Bot 70: 760–766.
45. Chapin FS, Kedrowski RA (1983) Seasonal changes in nitrogen and phosphorus fractions and autumn retranslocation in evergreen and deciduous taiga trees. Ecology 64: 376–391.
46. Pregitzer KS, Dickmann DI, Hendrick R, Nguyen PV (1990) Whole-tree carbon and nitrogen partitioning in young hybrid poplars. Tree Physiol 17: 79–93.
47. Paquin R, Margolis HA, Doucet R, Coyea MR (2000) Physiological responses of black spruce layers and planted seedlings to nutrient addition. Tree Physiol 20: 229–237.
48. Prior SA, Runion GB, Mitchell RJ, Rogers HH, Amthor JS (1997) Effects of atmospheric CO2 on longleaf pine: Productivity and allocation as influenced by nitrogen and water. Tree Physiol 17: 397–405.
49. Scholes RJ, Walker BH (1993) An African Savanna: Synthesis of the Nylsvley Study. Cambridge: Cambridge University Press. 306 p.
50. Rittenhouse LR, Roath LR (1987) Forage quality: primary chemistry of grasses. In: Capinera JL, editor. Integrated pest management on rangeland a shortgrass prairie perspective. Boulder: West-view. 25–37.
51. Treydte AC, Heitkonig IMA, Prins HHT, Ludwig F (2007) Trees improve grass quality for herbivores in African savannas. Perspect Plant Ecol Evol Syst 8: 197–205.
52. Tefera SB (2011) Soil seed bank dynamics in relation to land management and soil types in the semi-arid savannas of Swaziland. Afr J Agric Res 6: 2494–2505.
53. Scholes RJ, Bond W, Eckhardt H (2003) Vegetation dynamics in the Kruger ecosystem. In: du Toit J, Rogers K, Biggs H, editors. The Kruger experience. Ecology and Management of savanna heterogeneity. Washington DC: Island Press.pp 131–148.
54. Novozamsky I, Houba VJG, Eck RV, Vark VW (1983) A novel digestion technique for multi-element analysis. Commun Soil Sci Plant Nutr 14: 239–249.
55. Zuur AF, Ieno EN, Walker NJ, Saveliev AA, Smith GM (2009) Mixed Effects Models and Extensions in Ecology with R. New York: Springer-Verlag. 596 p.
56. Chapin FS (1980) The mineral nutrition of wild plants. Annu Rev Ecol Syst 11: 233–260.
57. Shipley B, Meziane D (2002) The balanced-growth hypothesis and the allometry of leaf and root biomass allocation. Funct Ecol 16: 326–331.
58. Chapin FS, Vitousek PM, Van Cleve K (1986) The Nature of Nutrient Limitation in Plant Communities. Am Nat 127: 48–58.
59. Garge BK, Burman U, Kathju S (2004) The influence of phosphorus nutrition on the physiological response of moth bean genotypes to drought. J Plant Nutr Soil Sci 167: 503–508.

60. Christensen N L, Peet RK (1981) Secondary forest succession on the North Carolina Piedmont. In: West D C, Shugart HH, Botkin DB, editors. Forest succession: concepts and applications. New York: Springer-Verlag. 230–245.

61. Tilman D (1986) Nitrogen-limited growth in plants from different successful stages. Ecology 67: 555–563.

62. Pessarakli M (2005) Handbook of photosynthesis, 2nd edition. Boca Raton: CRC Press. 928 p.

63. Portis AR (1992) Regulation of ribulose 1,5-bisphosphate carboxylase/oxygenase activity. Annu Rev Plant Physiol Plant Mol Biol. 43: 415–437.

64. Cheng L, Fuchigami LH (2000) Rubisco activation state decreases with increasing nitrogen content in apple leaves. J Exp Bot 51: 1687–1694.

65. Chapin FS, Schultz E, Mooney H (1990) The ecology and economics of storage in plants. Ann Rev Ecol Syst 21: 423–447.

66. Hu Y, Schmidhalter U (2005) Drought and salinity: A comparison of their effects on mineral nutrition of plants. J Plant Nutr Soil Sci 168: 541–549.

67. Grime JP (1979) Plant strategies and vegetation processes. New York: John Wiley and Sons. 222 p.

68. Fornara DA, Du Toit JT (2008) Responses of woody saplings exposed to chronic mammalian herbivory in an African savanna. Ecoscience 15: 129–135.

69. Sankaran M, Ratnam J, Hanan NP (2004) Tree-grass coexistence in savannas revisited - insights from an examination of assumptions and mechanisms invoked in existing models. Ecol Lett 7: 480–490.

70. Tomlinson KW, Poorter L, Sterck F, Borghetti F, Ward D, et al. (2013) Leaf adaptations of evergreen and deciduous trees of semi-arid and humid savannas on three continents. J Ecol 101: 430–440.

18

Total Nitrogen Concentrations in Surface Water of Typical Agro- and Forest Ecosystems in China, 2004-2009

Zhiwei Xu, Xinyu Zhang*, Juan Xie, Guofu Yuan, Xinzhai Tang, Xiaomin Sun, Guirui Yu

Key Laboratory of Ecosystem Network Observation and Modeling, Institute of Geographic Sciences and Natural Resources Research, Chinese Academy of Sciences, Beijing, China

Abstract

We assessed the total nitrogen (N) concentrations of 28 still surface water (lake and pond), and 42 flowing surface water (river), monitoring sites under 29 typical terrestrial ecosystems of the Chinese Ecosystem Research Network (CERN) using monitoring data collected between 2004 and 2009. The results showed that the median total N concentrations of still surface water were significantly higher in the agro- (1.5 mg·L^{-1}) and oasis agro- ecosystems (1.8 mg·L^{-1}) than in the forest ecosystems (1.0 mg·L^{-1}). This was also the case for flowing surface water, with total N concentrations of 2.4 mg·L^{-1}, 1.8 mg·L^{-1} and 0.5 mg·L^{-1} for the agro-, oasis agro- and forest ecosystems, respectively. In addition, more than 50% of the samples in agro- and oasis agro- ecosystems were seriously polluted (>1.0 mg·L^{-1}) by N. Spatial analysis showed that the total N concentrations in northern and northwestern regions were higher than those in the southern region for both still and flowing surface waters under agro- and oasis agro- ecosystems, with more than 50% of samples exceeding 1.0 mg·L^{-1} (the Class III limit of the Chinese National Quality Standards for Surface Waters) in surface water in the northern region. Nitrogen pollution in agro- ecosystems is mainly due to fertilizer applications, while the combination of fertilizer and irrigation exacerbates nitrogen pollution in oasis agro- ecosystems.

Editor: Manuel Reigosa, University of Vigo, Spain

Funding: This research was funded by the Key Direction in Knowledge Innovation Program of the Chinese Academy of Sciences (KZCX2-EW-310) and the National Natural Science Foundation of China (No. 41171153). The authors declare that no additional external funding was received for this study. The funders had no role in study design, data collection and analysis, decision to publish, or preparation of the manuscript.

Competing Interests: The authors have declared that no competing interests exist.

* E-mail: zhangxy@igsnrr.ac.cn

Introduction

Nitrogen (N) pollution, leading to eutrophication of inland waters, has resulted in an increase in global algal biomass and photosynthesis, such that primary production is approximately 60% higher than expected background levels in lakes [1], streams and rivers [2]. As a major contributor to eutrophication of water bodies, non-point losses of N (e.g. in runoff) have received particular attention. N pollution causes water eutrophication, which disrupts ecology and causes, among other problems, toxic algal blooms, loss of oxygen, loss of biodiversity (including species that are important for commerce and recreation) [3–5]. Eutrophication can also seriously affect our ability to use water for drinking, industry, agriculture, recreation, and other purposes.

Total N concentrations in surface water have been classified in China and other countries to control water eutrophication and improve water quality. To identify at-risk surface water bodies and protect them from eutrophication, the US EPA developed guidelines, which state that N concentrations should not exceed 0.3 mg·L^{-1} in streams and rivers or 0.1 mg·L^{-1} in lakes and reservoirs [6]. Water has been divided into 14 distinct aggregate nutrient ecoregions in the U.S. according to total P and total N concentrations, chlorophyll *a* and turbidity. In China surface water has been divided into five categories according to the Chinese National Quality Standards for Surface Water. Water categorized as class I to III can be used as drinking water, while class IV and V water is only suitable for industrial and agricultural uses. The total

N values for categories I – V are < 0.2 mg·L^{-1}, 0.2 – 0.5 mg·L^{-1}, 0.5 – 1.0 mg·L^{-1}, 1.0 – 1.5 mg·L^{-1} and 1.5 – 2.0 mg·L^{-1}, respectively [7].

In China, stream total N concentrations have tended to increase since the 1980s as a consequence of demographic, industrial and agricultural development [8,9]. Rivers and lakes in the Taihu Lake region are polluted to varying degrees by N [10], with 80% of samples having concentrations exceeding 1.0 mg·L^{-1}, meaning that the water is only suitable for industrial, agricultural and landscape uses according to the Chinese National Quality Standards for Surface Water [11]. Cai et al. [12] reported that eutrophication of lakes is serious in southern China, but that it is worse across a large part of northern China. Rural rivers in eastern China are severely polluted [13]. As a general rule, water pollution in China tends to intensify from tributaries to the main stems of river systems, from urban to rural areas, from surface water to groundwater and from the regional to the basin scale [12].

Natural and anthropogenic sources both contribute to surface water N inputs [14,15]. Now that point-source pollution has been controlled effectively, non-point source pollution, especially that which results from agricultural land management, has become the main influence on surface water and is an important environmental issue worldwide [16,17]. Crop production is by far the largest cause of human alteration of the global N cycle. Global industrial N fixation for fertilizers has increased rapidly to 80×10^6 Mg·year^{-1} [18]. In the United States (US) and Europe, only 18% of the N input in fertilizer is removed from farms in produce, leaving behind, on average, 174 kg·ha^{-1}·year^{-1} of surplus N [19,20]. This surplus may accumulate in soils, from where it may be either eroded and transported to surface water, leached to surface and ground water, or lost to the atmosphere [18]. Inorganic fertilizers now contribute 80 Tg N year^{-1} to the environment, while 32 – 45 Tg year^{-1} of inorganic fertilizers find their way into freshwater (from leaching and erosion) [21].

Fertilizer consumption has grown rapidly in China since 1978, and China recently became the biggest producer and consumer of N fertilizer in the world. The average annual application rate of N in China gradually increased to 130 kg·ha^{-1} in 1985, then rapidly increased to 236 kg·ha^{-1} in 1995 and then to 262 kg·ha^{-1} in 2001 [22]. However, less than half of the fertilizer N applied in China is taken up by crops [23]. The losses of total N through runoff and leaching were 5.6 – 9.0 kg·ha^{-1} from paddy fields and 12.5 – 19.5 kg·ha^{-1} from upland fields, accounting for 2.6 – 4.2% and 7.6 – 12.3% of the total applied N [24]. It is estimated that as much as 0.644 Tg of total N is leached, accounting for 2.8% of China's total N application [24].

Because of the increasing incidence of surface water N pollution, many countries are beginning to be concerned about sustainable water management. The European Union has implemented the Water Framework Directive, which aims to recover good status in water resources by 2015 [25]. Canada (both at the national and provincial levels) and the US have begun to take an effective and integrated approach to land-use management with respect to protection of drinking water sources. On a national level, Canada has established aboriginal water systems, while at the provincial level; British Columbia has established a water policy, similar to that developed by the US Environmental Protection Agency [26]. The Chinese government has published many water environmental protection measures such as the Water Pollution Prevention Program of Yangtze River basin (2011–2015) and other measures for important river basins (Yellow River, Huaihe River, Weihe River, Liaohe River, Haihe River, Songhuajiang River, Chaohu and Dianchi) [27]. Until now however, there have been few national scale assessments of N concentrations in surface water in China.

Using the Chinese Ecosystem Research Network (CERN) as the monitoring framework, total N concentrations at 28 still, and 45 flowing, surface water monitoring sites of 28 CERN monitoring stations were assessed from 2004 to 2009. The aims of this study were: (1) to assess surface water total N pollution under agro-, oasis agro- and forest ecosystems and (2) to identify sites vulnerable to N pollution in agro- and oasis agro- ecosystems. Findings from this study will provide key information that will be useful in controlling N pollution in surface water in China.

Materials and Methods

2.1 Ethics statement

The authors declare that no specific permits were required for the described field studies. The authors also declare that no specific permissions were required for these locations/activities. Locations are not privately-owned or protected in any way. We confirm that this study did not involve endangered or protected species and no protected species were sampled during the monitoring campaign.

2.2 Monitoring sites

Surface water monitoring of the CERN focuses on assessing the effects of typical terrestrial ecosystems, i.e. agro-, oasis agro- and forest ecosystems, on water quality. Twenty-eight monitoring stations of the CERN were chosen to represent typical agricultur-

al, oasis agricultural and forest ecosystems, geology, soils, and land use types found across a wide range of climatic zones in China (Figure 1). The annual average rainfall of the monitoring ecosystems (80°43′39″-133°18′03″E, 18°13′01″-47°27′15″N) varied from approximately 35 mm (Cele) to 1956 mm (Dinghushan), while the annual average temperature ranged from 1.5°C (Hailun) to 21.8°C (Xishuangbanna).

The 12 agro- ecosystems were located in (1) humid and sub-humid regions in the temperate zone of northeastern China (Hailun, Shenyang) and the warm temperate zone of northern China (Luancheng, Yucheng, Fengqiu), and (2) the Loess Plateau (Ansai and Changwu), (3) humid areas in the sub-tropical zone in the Yangtze River Delta (Changshu) and southern China (Huanjiang, Qianyanzhou, Yanting, Taoyuan,Yingtan), respectively (Figure 1). Based on geographical and climatic zones, we designated Hailun, Shenyang, Luancheng, Yucheng, Fengqiu, Ansai and Changwu agro- ecosystems as the northern group, while the Changshu, Huanjiang, Qianyanzhou, Yanting, Taoyuan and Yingtan agro- ecosystems were designated as the southern group. Land use in these agro- ecosystems was mainly crop growing, including wheat, maize, soybean, rice and cotton (Table 1).

Seven oasis agro- ecosystems (Akesu, Cele, Eerduosi, Fukang, Linze, Naiman, Shapotou) were located in the warm temperate zone of northwest and northern China in arid and semi-arid areas. The oasis agro- ecosystem is unique and found only in arid agricultural areas. Unlike agro- ecosystems in other areas represented in this study, crop cultivation depends heavily on irrigation. Land use in the oasis agro- areas was mainly cotton, maize and wheat growing (Table 1). Therefore, we classified these ecosystems as an independent group and called it oasis agro-ecosystem. We designated these seven oasis agro- ecosystems as the northwestern group.

Total N concentrations in surface water in forest ecosystems were assessed. The forest ecosystems were located along the north-south transect of eastern China (Figure 1), and were representative of native and secondary forests and were free from human fertilization and irrigation activities.

2.3 Monitoring and analysis methods

CERN surface water quality samples were collected according to the Water Monitoring Protocol of the Chinese Ecosystem Research Network [28]. 530 samples were collected from the still surface water sites, and 703 samples were collected from the 45 flowing surface water monitoring sites. The monitoring frequency ranged from 2 to 12 times per year, with sampling distributed evenly through the wet and dry seasons. The maximum sampling frequency was monthly (Ansai, Fengqiu, and Changshu), and the minimum sampling frequency was twice a year (for most of the other monitoring ecosystems), in both the dry and wet seasons. Water samples were analyzed at the Chinese Science Academy's laboratory following standard protocols and methods [28]. Total N was determined by spectrophotometry after potassium persulfate digestion. Information about crop rotation, soil type and fertiliser application rate was recorded for each sampling event at each monitoring station.

2.4 Statistical analysis

Data were analyzed with Matlab 7.11.0 (Massachusetts, USA). A Lilliefors test was conducted to test the normality of the data. Data were not normally distributed, so the nonparametric Kruskal-Wallis test was used to test for differences between the median total N concentrations for data grouped by ecosystem type (for agro-, oasis agro- and forest ecosystems) and geographical region (southern, northern and northwestern) for agro- and oasis

Figure 1. Distribution map of total nitrogen monitoring sites in agro- , oasis agro- and forest ecosystems of the Chinese Ecosystem Research Network (CERN).

agro- ecosystems. Where there were differences between data groups, the Kruskal-Wallis test was combined with a multi-comparison method in Matlab R 2010b to determine which groups were different ($p < 0.05$). We used linear regression to test for relationships between total N concentrations and soil N application rates using SPSS 19.0 for Windows. We used $p < 0.05$ as the significance level. We used Origin 8.0 software for box plots.

1.0 mg·L^{-1} (Class III limit of the Chinese National Quality Standards for Surface Waters) was used as the guideline to assess the exceedance frequency of total N concentrations at the monitoring sites in agro-, oasis agro- and forest ecosystems. Sites with high exceedance frequencies were identified as total N vulnerable zones.

Results

3.1 Total N concentrations under different ecosystems

Total N concentrations in the agro- and oasis agro- ecosystems were significantly higher than in the forest ecosystems. The typical total N concentrations (10th and 90th percentiles in box plots) in the agro-, oasis agro- and forest ecosystems ranged between 0.4 – 8.7 mg·L^{-1}, 0.7 – 15.2 mg·L^{-1}, and 0.2 – 6.6 mg·L^{-1}, respectively for still surface water (Figure 2(A)), and 0.4 – 10.9 mg·L^{-1}, 0.6 – 11.6 mg·L^{-1}, and 0.2 – 3.3 mg·L^{-1}, respectively for flowing surface water (Figure 2(B)). The median total N concentrations of still and flowing surface water in forest ecosystems were 1.1 mg·L^{-1} and 0.5 mg·L^{-1}, respectively. These concentrations were significantly lower than those of still (1.5 mg·L^{-1} and 1.8 mg·L^{-1}, respectively) and flowing surface water (2.4 mg·L^{-1} and 1.8 mg·L^{-1}, respectively) in the agro- and oasis agro-ecosystems ($p < 0.05$).

The surface water in agro- and oasis agro- ecosystems was seriously polluted by N. The median concentrations of total N under the agro- and oasis agro- ecosystems exceeded 1.0 mg·L^{-1}

[7] (Figure 2), indicating that more than 50% of surface water samples were heavily polluted by N in these ecosystems.

The total N concentrations in surface water were higher in Beijing, Huitong, Heshan and Dinghushan forest ecosystems than the other forest ecosystems. The exceedance frequencies (> 1.0 mg·L^{-1}) of flowing surface water total N concentrations in Huitong and Dinghushan were about 100%, which indicates that surface water in some forest ecosystems is eutrophic. The water quality standard was not exceeded for either still or surface water in Gongga, Ailao and Xishuangbanna (Table 2).

3.2 Total N concentrations of agro- and oasis agro- ecosystems in different regions

The surface water total N concentrations in northern and northwestern regions in China were higher than those in the southern region. The typical total N concentrations of still surface water in northern, southern and northwestern regions ranged from 0.6 – 4.3 mg·L^{-1}, 0.3 – 3.5 mg·L^{-1}, 0.6 – 15.6 mg·L^{-1} (Figure 3(A)) and 0.9 – 39.1 mg·L^{-1}, 0.3 – 4.4 mg·L^{-1}, 0.8 – 12.2 mg·L^{-1} for flowing surface water (Figure 3(B)). There were no significant differences between total N concentrations in still surface water for the different regions. However, the total N concentrations in flowing surface water in the northern region (3.8 mg·L^{-1}) were significantly higher than those in the southern (1.2 mg·L^{-1}) and northwestern regions (1.8 mg·L^{-1}) ($p < 0.05$).

Surface water in northern, southern and northwestern regions showed varying levels of N pollution, with median total N concentrations of still and flowing surface water all exceeding 1.0 mg·L^{-1} [7] (Figure 3). Results indicate that more than 50% of surface water samples were heavily polluted by N in northern agro- ecosystems, especially flowing surface water samples. The median total N concentrations for surface water in the southern and northwestern regions exceeded 1.0 mg·L^{-1} [7]. About 50% of

Table 1. TN concentrations in still and flowing surface water under agro- and oasis agro- ecosystems.

Spatial Regions	Station	Mean Precipitation (mm)	Soil type	Land use	N application rate (kg·ha⁻¹·year⁻¹)	Still surface water				Flowing surface water			
						n	Median	Max	>1.0 mg·L⁻¹ frequency(%)	n	Median	Max	>1.0 mg·L⁻¹ frequency(%)
North	Ansai	500	Loessial Soil	Soybean-Millet	120	—	—	—	—	56(1)	2.8	13.8	96
	Changwu	584	Loessial Soil	Maize-Wheat	345	24(1)	3.0	9.7	100	17(2)	1.6	3.9	65
	Fengqiu	597	Fluvo-aquic Soil	Maize- Wheat	345	—	—	—	—	57(1)	3.3	39.1	96
	Hailun	500-600	Black Soil	Wheat-Maizerotation	120	—	—	—	—	19(1)	57.0	84.8	100
	Shenyang	650-700	Aquic brown Soil	Maize	75	13(1)	0.8	9.4	23	11(1)	0.3	0.4	0
	Yucheng	582	Fluvo-aquic Soil	Maize-Wheat	510	—	—	—	—	75(2)	5.4	27.8	91
South	Changshu	1038	Red Soil	Paddy-Wheat	466	46(2)	1.9	7.9	80	96(1)	1.8	12.3	75
	Huanjing	1389	Calcareous soil	Maize-Soybean	—	5(1)	0.6	9.0	20	10(2)	0.3	10.3	20
	Qianyanhzou	1542	Red Soil	Paddy-Paddy	320	19(2)	1.0	7.3	53	9(1)	0.9	2.0	44
	Yanting	826	Purple Soil	Maize -Wheat	300	32(1)	1.7	4.5	84	—	—	—	—
	Taoyuan	1450	Red Soil	Paddy-Paddy	270	9(2)	1.0	4.4	44	20(2)	0.9	8.0	40
	Yingtan	1785	Red Soil	Peanut	150	44(2)	0.6	3.0	9	43(6)	0.8	34.8	35
Northwest	Akesu	45.7	Aeolian sandy soil	Cotton	160	15(2)	0.9	1.6	40	7(1)	0.8	1.8	43
	Cele	35	Aeolian sandy soil	Cotton-Maize rotation	468	18(1)	1.0	3.2	39	19(1)	2.8	13.4	84
	Eerduosi	348.3	Aeolian sandy soil	Cotton-Maize rotation	—	4(1)	1.0	1.2	50	21(1)	2.2	4.3	95
	Fukang	164	Aeolian sandy soil	Cotton-Maize rotation	275	9(1)	1.5	17.4	89	8(1)	1.6	91.0	63
	Linze	117	Aeolian sandy soil	Wheat-Maize rotation	122	8(2)	23.3	69.5	100	10(1)	12.2	72.0	100
	Naiman	340-450	Aeolian sandy soil	Wheat-Maize rotation	207	8(1)	1.4	8.3	75	14(2)	0.9	1.3	36
	Shapotou	180-220	Aeolian sandy soil	Wheat-Maize rotation	256	10(1)	11.5	43.0	100	—	—	—	—

Note: The "n" values represent the number of sampling sites and the monitoring sites (within brackets). "–"s illustrate that no detection data were available.

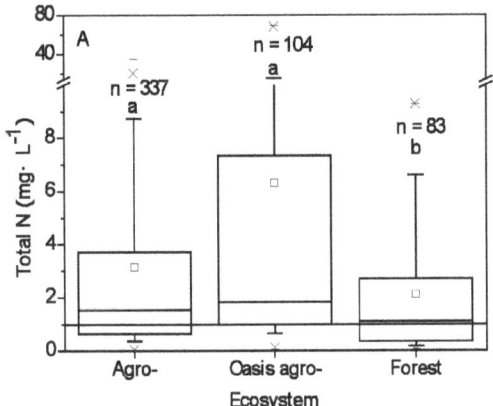 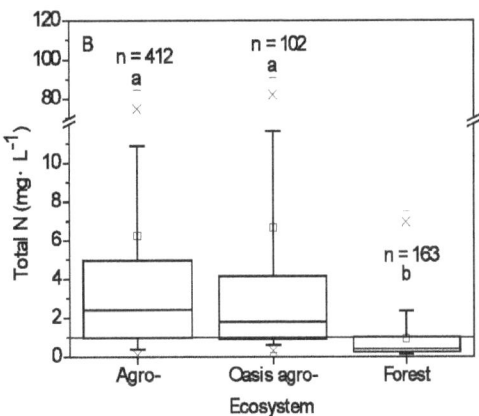

Figure 2. Total nitrogen concentrations in still surface water (A) and flowing surface water (B) of the agro-, oasis agro- and forest ecosystems between 2004 and 2009. Box plots illustrate the 25th, 50th, and 75th percentiles, the whiskers indicate the 10th, 90th percentiles, the "-"s indicate the maximum and minimum percentiles, the "×"s indicate the 1th and 99th percentiles, the "□"s indicate the mean values. Box plots labeled with different letters (a, b) indicate that differences in median values among the three ecosystem types are significant at $p < 0.05$. "—" illustrate the Class III guideline of 1.0 mg·L^{-1} for total nitrogen according to the national quality standards for surface waters of China.

surface water samples were seriously polluted by N in northwestern oasis agro- ecosystems.

The total N concentrations varied by station. The flowing surface water total N concentration was highest in the Hailun agro- ecosystem located in northern China, with a median value of 57.0 mg·L^{-1} (Table 1). The flowing surface water total N concentrations in the northern agro- ecosystems, except Shenyang, all exceeded 1.0 mg·L^{-1}. Exceedance frequencies at Ansai, Fengqiu, Hailun and Yucheng were 90 – 100%. In addition, the still surface water at Changwu was heavily polluted by N and had an exceedance frequency of 100%.

In southern agro- ecosystems, the two highest still and flowing surface water total N concentrations occurred in Changshu, the median values of which were 1.9 mg·L^{-1} (still) and 1.8 mg·L^{-1} (flowing). At this station about 80% of total N concentrations in still surface water and 75% of total N concentrations for flowing surface water exceeded 1.0 mg·L^{-1} (Table 1). The still and flowing surface water total N concentrations of other stations (Huanjiang, Qianyanzhou, Taoyuan, Yingtan) were about 1.0 mg·L^{-1}, and the exceedance frequencies varied from 20% to 50%.

The total N concentrations in the northwestern oasis agro-ecosystems were highest in Linze and Shapotou. The still and flowing surface water median total N concentrations were 23.3 mg·L^{-1} and 12.2 mg·L^{-1}, respectively, in Linze, while for still surface water in Shapotou, the median total N concentration was 11.5 mg·L^{-1}. The total N concentrations all exceeded 1.0 mg·L^{-1} in these two stations, indicating that the surface water was heavily polluted by N (Table 1).

3.3 Correlations between total N concentrations and N fertilizer application rates in agro- and oasis agro- ecosystems

Correlation analysis between the surface water total N concentrations and N application rates showed that the flowing surface water total N concentrations and N application rates were significantly correlated ($r^2 = 0.415$, $p = 0.009$) (Figure 4). Results indicated that total N concentrations tended to increase as N application rates increased. While there was a similar trend for still surface water and N application rates, the correlation was not significant ($r^2 = 0.225$, $p = 0.107$).

Discussion

4.1 Surface water nitrogen pollution in national and international

The median surface water total N concentrations under agro- and oasis agro- ecosystems of the CERN were lower than those in the Tai Lake region (6.4 mg·L^{-1}) [29], but similar to those in surface water in northern China [30,31]. Compared with studies in other countries, the total N concentrations for this study were higher than those of surface water for a Japanese agro- ecosystem [32], but similar to the total N concentrations of the Calapooia River Basin in a Western Oregon agro- ecosystem, where they ranged from 0.5 to 43 mg·L^{-1}[33]. Generally, total N concentrations in northern and northwestern regions were higher than in the southern region of China. A spatial assessment of lakes in China from 1990 to 2010 showed that total N concentrations decreased with rising latitude, but were not related to longitude [12]. Different farming practices, irrigation practices and crop rotations may influence N leaching [34]. For instance, total N losses due to leaching were 9.875 kg·ha^{-1} in the wheat growing season, but were only 1.868 kg N·ha^{-1} in the rice growing season [29]. Li et al [35]. Reported that nitrate-N concentrations in surface water from rice growing land were significantly higher than those in corn land. In addition, the dilution effect of precipitation could decrease the total N concentrations in the southern region, as precipitation is much higher than in the northern region.

The total N concentrations in surface water under the forest ecosystems were lower than in agro- and oasis agro- ecosystems. This result is similar to what was observed for surface water total phosphorus concentrations in the same study area [36]. Larned et al. [37] also indicated that dissolved N concentrations in pastoral and urban ecosystems were 2 – 7 times higher than in native and plantation forest ecosystems. Total N concentrations were higher in Chinese typical forest ecosystems than in surface water in Japanese forest ecosystems, where they ranged from 0.01 to 1.3, with a median value of 0.14 mg·L^{-1} [38]. In the US, total N concentrations in surface water in northern Californian national forests ranged from 0.03 to 0.1 mg·L^{-1} [39]; these concentrations are also lower than those in China.

Table 2. TN concentrations and background information of still and flowing surface water under the 9 forest ecosystems.

Ecotype	Station	Altitude (m)	Mean Precipitation (mm)	Soil type	Vegetation	Still surface water				Flowing surface water			
						n	Median	Max	>1.0 mg·L⁻¹ frequency (%)	n	Median	Max	>1.0 mg·L⁻¹ frequency (%)
Humid, Sub-humid Areas in Temperate Zone	Changbai	740	695	Dark Brown soil	Broad-leaved korean pine forest	—	—	—	—	9(1)	1.0	1.4	44
Humid, Sub-humid Areas in Warm Temperate Zone	Beijing	1248	612	Mountain brown soils	Man-made Pinus tabulaeformis Forests	—	—	—	—	9(1)	1.6	5.7	56
Humid Areas in North Sub-tropical Zone	Maoxian	1826	825	Cinnamon soil	Warm temperate coniferous forest	—	—	—	—	8(1)	0.7	1.0	13
	Gongga	2950	1974	Podzolic brown taiga soils	Sulbalpine dark coniferous forest	—	—	—	—	54(7)	0.3	0.8	0
	Huitong	541	1079	Yellow soil	Broad-leaved Trees Mixed Forests	11(1)	1.5	3.5	100	13(1)	1.8	3.7	92
Humid Areas in South Sub-tropical Zone	Ailao	2481	1880	Yellow Brown soil	Subtropical mid mountain humid evergreen broadleaved forest	18(1)	0.3	0.9	0	19(1)	0.2	0.5	0
	Heshan	90	1927	Ferrisols	Acacia mangium pure forests	12(1)	0.8	3.3	42	19(1)	1.6	7.5	63
	Dinghu	90	1700	Lateritic Red Soil	Broad-leaved Trees Mixed Forests	30(3)	5.2	9.3	100	10(1)	1.3	7.0	70
Humid Areas in tropical Zone	Banna	560	1539	Red soil	Tropical seasonal rain forest	12(1)	0.3	0.5	0	23(1)	0.2	0.9	0

Note: The "n" values represent the number of sampling sites and the monitoring sites (within brackets). "—"s illustrate that no detection data were available.

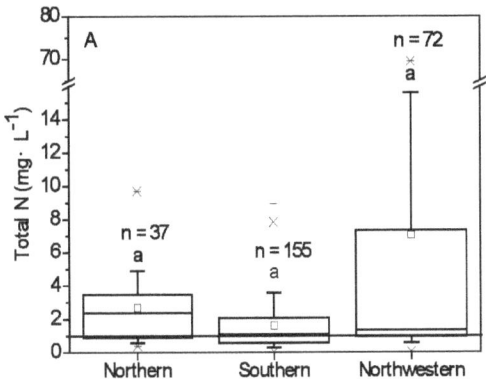 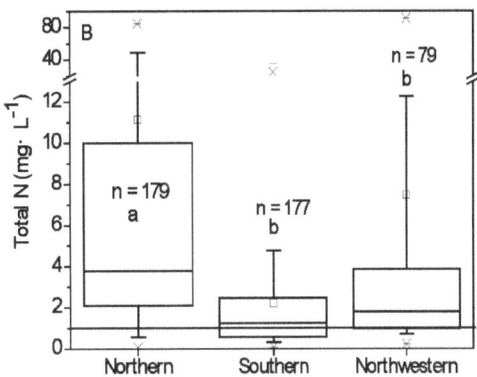

Figure 3. Total nitrogen concentrations of still surface water (A) and flowing surface water (B) under agro- and oasis agro-ecosystems in northern, southern and northwestern areas. The illustrations about Box plots are the same as those in figure 2.

Increasing N concentrations have been observed in many watersheds in the US, Europe, New Zealand and Canada because of land use change, atmospheric deposition, fertilizer application and burning fossil fuels [40–43]. Total N at the Konza Prairie Biological Station (Konza), located in the Flint Hills region of the Great Plains in Kansas, increased from 0.4 mg·L^{-1} to 1.2 mg·L^{-1} due to land use change [44]. There are concerns about N exports from streams in the US midwest because of excessive nutrient enrichment and eutrophication [45,46]. In agro- ecosystems in Baltimore, the annual total N exports were about 30 kg N ha^{-1} during 2005 [47]. The annual total N export rates from the world's rivers in different geographical areas between 76 °N and 43 °S ranged from 1 – 20 630 kg N ha^{-1} year^{-1} [48].

4.2 Factors influencing TN in surface water

Climate, hydrology, soil properties, geomorphology, topography, soil cover and land use are the main factors that influence nutrients in stream water [49-51]. As we were restricted by the sampling frequency, we did not explore relationships between mean annual precipitation and total N in surface waters. However, a significant inverse relationship between total N concentrations and precipitation was reported by a study which examined seasonal variation of total N in the Beijing urban ecosystem [52].

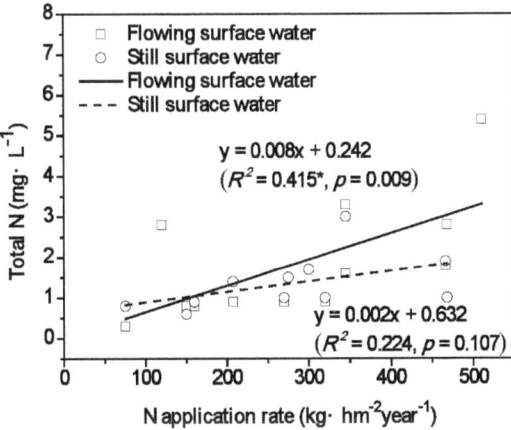

Figure 4. Correlations of total nitrogen concentrations and soil N fertilizer application rates in agro- and oasis agro- ecosystems ($p < 0.05$, n = 15). The total nitrogen concentrations in this figure were the median values of agro- and oasis agro- ecosystems.

Land use change can influence N concentrations. The N concentrations in rural residential areas and cultivated land were much higher than those in grass and forest ecosystems [53,54]. Kvítek et al. [55] reported that, as ploughed land in the catchment increased, nitrate contamination of surface water also increased. Studies of N pollution in the US and the Netherlands showed that 60% – 80% of N came from agricultural non-point sources [56–58]. Alvarez-Cobelas [48] also reported that exports of total N were four times higher from catchments dominated by crops than from forested catchments.

Nitrogen fertilizer, as a non-point source pollutant, is a critical source of pollution in agro- and oasis agro- ecosystems. Surface water N pollution in agro- ecosystems was closely related to agricultural activity. China surpassed the US and the European Union in its production and use of N fertilizers in approximately 2000. However, less than half of the fertilizer N applied in China was taken up by crops [23]. The rest was largely lost to the environment in gaseous (NH_3, NO, N_2O and N_2) or dissolved (NH_4^+ and NO_3^-) forms [59,60]. Leaching is the most important pathway for water eutrophication and poor N fertilizer management could lead to N leaching into the drainage water through run off and drainage [61,62]. In spite of considerable controls on point source pollution, water quality standards have not reached the criterion of the US and European countries due to the increasing contribution of non-point pollution [63]. In general, the waters under the agro- ecosystems have been widely polluted by total N, while only been polluted by total P in a few agro- ecosystem sites [36]. Therefore, the surface water eutrophication under the agro- ecosystems was P limited, which indicated that N fertilizer should generally be reduced and P fertilizer should only be controlled in the P polluted areas of the agro-ecosystems.

Soil textures differ according to location; therefore, losses of N fertilizer through leaching, drainage and runoff are also variable [64]. The relatively high total N concentrations in surface water in northwestern regions may be attributed to soil types. Studies in France have shown that leached N varied from 31 mg·L^{-1} in deep loamy soils to 92 mg·L^{-1} in shallow sandy soils [34]. Soils in the northwestern region of China are mainly aeolian sandy soils, and have low water-holding capacity, meaning that N fertilizer is likely to be lost from soil to surface water. Combined with frequent irrigation, serious soil losses in this region may contribute to the high total N concentrations [65].

Atmospheric N deposition may be an important source of surface water N [66]. While its effects may be weaker than other factors [69], it may still aggravate water eutrophication and affect ecosystem stability [67,68]. The 9 forest ecosystems in our study showed eutrophication tendencies, especially the Beijing, Huitong, Heshan and Dinghushan forest ecosystems. The reactive nitrogen (N_r) species emitted during fossil fuel combustion, have resulted in some of the most pronounced air pollution ever recorded in China, and the increased N_r emissions may have influenced atmospheric N deposition near the study areas. Research has shown that N deposition in Dinghushan and Beijing exceeds 30 kg·ha^{-1} year^{-1}. These rates are higher than those recorded for the other forest ecosystems in China [70–72]. Therefore, the higher atmospheric N deposition may contribute to the higher total N concentrations in the Beijing and Dinghu forest ecosystems.

Conclusions

The total N concentrations of surface water under agro- and oasis agro- ecosystems were much higher than those under forest ecosystems. About 50% of median total N concentrations exceeded 1.0 mg·L^{-1}. There was an obvious spatial pattern in surface water total N concentrations, with much higher total N concentrations in the northern region of China, especially in flowing surface water. The surface water of some agro- ecosystems (Ansai, Changwu, Fengqiu, Hailun) and oasis agro- ecosystems (Cele, Linze, Eerduosi) were severely polluted by N with exceedance (> 1.0 mg·L^{-1}) frequencies greater than 50%. Fertilizer applications were the main source of N pollution in the flowing surface water in agro- and oasis agro- ecosystems.

Acknowledgments

We gratefully acknowledge the many Chinese Ecosystem Research Network (CERN) personnel, who designed the monitoring networks and collected and analyzed the samples used in this study.

Author Contributions

Conceived and designed the experiments: ZX XZ XS GY. Performed the experiments: ZX XZ XS. Analyzed the data: ZX XZ JX. Contributed reagents/materials/analysis tools: JX. Wrote the paper: ZX XZ JX GY XT XS GY.

References

1. Lewis WM Jr (2011) Global primary production of lakes: 19th Baldi Memorial Lecture. Inland Waters 1(1): 1–28.
2. OECD (1982) Eutrophication of Waters-monitoring, Assessment and Control; Paris, France: Organization for Economic Co-operation and Development.
3. Tian YH, Yin B, Yang LZ, Yin SX, Zhu ZL (2007) Nitrogen runoff and leaching losses during rice-wheat rotations in Taihu Lake region, China. Pedosphere 17(4): 445–456.
4. Lassaletta L, García-Gómez H, Gimeno BS, Rovira JV (2009) Agriculture-induced increase in nitrate concentrations in stream waters of a large Mediterranean catchment over 25 years (1981–2005). Science of the Total Environment 407(23): 6034–6043.
5. Mian IA, Begum SM, Ridealg M, McClean CJ, Cresser MS (2010) Spatial and temporal trends in nitrate concentration in the river Derwent, North Yorkshire and its need for NVZ status. Science of the Total Environment 408(4):702–712.
6. US EPA (2002) National Recommended Water Quality Criteria: 2002. Office of Water, EPA-822-R-02-047, U.S. Environmental Protection Agency, Washington DC. http://www.epa.gov/waterscience/standards/wqcriteria.html.
7. China Ministry of Environmental Protection, China General Administration of Quality Supervision and Quarantine (2002) Environmental Quality Standards for Surface Water. GB3838-2002.
8. Yang AL, Zhu YM (1999) Study on nonpoint source pollution in surface water environment. Advance of Environmental Sciences 7(5):60–67.
9. Bao QS, Wang HD (1996) Perspective on water environment and nonpoint source pollution in China. Geographical Sciences 16(1): 66–71.
10. Xu H, Liu ZP, Jiao JG, Yang LZ (2008) Nitrogen pollution status of various types of passing-by water bodies in upper reaches of Taihu Lake. Chinese Journal of Ecology, 27(1): 43–49.
11. Liu ZD, Yu XG, Wang ZX (2003) The current water pollution of Taihu drainage basin and the new management proposals. Journal of Natural Resources 16(4): 467–474.
12. Cai LY, Li Y, Zheng ZH (2010) Temporal and spatial distribution of nitrogen and phosphorus of lake systems in China and their impact on eutrophication. Earth and Environment 38 (2):235–241.
13. Wang LX, Lü JL, Zhuang SY, Hu ZY (2009) Characteristics of phosphorus adsorption on paddy soil and river sediment in east China. Soils, 41(3):402–407.
14. He B, Oki K, Wang Y, Oki T (2009b) Using remotely sensed imagery to estimate potential annual pollutant loads in river basins. Water Science and Technology 60 (8): 2009–2015.
15. He B, Oki T, Kanae S, Mouri G, Kodama K, et al. (2009a) Integrated biogeochemical modelling of nitrogen load from anthropogenic and natural sources in Japan. Ecological Modelling 220(18): 2325–2334.
16. Galloway JN, Cowling EB (2002) Reactive nitrogen and the world: 200 years of change. Ambio 31(2): 64–71.
17. Galloway JN, Schlesinger WH, Levy II H, Michaels A, Schnoor JL (1995) Nitrogen fixation: anthropogenic enhancement-environmental response. Global Biogeochemical Cycles 9(2): 235–252.
18. Vitousek PM, Aber JD, Howarth RW, Likens GE, Matson PA, et al. (1997) Human alteration of the global nitrogen cycle: Sources and consequences. Ecological Applications 7(3):737–750.
19. Isermann K (1991) Share of agriculture in nitrogen and phosphorus emissions into the surface waters of Western Europe against the background of their eutrophication. Fertilizer Research 26(1–3):253–269.
20. National Research Council (1993b) Soil and water quality: an agenda for agriculture. National Academy Press, Washington, D.C., USA.
21. Smil V (1999) Nitrogen in crop production: An account of global flows. Global Biogeochemistry Cycles 13(2): 647–662.
22. Zhang D, Li G, Yang YS, Zhang X, Guo H (2009) Bio-geological processes of nitrogen transport and transformation in the aeration zone and aquifer. Hydrological Sciences Journal 54(2): 316–326.
23. Zhang FS, Wang YQ, Zhang WF, Cui ZL, Ma WQ, et al. (2008) Nutrient use efficiencies of major cereal crops in China and measures for improvement. Acta Pedologica Sinica 45(5):915–924.
24. Hu YT, Liao Qian JH, Wang SW, Yan XY (2011) Statistical analysis and estimation of N leaching from agricultural fields in China. Soils 43 (1): 19–25.
25. Letcher RA, Giupponi C (2005) Policies and tools for sustainable water management in the European Union. Environmental Modeling & Software 20(2):93–98.
26. Davies JM, Mazumder A (2003) Health and environmental policy issues in Canada: the role of watershed management in sustaining clean drinking water quality at surface sources. Journal of Environmental Management 68(3): 273–286.
27. The Central People's Government of the People's Republic of China (2012) Water pollution prevention program of Yangtze River basin (2011-2015). Gazette of the State Council of the People's Republic of China 11:51–60.
28. Yuan GF, Tang DY, Sun XM (2007) Water Monitoring Protocol of Chinese Ecosystem Research Network. Beijing: China Environmental Science Press.
29. Xie YX, Xiong ZQ, Xing GX, Sun GQ, Zhu ZL (2007) Assessment of nitrogen pollutant sources in surface waters of Taihu Lake region. Pedosphere 17(2): 200–208.
30. Chen LD, Peng HJ, Fu BJ, Qiu J, Zhang SR (2005) Seasonal variation of nitrogen-concentration in the surface water and its relationship with land use in a catchment of northern China. Journal of Environmental Science 17(2): 224–231.
31. Meng LY, Shi MW, Yan XX, Lu H, Zhu XL (2009) Evaluation of Nonpoint Source Pollution on Nitrogen and Phosphorus of Surface Water of Hutuo River Valley under Background State. Geography and Geo-information Science 25(3): 95–98.
32. Mouri G, Takizawa S, Oki T (2011) Spatial and temporal variation in nutrient parameters in stream water in a rural-urban catchment, Shikoku, Japan: Effects of land cover and human impact. Journal of Environmental Management 92(7):1837–1848.
33. Mueller-Warrant GW, Griffith SM, Whittaker GW, Banowetz GM, Pfender WF, et al. (2012) Impact of land use patterns and agricultural practices on water quality in the Calapooia River Basin of western Oregon. Journal of Soil and Water Conservation 67 (3): 183–201.
34. Beaudoin N, Saad JK, Laethem CV, Machet JM, Maucorps J, et al. (2005) Nitrate leaching in intensive agriculture in Northern France: Effect of farming practices, soils and crop rotations. Agriculture, Ecosystems and Environment 111(1–4): 292–310.
35. Li J, Lu WX, Zeng XK, Yuan JH, Yu FR (2010) Analysis of spatial-temporal distributions of nitrate-N concentration in Shitoukoumen catchment in northeast China. Environmental Monitoring and Assessment 169(1–4): 335–345.
36. Xie J, Zhang XY, Xu ZW, Sun XM, Ballantine D (2013) Total Phosphorus concentrations in surface water of Typical Chinese Ecosystems, 2004–2010. Frontiers of Environmental Science & Engineering, DOI: 10.1007/s11783-013-0601-5.
37. Larned ST, Scarsbrook MR, Snelder TH, Norton NJ, Biggs BJF (2004) Water quality in low-elevation streams and rivers of New Zealand: recent state and

trends in contrasting land-cover classes. New Zealand Journal of Marine and Freshwater Research 38(2): 347–366.

38. Zhang Z, Fukushima T, Shi PJ, Tao FL, Onda Y, et al. (2008) Baseflow concentrations of nitrogen in forested headwaters in Japan. Science of the Total Environment 402(1): 113–122.

39. Roche LM, Kromschroeder L, Atwill ER, Dahlgren RA, Tate KW (2013) Water quality conditions associated with cattle grazing and recreation on national forest lands. PLoS ONE 8(6): e68127.

40. Stoddard JL (1994) Long-term changes in watershed retention of nitrogen: Its causes and aquatic consequences. Environmental Chemistry of Lakes and Reservoirs 237: 223–284.

41. Boesch DF, Brinsfield RB, Magnien RE (2001) Chesapeake Bay eutrophication: Scientific understanding, ecosystem restoration, and challenges for agriculture. Journal of Environment Quality 30(2): 303–320.

42. Dayyani S, Prasher SO, Madani A, Madramootoo CA (2012) Impact of climate change on the hydrology and nitrogen pollution in a tile-drained agricultural watershed in eastern Canada. Transactions of the Asabe 55(2): 389–401.

43. Duggan IC, Collier KJ, Champion PD, Croker GF, Davies-Colley RJ, et al. (2002) Ecoregional differences in macrophyte and macroinvertebrate communities between Westland and Waikato: are all New Zealand lowland streams the same? New Zealand Journal of Marine and Freshwater Research 36(4): 831–845.

44. Kemp MJ, Dodds WK (2001) Spatial and temporal patterns of nitrogen concentrations in pristine and agriculturally-influenced prairie streams. Biogeochemistry 53(2): 125–141.

45. Dodds WK, Welch EB (2000) Establishing nutrient criteria in streams. Journal of the North American Benthological Society 19(1): 186–196.

46. US Environmental Protection Agency (USEPA) (2000) US Environmental Protection Agency (USEPA) Nutrient criteria technical guidance manual: rivers and streams, EPA-822-B-00-002. US Government Printing Office, Washington, DC.

47. Kaushal SS, Groffman PM, Band LE, Elliott EM, Shields CA, et al. (2011) Tracking nonpoint source nitrogen pollution in human-impacted watersheds. Environmental Science & Technology 45(19): 8225–8232.

48. Alvarez-Cobelas M, Angeler DG, Sánchez-Carrillo S (2008) Export of nitrogen from catchments: A worldwide analysis. Environmental Pollution 156(2): 261–269.

49. Ometto JPHB, Martinelli LA, Ballester MV, Gessner A, Krusche AV, et al. (2000) Effects of land use on water chemistry and macroinvertebrates in two streams of Piracicaba river basin, southeast Brazil. Freshwater Biology 44(2): 327–337.

50. McKee LJ, Eyre BD, Hossain S, Pepperell PR (2001) Influence of climate, geology and humans on spatial and temporal nutrient geochemistry in the subtropical Richmond River catchment, Australia. Marine and Freshwater Research 52(2): 235–248.

51. Likens GE, Buso, DC (2006) Variation in streamwater chemistry throughout the Hubbard Brook Valley. Biogeochemistry 78(1): 1–30.

52. Ren YF, Xu ZW, Zhang XY, Wang XK, Sun XM, et al. (2013) Nitrogen pollution and source identification of urban ecosystem surface water in Beijing. Frontiers of Environmental Science & Engineering, DOI: 10.1007/s11783-012-0474-z.

53. Zhu B, Wang T, Xu TP, Kuang FH, Luo ZX, et al. (2006) Nonpoint-source nitrogen movement and its environmental effects in a small watershed in hilly area of purple soil. Journal of Mountain Science 24(5):601–606.

54. Zhang XY, Xu ZW, Sun XM, Dong WY, Ballantine DJ (2013) Nitrate in shallow groundwater in typical agricultural and forest ecosystems in China, 2004–2010. Journal of Environmental Sciences 25(5): 1007–1014.

55. Kvítek T, Žlábek P, Bystřický V, Fučík P, Lexa M, et al. (2009) Changes of nitrate concentrations in surface waters influenced by land use in the crystalline complex of the Czech Republic. Physics and Chemistry of the Earth 34(8–9):541–551.

56. Boer PM (1996) Nutrient Emissions from agriculture in the Netherlands: causes and remedies. Water Science Technology 33(4–5): 183–189.

57. Kronvang B, Graesbøll P, Larsen SE, Svendsen LM, Andersen HE (1996) Diffuse nutrient losses in Denmark. Water Science technology 33(4–5):81–88.

58. Shepard R (2000) Nitrogen and phosphorus management on Wisconsin farms: Lessons learned for agricultural water quality programs. Journal of Soil and Watershed Conservation 55(1): 63–68.

59. Zhu ZL, Chen DL (2002). Nitrogen fertilizer use in China: contributions to food production, impacts on the environment and best management strategies. Nutrient Cycling in Agroecosystems 63(2–3): 117–127.

60. Ju XT, Xing GX, Chen XP, Zhang SL, Zhang LJ, et al. (2009). Reducing environmental risk by improving N management in intensive Chinese agricultural systems. Proceedings of the National Academy of Sciences of the United States of America 106(9): 3041–3046.

61. Barton L, Colmer TD (2006) Irrigation and fertilizer strategies for minimising nitrogen leaching from turfgrass. Agricultural Water Management 80(1–3):160–175.

62. Cang HJ, Xu LF, Li Z, Ren H (2004) Nitrogen losses from farmland and agricultural nonpoint source pollution. Tropical Geography, 24: 332–336.

63. Novotny V (2005) Diffuse pollution from agriculture in the world. In: Proceedings from the International Workshop on Where Do Fertilizers Go? Belgirate, Italy.

64. Nie SW, Gao W, Chen YQ, Sui P, Eneji AE (2009) Review of current status and research approaches to nitrogen pollution in farmlands. Agricultural Sciences in China 8(7):843–849.

65. Zhao SZ (2003) Ecological environmental problems resulting from exploitation of water resources in Northwest China. Northwestern Geology 36(3):92–96.

66. Krusche AV, de Camargo PB, Cerri CE, Ballester MV, Lara LBLS, et al. (2003). Acid rain and nitrogen deposition in a sub-tropical watershed (Piracicaba): ecosystem consequences. Environmental Pollution 121(3): 389–399.

67. Schindler DW (1988) Effects of acid rain on freshwater ecosystems. Science 239(4836): 149–157.

68. Bergström A, Jansson M (2006) Atmospheric nitrogen deposition has caused nitrogen enrichment and eutrophication of lakes in the northern hemisphere. Global Change Biology 12(4): 635–643.

69. Gorham E (1998). Acid deposition and its ecological effects: a brief history of research. Environmental Science & Policy 1(3): 153–166.

70. Huang ZL, Ding MM, Zhang ZP, Yi WM (1994) The hydrological processes and nitrogen dynamics in a monsoon evergreen broad-leafed forest of Dinghushan. Acta Phtoecologica Sinica 18(2): 194–199.

71. Zhou GY, Yan JH (2001) The influence of region atmospheric precipitation characteristics and its element inputs on the existence and development of Dinghushan forest ecosystems. Acta Ecologica Sinica 21(12): 2002–2012.

72. Zhang Y, Liu XJ, Zhang FS, Ju XT, Zhou GY, et al. (2006) Spatial and temporal variation of atmospheric nitrogen deposition in North China Plain. Acta Ecologica Sinica 26(6): 1633–1638.

19

Regime Shift in Fertilizer Commodities Indicates More Turbulence Ahead for Food Security

James J. Elser[1]*, Timothy J. Elser[2], Stephen R. Carpenter[3], William A. Brock[4,5]

1 School of Life Sciences, Arizona State University, Tempe, Arizona, United States of America, **2** Flyr, Inc., San Francisco, California, United States of America, **3** Center for Limnology, University of Wisconsin, Madison, Wisconsin, United States of America, **4** Department of Economics, University of Wisconsin, Madison, Wisconsin, United States of America, **5** Department of Economics, University of Missouri, Columbia, Missouri, United States of America

Abstract

Recent human population increase has been enabled by a massive expansion of global agricultural production. A key component of this "Green Revolution" has been application of inorganic fertilizers to produce and maintain high crop yields. However, the long-term sustainability of these practices is unclear given the eutrophying effects of fertilizer runoff as well as the reliance of fertilizer production on finite non-renewable resources such as mined phosphate- and potassium-bearing rocks. Indeed, recent volatility in food and agricultural commodity prices, especially phosphate fertilizer, has raised concerns about emerging constraints on fertilizer production with consequences for its affordability in the developing world. We examined 30 years of monthly prices of fertilizer commodities (phosphate rock, urea, and potassium) for comparison with three food commodities (maize, wheat, and rice) and three non-agricultural commodities (gold, nickel, and petroleum). Here we show that all commodity prices, except gold, had significant change points between 2007–2009, but the fertilizer commodities, and especially phosphate rock, showed multiple symptoms of nonlinear critical transitions. In contrast to fertilizers and to rice, maize and wheat prices did not show significant signs of nonlinear dynamics. From these results we infer a recent emergence of a scarcity price in global fertilizer markets, a result signaling a new high price regime for these essential agricultural inputs. Such a regime will challenge on-going efforts to establish global food security but may also prompt fertilizer use practices and nutrient recovery strategies that reduce eutrophication.

Editor: Luis Herrera-Estrella, Centro de Investigación y de Estudios Avanzados del IPN, Mexico

Funding: J.J.E. is grateful for support from the Arizona State University Sustainable Phosphorus Initiative and the US National Science Foundation (NSF). S.R.C. acknowledges support of the NSF from North Temperate Lakes LTER and the Water Sustainability and Climate program. W.B. is supported by the Vilas Trust and the NSF Center for Robust Decision Making on Climate Change and Energy Policy. The funders had no role in study design, data collection and analysis, decision to publish, or preparation of the manuscript.

Competing Interests: The authors have declared that no competing interests exist.

* E-mail: j.elser@asu.edu

Introduction

The human population has more than doubled during the past fifty years, during which time *per capita* food availability has nevertheless increased [1]. This was made possible by the "Green Revolution", the large-scale expansion and intensification of agricultural activity that included extension of cultivated lands, development of high yield crop varieties, increased irrigation, and heavy application of inorganic fertilizers to supply nitrogen (N), phosphorus (P), and potassium (K). However, many concerns about the present and future sustainability of these activities have been raised [2]; in this paper we focus on dimensions related to inorganic fertilizers [3]. Large amounts of N and P fertilizers are used in agriculture each year, with much of the added N and P being lost from farms and livestock operations and leading to widespread degradation of the quality of fresh waters as well as damage to coastal marine ecosystems [4]. Furthermore, concerns about the continued availability and affordability of inorganic fertilizers, especially those based on P, have recently been raised following the several-fold increase in the price of phosphate rock and fertilizers in 2007 and 2008 [5]. Such price increases are of particular importance because small conventional stakeholder farmers in developing countries often lack the financial resources

to buffer such increases in needed input commodities, such as fertilizers. As these farmers are a significant proportion of the global undernourished population, increases in inorganic fertilizer prices can diminish important components of food security [5]. Reserve estimates for phosphate rock have recently been revised substantially upward [6]. Furthermore, there is always potential for adaptive responses of technology (such as nutrient recovery or crop biotechnology approaches) and market systems to emerging geological scarcity [7–9]. Nevertheless, converging trends suggest that the fertilizer/food system will come under increased pressure in coming decades. First, to assure the food security of the global population in 2050, food production may need to double [10]. Second, growing worldwide affluence means that global diets will include increasing amounts of meat [11]; meat-intensive diets require disproportionately more nutrients (N, P) to produce [12]. Third, non-food uses of nutrient elements (e.g. N and P to produce biofuels [13]; P to produce lithium-phosphate batteries) are increasing dramatically and are beginning to impose novel market pressures on the fertilizer sector. While various approaches for diversifying the sources of P for fertilizer production are under investigation (e.g. P recycling from human waste and other waste streams; [14]), the time-scales of development and adoption of these strategies are not yet clear and thus concern remains about scenarios for future global P dynamics [15].

Because of the essential nature of fertilizers in supporting the food system, it is important to gain insight into overall trends and patterns of variability in fertilizer prices. This is an especially challenging task given how deeply N and P are embedded within an increasingly complex and globalized socioeconomic ecosystem. One way to achieve such insights is to examine time series for significant breakpoints in temporal trends as well as for symptoms of possible "critical transitions" that accompany, and sometimes presage, regime shifts in diverse domains from ecology (e.g.: fishery collapse) to medicine (e.g.: heart attack) [16,17] Regime shifts may be brought about by external perturbations but many of them can be explained by critical transitions in system dynamics [18,19]. In economic analyses, basic theory of asset markets (called the Efficient Market Hypothesis, EMH) argues that strong market forces make such price changes roughly unpredictable over time [20]. EMH theory does not rule out rapid up bursts or rapid declines nor does it limit volatility bursts or persistence in volatility; indeed, a posteriori analyses may detect significant change points. Here we focus on such change points and statistical symptoms of critical transitions near change points and provide the first such application of these approaches to time series related to agricultural input commodities (fertilizers) and crop commodities.

Our primary research goals were two-fold: first, to identify possible breakpoints in the data series for agricultural fertilizers that mark significant changes in market conditions and, second, to assess the possibility that these commodities are affected by nonlinear system dynamics and thus undergoing critical transitions. To do this we apply time series analytical methods [21] to phosphate rock price data along with data for two other fertilizer commodities, potassium (K) and urea. To assess if changes in fertilizer dynamics might be driven by associated regime shifts in key crops, we also used similar methods to evaluate the price time series for three major crop commodities: maize, rice, and wheat. To see if fertilizer dynamics might be associated with dynamics of energy prices we analyzed the time series for petroleum. We also considered the dynamics of nickel, as an indicator of non-agricultural volatility patterns, and of gold, as an indicator of overall financial volatility in the global system. Other investigators have previously considered volatility patterns in commodity prices with special attention on cross-sector price transmission [22–24]. However, our analysis focuses less on issues of transmission and elasticity (e.g. how do changes in fertilizer prices manifest in changes in food prices, or vice versa?) but more on issues related to characterizing the qualitative nature of underlying dynamics in the systems that underpin key commodities (e.g. linear vs nonlinear dynamics) and to identifying potential breakpoints in those dynamics, as these may signal regime shifts in those commodities. Our approach is mainly descriptive and assessments of all of the complex mechanistic inter-relations among these series await future analyses.

Methods

We analyzed several decades of monthly price time series for (i) three primary components of industrial fertilizer, phosphate rock, potassium (K), and urea (an organic molecule rich in N); (ii) three food commodities, maize, rice, and wheat; and (iii) three non-agricultural commodities, petroleum, gold, and nickel. The non-agricultural commodities were chosen in order to evaluate if possible changes identified in the agricultural realm were simply mirroring those ongoing in the energy (petroleum), financial (gold), or general industrial (nickel) sectors. Our analysis focuses on raw

input commodities (e.g. phosphate rock) instead of finished products (e.g. tri-, di-, sodium, potassium phosphate fertilizers) because we wished to more easily make comparisons among input commodities in a way that was less likely to be influenced by indirect effects that are felt for finished products (e.g. costs of transport, marketing, rents, overhead, etc). For each time series, we fit an autoregressive change point model to the series, filtered residuals using a general autoregressive conditional heteroskedasticity (GARCH) model, and subjected GARCH residuals to a test for linearity [27]. We also computed rolling-window variance and autocorrelation, indices that are commonly used as indicators of regime shifts [21,28]. Further details are provided in online Supplemental Information.

Monthly data for each commodity were obtained from the World Bank (data.worldbank.org/data-catalog/commodity-price-data). All series were adjusted for inflation using the US consumer price index time series for all urban consumers (downloaded 21 May 2013). Based on normal probability plots, time series were $\log_{10}$-transformed prior to analysis. All time series were subjected to 36-month rolling window calculations of variance and autocorrelation [21,27,28]. Autocorrelation statistics were transformed to autocorrelation time, $-1/\ln(AC)$ [29].

Critical transitions imply the operation of nonlinear or non-stationary mechanisms in the stochastic dynamical system, i.e. stochastic process, that underpins a time series [27]. In this paper, we define stochastic processes as "nonlinear-in-mean" if the underlying system is nonlinear in past values of "x" when external disturbances are set to zero. In addition, the stochastic process is "nonlinear-in-time trend" if the underlying dynamical system is a nonlinear function of time. To detect such systems, one can fit models that are linear but also have additional nonlinear terms as functions of past "x" and of time and evaluate the statistical significance of the models' nonlinear terms; the residuals of such models are also explored for further structure [27,30]. We followed this strategy here, fitting GARCH models [20] to the residuals of our model fits, and testing the standardized residuals for "extra structure" missed by the fitted GARCH models [27], including regime switches in variance.

Plots of autocorrelation and partial autocorrelation functions (PACF) indicated that all series required autoregression (AR) correction. Change point models with autoregressive terms were fit to first-differenced data by maximum likelihood. The best-fitting models were selected using Akaike's information criterion [31] as well as diagnostics for residual time series. Models with residuals exhibiting no predictable structure, especially non-significant PACFs, were selected for further analysis. Residuals of these models were fitted to a GARCH model. GARCH residuals were renormalized and analyzed by the bootstrap BDS (Brock-Dechert-Scheinkman) test using 1000 iterations. The BDS tests the null hypothesis that the renormalized GARCH residuals are identically distributed and independent [32]; rejection of this null hypothesis implies that some structure remains in the series, including potential nonlinearities.

Results

All nine of the commodities studied exhibited substantial dynamics (Figure 1). Indeed, we detected significant change points for each commodity except for gold (Table 1, Figure 1). All of the significant change points were associated with peaks in autocorrelation or variance (Figure 2). However, it is difficult to evaluate the statistical significance of the autocorrelation and variance time series. We used the BDS test (Table 1) for significance tests on the null hypothesis that each price time series followed linear

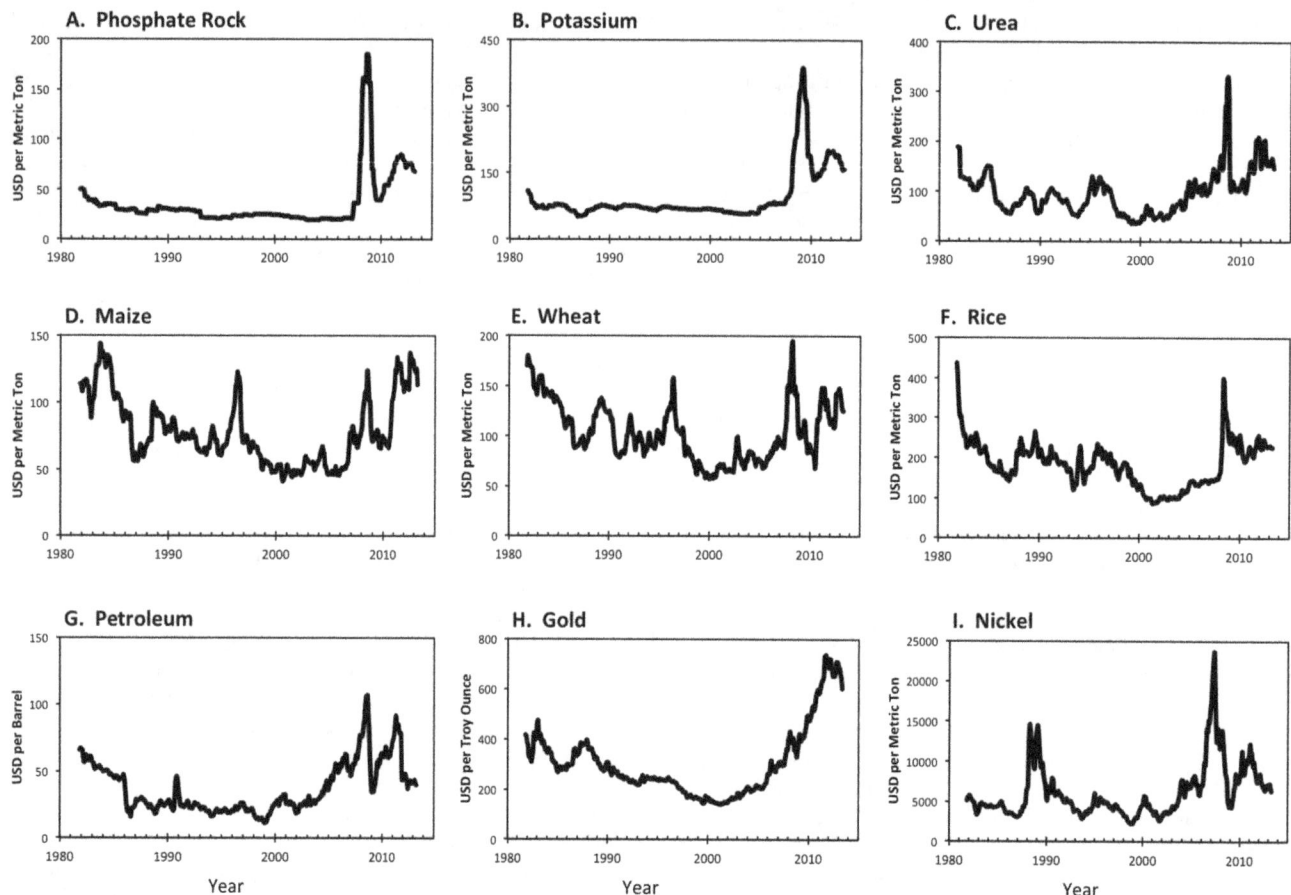

Figure 1. Commodity price time series from 1981 to 2011, corrected for inflation to 1982 price.

processes. Significant departures from linear behavior were detected for all three fertilizer components, as well as for rice (Table 1 and Table S1 and Table S2 in File S1).

Statistical analysis of the phosphate rock price time series (Figure 1A) identified three distinctive time intervals (Table 1): Regime 1 (before May 2007), Regime 2 (May 2007–March 2010),

and Regime 3 (after March 2010). In addition, for phosphate rock, both variance and autocorrelation rose and fell together during 2007–2010 (Figure 2A) and the BDS test for phosphate rock strongly rejects the hypothesis of linearity. The shift in 2007 represents a transition from a long phase of gradual change (Regime 1) to a sharp increase followed by a steep decrease

Table 1. Change points (if any) in years and results of the BDS test for the commodity time series.

Commodity	Change Points	BDS *P* values (3 values of epsilon)	Inference
Phosphate Rock	May 2007, March 2010	0, 0, 0	Change points, not linear
Potassium	January 2008, July 2009	0.015, 0.008, 0.014	Change points, not linear
Urea	January 2009	0.02, 0, 0	Change point, not linear
Rice	March 2008	0.054, 0.011, 0.032	Change point, not linear
Maize	September 2008	0.19, 0.14, 0.15	Change point, linear
Wheat	August 2007	0.61, 0.78, 0.91	Change point, linear
Petroleum	September 2008	0.23, 0.23, 0.25	Change point, linear
Gold	None	0.92, 0.92, 0.78	No change point, linear
Nickel	April 2007	0.61, 0.26, 0.42	Change point, linear

BDS tests the null hypothesis that the standardized residuals of the change point model come from a stationary stochastically independent process. A low *P* value rejects the hypothesis of stationary independence. 'Inference' is our interpretation of the statistics. Change point model fits, GARCH fits, and results of bootstrapped BDS *P* values are presented in Supplementary Information.

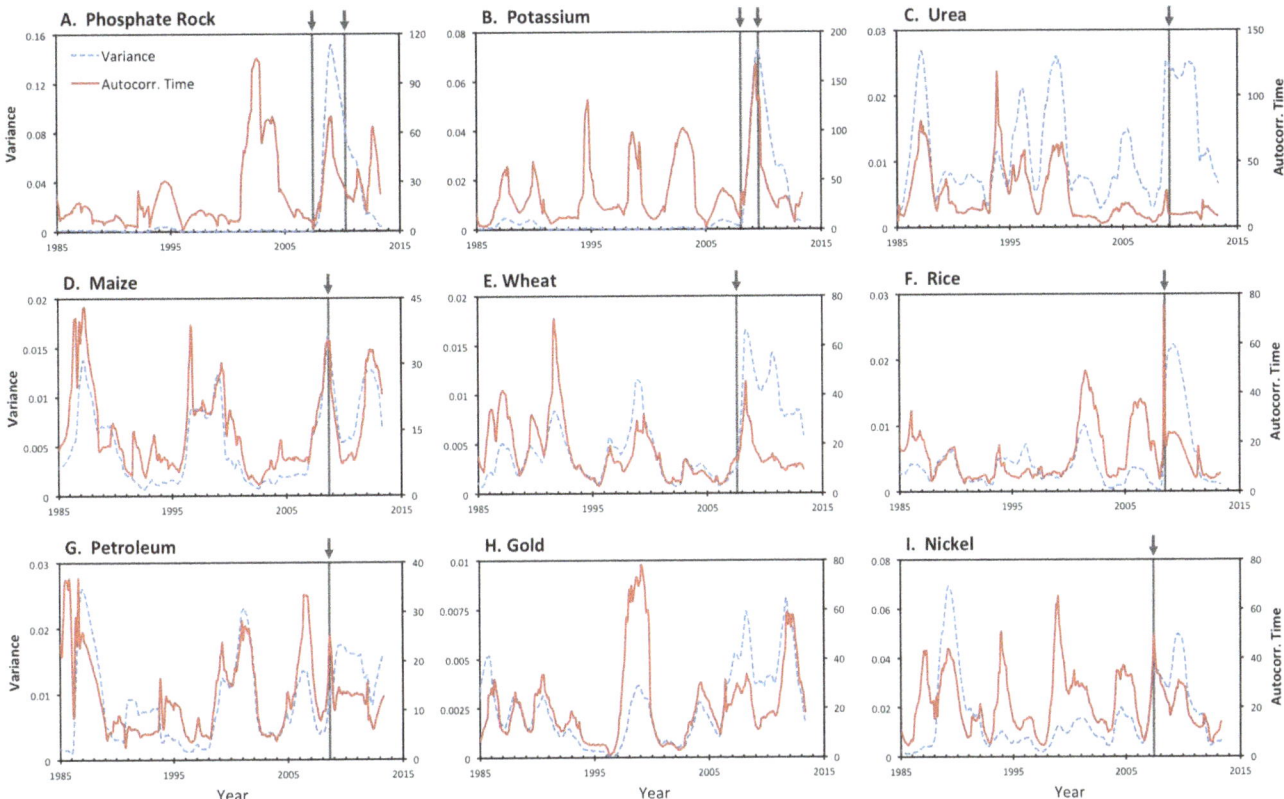

Figure 2. Temporal breakpoints and statistical indicators of critical transition for nine commodities. Arrows and vertical lines show statistically significant change points. Variance (blue, dashed) and autocorrelation time (red, solid) for log10-transformed data were computed for 36-month rolling windows. Autocorrelation time is the negative inverse of the natural logarithm of the autocorrelation coefficient [42].

(Regime 2) while the shift in 2010 is associated with the more gradual rise and subsequent recent decline near the end of the series (Regime 3). Thus, the behavior of the phosphate rock time series reveals two major breakpoints in recent years (including the earliest breakpoint detected of the commodities we studied) and statistical behavior that is consistent with a critical transition under nonlinear dynamics. However, we note that the distinctiveness of Regime 3 is somewhat uncertain given the limited amount of data following the March 2010 breakpoint.

Similar to phosphate rock dynamics, the K price time series also exhibits two change points (January 2008 and July 2009; Figure 1B) although the timing of these events is not coincident. As for phosphate rock, the BDS test for the K time series was consistent with nonlinear behavior, although the BDS phosphate rock values were somewhat higher (Table 1). The urea price time series contained a significant shift at the beginning of 2009 (Figure 1C) that was also marked by increases in variance (Figure 2C). The BDS test suggests that the urea time series cannot be explained entirely by linear processes (Table 1).

The three crop commodities considered also had significant change points (Table 1). For wheat, a significant disruption was detected in August 2007 while those for rice and maize came later (March 2008 and September 2008, respectively). In contrast to fertilizers, indications of nonlinear dynamics were weaker in the crop commodities. Indeed, only rice showed statistically significant BDS tests but the P values (0.011–0.054) were higher than those seen for the fertilizer commodities (Table 1 and Table S1 and Table S2 in File S1).

The non-agricultural commodities we evaluated showed behaviors distinct from those of the agricultural commodities,

and especially of urea, K, and phosphate rock. Petroleum price had a change point in September 2008 (Figure 1D). While this change point was associated with rising variance and autocorrelation time (Figure 2D), the BDS test for petroleum was not significant and thus the dynamics are consistent with linear processes. Gold fit a seasonal autoregressive model with no change points (Figure 1E, Table 1). While there were correlated changes in variance and autocorrelation after 2005 (Figure 2E), the BDS test for gold was not significant. Nickel dynamics exhibited a significant change point in 2007 (Table 1, Figure 1F). While nickel prices also exhibited coherent fluctuations in variance and autocorrelation time after 2006 (Figure 2F), the BDS result was consistent with linear processes (Table 1).

Based on these analyses, we argue that the evidence suggests that there was a temporary disruption in fertilizer markets during the 2007–2009 period that was especially notable in the phosphate rock market between March 2007 and July 2009, during which prices rose to very high levels and then dropped precipitously to levels somewhat higher than those preceding. This volatility in phosphate rock prices was followed by what appears to be a regime change to a phase of higher prices. Overall, high and increasing prices after 2009, along with signals in the variance structure, suggest the arrival of a new phosphate rock regime that is increasingly influenced by Hamilton's "scarcity price" [25], a scenario that will challenge attempts to achieve food security in coming decades [26].

Discussion

We assessed whether price dynamics of key agricultural input commodities {nutrient fertilizers: urea, potassium (K), and especially phosphorus (P) as phosphate rock} during recent decades contained evidence for significant change points as well as for critical transitions [17,33]. We also determined if such dynamics were distinct from the behavior of crop commodities as well as non-agricultural commodities, such as those connected to transportation (petroleum), financial (gold), or industrial (nickel) sectors. Our most significant finding is that the price dynamics of fertilizers appear to be unique among these time series. While all of the commodities showed considerable turbulence from 2006 to the present, all three fertilizer components, and especially phosphate rock, displayed multiple change points as well as significant departures from linearity during this period. Among the other commodities, only the dynamics of rice were suggestive of nonlinear dynamics associated with a significant change point but these indications were weaker than those seen for the fertilizers.

As a finite geological resource rises in price, it is likely that technological advances in resource extraction, conservation, or recycling will occur. Thus, the prospect of 'peak phosphorus' is a contested idea. Two recent articles [25,34] seek ways to detect evidence of "peaking" (e.g. "peak oil") or impending increased scarcity in an exhaustible resource using time series analyses. The basic Hotelling theory discussed by Smith [34] predicts a general uptrend in price due to an increasing scarcity component of the price above and beyond that which is set by shorter term forces of supply and demand [25]. Hamilton notes that this scarcity component is likely to increase as the resource becomes scarcer and harder to extract, as new geographic sources diminish, and as demand rises [25]. Here we do not attempt to evaluate the plausibility of 'peak phosphorus'. Instead, in the case of P we have focused on the price of phosphate rock in relation to the behavior of other key fertilizer components, rather than their geological abundances [35]. This focus is appropriate for the economics of industrial agriculture, and relevant in considerations of the food security of the world's poorer peoples who cannot independently afford inorganic fertilizers even at current prices.

For example, we find statistical evidence for three regimes in phosphate rock prices (Table 1, Figures 1 and 2). The first regime of stable and slightly declining prices is quite precisely fitted because it has a relatively large number of observations and rather consistent behavior over time; it lasted until approximately March 2007. The second regime (approximately March 2007 to February 2009) was a turbulent one with a rapid rise in prices followed by an equally rapid drop in prices. The final regime from February 2009 through 2011 exhibited a more moderate rise in prices. Because the sample size is limited after 2009 we cannot be certain if this third phase of gradually rising prices will be sustained or give way to something different. The dynamics of potassium (K) prices also were consistent with nonlinearity but the change points for potassium (K) followed those for phosphorus by about half a year. Urea underwent a change point in 2009, at the end of the turbulent period for phosphorus prices, and its dynamics also indicate nonlinearity. While evidence of change points was seen in the dynamics of several of the non-agricultural commodities we considered, no clear indications of nonlinearity were documented among these (only rice prices showed some signs of nonlinear behavior).

For an exhaustible resource like phosphate rock or oil, economists have developed the concept of a "scarcity price." This refers to the underlying price that a unit of the resource would fetch under average conditions of short run supply and demand (also sometimes called the "royalty price", [36]). This component of the current observed market price would be zero if known resources of the exhaustible resource were infinite. Though provisional, our analyses suggest that fertilizer markets, and especially that for phosphate rock, have entered a regime in which there is a scarcity price operating [25,34]. The future will reveal whether this scarcity regime is stable or only the prelude to yet another, as yet unknown, regime shift. These recent shifts, with their rapid bursts in price followed by rapid declines, can be viewed as a warning sign that similar large disruptions in fertilizer markets could occur in the future. In the case of phosphate rock, our analyses show that the volatility of residuals of fitted models for phosphate rock prices has increased in the last two regimes relative to the first. This is yet another piece of evidence that the underlying fundamentals of the phosphate rock market have changed, a change with significant implications for global food security.

The causes of these shifts cannot be identified via time-series analysis alone. Mechanistic inferences would instead require detailed macroeconomic modeling combined with fine-scale data about shifts in supply and demand for the various commodities. Such analyses are beyond the scope of this study, whose main goal was to characterize the volatility dynamics of these commodities and test for possible change points in the market systems connected to them. However, the differences we observed among the commodity time series are suggestive of possible contributing factors. First, the disconnect between the fertilizer commodity dynamics and those for the non-fertilizer commodities suggests that the underlying mechanisms are not tightly connected to demand-side drivers from major crops (as indexed by rice, maize, wheat), to shifts in energy costs (as indexed by petroleum), to general economic volatility (as indexed by gold), or to overall industrial activity (as indexed by an industrial material, nickel). This is somewhat surprising, given the previous documentation of price transmission between petroleum and "fertilizer" commodities [22,23]. It important to note, however that in these previously published analyses of price transmission, "fertilizer" corresponded to finished tri-sodium phosphate and not to phosphate rock, as in our analysis. Instead, our work suggests that market and sociopolitical forces localized within the agricultural sector itself (including the fertilizer industry) appear sufficiently strong to impose their own flavor on the emerging dynamics seen in P, K, and N commodity prices. Nevertheless, cross-sector analyses using bivariate and multivariate approaches (e.g. [24,37,38]) to evaluate connections among fertilizer, crop, and petroleum volatilities may be informative and await further analysis. Second, differences among the agricultural commodity series themselves provide some hints about potential contributing factors. Bearing in mind differences in how these other fertilizer components are sourced (e.g. beyond inorganic fertilizer, N inputs to farm fields can also involve rotation strategies that promote N fixation by legumes), if dynamics were driven largely by demand-side forces we would expect similar temporal behaviors either when comparing crops (maize, rice, wheat) to fertilizers (N, P, K) or when comparing the fertilizers to each other (since demand for these key fertilizer components is strongly coupled by their joint essentiality in supporting high crop yield). However, crop change points, as well as those for N and K, were significantly delayed with respect to those for P. Notably, the three fertilizer components and especially P showed signs consistent with nonlinear dynamics but, among the crop commodities, only rice prices showed evidence of nonlinearity and this was relatively weak. We speculate that local and regional factors associated with the supply sectors for each fertilizer component result in disjunct dynamics because the source countries are different for each element. For example, P supplies are increasingly dominated by activities and policies in

Morocco and China while K supplies are dominated by events in their main producers, Canada and Russia.

Time-series analyses are, of course, retrospective and it is problematic to forecast future trajectories from past events, as noted in particular case of the dynamical P system [7]. Nevertheless, the question naturally arises about the likelihood of a return to earlier regimes of low, relatively stable fertilizer prices. One natural market response to rising prices for an extractable resource such as phosphate rock is expansion of production and development of previously economically non-viable resources [25], a trend that is already underway for phosphate rock [7]. However, such responses beg the question if the focus is not on the trajectory of overall production (e.g. the possibility of "peak phosphorus") but instead is on the trajectory of price, as development of these "new" reserves involves exploitation of deposits that have lower resource concentrations and/or higher contaminant levels, lie under greater overburden, or are located further from fertilizer production facilities and markets. All of these imply higher costs and suggest that a near-term return to the earlier, low-price, regimes is unlikely, especially in light of projected increases in demand over the next several decades due to increasing human population size, rising meat consumption, and non-agricultural uses of P [10,12,13]. In the case of phosphate rock, this expectation of sustained high prices in the near-term (next decade or two) is shared by industry analysts despite documented expansion of exploration and exploitation [39]. Morocco's emerging oligopoly over global reserves of phosphate rock [40] may also help to sustain high P prices. However, transformational innovation in crop nutrient use efficiency and in nutrient recycling approaches in the food system [14,41] may be one pathway for a return to the earlier situation, a transition that would improve fertilizer access for all who need it and thus enhance global food security. Such a transition would also reduce nutrient exports from agricultural systems and improve water quality.

Put simply, the evidence presented here confirms for the first time that fertilizer prices have undergone unique patterns of recent volatility and have moved into a new high price regime. That is, an era (*the* era?) of cheap inorganic fertilizer appears to be over. The persistence of this regime will depend on hard-to-forecast transitions to new technologies and strategies for improved crop nutrient use and for nutrient recycling via reclamation from human, animal, and crop waste streams [14]. Importantly, our analysis indicates that the extreme price fluctuations seen in recent years are not merely the rise and fall of prices in "normal" (i.e. linear) market interactions. Instead, they are symptoms of the operation of a nonlinear system and portend the possibility of similar instabilities in the future. The potential continuation of this regime of rising and unpredictable prices has adverse implications for farmers and consumers in both developed and developing countries. Given the close connections between food security and national security, governmental and non-governmental institutions may wish to consider measures to head off further escalation of fertilizer prices and to stabilize their dynamics. Such measures could involve implementing fertilizer stockpiles analogous to the USA's strategic petroleum reserve (a concept first proposed by US President Franklin D. Roosevelt) or encouraging development of diversified local and regional sources of fertilizer from recycled sources (sewage, animal waste, food waste, etc.). A nutrient recycling strategy will have the added benefit of improving water quality in freshwater and marine ecosystems by reducing nutrient runoff from farms, livestock, and cities.

Supporting Information

File S1 This file contains Tables S1 and S2. **Table S1.** Parameter estimates for change point models. **Table S2.** Analysis of residuals from change point models by GARCH and BDS.

Acknowledgments

The views are those of the authors and do not represent views of the National Science Foundation. JM Anderies provided useful comments on an early draft of the manuscript; J Learned provided assistance with drafting the figures. We are grateful for the input of two anonymous reviewers whose comments improved the manuscript.

Author Contributions

Conceived and designed the experiments: JJE SRC WB. Analyzed the data: SRC TJE. Wrote the paper: JJE SRC WB TJE.

References

1. Alexandratos N, Bruinsma J (2012) World agriculture towards 2030/2050: the 2012 revision. Rome: FAO.
2. Foley J A, Ramankutty N, Brauman KA, Cassidy ES, Gerber JS, et al. (2011) Solutions for a cultivated planet. Nature 478: 337–42.
3. Elser JJ, Bennett E (2011) Phosphorus: a broken biogeochemical cycle. Nature 478: 29–31.
4. Bennett EM, Carpenter SR, Caraco NF (2001) Human impact on erodable phosphorus and eutrophication: A global perspective. Bioscience 51: 227–34.
5. Cordell D, Drangert J-O, White S (2009) The story of phosphorus: Global food security and food for thought. Glob Environ Chang 19: 292–305.
6. Van Kauwenbergh SJ (2010) World Phosphate Rock Reserves and Resources Technical Bulletin IFDC-T-75. Muscle Shoals, AL: International Fertilizer Development Center.
7. Scholz RW, Wellmer F-W (2013) Approaching a dynamic view on the availability of mineral resources: What we may learn from the case of phosphorus? Glob Environ Chang 23: 11–27.
8. Gaxiola R, Edwards M, Elser JJ (2011) A transgenic approach to enhance phosphorus use efficiency in crops as part of a comprehensive strategy for sustainable agriculture. Chemosphere 84: 840–5.
9. López-Arredondo DL, Herrera-Estrella L (2012) Engineering phosphorus metabolism in plants to produce a dual fertilization and weed control system. Nat Biotechnol 30: 889–9.
10. Tilman D, Balzer C, Hill J, Befort BL (2011) Global food demand and the sustainable intensification of agriculture. Proc Natl Acad Sci USA 108: 20260–4.
11. Kearney J (2010) Food consumption trends and drivers Philos. Trans R Soc B-Biological Sci 365: 2793–807.
12. Metson GS, Bennett EM, Elser JJ (2012) The role of diet in phosphorus demand. Environ Res Lett 7: 44043.
13. Childers DL, Corman JR, Edwards M, Elser JJ (2011) Sustainability challenges of phosphorus and food: Solutions from closing the human phosphorus cycle. Bioscience 61: 117–24.
14. Rittmann BE, Mayer B, Westerhoff P, Edwards M (2011) Capturing the lost phosphorus. Chemosphere 84: 846–53.
15. Wyant KA, Corman JR, Elser JJ, editors (2013) Phosphorus, food, and our future. Oxford: Oxford University Press.
16. Carpenter SR, Brock WA (2006) Rising variance: a leading indicator of ecological transition. Ecol Lett 9: 308–15.
17. Scheffer M, Bascompte J, Brock WA, Brovkin V, Carpenter SR, et al. (2009) Early-warning signals for critical transitions. Nature 461: 53–9.
18. Carpenter SR (2003) Regime shifts in lake ecosystems: pattern and variation. Oldendorf/Luhe, Germany: Ecology Institute.
19. Scheffer M (2009) Critical transitions in nature and society. Princeton: Princeton University Press.
20. Campbell J, Lo A, MacKinlay C (1997) The econometrics of financial markets. Princeton: Princeton University Press.
21. Dakos V, Carpenter SR, Brock W A, Ellison AM, Guttal V, et al. (2012) Methods for detecting early warnings of critical transitions in time series illustrated using simulated ecological data. PLoS One 7: e41010.
22. Baffes J (2007) Oil spills on other commodities. Resour Policy 32: 126–34.
23. Baffes J (2010) More on the energy/nonenergy price link Appl Econ Lett 17: 1555–8.
24. Serra T, Zilberman D, Gil JM, Goodwin BK (2011) Nonlinearities in the U.S. corn-ethanol-oil-gasoline price system. Agric Econ 42: 35–45.
25. Hamilton JD (2013) Oil prices, exhaustible resources, and economic growth. In: Fouquet R, editor. Handbook of energy and climate change. Cheltenham and Northampton: Edward Elgar Publications. 39–63.

26. Cordell D, Rosemarin A, Schroder JJ, Smit AL (2011) Towards global phosphorus security: A systems framework for phosphorus recovery and reuse options. Chemosphere 84: 747–58.

27. Brock WA, Hsieh D, LeBaron B (1991) Nonlinear dynamics, chaos, and instability: statistical theory and economic evidence. Cambridge: MIT Press.

28. Carpenter SR, Cole JJ, Pace ML, Batt R, Brock WA, et al. (2011) Early warnings of regime shifts: a whole-ecosystem experiment. Science 332: 1079–82.

29. Dai L, Vorselen D, Korolev KS, Gore J (2012) Generic indicators for loss of resilience before a tipping point leading to population collapse. Science 336: 1175–7.

30. Ashley R, Patterson D (2000) A nonlinear time series workshop: a toolkit for detecting and identifying nonlinear serial dependence. Dordrecht: Kluwer Academic Publishers.

31. Burnham KP, Anderson DR (1998) Model selection and inference. New York: Springer-Verlag.

32. Brock W, Dechert W, Scheinkman J, LeBaron B (1996) A test for independence based upon the correlation dimension. Econom Rev 15: 197–235.

33. Biggs R, Carpenter SR, Brock WA (2009) Turning back from the brink: detecting an impending regime shift in time to avert it. Proc Natl Acad Sci USA 106: 826–31.

34. Smith JL (2012) On the portents of peak oil (and other indicators of resource scarcity). Energy Policy 44: 68–78.

35. Burt D, Dumas M, Springer N, Vaccari DA (2013) Global phosphorus: geological sources and demand-driven production. In: Wyant KA, Corman JR, and Elser JJ, editors. Phosphorus, food, and our future. Oxford: Oxford University Press. 40–63.

36. Dasgupta PS, Heal GM (1979) Economic theory and exhaustible resources. Cambridge: Cambridge University Press.

37. Hassouneh I, Serra T, Goodwin BK, Gil JM (2012) Non-parametric and parametric modeling of biodiesel, sunflower oil, and crude oil price relationships. Energy Econ 34: 1507–13.

38. Figuerola-Ferretti I, Gilbert CL (2008) Commonality in the LME aluminum and copper volatility processes through a FIGARCH lens. J Futur Mark 28: 935–62.

39. Evans M (2012) Phosphate resources: future for 2012 and beyond. Sharm El-Sheikh, Egypt: AFA International Annual Fertilizer Forum & Exhibition.

40. Cooper J, Lombardi R, Boardman D, Carliell-Marquet C (2011) The future distribution and production of global phosphate rock reserves. Resour Conserv Recycl 57: 78–86.

41. Hering JG (2012) An end to waste? Science 337: 623.

42. Dakos V, Scheffer M, van Nes EH, Brovkin V, Petoukhov V, et al. (2008) Slowing down as an early warning signal for abrupt climate change. Proc Natl Acad Sci USA 105: 14308–12.

20

Influence of 20–Year Organic and Inorganic Fertilization on Organic Carbon Accumulation and Microbial Community Structure of Aggregates in an Intensively Cultivated Sandy Loam Soil

Huanjun Zhang[1], Weixin Ding[1]*, Xinhua He[2,3], Hongyan Yu[1], Jianling Fan[1], Deyan Liu[1]

1 State Key Laboratory of Soil and Sustainable Agriculture, Institute of Soil Science, Chinese Academy of Sciences, Nanjing, China, 2 Forests NSW, NSW Department of Primary Industries, West Pennant Hills, New South Wales, Australia, 3 School of Plant Biology, University of Western Australia, Crawley, Western Australia, Australia

Abstract

To evaluate the long–term effect of compost (CM) and inorganic fertilizer (NPK) application on microbial community structure and organic carbon (OC) accumulation at aggregate scale, soils from plots amended with CM, NPK and no fertilizer (control) for 20 years (1989–2009) were collected. Soil was separated into large macroaggregate ($>2,000$ μm), small macroaggregate (250–2,000 μm), microaggregate (53–250 μm), silt (2–53 μm) and clay fraction (<2 μm) by wet-sieving, and their OC concentration and phospholipid fatty acids (PLFA) were measured. The 20-year application of compost significantly ($P<0.05$) increased OC by 123–134% and accelerated the formation of macroaggregates, but decreased soil oxygen diffusion coefficient. NPK mainly increased OC in macroaggregates and displayed weaker influence on aggregation. Bacteria distributed in all aggregates, while fungi and actinobacteria were mainly in macroaggregates and microaggregates. The ratio of monounsaturated to branched (M/B) PLFAs, as an indicator for the ratio of aerobic to anaerobic microorganisms, increased inversely with aggregate size. Both NPK and especially CM significantly ($P<0.05$) decreased M/B ratios in all aggregates except the silt fraction compared with the control. The increased organic C in aggregates significantly ($P<0.05$) negatively correlated with M/B ratios under CM and NPK. Our study suggested that more efficient OC accumulations in aggregates under CM–treated than under NPK–treated soil was not only due to a more effective decrease of actinobacteria, but also a decrease of monounsaturated PLFAs and an increase of branched PLFAs. Aggregations under CM appear to alter micro-habitats to those more suitable for anaerobes, which in turn boosts OC accumulation.

Editor: Dafeng Hui, Tennessee State University, United States of America

Funding: This work was supported by the National Basic Research Program of China (2011CB100503), the Chinese Academy of Sciences (KSCX2-EW-N-08), and National Science Foundation of China (41171190). The funders had no role in study design, data collection and analysis, decision to publish, or preparation of the manuscript.

Competing Interests: The authors have declared that no competing interests exist.

* E-mail: wxding@issas.ac.cn

Introduction

Sequestration of carbon (C) in soils has gained increasing recognition in recent years for its role in global environmental change [1]. Soil aggregation affects soil properties such as aeration and water infiltration and alters micro–habitats and activities of microorganisms [2]. A growing body of work has revealed a close relationship between aggregation and soil organic C (OC) concentrations [3,4] as OC stabilization in aggregates is the principal mechanism controlling OC turnover and sequestration [5,6,7].

The potential of soils to sequester C depends on soil type, management practices, and soil aggregate structure [5,8,9]. Previous studies have suggested that OC is more effectively protected and more stable in microaggregates, whereas macroaggregates provide a niche for the storage of labile C [3,5,10]. Fertilization, a common agricultural practice, has been found to significantly influence aggregate formation and OC distribution in aggregates [11,12]. For example, increases in OC following organic manure application were observed in aggregates of all

sizes, especially in macroaggregates [11,13]. Kong et al. [14] found that the majority of OC derived from organic material was preferentially sequestered in microaggregates within small macro-aggregates. In contrast, Sarkar et al. [15] and Bhattacharyya et al. [16] found that the highest rate of increase in OC following repeated inorganic fertilizer applications primarily occurred in the silt and clay fractions of OC–poor soils.

The turnover of OC in soil aggregates is not only determined by the physical protection offered by that fraction, but also by the abundance and community of microorganisms, especially Gram–positive (G⁺) bacteria and actinobacteria [17,18,19]. Aggregates provide spatially heterogeneous habitats for microorganisms characterized by different substrates, oxygen concentrations and water contents [20]. In turn, microorganisms provide further ecosystem functions through their effects on soil structure by binding soil particles and organic matter to create aggregates [9,21]. Shifts in microbial community and function in response to different agricultural management practices can markedly alter rates of organic C loss from soils. Determining how fertilization affect the distribution of microbial functional groups among aggregates can lead to a better understanding of the differences in soil OC turnover under different management practices [2,22]. A previous study showed that different fertilizer treatments such as nitrogen fertilizer, green manure and sewage sludge did not dramatically alter the distribution pattern of bacteria in aggregates <200 μm [20].

The abundance and community structure of microorganisms are very responsive and can provide immediate and precise information on changes occurring in soil. In contrast, OC concentrations in soil and aggregates change at a slower rate [23]. To date, limited information has been available on the relationship between OC turnover and microbial community structure in soil aggregates. A long–term field experiment has been established in the Northern China plain to monitor changes in soil organic C under organic and inorganic fertilizer applications. In the present study, the abundance and community structure of microorganisms in aggregates were measured. The objectives of this study were to: (1) identify how repeat inorganic and organic fertilizer applications affect the distribution and community structure of microorganisms in aggregates; and (2) evaluate the relationship between OC accumulation and microbial community structure in aggregates.

Materials and Methods

Field Experiment and Soil Sampling

A long–term field experiment was set up in September 1989 to monitor the dynamic variation in OC following applications of compost and inorganic fertilizer to an intensively cultivated fluvo–aquic soil at the Fengqiu Agro–ecological Experimental Station, Chinese Academy of Sciences, Henan Province, China (35°00′N, 114°24′E). The soil was developed from alluvial sediments of the Yellow River and classified as Aquept [24]. Prior to the experiment, the field had been cultivated under a similar agricultural regime for at least 50 years, so the heterogeneity of soil fertility was minimal. The characteristics of the surface soil (0–20 cm) in September 1989 were listed in Table 1.

The experiment was based on a random design including three fertilization treatments with four replicate plots (9.5 m×5 m each): (1) no–fertilization control (control), (2) inorganic fertilizer NPK (NPK) and (3) compost (CM). Winter wheat (*Triticum aestivum*) was annually rotated with summer maize (*Zea mays*). A total of 150 kg N ha^{-1} for each crop was applied in the NPK and CM treatments. In the NPK treatment, urea was used at 60 or 90 kg N ha^{-1} as basal fertilizer in early June or early October for maize or wheat, respectively, and at 90 or 60 kg N ha^{-1} as supplement fertilizer in late July or late February for maize or wheat, respectively. In the NPK treatment, 75 kg P$_2$O$_5$ ha^{-1} (calcium superphosphate) and 150 kg K$_2$O ha^{-1} (potassium sulfate) were also added to each crop. In the CM treatment, a total of 2.76 tons ha^{-1} compost (made from wheat straw, oil cake and cotton cake, with a C/N ratio of approximately 8), which contained 1,164 kg C, 150 kg N, 51 kg P$_2$O$_5$ and 65 kg K$_2$O, was applied as basal fertilizer in early June for maize and early October for wheat. To match the same

amounts of P and K with the NPK treatment, 24 kg P$_2$O$_5$ ha^{-1} (calcium superphosphate) and 85 kg K$_2$O ha^{-1} (potassium sulfate) were added to the compost prior to the application. No fertilizer was applied in the control treatment. Fertilizers and composts were broadcasted onto the soil surface by hand and the surface soil (0–20 cm) was immediately tilled. Seeds of 195 kg wheat ha^{-1} and 37.5 kg maize ha^{-1} were immediately sown into the soil by hand after the basal fertilization. Seeding rows were 70 cm wide for maize and 15 cm wide for wheat cultivation. Over the entire 20–year period, field management followed local practices.

Ten soil cores (0–20 cm) from each replicate plot were collected with a 2.5 cm diameter auger and then mixed to form one composite sample on 7 June 2009 after wheat harvest. Fresh samples were stored at 4°C and immediately transported to the laboratory for wet–sieving and phospholipid fatty acids (PLFA) analyses, while air–dried and sieved soils were used to determine the OC content by the wet oxidation–redox titration method [25]. A further eight undisturbed cylinder soil samples (100 cm^3) were taken from each plot to establish the soil water retention curve and bulk density.

Soil Water Retention Curve and the Effective Diffusion Coefficient of Oxygen

Water retention curves were determined with a ceramic pressure plate assembly at equilibrium matric potentials of −0.1, −0.2, −1, −3.5, −6, −10, −33, −50, −100, −200, −500 and −1,500 kPa in a pressure chamber. The obtained data were used to calculate the soil water retention curves and to derive the van Genuchten parameters by RETC (RETention Curve) software [26] using the following equation:

$$\theta(\psi) = \theta_r + \frac{\theta_s - \theta_r}{[1 + (\alpha\psi)^n]^m}$$

where θ is the soil water content (cm^3 cm^{-3}), θ_s and θ_r are the saturated and residual water contents (cm^3 cm^{-3}), respectively, ψ is the matric potential (kPa) as indicated by pressure head, and the parameters of α, n and m ($m = 1 - 1/n$) are dimensionless.

According to the capillary rise theory, the pore size can be calculated using the following equation when the soil is hydrophilic at 20°C of water [27]:

$$d = \frac{3000}{|\psi|}$$

where d is the diameter of pore (μm), and $|\psi|$ is the absolute matric potential (kPa).

The effective diffusion coefficient (D) of oxygen through the pore space of the soil (m^2 s^{-1}) was calculated as follows [28]:

$$D = \frac{1}{N^2} \times [(D_{a0} \times Q_a^p) + (KH \times D_{w0} \times Q_w^p)]$$

where N is soil porosity, D_{a0} is the free diffusion coefficient in air

Table 1. Characteristics of the soil sampled in September 1989.

Texture (Sand, Silt, Clay) (%)	pH$_{H2O}$ (extract 1:5, w/v)	Organic C (g C kg^{-1})	Total N (g N kg^{-1})	Total P (g P kg^{-1})	Total K (g K kg^{-1})	Inorganic N (mg N kg^{-1})
52,33,15	8.65	4.48	0.43	0.5	18.6	9.51

$(1.8 \times 10^{-5} \text{ m}^2 \text{ s}^{-1}$ at 20°C), D_{w0} is the free diffusion coefficient in water $(2.2 \times 10^{-9} \text{ m}^2 \text{ s}^{-1}$ at 20°C), Q_a and Q_w are the proportion of soil porosity occupied by air and water, respectively, i.e. $Q_a + Q_w = 1$, and p is a power constant $(p = 3.4)$.

Q_w was calculated as follows:

$$Q_w = \rho \times \theta_m$$

where ρ is the soil bulk density, and θ_m is the soil moisture content.

Soil Fractionation

The water–stable aggregates of moist soils were separated using the wet–sieving protocol [29]. One hundred grams of moist soil samples (on an oven–dried basis) were submerged in deionized water for 5 min at room temperature on top of a 2,000–μm sieve. The sieve was manually moved up and down 3 cm, 50 times over a 2–min period. The fraction remaining on the 2,000–μm sieve was collected in a pre–weighed aluminum pan. Water plus the filtered soil was poured through a 250–μm sieve, and the sieving procedure repeated. Water plus the <250 μm fraction of soil was poured through a 53–μm sieve, and the sieving procedure repeated. To separate the silt (2–53 μm) from the clay (<2 μm) fraction, the remaining suspension was poured into centrifuge bottles and centrifuged at approximately 1,000 rpm for 3 min at 15°C. To obtain the clay fraction, all supernatants collected in several centrifuge bottles were centrifuged at 4,500 rpm for 30 min at 15°C. Large macroaggregate (>2,000 μm), small macroaggregate (250–2,000 μm), microaggregate (53–250 μm), silt (2–53 μm) and clay (<2 μm) fractions were obtained from the tested soils. An aliquot of each fraction was used to determine moisture content. Another subsample of each fraction was dried at 50°C to determine the OC content. The remaining sample was used to measure the microbial biomass and community structure using PLFA method.

PLFA Measurement

PLFAs were extracted using a modified Bligh–Dyer technique as described by Brant et al. [30]. In brief, three grams of soil sample (on an oven–dried basis) were incubated in a 2:1:0.8 solution of methanol, chloroform, and phosphate buffer. Soil extracts were centrifuged and the chloroform phases collected. Phospholipids were separated from glycolipids and neutral lipids using silicic acid bonded solid–phase–extraction columns. The phospholipid fractions were dried under a stream of nitrogen at 37°C. The fatty acid methyl esters (FAME) were produced through mild alkaline methanolysis, and the FAME were dried under nitrogen at 37°C. Finally, the FAME were dissolved in hexane, which contained a 19:0 (methyl nonadecanoate fatty acid) FAME standard. Samples were analyzed on a Shimadzu GC–MS QP 2010 PLUS (Shimadzu, Kyoto, Japan). Peaks were identified based on comparing retention times with known standards. Concentrations of each PLFA were obtained by comparing peak areas with the 19:0 FAME standards.

PLFAs have been used as biomarkers for various groups of microorganisms [31]. The PLFAs 16:3ω3, 18:1ω9, 18:2ω6,9 18:2ω9,12 and 20:1ω9 are used as biomarkers for fungi [19,32]; 16:1ω5 is generally attributed to arbuscular mycorrhizal fungi [33]; 10Me17:0, 10Me18:0, and 10Me20:0 for actinobacteria [34,35]; 3OH–15:0, 14:0, 15:0, 16:0, 17:0, 18:0, 19:0 and 20:0 for general bacteria; a15:0, i15:0, a16:0, i16:0, a17:0, i17:0 and i19:0 for G$^+$ bacteria [34]; and cy17:0, cy19:0, 16:1ω7c, 16:1ω7t, 18:1ω7c and 18:1ω7t for Gram–negative (G$^-$) bacteria [34,35]. According to Bossio et al. [36], 16:1ω7c, 16:1ω7t, 18:1ω7t,

18:1ω7t, 18:1ω9 and 16:1ω5 are monounsaturated PLFAs, while 3OH–15:0, a15:0, i15:0, a16:0, i16:0, a17:0, i17:0, i19:0, 10Me17:0, 10Me18:0 and 10Me20:0 are branched PLFAs. The ratio of monounsaturated to branched PLFAs is regarded as an indicator for the ratio of aerobic to anaerobic microorganisms and the development of aerobic conditions [36,37].

Statistical Analysis

The recovery rate of microbial PLFAs in all aggregates was calculated as follows:

$$\text{Re covery rate}(\%) = \sum \left(C_{aggregate} \times M_{aggregate} \right) \Big/ C_{soil} \times 100$$

where $C_{aggregate}$ is the concentration of microbial PLFAs in aggregates (nmol g^{-1} aggregate), $M_{aggregate}$ is the proportion of aggregates in whole soil by mass (kg kg^{-1} soil), and C_{soil} is the concentration of microbial PLFAs in soil (nmol g^{-1} soil).

Statistically significant differences among the treatments were determined using analysis of variance (one–way ANOVA) and Least Significant Difference (LSD) calculations at the 5% level with SPSS 18 for windows. Redundancy analysis (RDA) and regression analyses were used to test relationships between the increased organic C concentration in the CM and NPK treatments compared with the control and the abundance of microorganisms in aggregates in the CM and NPK treatments.

Results

Aggregate Mass Distribution and Aggregate–associated OC Concentration

Compost applications significantly $(P<0.05)$ increased the mass proportion of large and small macroaggregates by 175% and 44%, respectively, at the expense of reduction in the silt fraction or microaggregate compared with the control. Inorganic fertilizer NPK only significantly $(P<0.05)$ increased the proportion of small macroaggregates by approximately 30% (Fig. 1). The highest OC concentration was observed in small macroaggregates in all treatments, while the lowest occurred in the clay fraction (Table 2). Compost application more efficiently increased organic C in all aggregates than fertilizer NPK, and significantly $(P<0.05)$ increased the values by 123–134% compared with the control. NPK mainly increased organic C in large and small macroaggregates, and to a lesser extent those in microaggregate, silt and clay fractions.

Pore Size and Effective Diffusion Coefficient of Oxygen

The proportion of pores with a neck diameter <4 μm in the CM treatment amounted to 65.12%, which was significantly $(P<0.05)$ higher than in the control and NPK treatments, the latter being 55.87 and 57.28%, respectively (Fig. 2a). In contrast, the proportion of pores with a neck diameter of 15–60 μm showed a decrease in the order: control>NPK>CM and there was a significant difference between treatments. Pores greater than 300 μm were significantly $(P<0.05)$ increased by fertilization, but there was no obvious difference between CM and NPK treatments. The highest effective diffusion coefficient of oxygen was observed in the control treatment, peaking at $5.19 \times 10^{-6} \text{ m}^2 \text{ s}^{-1}$, while the lowest occurred in the CM treatment, measuring only $1.30 \times 10^{-6} \text{ m}^2 \text{ s}^{-1}$ (Fig. 2b).

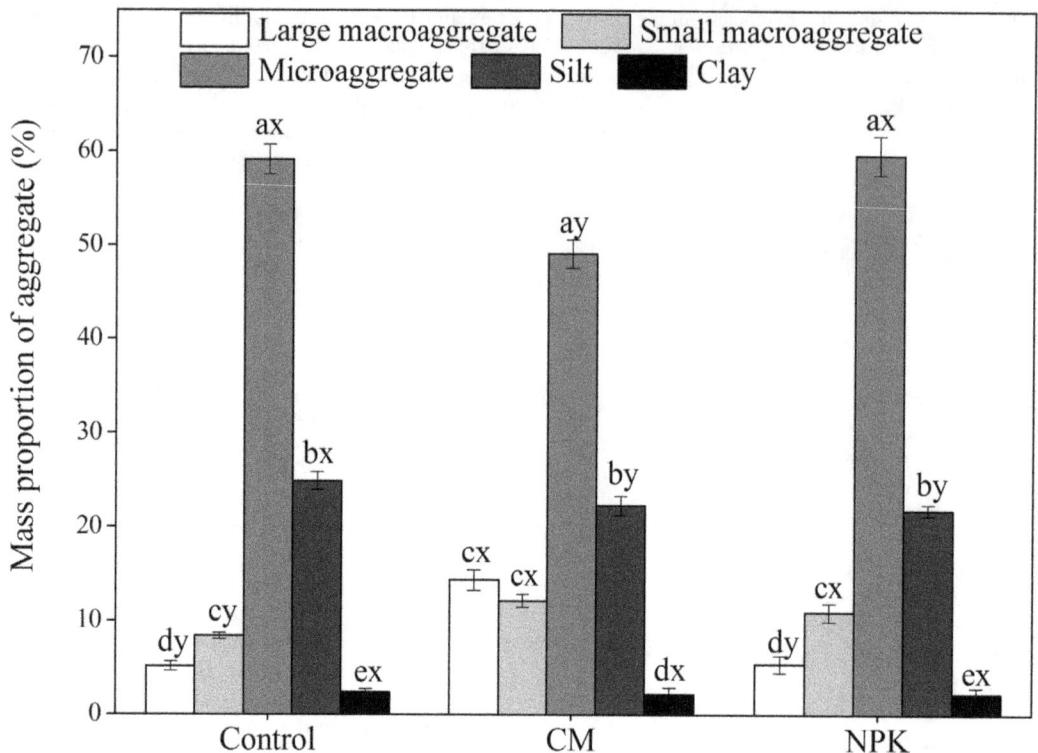

Figure 1. Effect of long-term applications of compost and fertilizer NPK on the mass proportion of aggregates in soil. Vertical bars indicate the standard error of the means ($n = 4$). Different letters denote significant differences between aggregates in the same treatment (a, b, c, d, e) and between treatments with the same aggregate (x, y, z) at $P < 0.05$.

Concentrations of Bacterial, Fungal and Actinobacterial PLFAs in Aggregates

The recovery rate of microbial PLFAs in all aggregates compared with soil was 95–100%. The highest concentrations of microbial PLFAs were found in small macroaggregates, ranging from 89.00 to 93.93 nmol g^{-1} aggregate, and the lowest was found in the clay fraction for all treatments (Fig. 3d). The NPK application did not significantly ($P > 0.05$) affect the concentration of total microbial PLFAs compared with the control. In contrast, compost significantly ($P < 0.05$) increased the concentration of total PLFAs in the microaggregates and clay fraction.

The concentrations of bacterial PLFAs in microaggregates in all treatments varied from 47.97 to 63.07 nmol g^{-1} aggregate, being significantly ($P < 0.05$) higher than in other aggregates, while the lowest was in the clay fraction (Fig. 3a). Compost significantly ($P < 0.05$) increased the concentrations of bacterial PLFAs in microaggregates, silt and clay fractions, but fertilizer NPK only significantly ($P < 0.05$) increased those in the silt fraction compared with the control.

The highest concentrations of fungi, varying from 20.39–21.53 nmol g^{-1} aggregate, were observed in the small macroaggregates, followed by large macroaggregates, while the lowest were measured in the clay fraction, being only 0.81–1.83 nmol g^{-1} aggregate in the three treatments. The concentrations of fungi in large macroaggregates were significantly ($P < 0.05$) reduced by NPK and compost application, but were markedly increased in the silt fraction after compost was applied (Fig. 3b).

No actinobacterial PLFAs were found in any of the clay fractions (Fig. 3c). The maximum concentration of actinobacteria was observed in small macroaggregates, and the minimum in the silt fraction. The application of compost significantly ($P < 0.05$)

reduced their concentration in all aggregates, whereas NPK fertilization principally lowered the concentration in macroaggregates and the silt fraction.

Concentrations of Monounsaturated and Branched PLFAs in Aggregates

The concentration of monounsaturated PLFAs, which were mainly composed of G$^-$ bacterial PLFAs, was highest in microaggregates and lowest in the silt fraction in all treatments (Fig. 4a). Compost significantly ($P < 0.05$) reduced the concentrations of monounsaturated PLFAs in all aggregates compared with the control, while fertilizer NPK reduced their levels in small macroaggregates but increased their concentrations in the silt and clay fractions. In contrast, fertilization, regardless of type, significantly ($P < 0.05$) elevated the concentrations of branched PLFAs including G$^+$ bacterial and actinobacterial PLFAs in large and small macroaggregates and clay fractions compared with the control (Fig. 4b). However, compost exerted a larger effect in large and small macroaggregates and weaker influences in the clay fraction than NPK. The ratio of monounsaturated to branched PLFAs (M/B ratio) increased as aggregate size reduced in all treatments, except for small macroaggregates and the silt fraction in the control treatment and small macroaggregates in the NPK treatment (Fig. 4c). Compost application reduced M/B ratios more efficiently in all aggregates than NPK, and the latter significantly ($P < 0.05$) increased the M/B ratio in the silt fraction.

Correlation between Microbial PLFAs and the Increase of OC in Aggregates

Compared with the control, the increase of organic C in aggregates was found to be closely related with concentrations of

Table 2. Effect of long-term applications of compost and fertilizer NPK on organic C concentrations (g C kg⁻¹ aggregate) in aggregates.

Treatment	Large macroaggregate (>2,000 μm) Concentration (g C kg⁻¹)	Increase (%)	Small macroaggregate (250–2,000 μm) Concentration (g C kg⁻¹)	Increase (%)	Microaggregate (53–250 μm) Concentration (g C kg⁻¹)	Increase (%)	Silt fraction (2–53 μm) Concentration (g C kg⁻¹)	Increase (%)	Clay fraction (<2 μm) Concentration (g C kg⁻¹)	Increase (%)
Control	4.32±0.04 cz	–	8.56±0.06 az	–	4.28±0.07 cz	–	4.57±0.22 bz	–	4.04±0.05 dy	–
CM	9.93±0.09 cx	130	20.01±0.29 ax	134	9.67±0.01 dx	127	10.37±0.06 bx	126	9.02±0.07 ex	123
NPK	8.16±0.10 by	89	17.82±0.11 ay	108	6.01±0.16 dy	47	6.72±0.10 cy	40	5.76±0.08 ey	43

Mean ± standard deviation (n = 4).
Increase (%) = (organic C in fertilization treatments – organic C in the control treatment)/organic C in the control treatment×100.
Different letters denote significant differences between aggregates with the same treatment (a, b, c, d, e) and between treatments with the same aggregate (x, y, z) at $P<0.05$, respectively.

branched PLFAs, fungi, and actinobacteria, as well as the ratio of M/B. The increased OC content of aggregates significantly ($P<0.05$) positively correlated with concentrations of fungal PLFAs for CM and NPK, and that of actinobacteria for CM (Fig. 5). The increase of OC was also found to be marginally ($P<0.1$) positively linked to the concentration of branched PLFAs, but significantly ($P<0.05$) and negatively correlated with the ratio of M/B in CM and NPK treatments.

Discussion

Distribution of Microbial PLFAs in Aggregates

Aggregates in soils provide a spatially heterogeneous micro–habitat for microorganisms characterized by different substrates and oxygen concentrations [38]. In this study, fungi and actinobacteria were mainly located in aggregates >250 μm. In contrast, although the highest abundance of bacteria was found in microaggregates, they were more even distributed across different aggregates than fungi and actinobacteria. However, bacteria in the silt and clay fractions accounted for 79–98% of total microbial PLFAs, but only 46–52% in large and small macroaggregates, indicating that bacteria were abundant across aggregate size classes in soil and dominant in the silt and clay fractions, while fungi and actinobacteria were mainly observed in macroaggregates [39,40]. Using low–temperature scanning electron microscopy, Chenu *et al.* [41] found that actinobacteria and fungi displayed usually on the surface of aggregates greater than 10 μm in size, while bacteria displayed as individual cells or small colonies in clay pores under 2 μm. Thus, it is likely that actinobacteria and fungi were physically prevented from accessing the interior of microaggregates, silt and clay fractions where the habitats' small spaces were limiting for fungi and actinobacteria [42]. In contrast, the protective habitats in microaggregates provided niches for bacteria through exclusion of predators (protozoa) and competition with fungi, because the predation regimen acts as a major structuring force for the bacterial community [20,43].

Sessitsch *et al.* [20] suggested that the availability of nutrients in finer aggregates like silt and clay fractions or in smaller aggregates within macroaggregates could stimulate bacterial growth and thus bacterial diversity. Fungi and actinobacteria have mainly been found in large and small macroaggregates because they have a broader range of extracellular enzymes and prefer the particulate organic matter (POM) or the recalcitrant organic C such as lignin and hemicellulose that is rich in macroaggregates [40,44].

Previous studies have shown that microbial biomass was lowest in the silt and clay fraction and highest in the 1,000–2,000 μm fraction [2,45]. In our study, the concentration of total microbial PLFAs was also highest in small macroaggregates and lowest in the clay fraction, showing a decreasing pattern with reduction of aggregate size from 250–2,000 μm to <2 μm fractions (Fig. 3d). We found that 20–year compost applications significantly ($P<0.05$) increased the concentration of microbial PLFAs in microaggregates and the clay fraction compared with the control, while NPK did not. However, compost–induced stimulation did not drastically alter the distribution pattern of microbial PLFAs in soil aggregates. Previous studies demonstrated that long–term fertilization and conversion from conventional tillage to no–tillage did not affect the distribution pattern of microorganisms in aggregates of a clay loam in Sweden [20] and a subtropical paddy soil in China [22]. These results indicated that the distribution pattern of microorganisms in aggregates was mainly controlled by aggregate size and responded to a lesser extent to management practices like fertilization.

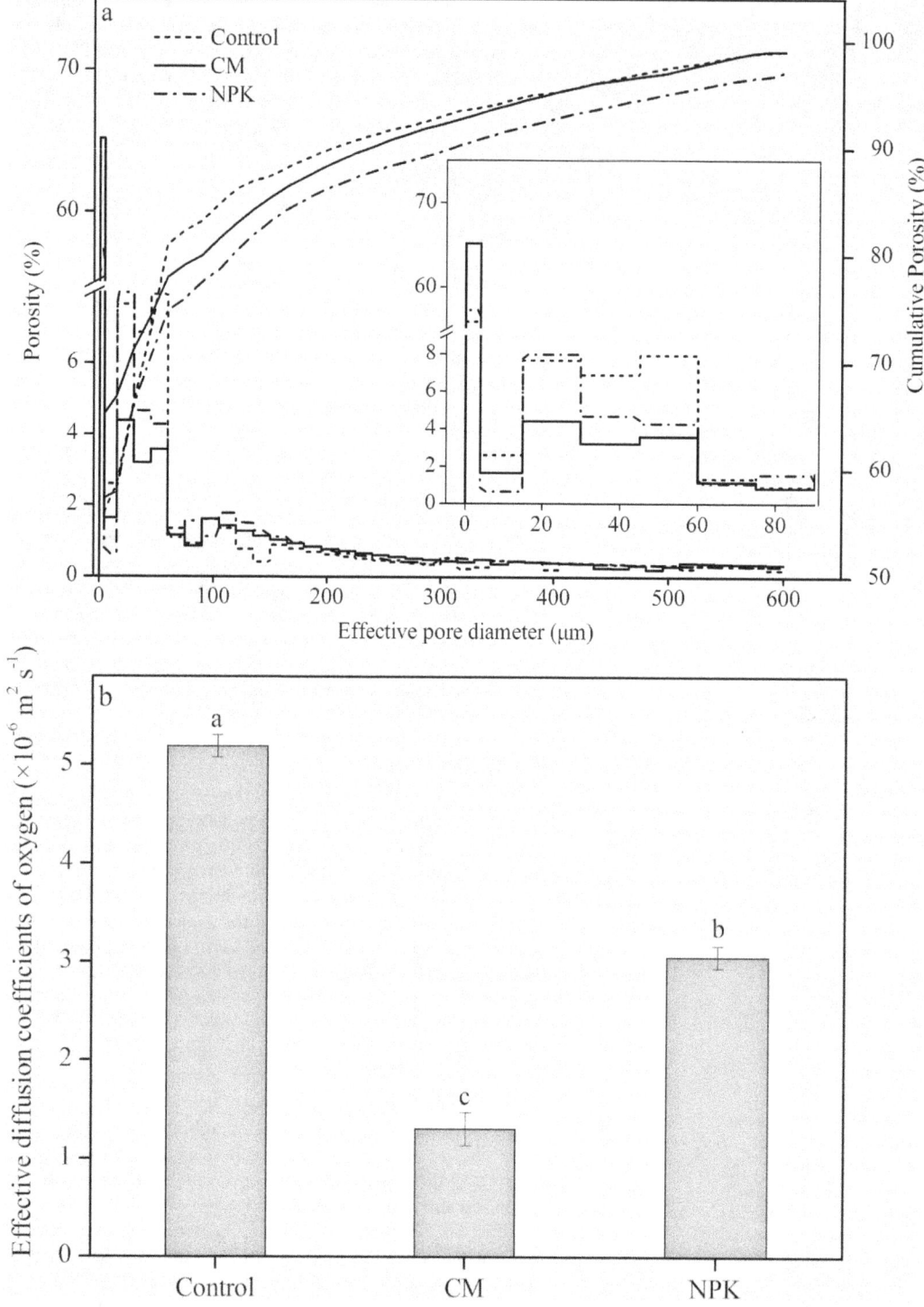

Figure 2. Effect of long–term applications of fertilizer NPK and compost on the proportions of pore volumes of different diameter (a) and effective diffusion coefficient of oxygen (b) in the soil. Vertical bars indicate the standard error of the means ($n = 4$). Different lowercase letters denote significant differences between treatments at $P < 0.05$.

Relationship between Microbial PLFAs and OC in Aggregates

The 20–year application of compost significantly increased organic C concentrations in all aggregates compared with the control, but fertilizer NPK exerted a much weaker effect (Table 2). The increase of organic C primarily depended on input levels of exogenous organic material and turnover of organic C [7,46]. Our previous study showed that the decomposition rate of per unit organic C in CM–treated soil was lower than in NPK–treated soil [47]. Thus, we argued that more efficient accumulation of organic C in CM–treated soil was not only attributable to higher annual inputs of organic C as compost or crop residues, but also to lower

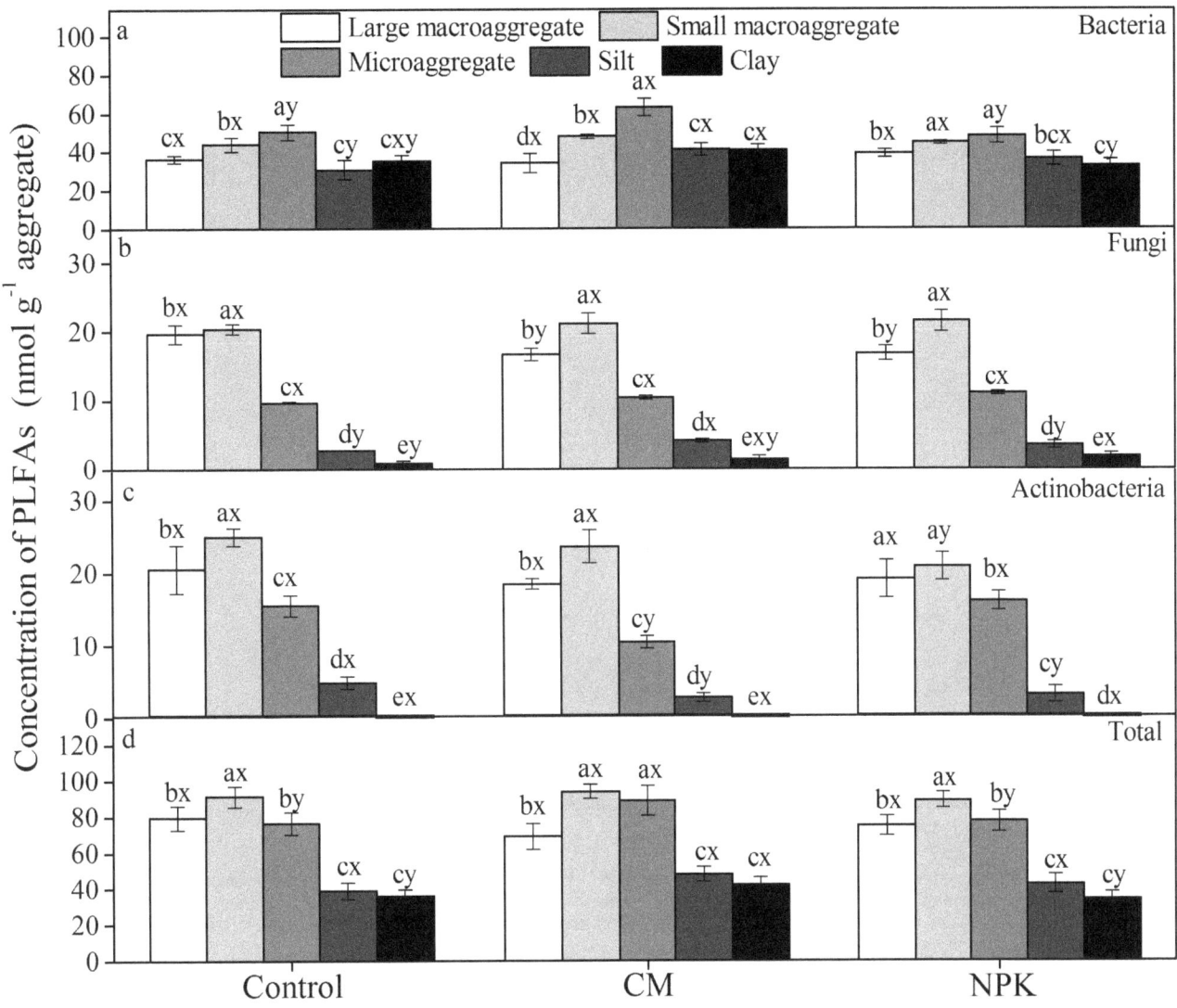

Figure 3. Effect of long–term applications of fertilizer NPK and compost on concentrations (nmol g⁻¹ aggregate) of bacterial, fungal, actinobacterial and total PLFA in aggregates. Values are means ($n = 4$) with standard error. Different letters denote significant differences between aggregates with the same treatment (a, b, c, d, e) and between treatments with the same aggregate (x, y, z) at $P < 0.05$, respectively.

decomposition rates of per unit organic C. In the present study, fungi and actinobacteria, which prefer particulate organic matter (POM), were reduced more efficiently in compost–treated soil, which might lead to the recalcitrant organic C accumulation in the CM treatment. However, correlation analysis showed that fungi and actinobacteria positively correlated with the increase in organic C in aggregates, possibly because they were mainly distributed in macroaggregates.

Killham *et al.* [48] and Dungait *et al.* [49] pointed out that the turnover of organic C depends on the accessibility of the OC to decomposer organisms or catalytic enzymes rather than its recalcitrance and location in the soil pore network. Organic C turnover was faster in pores with neck diameters greater than 4 µm than in those with smaller neck diameters [50]. Likewise, a previous study found the mineralization ratio of added fructose ¹³C in pores >291 µm to be 41.4% during a 13–day incubation, being significantly higher than in pores <97 µm [51]. In the present study, organic C in macroaggregates showed a more rapid rate of increase than in other aggregates in spite of the fact that

macroaggregates had higher abundances of microorganisms, especially fungi and actinobacteria as discussed above. Although it is possible that more exogenous organic material was accumulated in macroaggregates [14], we found that the M/B ratio declined in parallel with an increase in aggregate size (Fig. 4), and was negatively related to the rise in organic C in aggregates (Fig. 5). Our findings suggest that organic C turnover in aggregates is also dependent on the abundance and composition of the microbial community.

Ruamps *et al.* [51] suggested that the pore–scale distribution patterns of microorganisms did not occur by chance, but were the result of interactions between microorganisms and their habitat. Blagodatsky and Smith [52] found that the aeration of intra–aggregates changed with the formation of aggregates and could be completely anoxic. The increased organic C in microaggregates (free or within macroaggregates) caused an increase in pore–filling organic matter (mainly as POM or amorphous organic materials), and in turn reduced the proportion of large pores [53,54]. As a result, the proportion of pores <4 µm increased significantly in

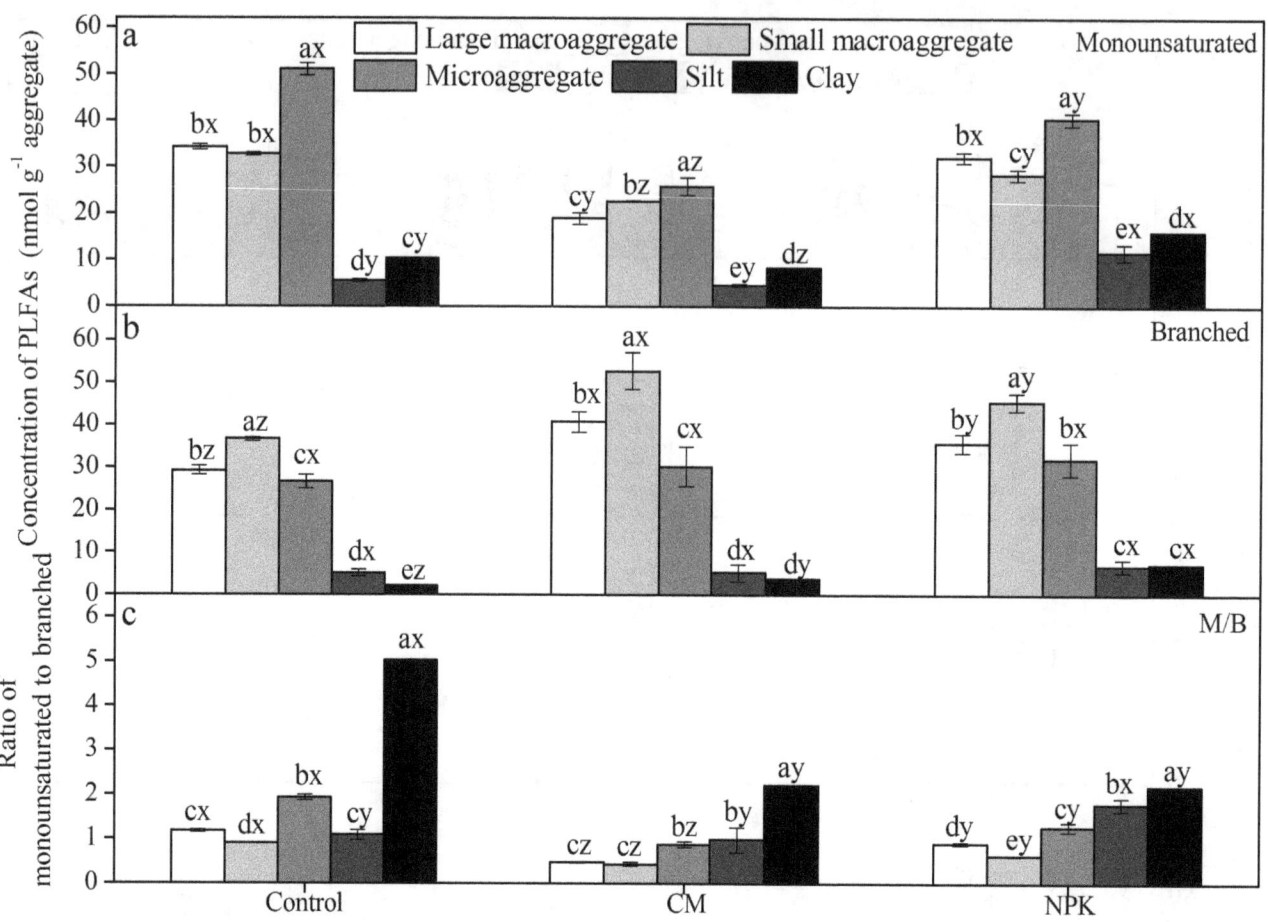

Figure 4. Effect of long–term applications of fertilizer NPK and compost on concentrations (nmol g⁻¹ aggregate) of monounsaturated and branched PLFA in aggregates. Values are means ($n=4$) with standard error. Different letters denote significant differences between aggregates with the same treatment (a, b, c, d, e) and between treatments with the same aggregate (x, y, z) at $P<0.05$, respectively.

compost–treated soil while the effective diffusion coefficient of oxygen in both treatments significantly decreased in our study (Fig. 2). Soil aeration and pore networks are known to play a large role in the sequestration and turnover of organic C [34], and when the oxygen concentration in soil air reduced to ≤10%, the accumulation of organic C increased by slowing the oxidation of soluble forms of organic C [55], because aerobes decompose organic C more efficiently than facultative or obligate anaerobes [56]. The M/B ratio has generally been found to decrease with the development of anaerobic conditions or reduction in oxygen availability in soils [36,57]. Monounsaturated PLFAs mainly represent G⁻ bacteria, which can utilize a variety of organic C sources and promote the decomposition of organic C under well–aerated conditions [58,59], while branched PLFAs being dominantly composed of G⁺ bacteria [36], which possess a greater proportion of peptidoglycan, which contains significant quantities of N–acetylglucosamine that is a precursor of relatively decay–resistant organic matter [60]. Thus, we considered that the aeration status would improve as aggregate size reduced [61], which in turn would control the activity and composition of microorganisms and organic C mineralization in aggregates. The more effective accumulation of organic C in macroaggregates than in other aggregates was at least partly due to the higher concentrations of branched PLFAs and low M/B ratios. However, the organic C accumulation was least in microaggregates in NPK

treatment, was mainly due to a low input of exogenous organic materials rather than the decomposition of organic C [54].

The increase in organic C in aggregates in NPK–treated soil was significantly ($P<0.05$) lower than in CM–treated soil. Yu et al. [54], in an incubation study using the same soils, demonstrated that compost more efficiently improved the stability of organic C, even labile organic C such as carbohydrate, in the silt plus clay fraction compared to NPK. John [62] suggested that the high turnover of organic C in the clay fraction was caused by a relatively high enrichment of organic C from fresh residues and/or microbial biomass. In NPK–treated soil, the newly increased organic C was mainly derived from root residues and exudates, while also from compost in CM–treated soil. Compost had been fermented for 2 months before use. During composting, the proportion of labile, hydrophilic, plant–derived organic compounds gradually reduced. In contrast, that of more stable hydrophobic moieties, including lignin–derived phenols and microbial–derived carbohydrates, increased [63]. When these microbial–derived carbohydrates are incorporated into the soil, they are not usually utilized by microorganisms and can become stabilized by mineral particles. It is likely that low annual input and high decomposability of exogenous organic material in NPK–treated soil resulted in low increases in organic C in the silt and clay fractions compared to CM. Furthermore, the mass proportion of macroaggregates was exponentially related to OC concentra-

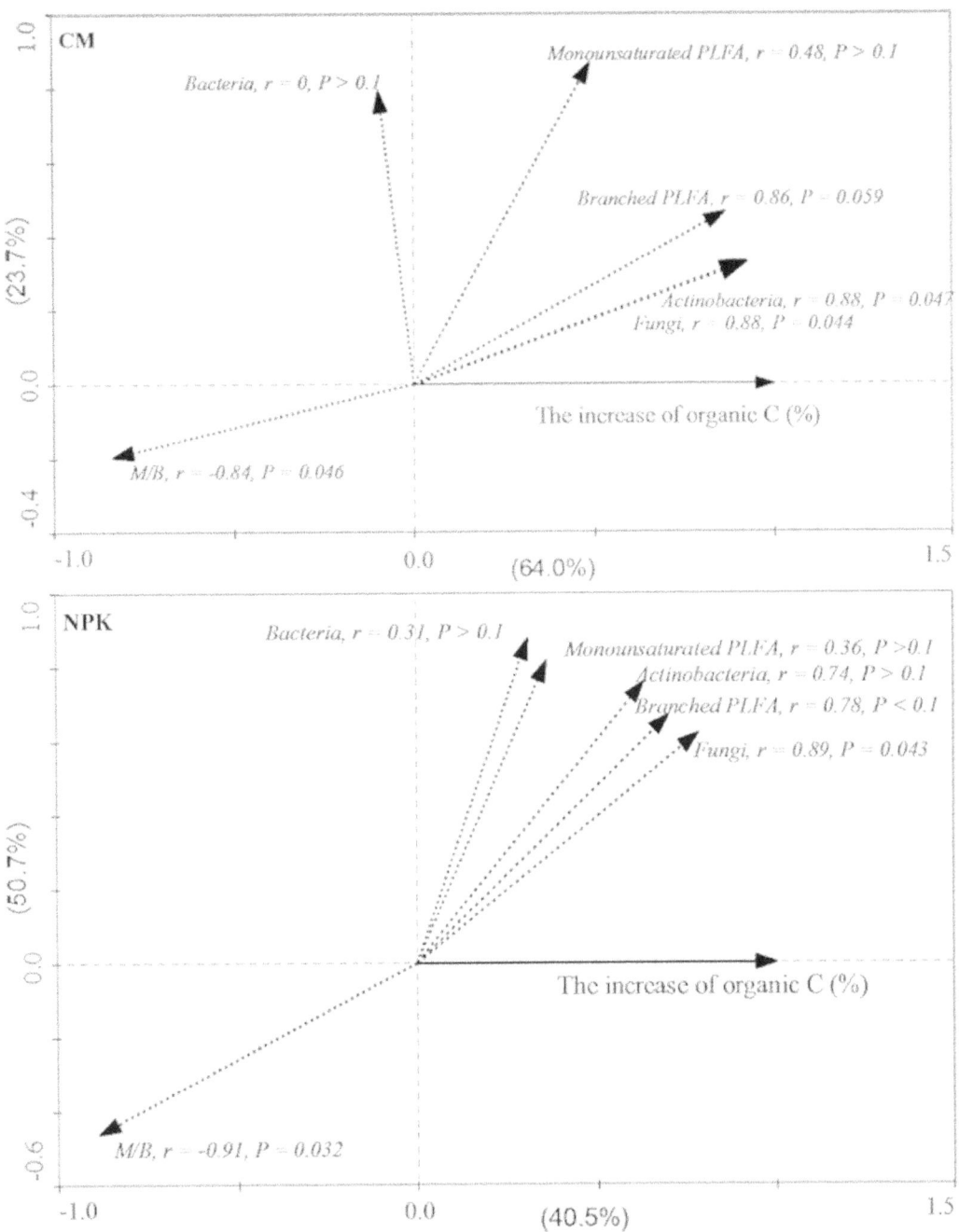

Figure 5. Relationship between increased organic C and the abundance of microbial PLFA (nmol g^{-1} aggregate) in soils. Redundancy analysis (RDA) and regression analyses were used to test relationships between the increased organic C concentration in the CM and NPK treatments compared with the control and the abundance of microorganisms in aggregates in the CM and NPK treatments.

tions in the silt plus clay fractions [54]. Greater input of recalcitrant OC and reduction in monounsaturated PLFAs is likely to have been critical for the observed increase in organic C in the silt and clay fractions. Our findings suggest that organic C accumulation in aggregates and aggregation increased the proportion of pores <4 μm and reduced the effective diffusion coefficient of oxygen, gradually leading to a change in microhabitats from those favorable for aerobes to those beneficial for facultative or obligate anaerobes in the CM–treated soil [37]; the shift in microbial community composition in turn altered organic C turnover and accumulation in aggregates.

Conclusions

Compost application more effectively increased organic C concentrations in aggregates, and accelerated the formation of macroaggregates than fertilizer NPK. The proportion of pores < 4 μm was significantly ($P<0.05$) increased in CM treatment, while the reduction effective diffusion coefficient of oxygen showed a decrease order: control>NPK>CM ($P<0.05$). The distribution pattern of microorganisms in aggregates was primarily controlled by aggregate size and responded to a lesser extent to management practices like fertilization. The concentration of actinobacteria were more significantly (P<0.05) reduced in all aggregates in CM–treated soil than in NPK–treated soil, whereas fungi were only reduced in macroaggregates under compost and NPK application compared with the control, resulting in recalcitrant organic C accumulated effectively in CM treatment. The M/B ratio, which was lowest in CM treatment in all aggregates, increased with the reduction of aggregate size in tested soils. Increased organic C in aggregates in CM– and NPK–treated soils was marginally (P<0.10) positively correlated with the concentration of branched PLFAs but significantly (P<0.05) negatively correlated with M/B ratios. More efficient accumulation of organic C in aggregates in CM–treated soil than in NPK–treated soil was probably due to the reduction of monounsaturated PLFAs and increase of branched PLFAs, which resulted from the change of micro–habitats in

aggregates from one more favorable for aerobes to that beneficial for facultative or obligate anaerobes with the formation of macroaggregates.

Acknowledgments

The authors thank Dr. Jingyu Huang from Hohai University for his assistance in sample analyses.

Author Contributions

Conceived and designed the experiments: WXD HYY HJZ. Performed the experiments: HJZ HYY. Analyzed the data: HJZ WXD XHH. Contributed reagents/materials/analysis tools: HJZ HYY JLF DYL. Wrote the paper: HJZ WXD XHH.

References

1. Anders MM, Beck PA, Watkins BK, Gunter SA, Lusby KS, et al. (2010) Soil aggregates and their associated carbon and nitrogen content in Winter Annual Pastures. Soil Sci Soc Am J 74: 1339–1347.

2. Helgason BL, Walley FL, Germida JJ (2010) No–till soil management increases microbial biomass and alters community profiles in soil aggregates. Appl Soil Ecol 46: 390–397.

3. Bossuyt H, Six J, Hendrix P (2002) Aggregate–protected carbon in no–tillage and conservation tillage agroecosystems using carbon–14 labeled plant residue. Soil Sci Soc Am J 66: 1965–1973.

4. Wright AL, Hons JM (2004) Soil aggregation and carbon and nitrogen storage under soybean cropping sequences. Soil Sci Soc Am J 68: 507–513.

5. Six J, Paustian K, Elliott ET, Combrink C (2000) Soil structure and organic matter. I. Distribution of aggregate–size classes and aggregate–associated carbon. Soil Sci Soc Am J 64: 681–689.

6. Bach EM, Baer SG, Meyer CK, Six J (2010) Soil texture affects soil microbial and structural recovery during grassland restoration. Soil Biol Biochem 42: 2182–2191.

7. Verchot LV, Dutaur L, Shepherd KD, Albrecht A (2011) Organic matter stabilization in soil aggregates: Understanding the biogeochemical mechanisms that determine the fate of carbon inputs in soils. Geoderma 161: 82–193.

8. Hassink J (1997) The capacity of soils to preserve organic C and N by their association with clay and silt particles. Plant Soil 191: 77–87.

9. Six J, Elliott ET, Paustian K (2000) Soil macroaggregate turnover and microaggregate formation: a mechanism for C sequestration under no–tillage agriculture. Soil Biol Biochem 32: 2099–2103.

10. Mikha MM, Rice CW (2004) Tillage and manure effects on soil aggregate associated carbon and nitrogen. Soil Sci Soc Am J 68: 809–816.

11. Aoyama M, Angers DA, N'Dayegamiye A (1999) Particulate and mineral–associated organic matter in water–stable aggregates as affected by mineral fertilizer and manure application. Can J Soil Sci 79: 295–302.

12. Manna MC, Swarup A, Wanjari RH, Singh YV, Ghosh PK, et al. (2006) Soil organic matter in a west Bengal inceptisol after 30 years of multiple cropping and fertilization. Soil Sci Soc Am J 70: 121–129.

13. Su YZ, Wang F, Suo DR, Zhang ZH, Du MW (2006) Long–term effect of fertilizer and manure application on soil–carbon sequestration and soil fertility under the wheat–wheat–maize cropping system in northwest China. Nutr Cycl Agroecosys 75: 285–295.

14. Kong AYY, Six J, Bryant DC, Denison RF, van Kessel C (2005) The relationship between carbon input, aggregation, and soil organic carbon stabilization in sustainable cropping systems. Soil Sci Soc Am J 69: 1078–1085.

15. Sarkar S, Singh SR, Singh RP (2003) The effect of organic and inorganic fertilizers on soil physical condition and the productivity of a rice–lentil cropping sequence in India. J Agric Sci 140: 419–425.

16. Bhattacharyya R, Prakash V, Kundu S, Srivastva AK, Gupta HS, et al. (2009) Long–term effects of fertilization on carbon and nitrogen sequestration and aggregate associated carbon and nitrogen in the Indian sub–Himalayas. Nutr Cycl Agroecosys 86:1–16.

17. Böhme L, Langer U, Böhme F (2005) Microbial biomass, enzyme activities and microbial community structure in two European long–term field experiments. Agri Ecosyst Environ 109: 141–152.

18. Billings SA, Ziegler SE (2008) Altered patterns of soil carbon substrate usage and heterotrophic respiration in a pine forest with elevated CO_2 and N fertilization. Global Change Biol 14: 1025–1036.

19. Kindler R, Miltner A, Thullner M, Richnow HH, Kästner M (2009) Fate of bacterial biomass derived fatty acids in soil and their contribution to soil organic matter. Org Geochem 40: 29–37.

20. Sessitsch A, Weilharter A, Gerzabek MH, Kirchmann H, Kandeler E (2001) Microbial population structures in soil particle size fractions of a long–term fertilizer field experiment. Appl Environ Microbiol 67: 4215–4224.

21. Jastrow JD, Miller RM, Lussenhop J (1998) Contributions of interacting biological mechanisms to soil aggregate stabilization in restored prairie. Soil Biol Biochem 30: 905–916.

22. Jiang XJ, Wright AL, Wang X, Liang F (2011) Tillage–induced changes in fungal and bacterial biomass associated with soil aggregates: A long–term field study in a subtropical rice soil in China. Appl Soil Ecol 48: 168–173.

23. Alagöz Z, Yilmaz E (2009) Effects of different sources of organic matter on soil aggregate formation and stability: A laboratory study on a Lithic Rhodoxeralf from Turkey. Soil Till Res 103: 419–424.

24. Soil Survey Staff (2010) Keys to soil taxonomy, 11th edn. USDA NRCS, Washington DC, USA 106–111.

25. Carter MR (1993) Soil samples and methods of analysis. Lewis Publishers, Boca Raton., 190–191.

26. van Genuchten MT (1980) A closed–form equation forpredicting the hydraulic conductivity of unsaturated soils. Soil Sci Soc Am J 44: 892–898.

27. Kutílek M, Nielsen DR (1994) Soil Hydrology. Catena Verlag, Cremlingen Destedt.

28. Davis GB, Power TR, Briegel D, Patterson BM (1998) BTEX vapour biodegradation rates in the vadose zone: initial estimates. Groundwater Quality: Remediation and Protection 250: 300–303.

29. Elliot ET (1986) Aggregate structure and carbon, nitrogen and phosphorus in native and cultivated soils. Soil Sci Soc Am J 50: 518–524.

30. Brant JB, Sulzman EW, Myrold DD (2006) Microbial community utilization of added carbon substrates in response to long–term carbon input manipulation. Soil Biol Biochem 38: 2219–2232.

31. Bossio DA, Scow KM, Gunapala N, Graham KJ (1998) Determinants of soil microbial communities: effects of agricultural management, season and soil type on phospholipid fatty acid profiles. Microb Ecol 36: 1–12.

32. Frostegard A, Bååth E (1996) The use of phospholipid fatty acid analysis to estimate bacterial and fungal biomass in soil. Biol Fertil Soils 22: 59–65.

33. Olsson PA (1999) Signature fatty acids provide tools for determination of the structure and interactions of mycorrhizal fungi in soil. FEMS Microbiol Ecol 29: 303–310.

34. Sundh I, Nilsson M, Borga P (1997) Variation in microbial community structure in two boreal peatlands as determined by analysis of phospholipid fatty acid profiles. Appl Environ Microbiol 63: 1476–1482.

35. Spring S, Schulze R, Overmann J, Schleifer KH (2000) Identification and characterization of ecologically significant prokaryotes in the sediment of freshwater lakes: molecular and cultivation studies. FEMS Microbiol Rev 24: 573–590.

36. Bossio DA, Fleck JA, Scow KM, Fujii R (2006) Alteration of soil microbial communities and water quality in restored wetlands. Soil Biol Biochem 38: 1223–1233.

37. Rajendran N, Matsuda O, Imamura N, Urushigawa Y (1992) Variation in microbial biomass and community structure in sediments of Eutrophic Bays as determined by phospholipid ester–linked fatty acids. Appl Environ Microbiol 58: 562–571.

38. Ladd JN, VanGestel M, Monrozier LJ, Amato M (1996) Distribution of organic C–14 and N–15 in particle–size fractions of soils incubated with C–14, N–15–labelled glucose/NH4, and legume and wheat straw residues. Soil Biol Biochem 28: 893–905.

39. Petersen SO, Debosz K, Schjønning P, Christensen BT, Susanne E (1997) Phospholipid fatty acid profiles and C availability in wet–stable macro–aggregates from conventionally and organically farmed soils. Geoderma 78: 181–196.

40. Kandeler E, Tscherko D, Bruce KD, Stemmer M, Hobbs PJ, et al. (2000) Structure and function of the soil microbial community in microhabitats of a heavy metal polluted soil. Biol Fertil Soils 32: 390–400.

41. Chenu C, Hassink J, Bloem J (2001) Short–term changes in the spatial distribution of microorganisms in soil aggregates as affected by glucose addition. Biol Fertil Soils 34: 349–356.

42. Beguin P, Aubert J (2006) The biological degradation of cellulose. FEMS Microbiol Rev 13: 25–58.

43. Jurgens K, Pernthaler J, Schalla S, Amann R (1999) Morphological and compositional changes in a planktonic bacterial community in response to enhanced protozoan grazing. Appl Environ Microbiol 65: 1241–1250.

44. Kölbl A, Kögel–Knabner I (2004) Content and composition of free and occluded particulate organic matter in a differently textured arable Cambisol as revealed by solid–state ^{13}C NMR spectroscopy. J Plant Nutr Soil Sci 167: 45–53.

45. Singh S, Singh JS (1995) Microbial biomass associated with water–stable aggregates in forest, savanna and cropland soils of a seasonally dry tropical region, India. Soil Biol Biochem 27: 1027–1033.

46. Ding WX, Meng L, Yin YF, Cai ZC Zheng XH (2007) CO_2 emission in an intensively cultivated loam as affected by long–term application of organic manure and nitrogen fertilizer. Soil Biol Biochem 39: 669–679.

47. Yu HY, Ding WX, Luo JF, Geng RL, Ghanni A, et al. (2012) Effects of long–term compost and fertilizer application on stability of aggregates–associated organic carbon in intensively cultivated sandy loam soil. Biol Fertil Soils 48: 325–336.

48. Killham K, Amato M, Ladd J (1993) Effect of substrate location in soil and soil porewater regime on carbon turnover. Soil Biol Biochem 25: 57–62.

49. Dungait JAJ, Hopkins DW, Gregory AS, Whitmore AP (2012) Soil organic matter turnover is governed by accessibility not recalcitrance. Global Change Biol 18: 1781–1796.

50. Strong DT, Wever HD, Merckx R, Recous S (2004) Spatial location of carbon decomposition in the soil pore system. Eur J Soil Sci 55: 739–750.

51. Ruamps LM, Nunan N, Chenu C (2011) Microbial biogeography at the soil pore scale. Soil Biol Biochem 43: 280–286.

52. Blagodatsky S, Smith P (2012) Soil physics meets soil biology: Towards better mechanistic prediction of greenhouse gas emissions from soil. Soil Biol Biochem 47: 78–92.

53. Zhuang J, McCarthy JF, Perfect E, Mayer LM, Jastrow JD (2008) Soil water hysteresis in water–stable microaggregates as affected by organic matter. Soil Sci Soc Am J 72: 212–220.

54. Yu HY, Ding WX, Luo JF, Geng RL, Cai ZC (2012) Long–term application of organic manure and mineral fertilizers on aggregation and aggregate–associated carbon in a sandy loam soil. Soil Till Res 124: 170–177.

55. Zibilske LM, Bradford JM (2007) Oxygen effects on carbon, polyphenols and nitrogen mineralization potential in soil. Soil Sci Soc Am J 71: 133–139.

56. Ding HB, Sun MY (2005) Biochemical degradation of algal fatty acids in oxic and anoxic sediment–seawater interface systems: effects of structural association and relative roles of aerobic and anaerobic bacteria. Marine Chem 93: 1–19.

57. Wixon DL, Balser TC (2013) Toward conceptual clarity: PLFA in warmed soils. Soil Biol Biochem 57: 769–774.

58. Ponder F, Tadros M (2002) Phospholipid fatty acids in forest soil four years after organic matter removal and soil compaction. Appl Soil Ecol 19: 173–182.

59. Feng XJ, Simpson MJ (2009) Temperature and substrate controls on microbial phospholipid fatty acid composition during incubation of grassland soils contrasting in organic matter quality. Soil Biol Biochem 41: 804–812.

60. Simpson AJ, Song G, Smith E, Lam B, Novotny EH, et al. (2007) Unraveling the structural components of soil humin by use of solution–state nuclear magnetic resonance spectroscopy. Environ Sci Technol 41: 876–883.

61. Rappoldt C, Crawford JW (1999) The distribution of anoxic volume in a fractal model of soil. Geoderma 88: 329–347.

62. John B (2003) Carbon turnover in aggregated soils determined by natural ^{13}C abundance. Ph.D. thesis, Georg–August–Universität Göttingen, Germany.

63. Said–Pullicino D, Kaiser K, Guggenberger G, Gigliotti G (2007) Changes in the chemical composition of water–extractable organic matter during composting: distribution between stable and labile organic matter pools. Chemosphere 66: 2166–2176.

21

Complex Effects of Fertilization on Plant and Herbivore Performance in the Presence of a Plant Competitor and Activated Carbon

Nafiseh Mahdavi-Arab[1,2], Sebastian T. Meyer[2]*, Mohsen Mehrparvar[3], Wolfgang W. Weisser[2]

1 Institute of Ecology, Friedrich Schiller University Jena, Jena, Germany, **2** Terrestrial Ecology Research Group, Department of Ecology and Ecosystem Management, School of Life Sciences Weihenstephan, Technische Universität München, Freising, Germany, **3** Department of Biodiversity, Institute of Science and High Technology and Environmental Sciences, Graduate University of Advanced Technology, Kerman, Iran

Abstract

Plant-herbivore interactions are influenced by host plant quality which in turn is affected by plant growth conditions. Competition is the major biotic and nutrient availability a major abiotic component of a plant's growth environment. Yet, surprisingly few studies have investigated impacts of competition and nutrient availability on herbivore performance and reciprocal herbivore effects on plants. We studied growth of the specialist aphid, *Macrosiphoniella tanacetaria*, and its host plant tansy, *Tanacetum vulgare*, under experimental addition of inorganic and organic fertilizer crossed with competition by goldenrod, *Solidago canadensis*. Because of evidence that competition by goldenrod is mediated by allelopathic compounds, we also added a treatment with activated carbon. Results showed that fertilization increased, and competition with goldenrod decreased, plant biomass, but this was likely mediated by resource competition. There was no evidence from the activated carbon treatment that allelopathy played a role which instead had a fertilizing effect. Aphid performance increased with higher plant biomass and depended on plant growth conditions, with fertilization and AC increasing, and plant competition decreasing aphid numbers. Feedbacks of aphids on plant performance interacted with plant growth conditions in complex ways depending on the relative magnitude of the effects on plant biomass and aphid numbers. In the basic fertilization treatment, tansy plants profited from increased nutrient availability by accumulating more biomass than they lost due to an increased number of aphids under fertilization. When adding additional fertilizer, aphid numbers increased so high that tansy plants suffered and showed reduced biomass compared with controls without aphids. Thus, the ecological cost of an infestation with aphids depends on the balance of effects of growth conditions on plant and herbivore performance. These results emphasize the importance to investigate both perspectives in plant herbivore interactions and characterize the effects of growth conditions on plant and herbivore performance and their respective feedbacks.

Editor: Daniel Doucet, Natural Resources Canada, Canada

Funding: This study was financially supported by grant WE 3081/15-2 and by a TUM Diversity Scholarship to NMA. Publication of this work was supported by the German Research Foundation (DFG) and the Technische Universität München within the funding program open access publishing. The funders had no role in study design, data collection and analysis, decision to publish, or preparation of the manuscript.

Competing Interests: The authors have declared that no competing interests exist.

* Email: sebastian.t.meyer@tum.de

Introduction

Understanding the mechanisms underlying the ecological interactions between insect herbivores and their host plants has been an important goal in ecology for a long time [1]. A key factor influencing these interactions is host plant quality [2]. Plant quality, defined as a general term, includes all physical, chemical or biological traits of a plant relevant for its herbivores (e.g. size and structure, phenology, secondary compounds, and nutritional status). Variation in plant quality influences herbivore-plant interactions [1,3–5] and consequently herbivore performance such as growth rates, fecundity, and survivorship [4,6–9]. Because the herbivores themselves influence plant growth, the fitness of a plant with herbivores is not only affected by factors such as fertilization or plant-plant-competition, but also by how these

growth conditions act on the feeding herbivore and resulting herbivore feedback effects on the plant. To disentangle direct from indirect effects, studies investigating the influence of growth conditions on plant and herbivore performance should therefore explicitly address these feedback effects, e.g. by rearing plants with and without the herbivore under identical conditions.

Nutrient availability is a major abiotic component of a plant's growth environment. Increasing the availability of nutrients, e.g. by the application of fertilizer, increases plant growth and affects plant quality [1,6]. Most studies that investigated the effects of host plant nutrient availability for insect herbivores have focused on effects of fertilization [6,10,11]. In general, the application of fertilizer is expected to increase the abundance of herbivores feeding on the plant, because a higher concentration of primary metabolites [12] or a reduction in plant anti-herbivore defenses

[13] will lead to increasing plant nutritional quality. In most studies, fertilization indeed increased insect abundance, due to changes in insect feeding preference and food consumption that resulted in shorter development times, higher rates of survival and higher fecundity [6,9,10,14,15]. However, there are some reports in which fertilization had no effect [16] or even a negative effect on herbivore performance, due to adverse changes in plant physiology at a high fertilizer level [17]. In addition, soil fertility can interact with other factors in the plant's growth environment, such as the presence of competitors or herbivores, making it a challenge to predict how addition of fertilizer affects plant and herbivore performance.

One of the main biotic factors that limits resource availability for plants is competition [18,19]. Individuals that suffer from competition typically show a reduction in nutrient uptake, growth rate, survival and fecundity. A decrease in the nutritional quality of a plant in competition can also decrease the performance of the herbivore feeding on it [20]. Herbivores generally reduce competitive abilities of attacked plants compared with unaffected neighbors [21], but their effects on the competitiveness of plants may depend on the identity of competing plant species, abiotic conditions, and the type, intensity and timing of herbivore damage [22]. Some experiments found that attacks by specialist herbivores had no [23], or even positive effects on plant competitive abilities [24]. For example, when aphids fed on *Poa* grasses, the effects of competition by forbs on above- and belowground *Poa* biomass were reduced compared with the situation without aphid infestation on *Poa* [20]. Despite many studies investigating the effects of herbivores and competition on plant performance [22–25], there is a lack of knowledge of how plant-plant competition interacts with herbivore effects on plants.

One mode of competition between plants is to release allelochemicals (toxic metabolites) into the environment [26]. Such allelopathic plants, often found among invasive species, can have strong effects on seed germination, growth or other fitness parameters of competitors [27–30]. For example, root exudates and root extracts of goldenrod, *Solidago canadensis*, have been documented to have an inhibitory effect on the growth of neighboring plants [31,32]. To separate negative effects of allelopathy from those of resource competition between two co-occurrence species, various studies used activated carbon (AC) [30,31,33]. AC can neutralize large organic compounds in the soil through adsorption, mechanical filtration, ion exchange, or surface oxidation [34,35].

In this study we investigated the direct and indirect effects of variation in plant growth condition on plant and herbivore performance, using the specialized aphid *Macrosiphoniella tanacetaria* on its host plant *Tanacetum vulgare*, tansy, and the plant competitor *Solidago canadensis*, goldenrod, as a model system. To manipulate plant growth conditions, tansy plants were fertilized and grown in competition with goldenrod. Importantly, the aphids used do not feed on goldenrod, thereby the effect of competition on the host plant was not confounded by the provision of an additional host for the herbivore. We added an AC-treatment to separate allelopathic from other competitive effects of goldenrod on tansy and measured both the performance of host plant and herbivore in response to manipulated growth conditions.

We had the following hypotheses (H1–H4): H1 - fertilization will increase plant (H1a) and aphid (H1b) performance by increasing plant biomass and aphid numbers. H2 - the addition of AC will increase plant (H2a) and aphid (H2b) performance in the presence of an allelopathic plant competitor, because it releases the host plant from competition. H3: plant-plant competition will decrease both host plant (H3a) and herbivore (H3b) performance.

H4: herbivory effects on plants depend on host plant growth conditions.

Materials and Methods

Experimental plants and aphids

The tansy aphid *Macrosiphoniella tanacetaria* ((Kaltenbach), Hemiptera: Aphididae) is a specialist herbivore on tansy. This species produces both sexual and asexual morphs (holocyclic) and spends its complete life-cycle on tansy (*Tanacetum vulgare*, L. Asteraceae). The aphid is not ant–attended [36] and feeds in loose colonies mainly on the tip of shoots.

Tansy is a perennial, herbaceous plant, native to Europe and Asia and has been introduced to America and Australia [37]. Natural habitats can be found in subalpine mountain river valleys in Siberia and Europe. Today, tansy is common in riverbanks, wastelands, along roadsides, and in rural and urban-industrial areas [38]. Tansy hosts more than 23 aphid species globally [39,40] among which *M. tanacetaria* is one of the most abundant [41].

Goldenrod (*Solidago canadensis*, L. Asteraceae) is native to North America and has become one of the most aggressive invaders in Europe occurring in the same habitats as tansy [38]. Both plants are comparable in size. There is evidence for allelopathic effects of goldenrod on co-occurring plants [31,42]. For *M. tanacetaria* goldenrod is not a suitable host plant.

Experimental design

In a fully factorial experimental design, a fertilizer treatment (three levels: control, inorganic fertilizer (F_{inorg}), inorganic and organic fertilizer ($F_{inorg+org}$)) was crossed with an AC treatment (two levels: with and without AC), a plant competition treatment (two levels: with and without competition by goldenrod) and an aphid treatment (two levels: with and without aphids), resulting in $3 \times 2 \times 2 \times 2 = 24$ treatment combinations.

The experiment was conducted in one liter pots ($11 \times 11 \times 12$ cm) in a greenhouse. As soil we used field soil excavated during maintenance of the Jena-Experiment and stored as pile on the field site (Jena, Thuringia, Germany; 50°55' N, 11°35' E) provided by the management of the Jena-Experiment (Anne Ebeling, Friedrich-Schiller-University, Jena). For the F_{inorg} treatment, the soil was mixed with 1 g of Osmocote (Hermann Meyer KG, Rellingen, Germany) per pot, a slow release NPK fertilizer. For the $F_{inorg+org}$ treatment, soil was mixed with unsterilized commercial peat soil containing organic humus (1:1) and 1 g of Osmocote was added. In the AC treatment, 8 g of finely ground AC were added, particle size <0.8 mm (Carl Roth Gmbh & Co. KG, Karlsruhe, Germany), as recommended by Abhilasha et al. [31]. We replicated every treatment combination 5 times, yielding a total of $2 \times 2 \times 3 \times 2 \times 5 = 120$ pots. Each set of replicates formed one of 5 blocks. No permits were required for this study. The study did not involve endangered or protected species.

Seeds where collected from naturally occurring tansy and goldenrod plants around the city of Jena, Germany, in 2010. Seeds from about 50 different plant individuals were mixed and seeds for the pots randomly drawn from this mixture. To grow the plants, about 20 seeds of tansy and/or goldenrod were sown directly into the experimental pots in May 2011 and maintained under controlled greenhouse conditions (temperature of ~25°C during the day and ~20°C at night; light regime of 16 h light: 8 h darkness). Plants germinated within the first two weeks after which the strongest individual from each species was kept in the pot and the rest was removed, resulting in pots with single tansy plants and pots with one tansy and one goldenrod plant. There was no treatment with competition by a second tansy plant as our study focused on the effects of interspecific competition with plants that are not suitable host plants to the herbivore.

In August, when plants were three months old, plants in the aphid treatment were infested with five unwinged adult aphids that were collected from the field. The adult females were removed after three days and 10 nymphs were left to grow and reproduce for two weeks. To avoid cross infection between plants by escaping aphids, all pots (with and without aphids) were covered with air-permeable perforated (<1 mm) polyethylene bags (20×35 cm) fixed to the pots with elastic bands. This transparent cover permitted the visual assessment of aphids on the plants without disturbing them. Two weeks after the start of the experiment, aphid numbers were counted, the aphids were removed, and aboveground parts of the plants were harvested, dried to constant mass at 70°C for 48 hours, and weighed.

For plant biomass, we calculated log response ratios (plant biomass LogRR) to directly quantify the effects of aphid infestation on plant growth, by calculating, for each particular combination of the fertilizer, competition and AC treatment, the log of the biomass of plants with aphids divided by the biomass of plants without aphids. Values of plant biomass LogRR <0 indicate that the biomass of plants without aphids is higher than the biomass of plants with aphids and thus aphid presence decreases plant biomass. In contrast, LogRR values >0 indicate that aphid presence increases plant biomass.

In addition, we calculated aphid load as aphid number per unit (g) plant biomass [43] to compare the effects of growth conditions on aphids relative to their effects on plants.

Statistical analysis

Linear mixed-effects models were used to analyze treatment effects on plant biomass. All treatments, i.e. the aphid, fertilizer, AC and competition treatment and their interactions were fitted as fixed factors in the model. Aphid numbers were analyzed in a similar way. In this model the biomass of plants not infected with aphids but exposed to the same other treatments in the same block was used as covariate because this biomass was not confounded with the aphid effect of reducing plant biomass. Also, plant biomass LogRR and aphid load were analyzed in a similar model, by fitting the fertilizer, AC and competition treatments together with their interactions. All models included block as a random factor and were estimated using the function "lme" in the nlme package [44] using version 2.14.1 of the R software [45]. Variables were log-transformed as necessary (indicated in Table 1). In all analyses, non-significant terms were removed during model simplification (in the order of least significance given in Table 1). All data are presented as means ± standard error (SE). Additional linear regressions were used to analyze the effects plant biomass (with or without aphids) and aphid numbers on the resulting plant biomass LogRR. These models we fit using the "lm" function and combined plants from all additional treatments (fertilizer, AC, competition).

Results

Tansy biomass

Effects of aphid infestation on tansy biomass. Infestation by aphids significantly decreased tansy biomass compared with uninfested control plants (Table 1, Fig. 1A). Consequently, average plant biomass LogRR was strongly negative (Table 1, Fig. 1B). Additional analyzes showed that plant biomass LogRR was independent of the biomass of tansy plants, i.e. the effect of

aphids on plants was not simply a function of the size of the host plant: neither the biomass at the end of the experiment of the plants with ($F_{1,57} = 2.28$; p = 0.136) nor of plants without aphids ($F_{1,57} = 2.304$; p = 0.134) explained variation in observed plant biomass LogRR. In contrast, plant biomass LogRR decreased with higher aphid numbers ($F_{1,57} = 3.97$; p = 0.05; Fig. 2). Thus, aphid effects on tansy plant biomass increased with the size of aphid colonies. None of the plants in any of the treatments flowered by the time the experiment was ended.

Effects of fertilizer on tansy biomass. In accordance with H1a, fertilization significantly increased tansy biomass (Table 1). Specifically, adding F_{inorg} or $F_{inorg+org}$ increased tansy biomass by two-fold compared with untreated control plants (Fig. 1A).

Effects of AC on tansy biomass. Results of AC addition were different than expected: adding AC to the soil generally increased tansy biomass but, in contrast to H2a, this effect was independent of the presence of goldenrod. Rather, the effect of AC depended on the fertilizer and aphid treatment. Addition of AC showed no effect in control soil, but increased tansy biomass when F_{inorg} was added. Adding AC together with $F_{inorg+org}$, did not increase tansy biomass in the absence of aphids and reduced biomass in the presence of aphids (Table 1, Fig. 1A) likely as a results of highest aphid numbers in this treatment combination (Table 1, Fig. 1A). As a consequence, plant biomass LogRR was more negative when AC was added in the soil with $F_{inorg+org}$ compared with the addition of just F_{inorg} (Table 1, Fig. 1B).

Effects of competition on tansy biomass. The presence of goldenrod decreased tansy biomass confirming H3a (Table 1, Fig. 1A), especially for plants without aphids. The reduction of tansy biomass by competition was reduced in aphid-infested plants (Table 1, Fig. 1A). Therefore, plant biomass LogRR was less negative for tansy plants with than without competition (Table 1, Fig. 1B, Fig. S1).

Summary of effects on tansy biomass. Infestation with aphids reduced plant performance and larger colonies had more negative effects on tansy biomass. As expected, fertilizer increased and competition decreased tansy biomass, and the negative effect of competition was smaller in the presence of herbivores. In contrast to our expectations there was no evidence for an allelopathic effect of goldenrod. AC increased tansy performance independent of the presence of the potentially allelopathic competitor.

Aphid performance

Relationship between plant biomass and aphid performance and aphid load. Aphid performance, measured as aphid numbers, increased with higher plant biomass (Table 1, Fig. 3) and depended on plant growth conditions, as detailed below. Aphid load, i.e. the ratio of aphid number and plant biomass, thus depended on the effects of growth conditions on both aphid performance (see below) and plant performance (see above).

Effects of fertilizer on aphid performance and aphid load. Adding fertilizers increased aphid numbers as predicted in H1b (Table 1, Fig. 1C, D). Interestingly, adding only F_{inorg} increased aphid numbers to intermediate levels, while adding $F_{inorg+org}$ to the soil increased aphid numbers by a factor of about two compared with controls. While both the number of aphids and plant biomass increased with fertilization, the response of the plant was stronger and consequently, aphid load was lower for fertilized plants compared with control plants (Table 1, Fig. 1).

Effects of AC on aphid performance. Adding AC to the soil tended to increased aphid numbers and, in contrast to H2b, this effect was independent of the presence of goldenrod (Table 1).

Table 1. Results from linear mixed-effects models for plant and aphid performance as depending on host plant growth conditions.

Treatment	Tansy biomass* N = 120	Plant biomass LogRR N = 60	Aphid number* N = 60	Aphid load* N = 60
AP/TB°	$F_{2,101} = 87.0$; p<<0.001	–	$F_{1,42} = 426$; p<<0.001	–
Fertilizer	$F_{2,101} = 812$; p<<0.001	$F_{2,52} = 0.28$; p = 0.759	$F_{2,42} = 16.8$; p<<0.001	$F_{2,49} = 83.4$; p<<0.001
AC	$F_{2,101} = 31.3$; p<<0.001	$F_{1,52} = 2.73$; p = 0.104	$F_{1,42} = 12.0$; p = 0.001	$F_{1,49} = 0.53$; p = 0.470
Competition	$F_{2,101} = 44.1$; p<<0.001	$F_{1,52} = 9.04$; p = 0.004	$F_{1,42} = 0.64$; p = 0.428	$(F_{1,48} = 0.40$; p = 0.527)^4
AP/TB×Fertilizer	$F_{2,101} = 0.20$; p = 0.817	–	$F_{2,42} = 3.18$; p = 0.051	–
AP/TB×AC	$F_{1,101} = 2.52$; p = 0.115	–	$F_{1,42} = 51.1$; p<<0.001	–
Fertilizer×AC	$F_{2,101} = 51.7$; p<<0.001	$F_{2,52} = 9.66$; p = 0.001	$F_{2,42} = 6.98$; p = 0.002	$F_{2,49} = 45.7$; p<<0.001
AP/TB×Competition	$F_{1,101} = 7.54$; p = 0.007	–	$(F_{1,41} = 0.05$; p = 0.826)^7	–
Fertilizer×Competition	$(F_{2,99} = 1.61$; p = 0.205)^6	$(F_{2,50} = 0.25$; p = 0.782)^3	$F_{2,42} = 3.77$; p = 0.031	$(F_{2,46} = 2.08$; p = 0.136)^3
AC×Competition	$(F_{1,98} = 0.94$; p = 0.334)^5	$(F_{1,47} = 0.44$; p = 0.510)^2	$(F_{1,40} = 0.80$; p = 0.376)^6	$(F_{1,45} = 0.64$; p = 0.426)^2
AP/TB×Fertilizer×AC	$F_{2,101} = 7.89$; p = 0.001	–	$(F_{2,36} = 0.18$; p = 0.838)^4	–
AP/TB×Fertilizer×Competition	$(F_{2,96} = 0.22$; p = 0.806)^4	–	$(F_{2,34} = 0.84$ p = 0.439)^3	–
AP/TB×AC×Competition	$(F_{1,95} = 0.38$; p = 0.539)^3	–	$(F_{1,33} = 0.11$; p = 0.738)^2	–
Fertilizer×AC×Competition	$(F_{2,93} = 2.18$; p = 0.118)^2	$(F_{2,47} = 0.42$; p = 0.658)^1	$(F_{2,38} = 1.67$; p = 0.207)^5	$(F_{2,43} = 1.22$; p = 0.304)^1
AP/TB×Fertilizer×AC×Competition	$(F_{2,91} = 0.35$; p = 0.704)^1	–	$(F_{2,31} = 2.05$; p = 0.145)^1	–

Given are separate models for tansy biomass, the log response ratio (plant biomass LogRR) of tansy biomass infested with aphids compared with control plants, aphid numbers, and aphid load (aphid number per unit plant biomass) from a greenhouse experiment using *Macrosiphoniella tanacetaria* on *Tanacetum vulgare* in competition with *Solidago canadensis*. Minimum adequate models are presented together with terms removed from the model given in brackets. Superscripts give the order in which terms have been removed from the model starting with highest order interactions based on least significance. Significant terms in the final models are given in bold. A random effect for 5 blocks which was included in all models is not shown. °In the model for tansy biomass, aphid presence (AP) was included as explanatory variable, while in the model for aphid number, tansy biomass (TB) was used as a covariate. The biomass of the control plants not infected with aphids but exposed to the same other treatments was used as covariate because this biomass was not confounded with the aphid effect of reducing plant biomass. *indicates data was log transformed.

The effect of AC depended again on the fertilizer treatment. In control soils, AC decreased aphid numbers while in both fertilizer treatments, AC increased aphid numbers which were highest in the combination with $F_{inorg+org}$ (Table 1, Fig. 1C). Thus, the strength of the response of aphid numbers to fertilization and AC was different from the responses of plant biomass to the same treatments, and this was reflected in how the treatments affected aphid load. In general, fertilization led to a decrease in aphid load as the increase in plant biomass was stronger than the increase in aphid numbers (Fig. 1A, C). This was true for control soil and soil with F_{inorg}, where, adding AC decreased aphid loads (Table 1, Fig. 1D). In contrast, plant biomass in the AC+$F_{inorg+org}$ treatment was not higher than in the AC+F_{inorg} treatment while aphid numbers were highest in AC+$F_{inorg+org}$ treatment (Fig. 1 A, C) and hence, aphid load was higher when both fertilizers were added together with AC than when just $F_{inorg+org}$ was added.

Addition of AC consequently also affected the relationship between aphid number and plant biomass (Table 1). The positive correlation between aphid numbers and plant biomass was stronger in the presence of AC than in the absence of AC (Fig. 3).

Effects of competition on aphid performance. On tansy plants in competition with goldenrod, aphid numbers were generally lower than on plants without competition, as predicted in H3b, with the one exception that aphid numbers increased in competition when both $F_{inorg+org}$ and AC were added to the soil (Table 1, Fig. 1C). Thus, in contrast to hypothesis H3b, plant-plant competition did not simply decrease aphid numbers, but the response of aphid numbers to competition depended on an interaction with fertilizer. Aphid load was not significantly influenced by the competition treatment (Table 1, Fig. 1D).

Summary of effects on aphid performance. Fertilizer increased and competition decreased aphid numbers, as expected.

In contrast to the expectations, AC increased aphid number independent of the potentially allelopathic competitor. Aphid load, which quantifies the number of aphids relative to plant biomass, depended on the magnitude of effects on both aphid and plant performance. Consequently, the effects of aphids on plant performance interacted with plant growth conditions confirming H4.

Discussion

Our results show that host plant growth conditions affect the performance of both tansy and the specialized aphid herbivore feeding on it. Fertilization increased plant and aphid performance, and competition decreased tansy biomass and aphid number, confirming our first and third hypothesis, respectively. AC increased plant and aphid performance both in the presence and absence of a potentially allelopathic competitor, thus contrary to hypothesis 2, there was no indication for goldenrod competing via allelochemicals. Rather, the addition of AC acted more like an additional fertilizer by increasing plant and aphid performance. As predicted by hypothesis 4, herbivore effects on plants depended on host plant growth conditions because the magnitude, and partly also the direction of the response to the fertilizer, AC and competition treatments differed between the plant and the herbivore. Analysis of the feedback effects of herbivores on plants as a function of the different treatments was only possible because of our full factorial design, in particular by rearing plants with and without the herbivore under identical conditions.

Effects of fertilizer on plant and aphid performance

Resource availability and resource quality, e.g. N-content may have different and independent effects on herbivores [46]. In our

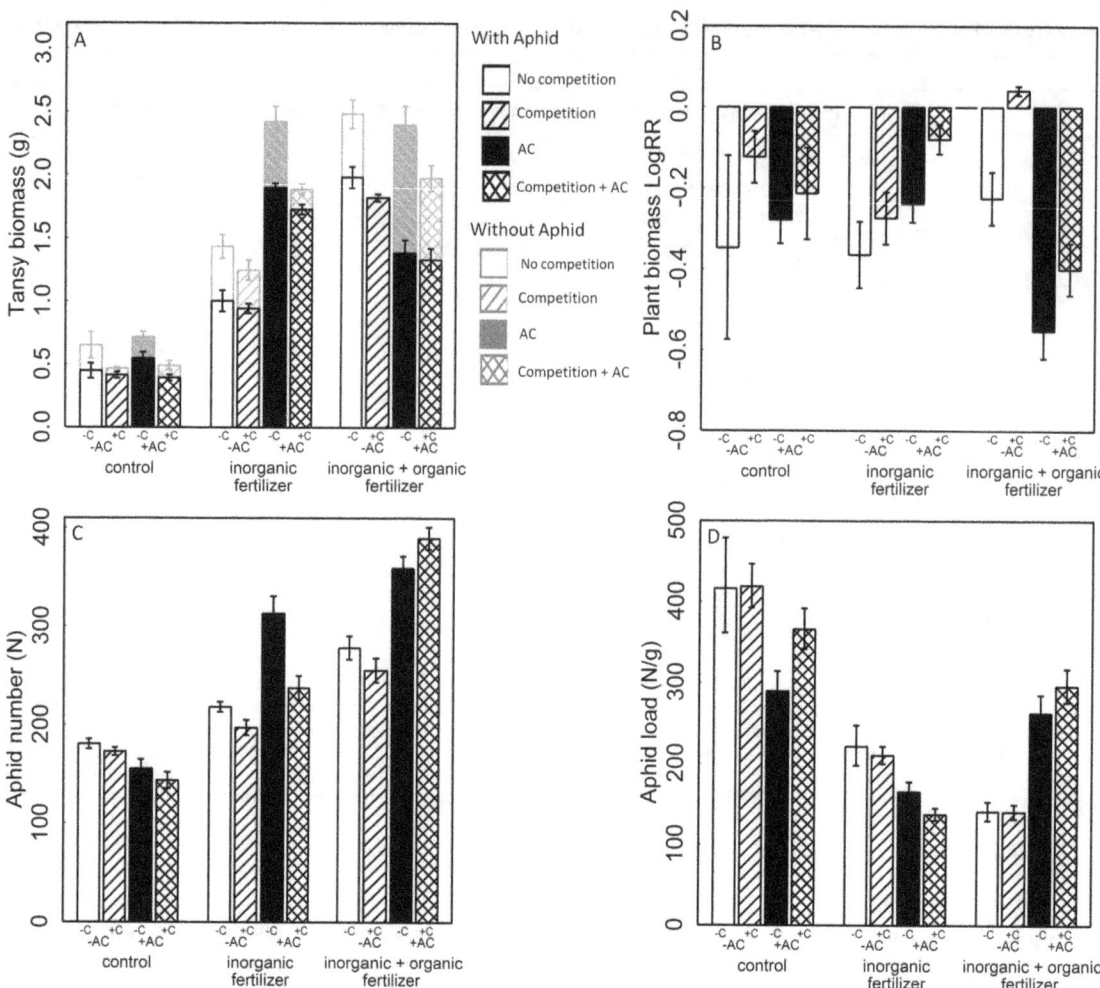

Figure 1. Response of plant and aphid performance on host plant growth conditions. Tansy growth conditions were manipulated in experimental treatments with three levels of fertilizer crossed with two levels of activated carbon (AC), competition and infestation with aphids (*Macrosiphoniella tanacetaria*). We replicated every treatment combination 5 times, yielding a total of $2\times2\times3\times2\times5=120$ pots. (A) tansy biomass, (B) plant biomass log response ratio (plant biomass LogRR), i.e. biomass of plants infested with aphids divided by biomass of control plants subjected to the same fertilization, AC and competition treatment, (C) aphid number, and (D) aphid load (aphid number per unit plant biomass). The colour of bars indicates aphid treatment (black bars: with aphids, gray bars: without aphids) while bar patterns indicate the AC treatment (+AC: black and hatched columns; −AC: white and slanted columns) and the competition treatment (with competition (+C): hatched and slanted and without competition (−C): white and black columns). Means ± SE are shown. For statistical tests see Table 1.

experiment, we specifically manipulated plant quality and the effects of fertilization on aphids were due to changes in host plant quality, not host plant biomass, for various reasons. First, the total number of aphids was always relatively small, even when aphid population size reached its maximum of 430 individuals, only a small part of the shoot was covered with aphids. Second, even though there were clear detrimental effects of aphids on plant growth, there were no visible signs of damage and aphids were not observed walking on the plant searching for feeding sites. Third, and most importantly, the effects of aphids on tansy plant biomass were independent of plant biomass.

As expected, both fertilizer treatments increased tansy biomass. Application of fertilizer also improved the performance of the studied aphid; these results are consistent with previous studies that showed that nitrogen application increased population growth of aphid species [47–49]. These results have direct field relevance, as for *M. tanacetaria* aphid densities in the field are higher on fertilized than on less fertilized plants [36]. Kleine and Muller [13]

fertilized tansy and found that the C:N ratio decreased, i.e. that more nitrogen was available in fertilized tansy plants, and the increased aphid performance on fertilized plants was thus likely a result of increased levels of nutrients in the plant [3,47,49]. Aphids respond to such differences in plant quality as indicated by a laboratory experiment by Nowak and Komor [50] who found that *M. tanacetaria* is more likely to settle and start feeding on tansy plants with higher amino acid concentrations in the phloem sap which increased with fertilization. In addition to the increase in host nutrient status, fertilization may also affect other aspects of host quality: e.g. Kleine and Muller [13] found that fertilized tansy plants show lower levels of terpenoid defence chemicals.

While the positive effects of fertilizer on plant and aphid performance was not surprising and in line with our hypothesis H1, the feedback effect of the aphid on the plant were not as straightforward as may have been expected, due to subtle differences in how fertilization affected the plant and the aphid. To understand the fertilization effects on the plant-aphid

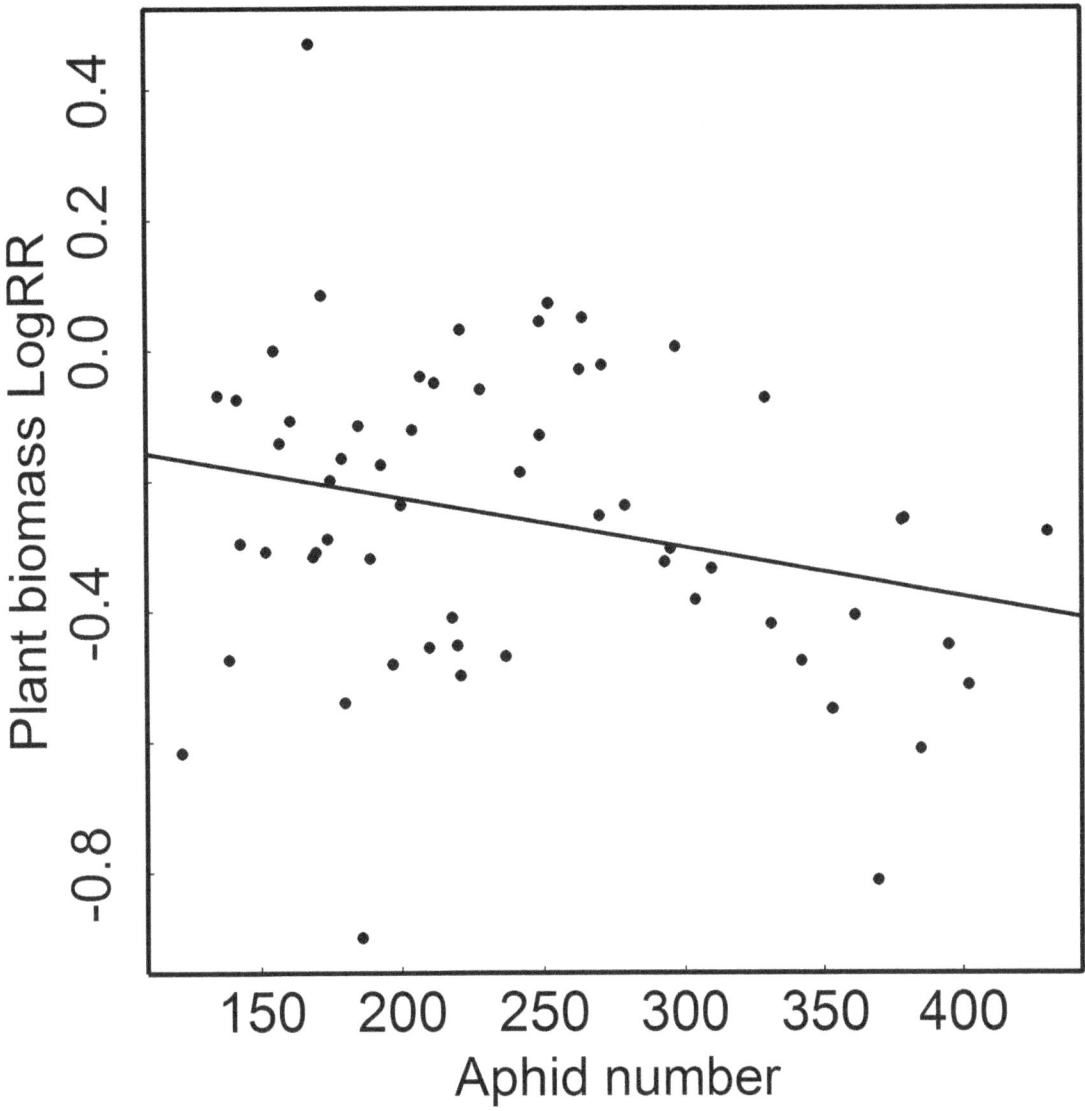

Figure 2. Dependence of plant biomass log response ratio on aphid numbers. Tansy plant biomass LogRR was measured as tansy biomass in the presence of aphids (*Macrosiphoniella tanacetaria*) divided by control plant biomass. This analysis combined host plants of all additional treatments imposed in the experiment.

interactions, we first calculated aphid load that quantifies aphid performance relative to plant performance. In our experiment, the increase in plant biomass and aphid numbers with fertilization was not proportional, i.e. with each additional unit plant biomass less than one additional unit of aphids occurred. As a consequence, the highest aphid loads occurred on plants growing in the control soil, even though the absolute number of aphids was lower compared with fertilized treatments. Aphid load declined with increasing host plant biomass, causing the largest plants to show the largest aphid numbers but not the highest aphid load. Organic fertilizer further complicated the issue: aphid load was higher in the treatment with $F_{inorg+org}$ than in the treatment with F_{inorg} alone. This was because biomass of tansy was not increased further when F_{org} was added in addition to F_{inorg}, as shown for plants without aphids. In contrast, in the aphid treatment, the number of aphids responded positively to an increased level of fertilizer even at the highest level.

The second measure for feedback effects was plant biomass LogRR that measured the reduction in plant biomass due to aphids for all combinations of the fertilizer, AC and competition treatments. While aphid load is a measure of the potential impact of the herbivore on the plant, LogRR directly quantifies this impact. Because aphids reduced plant biomass, as has been found in other studies [51–54], plant biomass LogRR was always negative and increased with the number of aphids. While aphid load increased with the addition of fertilizer, tansy plants apparently profited more from increases in nutrient availability by accumulating more biomass than they suffered from the higher number of aphids, resulting in less negative plant biomass LogRR. The exception was at the highest nutrient levels, where aphid numbers were even higher and had a stronger negative effect on tansy biomass, resulting in a decrease in plant biomass LogRR. Generally, patterns of aphid load and plant biomass LogRR were

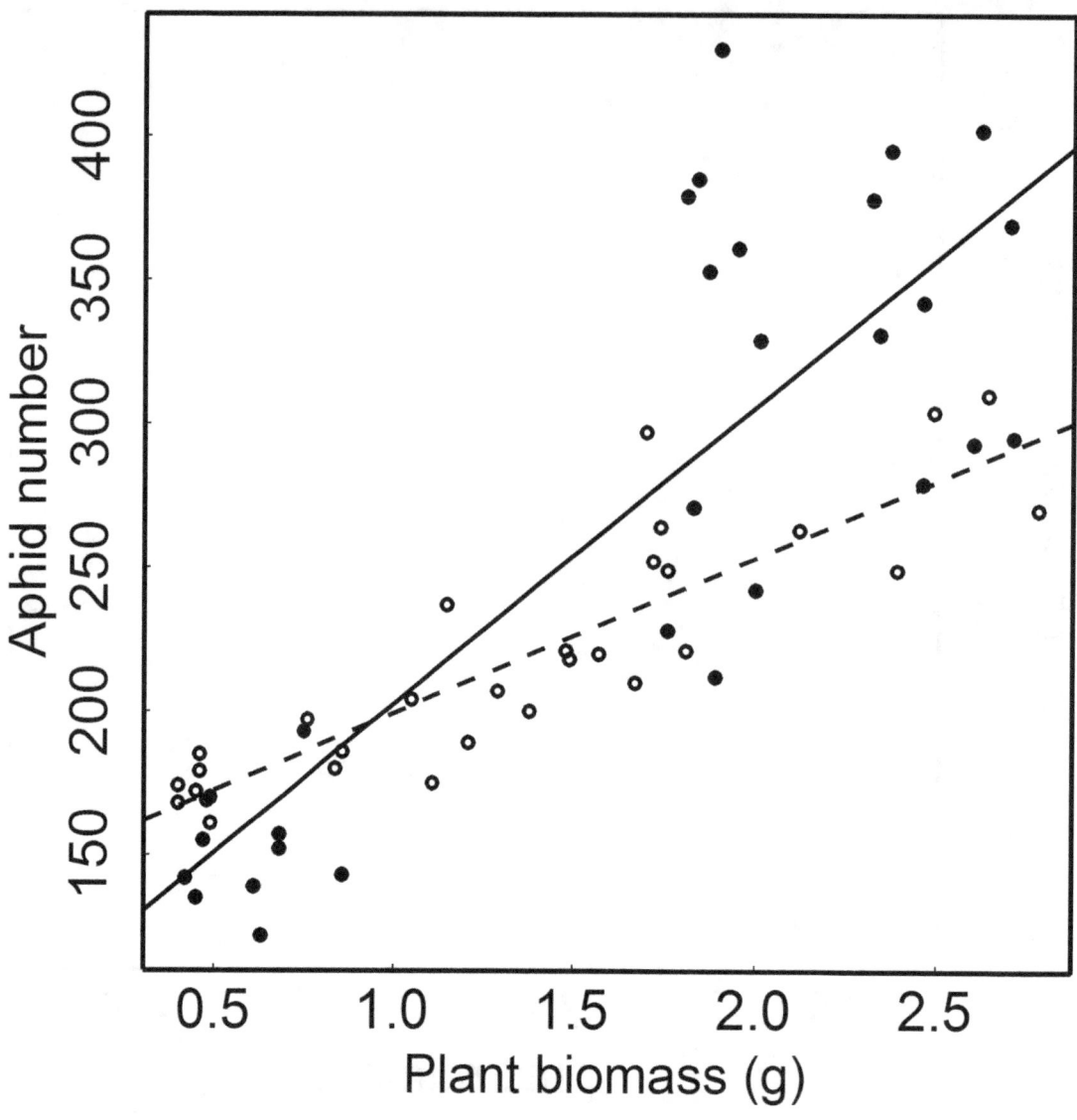

Figure 3. Dependence of aphid numbers on the biomass of the host plant. Open circles and dashed line represent the number of *Macrosiphoniella tanacetaria* aphids on tansy plants in soil without activated carbon. Closed circles and solid line are aphid numbers on plants in soil treated with activated carbon. The biomass of the control plants not infected with aphids but exposed to the same other treatments was used as explanatory variable because this biomass represents the potential size of the plants without the reduction of plant size due to the aphid infection.

in opposite directions, showing that for understanding feedback effects of the herbivore on plants it is necessary to calculate both.

Our results emphasize the subtleties in the plant-herbivore interactions that depend on the exact way in which plant and herbivore can exploit an increase in plant nutrient availability. Fertilization benefits both plants and herbivore and some fertilization appears to benefit the plant more than the herbivore, such that fertilization decreases the herbivore impact on the plant. However, our results suggest that at some level of fertilization, it is mainly the herbivore that benefits so that any additional increase in nutrients will aid the build-up of herbivore populations rather than further benefitting their hosts. This may not be the case for all plant-herbivore systems and, if such effects occur, at different levels of nutrient availability depending on the specific interaction. In agricultural systems, an increase fertilization is generally considered to be beneficial for plants, and the potential negative

consequences of feeding a herbivore population are possibly underexplored.

Plant competition and plant-herbivore interactions

As expected and in line with our hypothesis H3, tansy plants growing in competition with goldenrod were smaller in both treatments, with and without aphids, compared with control plants without competition. This was most likely a result of competition for limiting nutrient availability in the soil [55]. While aboveground competition for light cannot be ruled out completely, tansy plants were generally taller than goldenrod (tansy height mean 35.94 ± 0.98 cm, goldenrod height mean 31.13 ± 1.18 cm, respectively) making it unlikely that goldenrod affected tansy by shading.

Plant-plant competition also had negative effects on the aphid herbivore (H3), as aphid numbers were generally lower on tansy plants in competition. A similar reduction of aphid performance in the presence of a competitor of the host plant was documented by

Schädler et al. [20], who discussed differences in the quality (rather than quantity) of N-containing compounds as a mechanism for the reduced herbivore performance, because aphid numbers were less reduced in the presence of a N-fixing competitor than in the presence of another forb [20].

Previous studies demonstrated an allelopathic potential of goldenrod by finding reduced germination of potential plant competitors in the presence of goldenrod root or leaf extracts [31,42], which was the main motivation for our AC treatment. However, we did not find any evidence for allelopathic effects; adding AC did not alleviate effects of competition. There are several potential explanations for this result. Beside experimental differences between our study and the previous studies [31,42], it may be that allelopathic effects of goldenrod are restricted to seed germination or seedling establishment. This interpretation is in line with a study of Pisula and Meiners [42] who found evidence for allelopathy of goldenrod in germination essays with lettuce (*Lactuca sativa* L.) and radish (*Raphanus sativus* L.), but did not observe any allelopathy when investigating successional dynamics in old fields with high densities of goldenrod. While the AC treatment did not reduce competitive effects in our study, it did indicate potential side effects of AC that have been described before [34,56,57]. The observed increases in plant and aphid performance in the AC treatment are consistent with observations that addition of AC can act as a fertilizer [34,56]. There were also interactions between AC and the organic fertilizer that point to complex soil processes when both AC and dead organic matter are added to the soil. Such unclear interactions may be behind unexpected effects such as the increase in aphid numbers under plant-plant competition when both $F_{inorg+org}$ and AC were added to the soil.

Interactions between herbivory and competition and implications for the field

Herbivory and competition are expected to be synergistic in their negative effect on the focal plant [21,23], but they may also act antagonistically [22,25]. For example, Haag et al. [25] found that the effect of herbivory and interspecific plant competition can be antagonistic when herbivory affects all plants within the community. Also, Schädler et al. [20] performed an experiment similar to ours where herbivory was restricted to the focal plant, using cereal aphids on *Poa* as a model system. They found that also when herbivory is restricted to the focal plant, competition and herbivory can interact antagonistically as long as interspecific plant competition decreases the population growth of herbivores on the focal plant [20]. In our study, the effect of competition on tansy biomass was stronger in the absence than in the presence of aphids, in other words plant biomass LogRR was less negative for tansy plants in competition. This indicates that aphid effects were less detrimental for plants in competition compared with plants that did not suffer from competition. This effect resulted most likely from competition decreasing the number of aphids on the tansy plant, which in turn reduced the negative effect of competition on tansy plants. As a result, the biomass of plants without competition was more strongly reduced by larger colonies of aphids than the biomass of plants in competition, which were infected by smaller colonies.

The results of our greenhouse study are of relevance to understand plant herbivore interactions under field conditions. In non-agricultural systems, aphid colonies are mostly small and only few are very large, with several hundred to several thousand individuals per plant [58,59]. Thus, under natural conditions, it is also mostly resource quality rather than resource quantity that limits aphid population growth. An additional critical factor

limiting colony size in the field is predation by a large guild of predators [60]. While our study has not considered how plant growth conditions affect higher trophic levels and their feedback effects on the herbivores (and possibly the plant), such interactions can be also affected by resource availability [61]. Because of high rates of predation in the field, aphid colony growth critically depends on the balance between reproduction that is influenced by host plant growth conditions and mortality due to predation. Thus, any small negative effect on aphid growth rates, due to plant-plant competition, and every positive effect on growth rates, due to higher nutrient availability in the soil, is likely to critically affect local aphid persistence. Such small scale effects on the dynamics of aphid populations are underexplored [36]. Aphids are special herbivores in the sense that they produce many generations per year and may quickly build up large populations. Consequently effects of host plant growth conditions might be especially apparent for aphids. Yet, spatial variation in the competitive situation and nutrient availability of the host plant will generally create variation in the local growth rates of plants and herbivores and such spatial heterogeneity has been shown to have dynamical consequences for plant-herbivore systems and also the dynamics of predators and parasitoids feeding on the herbivores [62].

Conclusions

Our study has found that plant and herbivore growth and the feedback effects of the herbivore on plants are affected by both the abiotic and biotic plant growth conditions, in our case fertilization and plant-plant competition. While generally effects of growth condition on plant biomass and aphid numbers mirrored each other, our results emphasize the shifts in the plant-herbivore interactions that depend on the exact way in which plant and herbivore can exploit an increase in plant nutrient availability and react to competition. Fertilization benefited both plants and herbivores. Yet, our results suggest that at some levels of fertilization it is mainly the plant at others the herbivore that benefits from additional nutrients. Consequently, aphid impacts can decrease under fertilization even when absolute aphid numbers increase. The ecological costs of an infestation with herbivores, thus, depend on the balance of effects of growth conditions on plant and herbivore performance. Mechanistic insight into the feedback effects can be reached when the responses of all partners to a manipulation in plant growth conditions are studied for each partner both in isolation and their interactions.

Supporting Information

Figure S1 Comparison of plant biomass log response ratio (LogRR) of tansy with and without competition by goldenrod. Plant biomass LogRR (mean ± SE) was less negative for tansy plants in competition, thus infestation with *Macrosiphoniella tanacetaria* aphids was less detrimental for plants in competition compared with control plants. Means ± SE are shown. For statistical tests see Table 1.

Acknowledgments

We wish to thank Sylvia Creutzburg, Robert Schaelike and Eric Eckmann for their assistance in conducting the experiment and the management of the Jena-Experiment, especially Anne Ebeling, for provision of the soil used in the experiment. We would like to thank John Guess and three anonymous reviewers for their valuable comments on an earlier drafts of this paper.

Author Contributions

Conceived and designed the experiments: WWW NMA. Performed the experiments: NMA MM. Analyzed the data: NMA STM. Contributed to the writing of the manuscript: NMA STM WWW.

References

1. Sarfraz RM, Dosdall LM, Keddie AB (2009) Bottom-up effects of host plant nutritional quality on *Plutella xylostella* (Lepidoptera: Plutellidae) and top-down effects of herbivore attack on plant compensatory ability. European Journal of Entomology 106: 583–594.

2. Louda SM, Collinge SK (1992) Plant-resistance to insect herbivores: A field test of the environmental stress hypothesis. Ecology 73: 153–169.

3. Awmack CS, Leather SR (2002) Host plant quality and fecundity in herbivorous insects. Annual Review of Entomology 47: 817–844.

4. Edelsteinkeshet L (1986) Mathematical-theory for plant herbivore systems. Journal of Mathematical Biology 24: 25–58.

5. Mcnaughton SJ, Chapin FS (1985) Effects of phosphorus nutrition and defoliation on C^4 graminoids from the serengeti plains. Ecology 66: 1617–1629.

6. Zehnder CB, Hunter MD (2008) Effects of nitrogen deposition on the interaction between an aphid and its host plant. Ecological Entomology 33: 24–30.

7. Agrawal AA (2004) Plant defense and density dependence in the population growth of herbivores. American Naturalist 164: 113–120.

8. Ladner DT, Altizer S (2005) Oviposition preference and larval performance of North American monarch butterflies on four *Asclepias* species. Entomologia Experimentalis Et Applicata 116: 9–20.

9. Tsai JH, Wang JJ (2001) Effects of host plants on biology and life table parameters of *Aphis spiraecola* (Homoptera : Aphididae). Environmental Entomology 30: 44–50.

10. Nevo E, Coll M (2001) Effect of nitrogen fertilization on *Aphis gossypii* (Homoptera : Aphididae): Variation in size, color and reproduction. Journal of Economic Entomology 94: 27–32.

11. Perkins MC, Woods HA, Harrison JF, Elser JJ (2004) Dietary phosphorus affects the growth of larval *Manduca sexta*. Archives of Insect Biochemistry and Physiology 55: 153–168.

12. Meyer GA, Root RB (1996) Influence of feeding guild on insect response to host plant fertilization. Ecological Entomology 21: 270–278.

13. Kleine S, Muller C (2013) Differences in shoot and root terpenoid profiles and plant responses to fertilisation in *Tanacetum vulgare*. Phytochemistry 96: 123–131.

14. Mattson WJ (1980) Herbivory in relation to plant nitrogen-content. Annual Review of Ecology and Systematics 11: 119–161.

15. Cisneros JJ, Godfrey LD (2001) Midseason pest status of the cotton aphid (Homoptera : Aphididae) in California cotton: Is nitrogen a key factor? Environmental Entomology 30: 501–510.

16. Müller CB, Fellowes MDE, Godfray HCJ (2005) Relative importance of fertiliser addition to plants and exclusion of predators for aphid growth in the field. Oecologia 143: 419–427.

17. Bethke JA, Redak RA, Schuch UK (1998) Melon aphid performance on chrysanthemum as mediated by cultivar, and differential levels of fertilization and irrigation. Entomologia Experimentalis Et Applicata 88: 41–47.

18. Casper BB, Jackson RB (1997) Plant competition underground. Annual Review of Ecology and Systematics 28: 545–570.

19. Fowler N (1986) The role of competition in plant-communities in arid and semiarid regions. Annual Review of Ecology and Systematics 17: 89–110.

20. Schädler M, Brandl R, Haase J (2007) Antagonistic interactions between plant competition and insect herbivory. Ecology 88: 1490–1498.

21. Belsky AJ (1986) Does herbivory benefit plants - a review of the evidence. American Naturalist 127: 870–892.

22. Newingham BA, Callaway RM (2006) Shoot herbivory on the invasive plant, *Centaurea maculosa*, does not reduce its competitive effects on conspecifics and natives. Oikos 114: 397–406.

23. Lee TD, Bazzaz FA (1980) Effects of defoliation and competition on growth and reproduction in the annual plant *Abutilon theophrasti*. Journal of Ecology 68: 813–821.

24. Mcnaughton SJ (1986) On plants and herbivores. American Naturalist 128: 765–770.

25. Haag JJ, Coupe MD, Cahill JF (2004) Antagonistic interactions between competition and insect herbivory on plant growth. Journal of Ecology 92: 156–167.

26. Inderjit, Callaway RM, Vivanco JM (2006) Can plant biochemistry contribute to understanding of invasion ecology? Trends in Plant Science 11: 574–580.

27. Callaway RM, Aschehoug ET (2000) Invasive plants versus their new and old neighbors: A mechanism for exotic invasion. Science 290: 521–523.

28. Callaway RM, Ridenour WM (2004) Novel weapons: invasive success and the evolution of increased competitive ability. Frontiers in Ecology and the Environment 2: 436–443.

29. Cappuccino N, Arnason JT (2006) Novel chemistry of invasive exotic plants. Biology Letters 2: 189–193.

30. Inderjit, Callaway RM (2003) Experimental designs for the study of allelopathy. Plant and Soil 256: 1–11.

31. Abhilasha D, Quintana N, Vivanco J, Joshi J (2008) Do allelopathic compounds in invasive *Solidago canadensis* s.l. restrain the native European flora? Journal of Ecology 96: 993–1001.

32. Butcko VM, Jensen RJ (2002) Evidence of tissue-specific allelopathic activity in *Euthamia graminifolia* and *Solidago canadensis* (Asteraceae). American Midland Naturalist 148: 253–262.

33. Hierro JL, Callaway RM (2003) Allelopathy and exotic plant invasion. Plant and Soil 256: 29–39.

34. Lau JA, Puliafico KP, Kopshever JA, Steltzer H, Jarvis EP, et al. (2008) Inference of allelopathy is complicated by effects of activated carbon on plant growth. New Phytologist 178: 412–423.

35. Wurst S, Vender V, Rillig MC (2010) Testing for allelopathic effects in plant competition: does activated carbon disrupt plant symbioses? Plant Ecology 211: 19–26.

36. Stadler B (2004) Wedged between bottom-up and top-down processes: aphids on tansy. Ecological Entomology 29: 106–116.

37. Mitch LW (1992) Intriguing world of weeds – tansy. Weed Technology 6: 242–244.

38. Rebele F (2000) Competition and coexistence of rhizomatous perennial plants along a nutrient gradient. Plant Ecology 147: 77–94.

39. Blackman RL, Eastop VF (2006) Aphids on the world's herbaceous plants and shrubs. John Wiley & Sons, London, UK.

40. Holman J (2009) Host plant catalogue of aphids: Palaearctic region. Springer, Berlin.

41. Mehrparvar M, Mahdavi Arab N, Weisser WW (2013) Diet-mediated effects of specialized tansy aphids on survival and development of their predators: Is there any benefit of dietary mixing? Biological Control 65: 142–146.

42. Pisula NL, Meiners SJ (2010) Allelopathic effects of goldenrod species on turnover in successional communities. American Midland Naturalist 163: 161–172.

43. Petermann JS, Müller CB, Weigelt A, Weisser WW, Schmid B (2010) Effect of plant species loss on aphid-parasitoid communities. Journal of Animal Ecology 79: 709–720.

44. Pinheiro JC, Bates DM, DebRoy S, Sarkar D, R Core Team (2012) nlme: linear and nonlinear mixed effects models. R package version 3.1–106.

45. R Development core team (2010) A language and environment for statistical computing. R Foundation for Statistical Computing,. Vienna, Austria.

46. de Sassi C, Lewis OT, Tylianakis JM (2012) Plant-mediated and nonadditive effects of two global change drivers on an insect herbivore community. Ecology 93: 1892–1901.

47. Duffield SJ, Bryson RJ, Young JEB, SylvesterBradley R, Scott RK (1997) The influence of nitrogen fertiliser on the population development of the cereal aphids *Sitobion avenae* (F) and *Metopolophium dirhodum* (Wlk) on field grown winter wheat. Annals of Applied Biology 130: 13–26.

48. Garratt MPD, Wright DJ, Leather SR (2010) The effects of organic and conventional fertilizers on cereal aphids and their natural enemies. Agricultural and Forest Entomology 12: 307–318.

49. Gash AFJ (2012) Wheat nitrogen fertilisation effects on the performance of the cereal aphid *Metopolophium dirhodum*. Agronomy Journal 2: 1–13.

50. Nowak H, Komor E (2010) How aphids decide what is good for them: experiments to test aphid feeding behaviour on *Tanacetum vulgare* (L.) using different nitrogen regimes. Oecologia 163: 973–984.

51. Crawley MJ (1989) Insect herbivores and plant-population dynamics. Annual Review of Entomology 34: 531–564.

52. Foster WA (1984) The distribution of the sea-lavender aphid *Staticobium staticis* on a marine saltmarsh and its effect on host plant fitness. Oikos 42: 97–104.

53. Zvereva EL, Lanta V, Kozlov MV (2010) Effects of sap-feeding insect herbivores on growth and reproduction of woody plants: a meta-analysis of experimental studies. Oecologia 163: 949–960.

54. Choudhury D (1984) Aphids and plant fitness - a test of the Owen and Wiegert hypothesis. Oikos 43: 401–402.

55. Louda SM, Keeler KH, Holt RD (1990) Herbivore influence on plant performance and competitive interactions. In: Grace JB, Tilman D, editors. Perspective on Plant Competition. New York: Academic Press. 413–444.

56. Weisshuhn K, Prati D (2009) Activated carbon may have undesired side effects for testing allelopathy in invasive plants. Basic and Applied Ecology 10: 500–507.

57. Pietikainen J, Kiikkila O, Fritze H (2000) Charcoal as a habitat for microbes and its effect on the microbial community of the underlying humus. Oikos 89: 231–242.

58. Addicott JF (1978) The population dynamics of aphids on fireweed: A comparison of local populations and metapopulations. Canadian Journal of Zoology 56: 2554–2564.

59. Weisser WW (2000) Metapopulation dynamics in a aphid-parasitoid system. Entomologia Experimentalis et Applicata 97: 83–92.

60. Dixon AFG (1998) Aphid Ecology: An optimization approach. London: Chapman & Hall. 300 p.

61. Hartvigsen G, Wait D, Coleman J (1995) Tri-trophic interactions influenced by resource availability: predator effects on plant performance depend on plant resources. Oikos: 463–468.

62. Hassell MP (1978) The dynamics of arthropod predator-prey systems. Princeton: Princeton University Press.

22

Linking Annual N$_2$O Emission in Organic Soils to Mineral Nitrogen Input as Estimated by Heterotrophic Respiration and Soil C/N Ratio

Zhijian Mu[1]*, Aiying Huang[2], Jiupai Ni[3], Deti Xie[3]

1 Chongqing Key Laboratory of Soil Multi-scale Interfacial Processes, College of Resources & Environment, Southwest University, Chongqing, China, **2** College of Agronomy & Biotechnology, Southwest University, Chongqing, China, **3** Chongqing Engineering Research Center for Agricultural Non-point Source Pollution Control in Three -Gorges Region, College of Resources & Environment, Southwest University, Chongqing, China

Abstract

Organic soils are an important source of N$_2$O, but global estimates of these fluxes remain uncertain because measurements are sparse. We tested the hypothesis that N$_2$O fluxes can be predicted from estimates of mineral nitrogen input, calculated from readily-available measurements of CO$_2$ flux and soil C/N ratio. From studies of organic soils throughout the world, we compiled a data set of annual CO$_2$ and N$_2$O fluxes which were measured concurrently. The input of soil mineral nitrogen in these studies was estimated from applied fertilizer nitrogen and organic nitrogen mineralization. The latter was calculated by dividing the rate of soil heterotrophic respiration by soil C/N ratio. This index of mineral nitrogen input explained up to 69% of the overall variability of N$_2$O fluxes, whereas CO$_2$ flux or soil C/N ratio alone explained only 49% and 36% of the variability, respectively. Including water table level in the model, along with mineral nitrogen input, further improved the model with the explanatory proportion of variability in N$_2$O flux increasing to 75%. Unlike grassland or cropland soils, forest soils were evidently nitrogen-limited, so water table level had no significant effect on N$_2$O flux. Our proposed approach, which uses the product of soil-derived CO$_2$ flux and the inverse of soil C/N ratio as a proxy for nitrogen mineralization, shows promise for estimating regional or global N$_2$O fluxes from organic soils, although some further enhancements may be warranted.

Editor: Shuijin Hu, North Carolina State University, United States of America

Funding: This research was supported by the National Natural Science Foundation of China (grant number 41371211) and the National Major Science and Technology Projects for Water Pollution Control and Management (grant number 2012ZX07104-003). The funders had no role in study design, data collection and analysis, decision to publish, or preparation of the manuscript.

Competing Interests: The authors have declared that no competing interests exist.

* E-mail: muzj01@gmail.com

Introduction

Although organic soils occupy only 3% of the Earth's land area, they contain approximately 40% (610 Pg) of the terrestrial soil organic carbon (SOC) [1]. Climate warming and human disturbance such as drainage and cultivation are expected to accelerate carbon decomposition in organic soils, and the decomposition of SOC can facilitate the release of mineral nitrogen which can then be utilized by denitrifying and nitrifying bacteria to produce the potent greenhouse gas N$_2$O [2,3]. N$_2$O emissions from organic soils under agricultural use in Nordic countries were on average four times higher than those from mineral soils, indicating that N$_2$O derived from SOC decomposition dominates overall fluxes [4]. However, no consistent and quantitative relationship has been reported for N$_2$O emission and organic carbon decomposition in organic soils.

Organic carbon and nitrogen in soils, plant and microbial biomass are usually covalently bonded at relatively constant ratios. It is thus logical to expect that N$_2$O and CO$_2$ originated from SOC decomposition should be closely linked. Some studies have indeed found a significant relationship between soil N$_2$O and CO$_2$ emissions at the site level [5,6]. This relationship, however, was weaker when data were pooled across sites or ecosystems[7,8]. The

variability of soil C/N ratio may be one of the important factors undermining the correlation for organic soils. The C/N ratio in organic soils ranges from 50~100 in weakly decomposed peat to 12~35 in highly decomposed peat [9]. The supply of mineral nitrogen from SOC decomposition is the outcome of two concurrent and oppositely directed microbial processes – nitrogen mineralization and immobilization [10]. Soils with a high C/N ratio may be characterized by rapid immobilization of nitrogen and soils with a low C/N ratio by higher net nitrogen mineralization and a surplus of available NH$_4^+$ and NO$_3^-$ [11]. A negative relationship has accordingly been shown for C/N ratio of soils and N$_2$O fluxes [9]. Similar to the relationship between N$_2$O and CO$_2$ emissions, the correlation of N$_2$O emission with soil C/N ratio tended to be weak when the data from different sites at larger scales were included [4,12], which makes it difficult to scale up N$_2$O fluxes by CO$_2$ emissions or C/N ratio alone from individual sites to regional scales. In view of the coupling of soil carbon and nitrogen processes and the bridging function of C/N ratio, we hypothesized that a combination of soil CO$_2$ emission and C/N ratio would likely provide better measurements of N$_2$O emission at larger scales. In fact, Mu et al. [13] have linked N$_2$O flux to soil mineral nitrogen as estimated by CO$_2$ emission and C/N ratio for agricultural mineral soils. To our knowledge, no such kind of attempt has ever been made for organic soils. The aim of this study was therefore to determine: 1) if N_2O flux from organic soils is related to soil mineral nitrogen input estimated from heterotrophic respiration divided by soil C/N ratio (a derived measure of soil nitrogen mineralization) plus fertilizer nitrogen; and 2) whether or not the relationship is sufficiently robust to serve as an approach for estimating N_2O flux from organic soils.

Materials and Methods

Data source

To test the hypothesis, we collected journal-published data of N_2O and CO_2 emissions measured simultaneously in the fields on peatlands or histosols for which the carbon and nitrogen content or ratio of the organic matter in the upper layers of the soil has been reported. Occasional and short-period flux measurements were not used and only data on annual emissions were considered. For long-term measurements, we used annual estimates rather than multi-year averages to reflect temporal variability. Annual emissions were directly reported by authors or estimated from points in the figures of publications. The final dataset comprised of 122 field measurements from 28 geographical sites (Table S1). Of all data, only 12 measurements at 9 sites were from the tropical regions and the rest were from the temperate regions. Most of the flux measurements were made using closed chamber technique with sampling frequency varying from 1–3 times per week to once per month. Other factors such as soil pH and water table level, if reported, were also recorded in the database. Readers should refer to the original papers for a more complete presentation of the data.

Estimation of soil mineral nitrogen input

The CO_2 emission measured in bare soils can be taken as the proxy of SOC decomposition or heterotrophic respiration [14]. There are limited studies in which CO_2 emission was measured in bare soils (Table S1). For the CO_2 emissions measured in soils with plants, the contribution of heterotrophic respiration or SOC decomposition was estimated using the following equation adapted from Bond-Lamberty and Thomson [15]:

$$R_h = 10e^{[0.22 + 0.87\ln(R_t/10)]} \qquad (1)$$

where R_h is heterotrophic respiration and R_t is total soil respiration (kg C ha^{-1} yr^{-1}).

The nitrogen mineralization rate from soil organic matter was then calculated using the following equation [13]:

$$N_m = R_h/S_{CN} \qquad (2)$$

where N_m is the gross nitrogen mineralization (kg N ha^{-1} yr^{-1}) and S_{CN} is soil C/N ratio.

The mineralized nitrogen from soil organic matter decomposition and the inorganic nitrogen from chemical fertilizers constitute the total input of soil mineral nitrogen (N_{mf}). Atmospheric nitrogen deposition, as another important external source of soil mineral nitrogen, was not considered for our study since there were few papers reporting it.

Statistical analysis

The dataset in the current study is of unbalanced nature with observations collected from peer-reviewed papers rather than from systematically designed experiments. Accordingly, the effects of soil mineral nitrogen input and other variables on N_2O flux were analyzed using the mixed model-REML estimation method of SAS/MIXED procedure (version 9.3), which is suitable for handling unbalanced data. The values of N_2O flux were first natural-log transformed to normalize their distribution and then analyzed by the following model:

$$\ln(f_{N2O}) = \text{constant} + \ln(N_{mf}) + pH + WT + NS_i + E\cos ys_j$$
$$+ NS_i \times \ln(N_{mf}) + E\cos ys_j \times \ln(N_{mf}) + E\cos ys_j \times WT$$

where f_{N2O} is the N_2O flux; N_{mf}, pH, WT, NS_i and $Ecosys_j$ are the fixed effects of mineral nitrogen input, soil pH, water table level, nitrogen source (i is mineralized nitrogen only or a combination of mineralized nitrogen and inorganic nitrogen from chemical fertilizers), and ecosystem type (j is forest or non-forest type), respectively. A preliminary check of the data showed that the general trend of N_2O flux in forest system differed from grass and cropland, so the ecosystems were simply classified into two subclasses as forest and non-forest. Some two-factor interactions were also included in the model. A significant level of $p = 0.05$ was used to determine if a given variable or interactive effect was kept in the model to further seek solutions for fixed effects. Four negative values of N_2O flux reported by Inubushi et al. [16] and Mojeremane et al. [17] can not be subjected to log-transformation and were not included in the analysis. In addition to determination coefficient (i.e., R^2 value), concordance between observed N_2O fluxes and model fits was also analyzed using Lin's concordance correlation coefficient (CCC, Stata SE 12.0) to assess the goodness-of-fit of the finalized models. The resulting CCC was interpreted using the benchmarks described by Klevens et al. [18] as follows: <0.20 is considered virtually no agreement; 0.21–0.40 is considered slight; 0.41–0.60 is considered fair; 0.61–0.80 is considered moderate; and 0.81–0.99 is substantial.

Results

As shown in Table 1, soil pH, soil mineral nitrogen source (NS) and ecosystem type did not affect the annual N_2O flux ($p>0.05$), while the input of soil mineral nitrogen (N_{mf}) and water table level (WT) had significant effects on N_2O flux ($p<0.01$). The F value of N_{mf} was the biggest, indicating the input of soil mineral nitrogen was the main factor controlling N_2O emission in organic soils. The two-factor interactive effects between NS, N_{mf}, WT and ecosystem type on N_2O flux were not statistically significant ($p>0.05$).

Only the significant variables were then kept in the model to solve the estimates for their effects. Two models with different combinations of independent variables are shown in Table 2. The first model was the simplest one with N_{mf} as the single independent variable. The second model was expanded by adding the effect of water table level. The 95% confidence intervals of the estimated effect of N_{mf} were overlapped for different models. The models indicated that N_2O flux was positively correlated with N_{mf} and negatively with water table level. Using the estimated effects and the variables in the dataset allowed a comparison between predicted and observed annual N_2O fluxes from organic soils. The variable N_{mf} explained up to 69% of the variability in the overall data of observed N_2O fluxes (Fig. 1), while the addition of water table level increased the explanatory ability to 75% (Fig. 2). When the overall data were further divided by ecosystem types, the performance of models was somewhat different (Fig. 1 & 2). For forest, the determination coefficient (R^2) was nearly stable at the value of 0.63 for both models. In contrast, the introduction of water table level into models slightly improved the fitted results for

Table 1. Results of type III tests of fixed effects.

Effect	Numerator DF	Denominator DF	F Value	Pr>F
N_{mf}	1	96	13.16	0.0005
pH	1	96	1.43	0.2344
WT	1	96	5.15	0.0255
NS	1	96	0.10	0.7472
Ecosystem	1	96	0.70	0.4040
NS$\times$N$_{mf}$	1	96	0.11	0.7426
WT$\times$N$_{mf}$	1	96	3.20	0.0767
Ecosystem$\times$N$_{mf}$	1	96	0.21	0.6506
Ecosystem$\times$WT	1	96	2.17	0.1437

N_{mf}, the mineral nitrogen input to soil; WT, water table level; NS, the source of soil mineral nitrogen.

non-forest systems with R^2 values increasing from 0.59 to 0.69. This indicated that the input of mineral nitrogen was the most important predictor of N_2O flux, while water table level was a weak predictor of N_2O flux and appeared to be dependent on ecosystem type.

The slope of regression lines in Fig. 1 & 2 ranged from 0.50 to 0.75, indicating that the relationship strays from the ideal 1:1 line. Therefore the concordance correlation coefficient (CCC) between observed and predicted N_2O fluxes was calculated to measure robustness of the models. For the overall data with log-transformation, the concordance was substantial with the CCC ranging from 0.82 to 0.86 for the two models. When the log-transformed data were converted to actual N_2O fluxes, however, the cluster of fluxes greater than 15.0 kg N ha^{-1} yr^{-1} was found to be distinctly underestimated. The CCC for this cluster of data

ranged from -0.002 to 0.16 and showed virtually no agreement, suggesting that some important factors responsible for these high fluxes were not accounted for by the models. For the rest of the data (103 fluxes out of 118), the CCC (ranging from 0.63 to 0.68) still showed a moderate concordance.

The variable N_{mf} in the models can be decomposed into soil heterotrophic respiration (R_h), C/N ratio and inorganic nitrogen rate from chemical fertilizer (N_f). The mixed procedure analysis indicated that each of these components of N_{mf} had a significant influence on N_2O flux ($p<0.001$), with R_h and N_f being positively related to N_2O flux and C/N ratio negatively related to N_2O flux. Soil carbon and nitrogen contents, which could replace the variable of C/N ratio, were also significantly negatively or positively correlated with N_2O flux ($p<0.001$). The fitting efficiency between observed and predicted N_2O fluxes by models

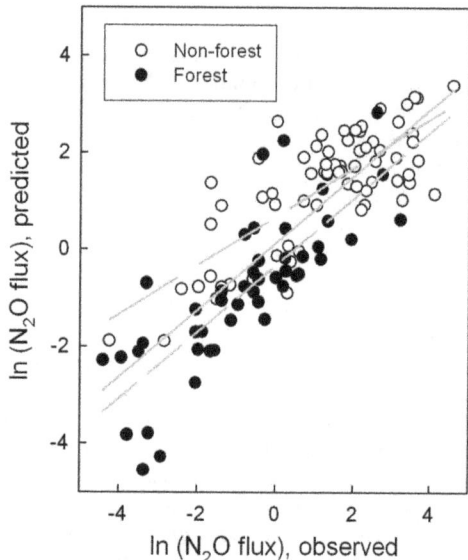

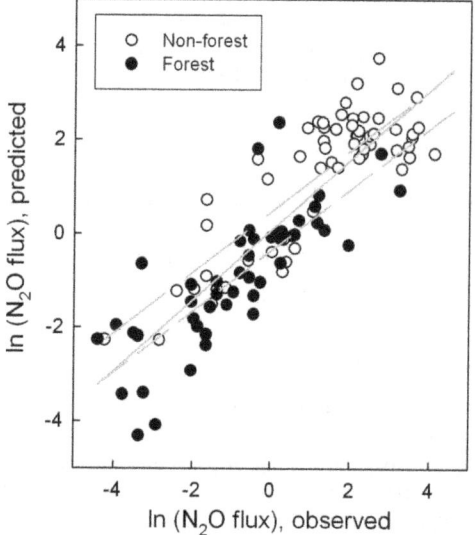

Figure 1. Correlation of observed fluxes of N_2O from organic soils and predicted values (kg N_2O-N ha^{-1}) by model 1 as presented in Table 2: ln(N_2O flux) = 1.8685 ln(N_{mf})−9.0314. Solid line shows linear regression fit for the overall data: y = 0.69x+0.13, R^2 = 0.69. Long-dashed line shows linear regression fit for the non-forest system: y = 0.50x+0.68, R^2 = 0.59. Short-dashed line shows linear regression fit for the forest system: y = 0.69x−0.33, R^2 = 0.63.

Figure 2. Correlation of observed fluxes of N_2O from organic soils and predicted values (kg N_2O-N ha^{-1}) by model 2 as presented in Table 2: ln(N_2O flux) = 1.5374 ln(N_{mf})−0.0221 WT−8.2334. Solid line shows linear regression fit for the overall data: y = 0.75x+0.08, R^2 = 0.75. Long-dashed line shows linear regression fit for the non-forest system: y = 0.65x+0.48, R^2 = 0.69. Short-dashed line shows linear regression fit for the forest system: y = 0.65x−0.35, R^2 = 0.63.

Table 2. Solutions for fixed effects of the models with log-transformed N_2O flux as dependent variable.

Model	Effect	Estimate	SE	DF	t Value	Pr>|t|	95% confidence	
							Lower	Upper
1	Intercept	−9.0314	0.5952	116	−15.17	<.0001	−10.2102	−7.8526
	ln (N_{mf})	1.8685	0.1157	116	16.15	<.0001	1.6393	2.0976
2	Intercept	−8.2334	0.6222	103	−13.23	<.0001	−9.4675	−6.9994
	ln (N_{mf})	1.5374	0.1479	103	10.39	<.0001	1.2441	1.8308
	WT	−0.0221	0.0053	103	−4.14	<.0001	−0.0326	−0.0115

N_{mf}, the mineral nitrogen input to soil (kg N ha^{-1} yr^{-1}); WT, water table level (cm).

using the above-mentioned components of N_{mf} as inputs were nearly the same as those of models using N_{mf} itself (data not shown).

Discussion

Previous studies have linked N_2O flux directly to either CO_2 flux or soil C/N ratio [5,8,9]. In this study, soil CO_2 emission and C/N ratio were combined to estimate mineral nitrogen input, and the latter accounted for up to 69% of the variability of N_2O fluxes from organic soils with various properties, land management practices and climates. Soil CO_2 flux or C/N ratio alone explained only 49% and 36% of the overall variability of N_2O fluxes, respectively (Fig. 3). This suggests the necessity of combining soil CO_2 flux and C/N ratio for predicting N_2O flux on a large scale. Of course, soil CO_2 flux and C/N ratio can be independently incorporated into the same models, but the interpretation of such models would be relatively complicated and evasive since there are various mechanisms which may explain the control of CO_2 flux and C/N ratio over N_2O flux [8,9,19]. In contrast, the quotient of soil CO_2 flux and C/N ratio can well represent in theory the gross nitrogen mineralization [20], and the implication of models using such a quotient as input is straightforward and self-evident in the importance of mineral nitrogen input for regulating soil N_2O flux. There is no significant difference in the influence of different sources of mineral nitrogen on N_2O flux (Table 1), suggesting that the simplified models might also be suitable for evaluating the effect of mineral nitrogen from other sources such as atmospheric deposition, though this idea needs further verification.

A negative relationship between N_2O flux and groundwater level has been observed for individual sites [21,22], and still holds at a large scale as shown in this study. This is logical simply because high moisture with increasing water table level can limit N_2O emission from soils due to the low availability of nitrate and/or efficient reduction of N_2O to N_2 through denitrification [16,23], while the lowering of water table increases oxygen penetration into the peat and enhances the decomposition of organic matter, as indicated by the negative relationship between heterotrophic respiration and water table level ($R^2 = 0.31$, $p<0.0001$). It has been reported that the control of soil water content or water table level over N_2O flux is important only when soil is not nitrogen limiting [24,25]. In this study, the percentage of observations with N_{mf} greater than 150 kg N ha^{-1} was only 19% for forest, but up to 87% for non-forest systems (Table S1). This suggests that forest soil is nitrogen limiting when compared with non-forest systems, which may be responsible for the insensitivity of N_2O flux to water table level for forest systems (Fig.1& 2). Besides the input of mineral nitrogen, forest differs from non-forest systems in many other factors, such as vegetation, below-/above-ground biomass, litter fall, soil compaction, and land management practices, all of which can influence N_2O flux but are not considered here due to limited and unsystematic information in literature sources of the current dataset. To fill the gap, ecosystem type was used as a proxy variable that we tried to incorporate into models; however, statistical analysis showed that its effect was not significant (Table 1).

It should be acknowledged that the models described here were dependent on simplifying assumptions that can introduce error. That is, the gross nitrogen mineralization was estimated from carbon mineralization and soil C/N ratio by assuming that the rate of carbon mineralization is the same as the rate of respiration and the C/N ratio of mineralized organic matter is the same as that of the bulk soil organic matter. In fact, carbon and nitrogen mineralization from soils originates from decomposable fractions

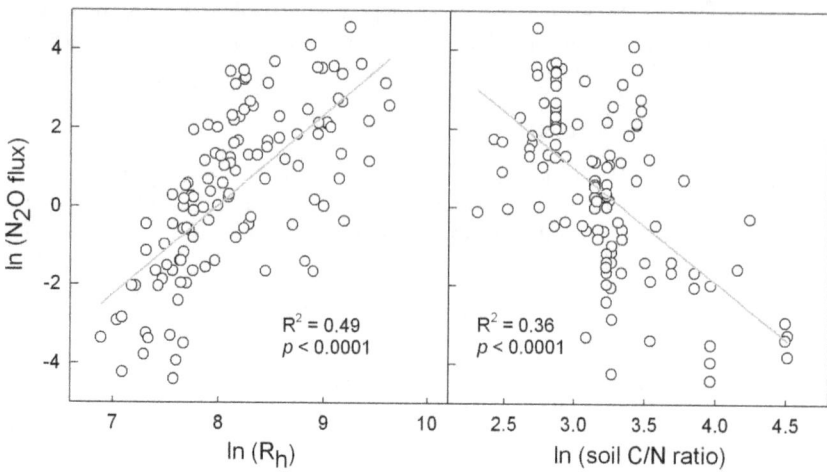

Figure 3. Correlation of observed fluxes of N₂O with estimated soil heterotrophic respiration (Rₕ, left panel) and C/N ratio (right panel).

of organic matter with different C/N ratios [26]. Most likely, the ratio of carbon evolved/nitrogen mineralized is much wider than the bulk soil carbon to nitrogen ratio [27,28]. This indicates that gross nitrogen mineralization might be over- or under-estimated if bulk soil C/N ratio was used in equation 2. The respiration process is also not exactly identical to carbon mineralization. The amount of carbon that is ultimately lost through respiration depends on how effectively the decomposer community converts mineralized carbon to biomass [29]. Similarly, the amount of nitrogen that is ultimately available to denitrifier or nitrifier for producing N_2O depends on how effectively the decomposer community converts mineralized nitrogen to biomass and plants compete with microbes for mineral nitrogen [10,20]. Empirical relationships have been established between nitrogen and carbon mineralization in studies performed usually under laboratory conditions [30]. Different organic matter fractions or their C/N ratios, and varying microbial use efficiency of carbon and nitrogen have also been proposed to predict nitrogen release [20,29]. However, these relationships are strongly dependent on the experimental conditions in which they have been established. Moreover, the current dataset is based on the in situ measurements in the field environment and contains only the basic information of respiratory carbon and bulk soil C/N ratio, thus necessitating the above-mentioned assumptions to estimate mineralized nitrogen. Such simplifications and assumptions may bring uncertainties, but it is necessary in some cases to understand the general trends and probabilistic nature of the environment [31].

N_2O emission from soils is of small magnitude and highly variable in space and time, and is thus very difficult to estimate. The measurement of soil N_2O flux also requires intricate techniques along with a lot of time and labor. In contrast, soil CO_2 emission is controlled primarily by soil temperature and moisture, and is relatively easy to measure or predict [32,33]. In addition, the estimates of soil respiration are currently more widely available than those of soil N_2O emission. The models developed in this study showed a promising approach to estimating N_2O emission from organic soils by using soil C/N ratio and CO_2 emission data derived from measurements or biogeochemical modeling. It should be mentioned, however, that several aspects of the information in the current dataset might impose uncertainties on these models. First, soil heterotrophic respiration was simply estimated from total soil respiration using a universal relationship

between them [15], but the relative contribution of organic matter decomposition or heterotrophic respiration would vary over time and depend on root respiration of the growing plants [8]. Second, the majority of the global organic soils are distributed in the boreal and sub-arctic regions and about 10%–15% in the tropical countries [1,3], but most of the current data came from northern Europe, indicating that the models developed in the present study might be biased to the temperate regions.

Conclusion

A fairly large number of data were collected to explore the relationship between annual N_2O emission and multiple variables for organic soil by a mixed-model analysis, and the input of soil mineral nitrogen was found to be the most useful predictor for N_2O flux. Soil mineral nitrogen was supposed to be composed of organic nitrogen mineralization as estimated by CO_2 emission and soil C/N ratio, thus providing a possibility for upscaling N_2O emission from organic soils by use of regional soil databases including information on C/N ratio and carbon storage change or CO_2 emission data. The approach proposed here may have validity as a whole, but needs further evaluation and advancement before practical application due to uncertainties associated with simplifying assumptions and a regionally unbalanced data source. A better understanding of the processes of carbon and nitrogen mineralization and their stoichiometric relationship as well as additional experimental data from organic soils outside of temperate Europe regions will help to improve the relationship established in this study.

Author Contributions

Conceived and designed the experiments: ZJM. Performed the experiments: ZJM AYH. Analyzed the data: ZJM. Contributed reagents/materials/analysis tools: JPN DTX. Wrote the paper: ZJM.

References

1. Page SE, Rieley J, Banks CJ (2011) Global and regional importance of the tropical peatland carbon pool. Global Change Biol 17: 798–818.
2. Goldberg SD, Knorr KH, Blodau C, Lischeid G, Gebauer G (2010) Impact of altering the water table height of an acidic fen on N_2O and NO fluxes and soil concentrations. Global Change Biol 16: 220–233.
3. Frolking S, Talbot J, Jones MC, Treat CC, Kauffman JB, et al. (2011) Peatlands in the Earth's 21st century climate system. Environ Rev 19: 371–396.
4. Maljanen M, Sigurdsson BD, Guðmundsson J, Óskarsson H, Huttunen JT, et al. (2010) Greenhouse gas balances of managed peatlands in the Nordic countries – present knowledge and gaps. Biogeosciences 7: 2711–2738.
5. Garcia-Montiel DC, Melillo JM, Steudler PA, Neill C, Feigl BJ, et al. (2002) Relationship between N_2O and CO_2 emissions from the Amazon Basin. Geophys Res Lett 29: Art.No. 1090.
6. Chatskikh D, Olesen JE (2007) Soil tillage enhanced CO_2 and N_2O emissions from loamy sand soil under spring barley. Soil Till Res 97: 5–18.
7. Keller M, Varner R, Dias JD, Silva H, Crill P, et al. (2005) Soil–atmosphere exchange of nitrous oxide, nitric oxide, methane, and carbon dioxide in logged and undisturbed forest in the Tapajos national forest, Brazil. Earth Interact 9: Art. No. 23.
8. Xu XF, Tian HQ, Hui DF (2008) Convergence in the relationship of CO_2 and N_2O exchanges between soil and atmosphere within terrestrial ecosystems. Global Change Biol 14: 1651–1660.
9. Klemedtsson L, Von Arnold K, Weslien P, Gundeersen P (2005) Soil CN ratio as a scalar parameter to predict nitrous oxide emissions. Global Change Biol 11: 1142–1147.
10. Luxhøi J, Bruun S, Stenberg B, Breland TA, Jensen LS (2006) Prediction of gross and net nitrogen mineralization-immobilization-turnover from respiration. Soil Sci Soc Am J 70: 1121–1128.
11. Bengtsson G, Bengtson P, Månsson KF (2003) Gross nitrogen mineralization-, immobilization-, and nitrification rates as a function of soil C/N ratio and microbial activity. Soil Biol Biochem 35: 143–154.
12. Ojanen H, Minkkinen K, Alm J, Penttila T (2010) Soil – atmosphere CO_2, CH_4 and N_2O fluxes in boreal forestry-drained peatlands. Forest Ecol Manage 260: 411–421.
13. Mu ZJ, Huang AY, Kimura SD, Jin T, Wei SQ, et al. (2009) Linking N_2O emission to soil mineral N as estimated by CO_2 emission and soil C/N ratio. Soil Biol Biochem 41: 2593–2597.
14. Hanson PJ, Edwards NT, Garten CT, Andrews JA (2000) Separating root and soil microbial contributions to soil respiration: A review of methods and observations. Biogeochemistry 48: 115–146.
15. Bond-Lamberty B, Thomson A (2010) A global database of soil respiration data. Biogeosciences 7: 1915–1926.
16. Inubushi K, Furukawa Y, Hadi A, Purnomo E, Tsuruta H (2003) Seasonal changes of CO_2, CH_4 and N_2O fluxes in relation to land-use change in tropical peatlands located in coastal area of South Kalimantan. Chemosphere 52: 603–608.
17. Mojeremane W, Rees RM, Mencuccini M (2012) The effects of site preparation practices on carbon dioxide, methane and nitrous oxide fluxes from a peaty gley soil. Forestry 85: 1–15.
18. Klevens J, Trick WE, Kee R, Angulo F, Garcia D, et al. (2011) Concordance in the measurement of quality of life and health indicators between two methods of computer-assisted interviews: self-administered and by telephone. Qual Life Res 20: 1179–1186.
19. Rochette P, Tremblay N, Fallon E, Angers DA, Chantigny MH, et al. (2010) N_2O emissions from an irrigated and non-irrigated organic soil in eastern Canada as influenced by N fertilizer addition. Eur J Soil Sci 61: 186–196.
20. Murphy DV, Recous S, Stockdale EA, Fillery IRP, Jensen LS, et al. (2003) Gross nitrogen fluxes in soil: Theory, measurement and application of ^{15}N pool dilution techniques. Adv Agron 79: 69–118.
21. Regina K, Silvola J, Martikainen PJ (1999) Short-term effects of changing water table on N_2O fluxes from peat monoliths from natural and drained boreal peatlands. Global Change Biol 5: 183–189.
22. Danevcic T, Mandic-Mulec I, Stres B, Stopar D, Hacin J (2010) Emissions of CO_2, CH_4 and N_2O from southern European peatlands. Soil Biol Biochem 42: 1437–1446.
23. Maljanen M, Shurpali N, Hytönen J, Mäkiranta P, Aro L, et al. (2012) Afforestation does not necessarily reduce nitrous oxide emissions from managed boreal peat soils. Biogeochemistry 108: 199–218.
24. Smith KA, Thomson PE, Clayton PE, McTaggart IP, Conen F (1998) Effects of temperature, water content and nitrogen fertilization on emissions of nitrous oxide by soil. Atmos Environ 32: 3301–3309.
25. Weslien P, Klemedtsson AK, Borjesson G, Klemedtsson L (2009) Strong pH influence on N_2O and CH_4 fluxes from forested organic soils. Eur J Soil Sci 60: 311–320.
26. Springob G, Kirchmann H (2003) Bulk soil C to N ratio as a simple measure of net N mineralization from stabilized soil organic matter in sandy arable soils. Soil Biol Biochem 35: 629–632.
27. Sollins P, Spycher G, Glassman CA (1984) Net nitrogen mineralization from light- and heavy-fraction forest soil organic matter. Soil Biol Biochem 16: 31–37.
28. Kader MA, Sleutel S, Begum SA, D'Haene K, Jegajeevagan K, et al. (2010) Soil organic matter fractionation as a tool for predicting nitrogen mineralization in silty arable soils. Soil Use Manage 26: 494–507.
29. Manzoni S, Taylor P, Richter A, Porporato A, Ågren GI (2012) Environmental and stoichiometric controls on microbial carbon-use efficiency in soils. New Phytol 196: 79–91.
30. Nicolardot B, Recous S, Mary B (2001) Simulation of C and N mineralization during crop residue decomposition: a simple dynamic model based on the C:N ratio of the residues. Plant Soil 228: 83–103.
31. Yan XY, Yagi K, Akiyama H, Akimoto H (2005) Statisical analysis of the major variables controlling methane emission from rice fields. Global Change Biol 11: 1131–1141.
32. Lloyd J, Taylor JA (1994) On the temperature-dependence of soil respiration. Funct Ecol 8: 315–323.
33. Raich JW, Potter CS, Bhagawati D (2002) Interannual variability in global soil respiration, 1984-94. Global Change Biol 8: 800–812.

23

Effects of Oscillatory Flow on Fertilization in the Green Sea Urchin *Strongylocentrotus droebachiensis*

Louise T. Kregting[1,2*¤a], **Anna L. Bass**[1], **Òscar Guadayol**[2¤b], **Philip O. Yund**[1¤c], **Florence I. M. Thomas**[2]

1 Marine Science Center, University of New England, Biddeford, Maine, United States of America, **2** Hawai'i Institute of Marine Biology, University of Hawai'i, Kane'ohe, Hawai'i, United States of America

Abstract

Broadcast spawning invertebrates that live in shallow, high-energy coastal habitats are subjected to oscillatory water motion that creates unsteady flow fields above the surface of animals. The frequency of the oscillatory fluctuations is driven by the wave period, which will influence the stability of local flow structures and may affect fertilization processes. Using an oscillatory water tunnel, we quantified the percentage of eggs fertilized on or near spawning green sea urchins, *Strongylocentrotus droebachiensis*. Eggs were sampled in the water column, wake eddy, substratum and aboral surface under a range of different periods ($T = 4.5 - 12.7$ s) and velocities of oscillatory flow. The root-mean-square wave velocity ($\mathrm{rms}(u_w)$) was a good predictor of fertilization in oscillatory flow, although the root-mean-square of total velocity ($\mathrm{rms}(u)$), which incorporates all the components of flow (current, wave and turbulence), also provided significant predictions. The percentage of eggs fertilized varied between $50 - 85\%$ at low flows ($\mathrm{rms}(u_w) < 0.02$ m s^{-1}), depending on the location sampled, but declined to below 10% for most locations at higher $\mathrm{rms}(u_w)$. The water column was an important location for fertilization with a relative contribution greater than that of the aboral surface, especially at medium and high $\mathrm{rms}(u_w)$ categories. We conclude that gametes can be successfully fertilized on or near the parent under a range of oscillatory flow conditions.

Editor: Chris D. Wood, Universidad Nacional Autónoma de México, Mexico

Funding: Funding was provided by the National Science Foundation, Ocean Science Division, OCE-04-25088 and OCE-12-33868 to POY and OCE-04-24978 to FIMT. This is contribution number 1563 from the Hawaii Institute of Marine Biology and 8858 from University of Hawaii, School of Ocean and Earth Science and Technology. The funders had no role in the study design, data collection and analysis, decision to publish, or preparation of the manuscript.

Competing Interests: The authors have declared that no competing interests exist.

* E-mail: l.kregting@qub.ac.uk

¤a Current address: School of Planning, Architecture and Civil Engineering, Queens University of Belfast, Northern Ireland, United Kingdom
¤b Current address: School of Biological, Biomedical and Environmental Sciences, University of Hull, Kingston-upon-Hull, United Kingdom
¤c Current address: The Downeast Institute, Beals, Maine, United States of America

Introduction

Many free-spawning benthic invertebrates live in energetic coastal environments where variation in water motion will exert a major influence on fertilization processes. In the simplest sense, water motion experienced by an organism is composed of underlying currents, ocean and local wind driven oscillations and turbulence. In unidirectional flow, gametes of a spawning animal are transported by the currents and turbulent diffusion, which create a spatially spreading plume of gametes downstream of the animal [1], [2]. Oscillatory flow is characterized by orbital motion. As flow approaches the seabed the vertical portion of this flow is damped creating horizontal oscillatory water motion. Thus animals spawning in oscillatory flow experience reversing horizontal flows in addition to currents and turbulent diffusion. Further, as water moves past an organism, vortices are created in the downstream wake [3]. The extent of vortex shedding depends on velocity of the flow, as well as the period of reversals. Thus the time scales of variation in water motion that affect the movement of gametes of free-spawning benthic organisms in unidirectional vs. oscillatory motion are quite different and this difference is likely to have important implications for the success of fertilization processes.

Investigations to date have focused largely on the influences of unidirectional flow on fertilization processes of benthic organisms that spawn externally [4–6]. In steady unidirectional flow the shear created at the surface of a spawning animal is constant and depends on flow velocity, animal size and morphology. Furthermore, the wake downstream of the animal potentially constitutes an important site for fertilization [5]. Past work [4], [5], [7] has demonstrated that as velocity increases, fertilization levels in benthic organisms decrease owing to increased advection and turbulent diffusion of gametes, which decreases concentrations of gametes to levels below that required for effective fertilization [7–9].

While studies based on unidirectional flow regimes have significantly increased our understanding of fertilization processes for organisms that live in environments dominated by tidal currents, many broadcast-spawning invertebrates live in high-energy shallow coastal environments where oscillatory water motion from locally generated or oceanic wind waves is the norm. In spite of the prevalence of this habitat, no quantitative studies have determined the influence of oscillatory flow on fertilization in benthic organisms (but *see* Denny et al. [10] for work in surge channels on wave-swept shores). In oscillatory flow, velocity gradients formed as water flows in one direction are quickly

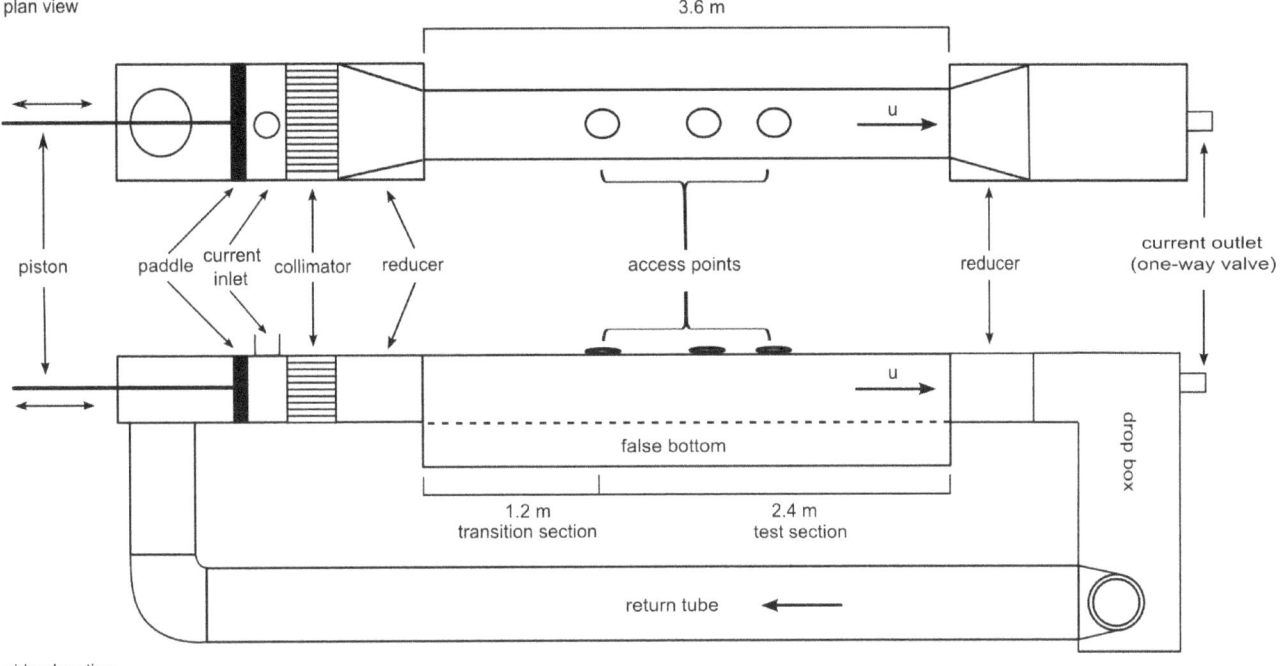

Figure 1. Diagram of the oscillatory water tunnel (OWT). The OWT chamber was used to determine the effects of oscillatory flow on fertilization of the green sea urchin *Strongylocentrotus droebachiensis*. The drive mechanism that attaches to the paddle (via two coupled hydraulic pistons driven by a flywheel attached to an electric motor powered by an adjustable frequency drive) has been omitted from the figure.

disrupted as the flow reverses. Consequently, the shear layers are constantly forming and shedding vortices; high shear stresses are potentially produced close to the surface of the animal, with the frequency of these perturbations determined by the magnitude of flow, structure size and period of reversals [11]. Thus it may be expected that fertilization in oscillatory flow will be dependent in part on the magnitude and period of the oscillatory wave.

The interaction between hydrodynamics and the biology of the organism is also important. Gametes are typically released in a viscous fluid that, under steady flow conditions, can form piles on the surface of a spawning animal or clumps and strings that drift into the water column, thus releasing eggs and sperm over time [12], [13]. Because velocity is reduced in the boundary layer at the surface of an organism, the presence of gamete piles can enhance fertilization by reducing the rate of diffusion of sperm from a male and providing sperm to the eggs of a downstream female over a period of several hours [14]. At a critical stage in the flow regime, however, a shear threshold will be reached where gametes are ablated away from the surface of the parent [5]. In sea urchins, the negatively buoyant eggs can be entrained in an eddy downstream of a female or fall to the substrate, with many eggs fertilized in these locations [5]. Therefore under steady unidirectional flow conditions, fertilization in sea urchins can take place at a number of identifiable locations near the parent. In an oscillatory flow regime, however, the continuously accelerating and decelerating flow over and around the organism generates vortices that are likely to have a major influence on gamete dispersion and hence the location and level of fertilization.

Here we present results from a laboratory experiment on fertilization in green sea urchins, *Strongylocentrotus droebachiensis*, subjected to different combinations of velocities and periods of oscillatory motion. Our objectives were to 1) quantify the percentage of eggs fertilized in the water column, in the wake

eddies, on the substratum, and on the aboral surface of a spawning sea urchin for a range of periods and velocities relevant to local field populations, 2) determine an appropriate hydrodynamic parameter for relating the water flow to the fertilization processes, and 3) determine the relative contribution to overall fertilization rates of different spatial areas adjacent to individual animals.

Materials and Methods

Ethics Statement

The sea urchins were purchased from commercial sea urchin divers who hand collected them under a commercial license. The species *Strongylocentrotus droebachiensis* is not an endangered species.

Oscillatory water tunnel

Experiments were conducted in a fully enclosed acrylic oscillatory water tunnel (OWT) with a square working section of 3.6 length (L) ×0.31 width (W) ×0.31 height (H) m (Fig. 1) designed to mimic the oscillatory motion of breaking ocean swells. Oscillatory motion was produced by a flywheel attached to a motor, with motor rotational velocity controlled by an adjustable frequency drive (the flywheel, motor and adjustable frequency drive are not shown in Fig. 1). This oscillatory motion was transferred to the water in the tunnel via a paddle driven by two coupled hydraulic piston cylinders. The resulting wave forms were sinusoidal with essentially no vertical or transverse water motion (*see* example in Fig. 2A). The driving cylinder was attached to the flywheel at various points that were different distances from the center so that both the period and velocity could be independently manipulated to produce a range of oscillatory conditions (Fig. 2B). Because of size limitations and constrained flow, this OWT primarily reproduces reversals of flow and the development of shear layers on the urchin, but does not allow for full development of boundary conditions above the substratum, damps the vertical components of flow, and may not provide small-scale turbulence conditions that scale appropriately with wave period and velocity.

To prevent saturation of the water tunnel with gametes due to back and forth mixing, and to simulate a unidirectional current underlying the waves (e.g., a tidal current oriented parallel to waves), a flow-through current of approximately 0.03 m s^{-1} was introduced in the water tunnel. This current was evenly distributed along the vertical axis of the water column (Fig. 2C). Ambient seawater entered the OWT via a dispersing inlet pipe positioned perpendicular to the long axis of the OWT between the paddle and the collimator (Fig. 1) and exited through a one-way outlet valve at the end of the OWT. Seawater not exiting the OWT flowed into a drop-box connected to the upstream end of the OWT with a pipe to create a continuous water connection to balance the flow on both sides of the paddle. However, any remaining sperm were unable to pass the paddle and re-enter the working section. To create orderly flow through the test section, a collimator was placed at the entrance of the working section followed by a reducer (Fig. 1). Another reducer was located at the end of the working section. The floor of the OWT was covered with Lego^{TM} base plates (0.25×0.25 m) glued to a 0.009 m thick acrylic sheet that provided a uniform rough surface as well as a substrate for securing the animals.

Sea urchins were secured in the center of the OWT located 1.24 and 2.24 m (male and female respectively) downstream from the reducer section in the test section of the OWT. Each sea urchin was secured by first holding the individual in place with rubber bands on separate small modified Lego^{TM} plates (0.065 (L) ×0.045 (W) ×0.003 (H) m), which attached easily and quickly to the floor of the OWT with minimal disruption to the animal. Portholes on the otherwise sealed top of the working section provided access to the animals. Preliminary dye experiments were carried out to ensure that gamete concentrations were not constrained by diffusing to the sidewalls until well downstream of the sampling locations. These preliminary tests were conducted under the full range of hydrodynamic conditions used in the trials.

Water velocity measurements were obtained using an Acoustic Doppler Velocimeter (SonTek ADVField) to characterize flow at the height of the urchin when the urchin was not present. After each trial the sea urchins were removed from the OWT and the ADV was positioned vertically above where the female had been secured, with the sample volume 0.04 m above the substrate (corresponding to the average height of the female sea urchins used in the trials: 0.037 m ± 0.001 SE). The ADV sample volume was located 5 cm from the probe head and occupied <0.1 cm^3. The vertical position of the probe minimized both the area perpendicular to any vertical flow and the volume of instrument submersed, thus minimizing the effect of any vortices shed from the instrument. The very low vertical velocities observed in all cases ensure that any small vortices shed from the probe head did not significantly affect the sampling volume. The x sensor was oriented with the dominant current direction. After the underlying current and oscillatory motion were established at the trial hydrodynamic condition for 1 minute, velocity was recorded for 2 minutes at a sampling frequency of 16 Hz. Velocimeter signal-to-noise ratios (SNR) were consistently above 20 dB and the correlation values for the 3 sensors remained above 90%. Before flow parameters were calculated, spikes were removed following 3D phase space threshold techniques [15], [16]. Values with correlations <60 and SNR <20 were also removed [17].

Previous fertilization experiments based on steady flow conditions have primarily used the mean flow ($\bar{u}$) to determine the sensitivity of fertilization to velocity [5], [18], [19], which in predominantly unidirectional flow is a measure of the flow magnitude and is composed of both the current and turbulence components. Because we were interested in understanding the effect of the various components of oscillatory flow on fertilization, the flow signal was decomposed. In oscillatory flow, a measurement of water velocity contains the current, the magnitude of wave driven flow, the turbulence, and instrumental noise. These components can be distinguished by extending Reynolds decomposition of the signal so that the total flow signal in the predominant direction (u) may be decomposed as:

$$u = U + u_w + u_t + u_{noise} \tag{1}$$

where U represents the low frequency component of the signal (that is, the underlying current), u_w is the oscillatory component (wave component), u_t is the turbulent component and u_{noise} represents the instrumental white noise. Squaring and averaging the terms yields:

$$\overline{u^2} = \overline{U^2} + \overline{u_w^2} + \overline{u_t^2} + \overline{u_{noise}^2} + \overline{2Uu_w} + \overline{2Uu_t} + \overline{2Uu_{noise}} + \overline{2u_wu_t} + \overline{2u_wu_{noise}} + \overline{2u_tu_{noise}} \tag{2}$$

Assuming that all components are uncorrelated among themselves (that is, that all double products are 0), and that $\overline{U^2} = U^2$ the equations simplifies to:

$$\overline{u^2} = \overline{U^2} + \overline{u_w^2} + \overline{u_t^2} + \overline{u_{noise}^2} \tag{3}$$

The square roots of the elements in this equation are actually the root-mean-square (rms) velocities of the different components. It is possible to estimate these components directly from the energy density spectrum (Fig. 3) [20], since the integral of the spectral density equals the variance of the signal as long as the time series is sufficiently long. This process can be repeated for flows perpendicular to the mainstream (v) and vertical (w).

For each time series, the white noise was identified as a flat segment appearing at the highest frequencies in the power density spectrum, and its variance was computed integrating this noise level over all frequencies of the spectral density. The turbulent component (rms(u_t)) was obtained by the integration of the inertial subrange (i.e., the segment of the spectral density that follows a −5/3 slope), after subtraction of the white noise. The components were calculated from vertical velocities only because, due to the geometry of the probe, this component has the lowest noise level and the observed inertial subrange is wider and more easily identifiable [21]. In addition, since the oscillating motion was generated mostly along the horizontal long axis of the OWT, the vertical spectra showed little or no peak at the oscillation frequency.

The wave component (rms(u_w) was not calculated from the spectral analyses because the time series were not sufficiently long to adequately resolve spectral densities at the frequencies of oscillation, particularly for experimental conditions with longer periods. Therefore, an alternative approach was followed to calculate rms(u_w). For each oscillatory experiment, the period of oscillation (T, s^{-1}) was estimated as the first local maximum in the autocorrelation function. The signal was ensemble averaged using period as the fundamental length. Then rms(u_w) was estimated as the root-mean-square of the resulting ensembled wave [22].

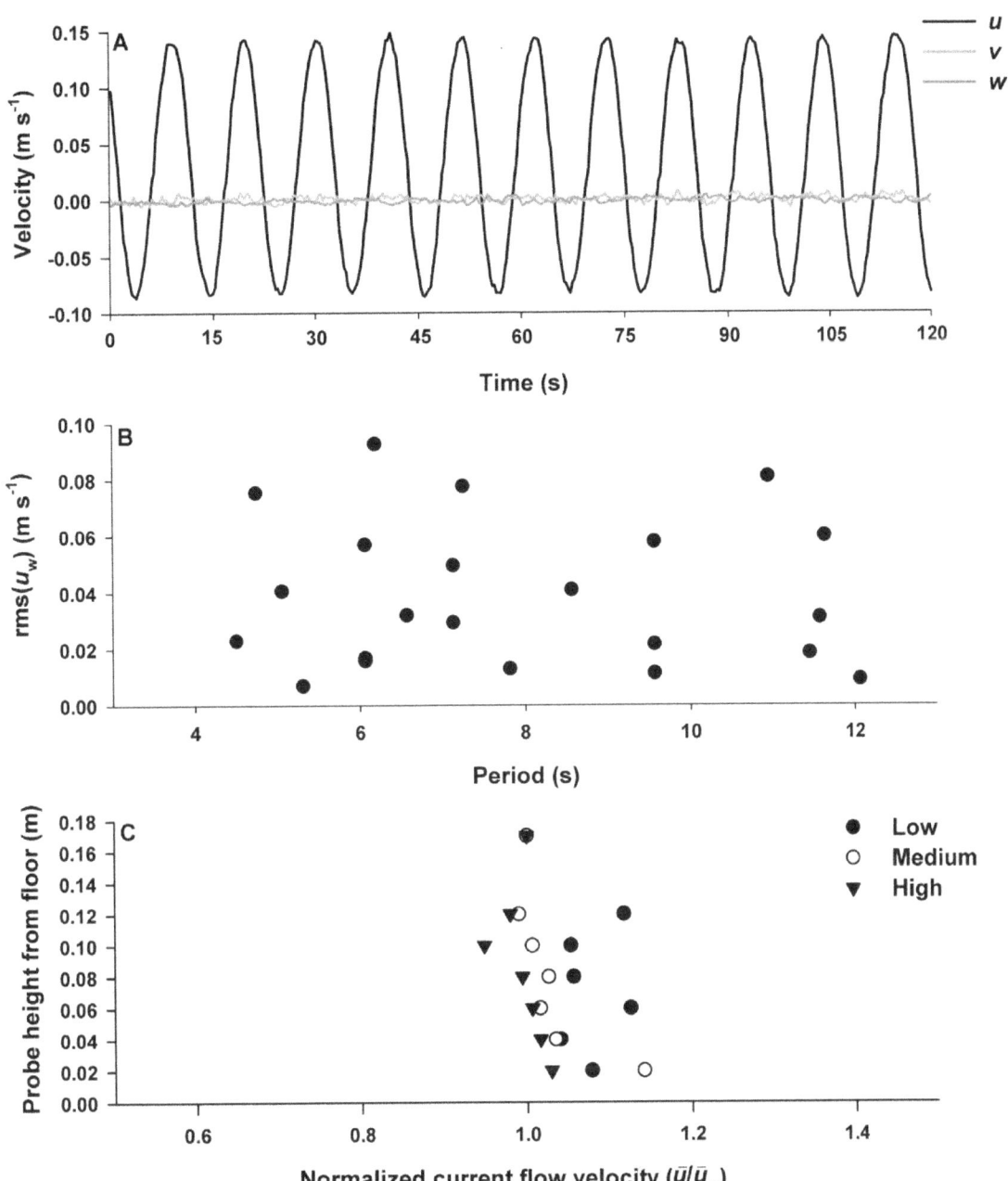

Figure 2. Hydrodynamic characterisation of the oscillatory water tunnel. (A) Sample wave form with underlying current (u, v, and w velocities (m s^{-1})) from an ADV positioned with the x probe oriented with the dominant current direction and the sample volume 0.04 m above the substrate (corresponding to the average height of the female sea urchins used in the trials). (B) Range of hydrodynamic conditions explored in the experimental trials, with both rms(u_w) (m s^{-1}) and period (s) manipulated independently. (C) Test section velocity profiles in the absence of the female sea urchin. Mean longitudinal velocity ($\bar{u}$) normalized by mean longitudinal velocity of the maximum probe height ($\bar{u}_m$) from the tank floor (0.18 m) from three velocity profiles representative from each of the rms(u_w) categories: LOW [< 0.03 m s^{-1}], MEDIUM [$> 0.03 < 0.06$ m s^{-1}], HIGH [> 0.06 m s^{-1}]

In addition, to these flow parameters, we calculated a Keulegan-Carpenter number which relates the wave orbital excursion length to the object size (KC; Keulegan and Carpenter [23]):

$$KC = V \bullet T / L \qquad (4)$$

where V is a flow parameter, here we used rms(u_w) as it was the best predictor for fertilization (*see* below) and L is the relevant length scale (sea urchin width, m). KC is used widely in civil and coastal engineering as a dimensionless number to describe the importance of drag forces relative to inertial forces for solid objects [24] and is related to the tendency of an object to shed vortices [11].

Fertilization experiments were conducted under a range of hydrodynamic conditions with both period ($T = 4.5 - 12.1$ s^{-1}) and rms(u_w) ($0.007 - 0.093$ m s^{-1}) manipulated independently in a total of 22 hydrodynamic trials (Fig. 2B). Flow conditions did not

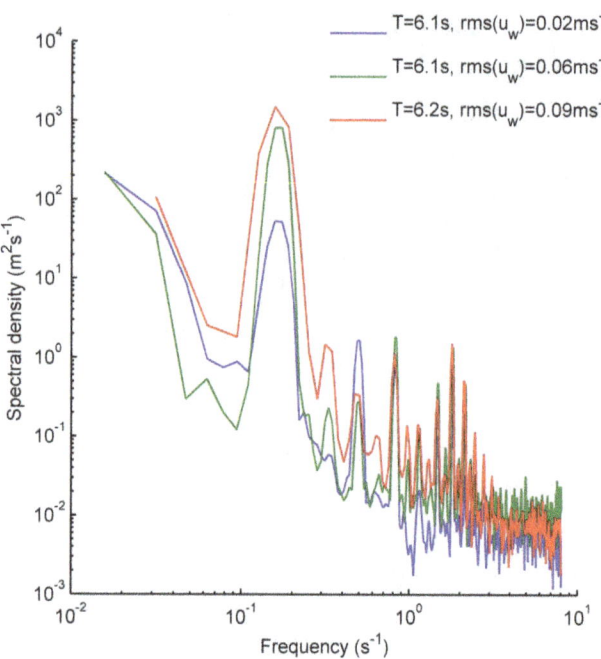

Figure 3. Example spectral density plots. Spectral density plots calculated from vertical velocity components representative from each of the rms(u_w) categories: LOW [< 0.03 m s^{-1}, blue], MEDIUM [> 0.03<0.06 m s^{-1}, green], HIGH [> 0.06 m s^{-1}, red]

vary during an experiment. These conditions were within the range where dye experiments showed that gametes would not reach the walls and become constrained over the experimental distances. The periods were representative of those observed from a buoy located in Penobscot Bay (NERACOOS F01), and flows were representative of those measured in the field at the height of *S. droebachiensis* sea urchins in Pemaquid Point (rms(u_w) (0.003 – 0.164 m s^{-1})) (unpublished data) on the north-east coast of Maine. *Strongylocentrotus droebachiensis* populations are exposed to a wide range of flow conditions as they are found in depths ranging from 0 – 300 m. Prior to recent harvesting, however, the bulk of the population was located below the surge zone in depths of 7–16 m [25].

Assessing fertilization several

To assess the effect of period and flow velocity on fertilization in the green sea urchin *S. droebachiensis* at different sampling locations near a spawning female (water column, wake eddy, substrate and aboral surface), we conducted a series of experiments during the spawning season, March to April 2007. Sea urchins were hand collected from the coast of Maine and housed in flow-through seawater tanks in ambient temperatures ranging between 2 – 4.5°C. For each hydrodynamic trial (n = 22), several sea urchins with an average test width of 0.065 ± 0.001 SE m were initially selected and induced to spawn by injection of ~ 3 mL of 0.5 mol L^{-1} potassium chloride (KCl) into the coelomic cavity. Of these urchins induced to spawn, only one female and one male animal that were spawning from at least 4 out of 5 gonopores were then selected and secured in the OWT, which was already filled with ambient seawater with a velocity and wave period selected. The trial was initiated when the motor and underlying current were restarted. Time from placement of the urchins in the OWT to when the trial was initiated was less than 2 minutes.

Prior to each experiment, a control sample of eggs and sperm were assayed to: 1) determine the absence of abnormal (e.g., immature) eggs, 2) test whether the male and female pair were compatible [26], [27], and 3) ensure that no eggs had raised membranes that might represent either sperm contamination or mechanical damage. In the compatibility test, if less than 95% of the eggs were fertilized for any pair, new animals were selected. The OWT was housed in a marine lab at the University of New England in Biddeford, Maine, and seawater was pumped from an estuary in a region that was essentially devoid (due to past harvesting) of adult sea urchins. Nevertheless, to ensure there were no background sperm in the seawater, control trials with no male urchin were conducted intermittently during the experimental period (n = 5).

Fertilization was assessed as a time integrated measure during a 1 h period [5]. Eggs were sampled at 5 time points (0, 16, 28, 40, and 60 min), and trials were terminated when either 5 samples had been collected or the male or female gamete piles were exhausted. At each time point, eggs were sampled from the water column, wake eddy, substrate and aboral surface respectively. All samples were collected on the downstream side of the female in relation to the underlying current except for the aboral surface sample. For the water column, a 1 L sample was siphoned 0.7 m downstream of the female 0.1 m above the substrate. The wake eddy was sampled with a 50 mL syringe with a narrow extension tube to ensure minimal disruption to the wake eddy 0.05 m downstream of the female. Both the substrate and the aboral surface were carefully sampled with a long 5 mL Pasteur pipette. The substrate was sampled immediately beside the female. Samples from each location were immediately filtered through a 45 μm mesh sieve and the eggs were rinsed with aged seawater (24 – 48 h to eliminate ambient sperm) into separate 20 mL scintillation vials containing a mixture of 2 mL 37% formaldehyde and 10 mL aged seawater within two minutes of collection. During each trial, observations were also recorded of the female and male spawning.

The percentage of eggs fertilized (PF) for each sample was calculated from a random sub-sample of at least 100 eggs or from all the eggs collected in a sample if the number of eggs collected was <100, but >10. Samples with less than 10 eggs were not included in the analysis.

Effect of hydrodynamics on fertilization

To determine how water motion influences the percentage of eggs fertilized (PF) at each location (water column, wake eddy, substrate, aboral surface) we calculated a mean, weighted by sampling time, at each of 5 sample periods. These numbers represent the percentage of eggs found at a location that were fertilized and do not account for the number of eggs at each location. To examine how different components of flow affect PF we used a stepwise regression (IBM SPSS Statistics V.19) of arcsine-transformed values of PF against the components of flow (current (U), wave component (rms(u_w)), turbulent component (rms(u_t)) and period (T)). We also conducted these regressions using the turbulent kinetic energy dissipation rate (ε), a potentially better predictor of water column processes, instead of rms(u_t). We then compared the resultant regressions to those obtained with the total flow signal (rms(u)) and KC alone to (i) determine whether using the easily measured rms(u) impacts our ability to predict fertilization and (ii) determine if KC, a dimensionless parameter that incorporates the period would improve our predictions of fertilization. Before running the analyses all parameters were tested for normality. If the parameters were found to be non-normally distributed they were normalized using a square root transformation. Current (U) and the wave component of flow

$(\mathrm{rms}(u_w))$ were highly collinear. Rather than using both of these terms in the model, we regressed current versus the wave component and used the residuals of the regression in the model, thus we are asking whether aspects of the underlying current, in addition to those that are directly collinear with the wave component, have effects on fertilization. Our intent was to keep current constant in these experiments thus it did not vary a great deal, additionally the OWT design did not allow for high currents, thus we are simply testing the impacts of small variations in current on fertilization.

Since the percent of eggs fertilized does not provide information about the relative contribution each location makes to total fertilization, we calculated the percentage of the total number of eggs spawned that were fertilized (PFT = [PF * # of eggs in a location]/total # of eggs spawned) and the relative contribution of each location to the total number of eggs fertilized (RCO). Eggs did not accumulate on the substratum and were not traceable in the eddy. Therefore, we only estimated the relative contribution of the aboral surface and water column.

To perform these calculations we required an estimate of the total number of eggs spawned by an average female *S. droebachiensis*. Six replicate females were allowed to spawn for one hour in flowing seawater at a low flow rate of $0.02 \mathrm{~m~s}^{-1}$. At this flow speed, large egg piles form on the aboral surface of female sea urchins and loss of eggs by advection is low [5]. After one hour, the egg piles were gently siphoned off each female's aboral surface and counted. The mean number of eggs spawned by the six replicate females was 2.2×10^7, SD $= 4.5 \times 10^5$. This number may be an underestimate of the total number of eggs because of some loss by advection during the hour; however at this low flow loss was minimal.

To establish a nondestructive method of estimating the numbers of eggs in the piles retained on the aboral surface of the urchins at the end of each trial, we established a relationship between the number of eggs in the piles and the volume of the piles. For each individual spawned (see above), we photographed the piles from the top and side. The volumes of piles were estimated from the photographs using Image-Pro® Plus (Media Cybernetics, version 4.1 for Windows^{TM}), based on equations for the closest simple geometric shapes (i.e., cones and ellipses) effectively providing a 3D representation of the egg piles. We then compared these estimated volumes to the numbers of eggs that were counted in each pile. Regressions of number of eggs enumerated versus volume of piles estimated from photographs yielded good approximations (Aboral = 13×10^4 eggs mL^{-1}, $R^2 = 0.92$, $n = 6$). For any egg piles that did remain on the surface of the female urchin at the end of a trial we could thus use the regression equation and photographs to quantify the number of eggs. The eggs remaining on the female were then subtracted from the total number of eggs that a female urchin could produce to obtain an estimate of the number of eggs transported unfertilized into the water column (total number spawned – number left at end = total number of eggs transported off of the aboral surface).

We assumed that the remainder of the eggs that were not fertilized on the aboral surface passed into the water column and determined the number of these eggs fertilized based on fertilization success found in the water column. Our assumptions, however, overestimate the contribution of the water column to fertilization, especially at low flows, as we know that the substrate and wake eddy can be important sites for fertilization [5]. For each location we then calculated the mean and standard error of the number of eggs that were fertilized in three $\mathrm{rms}(u_w)$ categories: LOW [$<0.03 \mathrm{~m~s}^{-1}$ ($n = 10$)], MEDIUM [$>0.03 < 0.06 \mathrm{~m~s}^{-1}$ ($n = 7$)], HIGH [$>0.06 \mathrm{~m~s}^{-1}$ ($n = 5$)].

Results

As noted in earlier investigations [5], visual observations at the start of each experiment confirmed the accumulation of eggs and sperm in piles on the aboral surface of the spawning sea urchins. Our observations suggested that the speed at which the piles were depleted was dependent on the velocity and period of the experimental flow conditions, with higher $\mathrm{rms}(u_w)$ values associated with increased rates of gamete erosion. Irrespective of the presence of piles of eggs and sperm on the parent surface, in all experimental trials both male and female animals were observed to continue spawning throughout the duration of the experimental period of 1 hour.

These visual observations also provided general information on the movement of eggs around female animals under the various experimental conditions. Over the range of flow conditions generated, eggs were consistently entrained from the aboral surface of the female into the vortex that formed downstream of the body of the animal. Immediately following reversal of the flow in the oscillating flow-field, the vortex structure broke down and the eggs were carried back over the animal and became entrained in the newly formed vortex on the opposite side of the animal. The transfer of eggs from one side of the animal to the other also resulted in some transport of the eggs into the overlying water column. We also observed that except for the lowest $\mathrm{rms}(u_w)$ $<0.01 \mathrm{~m~s}^{-1}$, no piles of eggs were formed directly beside the female, although even at the lowest $\mathrm{rms}(u_w)$ values eggs could be seen rolling along the substrate (essentially transported as bed load).

At low flows ($\mathrm{rms}(u_w)$ $<0.02 \mathrm{~m~s}^{-1}$) for the majority of the sampling locations (water column, wake eddy and substrate), PF were high with 60 – 85% of the eggs fertilized (Fig. 4). The exception was the aboral surface where maximum PF was only ~50%. As $\mathrm{rms}(u_w)$ increased, fertilization decreased to below 10% for all sampling locations except the water column where fertilization remained at 20% (Fig. 4).

A stepwise regression of the components of flow (current (U), wave component ($\mathrm{rms}(u_w)$), turbulence component ($\mathrm{rms}(u_t)$) and period (T)) indicated that for all the sampling locations except the substrate only $\mathrm{rms}(u_w)$ had a significant relationship to the percentage of eggs found fertilized at each location (Table 1). For the substrate, including period in the regression improved model fit. Interestingly, including the turbulence component ($\mathrm{rms}(u_t)$) or energy dissipation (ε) in the models did not improve the fit. Thus larger scale measures of flow that are correlated with smaller scale fluctuations are more predictive of fertilization under these flow conditions than measures of small scale turbulent mixing. A comparison of the stepwise regressions fits of $\mathrm{rms}(u_w)$ and those using $\mathrm{rms}(u)$ and KC alone as predictors of fertilization indicated that the stepwise regressions provided slightly better fits than either $\mathrm{rms}(u)$ or KC at most sampling locations (Table 1 & 2). Using $\mathrm{rms}(u)$ does provide significant predictions, however, and is more easily calculated than the separate components.

The percentage of total number of eggs fertilized (PFT) on the aboral surface and water column samples declined between low and medium $\mathrm{rms}(u_w)$ categories and remained constant between the medium and high $\mathrm{rms}(u_w)$ categories (Fig. 5A). Similar PFT were observed between the aboral surface and water column (~30%) at low flows, however as $\mathrm{rms}(u_w)$ categories increased, only 6% of the eggs on the aboral surface were fertilized compared to ~25% in the water column (Fig. 5A). The relative contribution (RCO) was similar between the aboral surface and water column at low $\mathrm{rms}(u_w)$ categories. At medium and high categories, fertilization was lower at the aboral surface than in the water

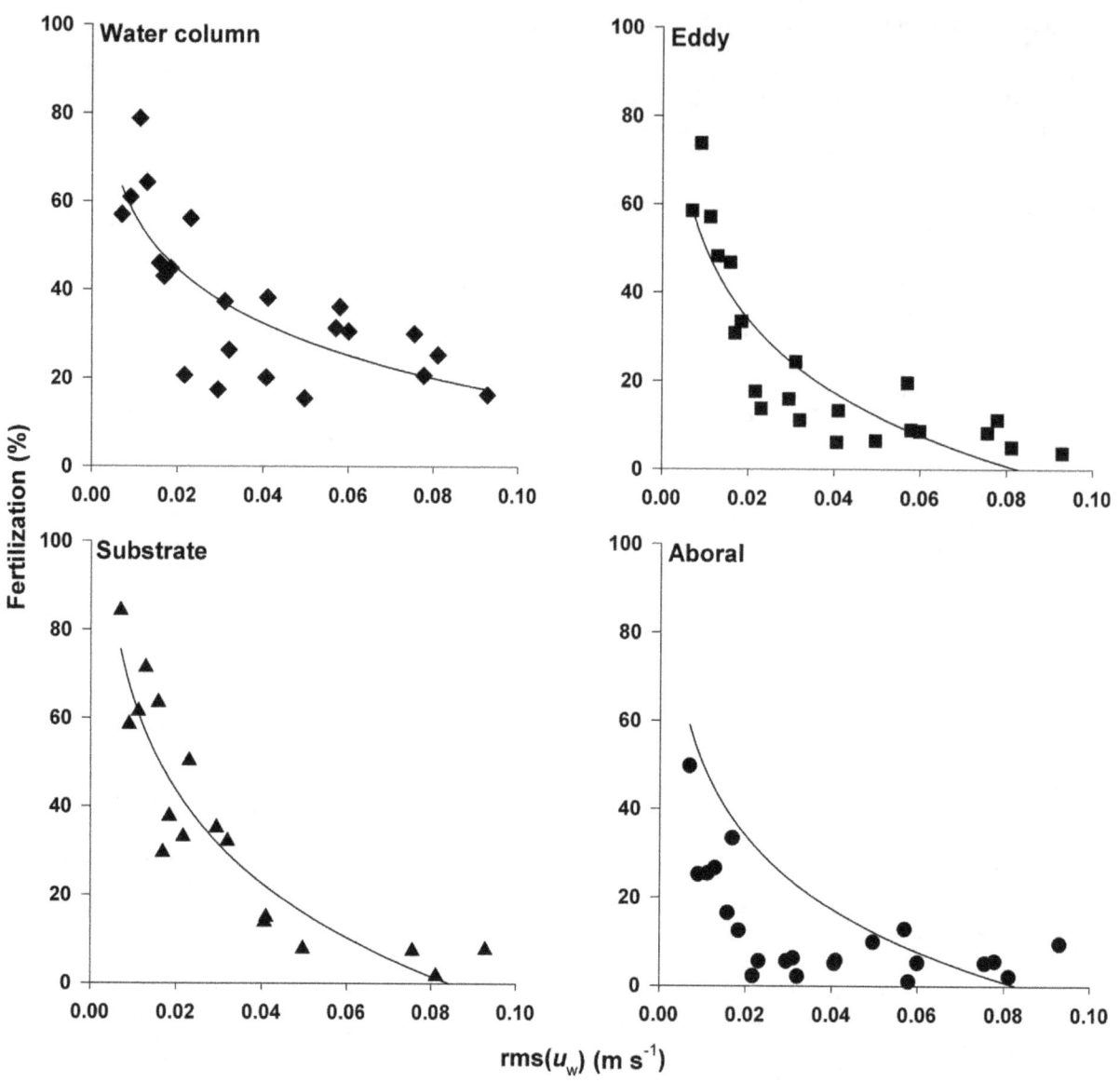

Figure 4. Variation in fertilization as a function of flow. Mean percent fertilization (PF) is plotted as a function of rms(u_w) (m s^{-1}) in the four sampling locations: water column, wake eddy, substrate, and aboral surface. Symbols represent weighted mean of all the time points. Lines represent a nonlinear curve fit to the untransformed data. PF was strongly dependent on rms(u_w) (m s^{-1}) for all sampling locations (Table 1).

Table 1. Stepwise regression analyses.

Location	Step	r^2	Delta r^2	T	p
Water column	1. Waves	57.8	57.8	−5.24	0.0001
Eddy	1. Waves	81.8	81.8	−9.48	0.0001
Substrate	1. Waves	77.7	77.7	−8.35	0.0001
	2. Period	84.6	6.9	−2.91	0.009
Aboral	1. Waves	55.2	55.2	−4.96	0.0001

Results of stepwise regression analysis for the four components of flow: current (U m s^{-1}), wave (rms(u_w) m s^{-1}), turbulent (rms(u_t) m s^{-1}), and period (T (s)), vs. arcsine-transformed values of PF at the four sampling locations. Stepwise regressions were done with $p<0.05$ to enter and $p<0.10$ to remain.

Table 2. Regression analyses for percent eggs fertilized (PF).

Location	rms(u) (m s^{-1})	KC
Water column	55.1 ($F_{1,20} = 24.56$)	49.5 ($F_{1,20} = 19.66$)
Eddy	76.8 ($F_{1,20} = 66.0$)	64.0 ($F_{1,20} = 35.64$)
Substrate	84.0 ($F_{1,15} = 78.99$)	88.3 ($F_{1,15} = 113.53$)
Aboral	49.2 ($F_{1,20} = 19.38$)	61.5 ($F_{1,20} = 31.94$)

Results of regression analysis for the two flow parameters: Total flow signal (rms(u)) and KC, vs. arcsine-transformed values of PF at the four sampling locations. Values are r^2 with df and F-ratio in parentheses. All predictors were significant to the 0.0001 level.

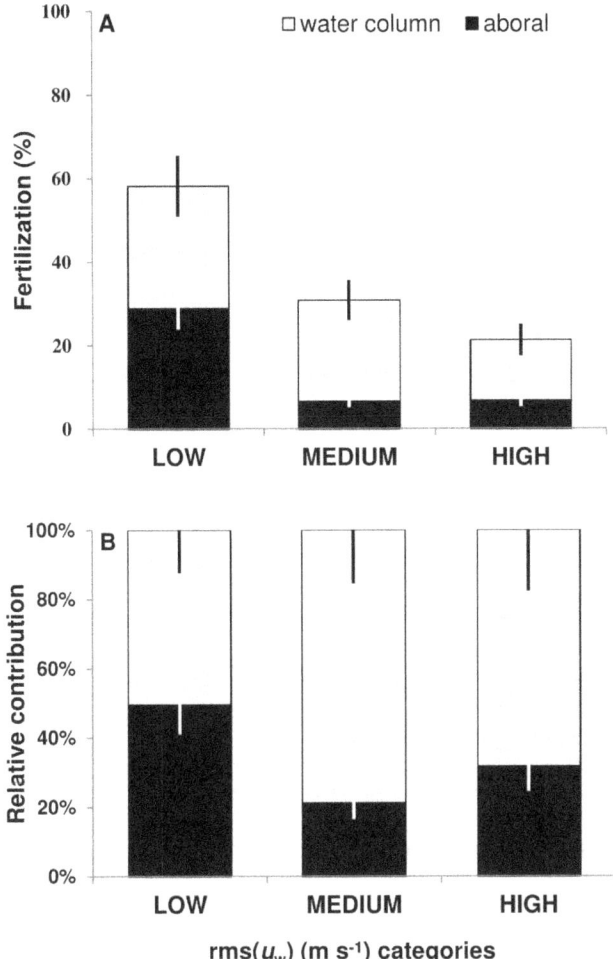

Figure 5. Importance of location for egg fertilization. Estimate of (A) the percentage of the total number of eggs spawned that were fertilized (PTF) and (B) the relative contribution to overall fertilization (RCO) at the aboral surface and water column as a function of rms(u_w) categories: LOW (<0.03 m s^{-1}), MEDIUM (>0.03<0.06 m s^{-1}) and HIGH (>0.06 m s^{-1}).

column with approximately 25% of the eggs being fertilized compared to ~75% for the water column (Fig. 5B).

Controls to determine background sperm levels in the OWT indicated that no significant sperm concentrations were present within the OWT with >98% of the eggs remaining unfertilized.

Discussion

The root-mean-square of the total flow signal (rms(u)) was a relatively good predictor of fertilization at all sampling locations around a spawning female (p<0001). This parameter is composed of the underlying currents (U), wave generated fluctuations (rms(u_w)), and turbulent fluctuations (rms(u_t)). In our analyses the wave component rms(u_w) provided a better fit to our data than did rms(u) (Table 2). These results indicate that rms(u_w) captures most of the variance associated with processes driving fertilization under the flow conditions tested in our OWT. The small effects attributed by period for the substrate locations are intriguing and suggest that period plays some role in controlling fertilization processes and could suggest that more complex experiments with varying period could be informative. Regardless, from a practical

perspective, rms(u_w) can be used as a predictor of fertilization in complex flows and captures the important components of flow.

Due to the potential influence of period on fertilization we compared fertilization success to a non-dimensional parameter KC. This parameter represents the ratio of orbital excursion length (length of flow movement in one direction) to animal size. Thus it may be an important concept for analyzing fertilization for a range of animals of different sizes experiencing a range of flow excursion lengths. In our study, using KC did not improve our predictions of fertilization, but this may be due to minimal size variation in our urchins. We included KC in our analysis, however, as it is a good predictor of fertilization and may prove useful for studies where animals exhibit greater variation relative to excursion length.

Similar to previous studies assessing fertilization in unidirectional flow conditions, fertilization declined with increasing flow speeds [7-9]. The percentage of eggs fertilized at low flows (rms(u_w) <0.02 m s^{-1}) varied between 50 – 85%, depending on the location sampled (water column, wake eddy, substrate or aboral surface), but declined to levels below 10% for most locations as rms(u_w) increased. Water column fertilization remained at 20% even at these higher flow velocities.

Despite continual fluctuations in flow velocity around the female sea urchin, fertilization occurred in various localities on or near the female. These results extend previous work of Yund and Meidel [5] showing that fertilization is not just a water column process in unidirectional flows. The lowest levels of fertilization occurred on the aboral surface where PF reached a maximum of 50% and was reduced to <10% at rms(u_w) >0.04 m s^{-1}. Nonetheless these results indicate that eggs can already be fertilized by the time they leave the female. Eggs leaving the aboral surface unfertilized may still be fertilized for some time, though this concept has received little attention in experimental design [5]. Eggs can become entrained in the vortices on either side of the sea urchin in the oscillating flow-field or transferred to the substrate by the pressure forces acting on the parent. Fertilization can occur in these locations as evidenced by higher fertilization levels in the substrate and wake eddy samples (Fig. 4).

Our results indicate that fertilization can remain upwards of 20% even at relatively fast flows and short periods of oscillation. Furthermore, our results indicate that the water column is an important location for fertilization in oscillatory flow. The total number of eggs fertilized and the relative contribution of the water column is greater than the aboral surface, especially at medium and high rms(u_w) categories. It is important to note, however, that fertilization can and does occur at the aboral surface location and may be a significant contributor to overall fertilization success. Oscillatory flow suspended the eggs in the overlying water column and inhibited the formation of discernible piles of gametes on the substrate. Although we were not able to determine the contribution of fertilized eggs from the wake eddy and substrate to the water column sample, the eggs found in the water column are likely to include eggs already fertilized from all three locations near or on the parent (aboral surface, substrate and wake eddy). Because of the additional contribution of these locations to water column fertilization estimates we may be overestimating the water column contribution. Nevertheless, sea urchin gametes can be successfully fertilized when spawning under a range of oscillatory flow conditions.

Evaluating different hydrodynamic parameters for relating fluid dynamics to fertilization processes is an essential step in establishing possible ecological effects on fertilization [28], [29]. Because of its size, our oscillatory OWT was unable to properly mimic boundary layer effects. The thickness of a boundary layer grows as flow moves over a surface, and the relatively short length of our OWT probably produced boundary layers that were thinner (at the sites of the spawning sea urchins) than in nature. A further consideration of the OWT was that we captured large scale oscillatory processes over a time integrated measure rather than the small scale processes at the scale of fertilization. Shear stress is an important process influencing fertilization at the scale of sperm encountering eggs [30–32], and it is possible that the OWT did not correctly capture these small scale processes. Nevertheless we expect that our understanding of fertilization processes should be correct for the $rms(u_w)$ values actually measured in the OWT, but recognize that in the field, similar $rms(u_w)$ values at the sea urchin height above the substrate may occur with different combinations of current and turbulence. Consequently, further work is necessary to establish how parameters measured in our OWT correspond to real-world wave forces. We also held the unidirectional current constant. In nature underlying currents may vary a great deal. Our results indicate that currents contribute to the fertilization occurring within eddies near the spawning female thus underlying currents may influence the relative importance of different locations to fertilization in natural systems.

Our experiments only address the simple spawning scenario of a single male and female (female positioned directly downstream of the male) and therefore are not representative of population-wide spawning events. *Strongylocentrotus droebachiensis* can be found in densities from 0.1 to 250 urchins m^{-2} [9] and are expected to aggregate before a mass synchronous spawning event [6], [33], [34]. Aggregation behavior has been observed in this species and other mobile invertebrates [35–37] and can only increase fertilization rates and mitigate the effects of sperm limitation. When individuals are densely packed in oscillatory flow, eggs and sperm may pass over neighbors to the front and back of the spawning individual several times depending on the period and underlying current as observed by [36]. Oscillatory flow may therefore enhance gamete coalescence and increase the number of eggs fertilized in mass spawns relative to unidirectional flow under similar density conditions.

The nature of flow is a fundamental component influencing the success of fertilization and some form of water motion is required to bring sperm into contact with eggs, but there appears to be a fine balance between excessive and sufficient. We know that small scale turbulent processes can result in regions of high concentration of sperm and eggs thereby enhancing fertilization [2], however, turbulence did not explain the variance in fertilization beyond that captured by $rms(u)$.

Whilst there are numerous investigations into the cues that trigger large-scale broadcast spawning events [38], [39] and biological conditions, e.g., density and number of spawning individuals [7], [9], [40], we know little about the flow conditions during natural spawning occasions [41]. Sea urchins have been observed to spawn when water flow rates were minimal [34] or in weak to moderate currents [36]. Other taxa such as the anemone *Oulactis mucosa* or fucoid algae have also been observed to either release spores or spawn in the intertidal at low velocities (< 0.2 m s^{-1}) or at low tide in rock pools [42–44]. Quantitative hydrodynamic information during natural spawning events for subtidal species is therefore essential if we are to understand and model fertilization success in the environment. Based on our experience the most adaptive strategy would be for urchins to synchronize spawning events during low flow conditions.

In summary, although we perceive the wave environment to be extremely turbulent and therefore expect fertilization levels to be low, our expectations are not always valid. Our results demonstrate that under certain oscillatory flows, the percentage of eggs fertilized can attain high levels (50–80%). These results highlight the fact that gametes can attain successful fertilization in energetic coastal wave environments. In addition, fertilization can occur on the aboral surface before eggs mix into the water column and fertilization rates can be quite high depending on flow conditions.

Acknowledgments

We thank Emily Zimmermann and Jessica Zima for assistance in the lab and Tim Arienti for construction of the oscillatory water tunnel.

Author Contributions

Conceived and designed the experiments: LTK ALB POY FIMT. Performed the experiments: LTK ALB. Analyzed the data: LTK OG POY FIMT. Wrote the paper: LTK ALB OG POY FIMT.

References

1. Benzie JAH, Black KP, Moran PJ, Dixon P (1994) Small-scale dispersion of eggs and sperm of the crown-of-thorns starfish (*Acanthaster planci*) in a shallow coral reef habitat. Biol Bull 186: 153–167. doi: 10.2307/1542049.

2. Crimaldi JP, Browning HS (2004) A proposed mechanism for turbulent enhancement of broadcast spawning efficiency. J Mar Syst 49: 3–18. doi: 10.1016/j.jmarsys.2003.06.005.

3. Guichard F, Bourget E (1998) Topographic heterogeneity, hydrodynamics, and benthic community structure: A scale-dependent cascade. Mar Ecol Prog Ser 171: 59–70.

4. Pennington JT (1985) The ecology of fertilization of echinoid eggs: the consequences of sperm dilution, adult aggregation, and synchronous spawning. Biol Bull 169: 417–430. doi: 10.2307/1541492.

5. Yund PO, Meidel SK (2003) Sea urchin spawning in benthic boundary layers: are eggs fertilized before advecting away from females? Limnol Oceanogr 48: 795–801.

6. Simon TN, Levitan DR (2011) Measuring fertilization success of broadcast-spawning marine invertebrates within seagrass meadows. Biol Bull 220: 32–28.

7. Levitan DR, Sewell MA, Chia F (1992) How distribution and abundance influence fertilization success in the sea urchin *Strongylocentrotus franciscanus*. Ecology 73: 248–254. doi: 10.2307/1938736.

8. Levitan DR, Young CM (1995) Reproductive success in large populations: empirical measures and theoretical predictions of fertilization in the sea biscuit *Clypeaster rosaceus*. J Exp Mar Biol Ecol 190: 221–241. doi: 10.1016/0022-0981(95)00039-T.

9. Wahle RA, Peckham SH (1999) Density-related reproductive trade-offs in the green sea urchin, *Strongylocentrotus droebachiensis*. Mar Biol 134: 127–137. doi: 10.1007/s002270050531.

10. Denny M, Dairiki J, Distefano S (1992) Biological consequences of topography on wave-swept rocky shores: I. Enhancement of external fertilization. Biol Bull 183: 220–232. doi: 10.2307/1542209.

11. Lam KM, Liu P, Hu JC (2010) Combined action of transverse oscillations and uniform cross-flow on vortex formation and pattern of a circular cylinder. J Fluid Struc 26: 703–721. doi: 10.1016/j.jfluidstructs.2010.03.007.

12. Thomas FIM (1994) Physical properties of gametes in three sea urchin species. J Exp Biol 194: 263–284.

13. Marshall DJ (2002) In situ measures of spawning synchrony and fertilization success in an intertidal, free-spawning invertebrate. Mar Ecol Prog Ser 236: 113–119. doi: 10.3354/meps236113.

14. Meidel SK, Yund PO (2001) Egg longevity and time-integrated fertilization in a temperate sea urchin (*Strongylocentrotus droebachiensis*). Biol Bull 201: 84–94. doi: 10.2307/1543529.

15. Goring D, Nikora V (2002) Despiking acoustic doppler velocimeter data. J Hydraul Eng 128: 117–126. doi: 10.1061/(ASCE)0733-9429(2002)128:1(117).

16. Wahl TL (2003) Discussion of "Despiking acoustic doppler velocimeter data" by Derek G. Goring and Vladimir I. Nikora. J Hydraul Eng 129: 484–487. doi: 10.1061/(ASCE)0733-9429(2003)129:6(484).

17. McLelland SJ, Nicholas AP (2000) A new method for evaluating errors in high-frequency ADV measurements. Hydrol Processes 14: 351–366. doi: 10.1002/(SICI)1099-1085(20000215)14:2<351::AID-HYP963>3.0.CO;2-K.

18. Levitan DR (1991) Influence of body size and population density on fertilization success and reproductive output in a free-spawning invertebrate. Biol Bull 181: 261–268. doi: 10.2307/1542097.

19. Babcock RC, Mundy CN, Whitehead D (1994) Sperm diffusion models and in situ confirmation of long-distance fertilization in the free-spawning Asteroid *Acanthaster planci*. Biol Bull 186: 17–28. doi: 10.2307/1542033.

20. Stiansen JE, Sundby S (2001) Improved methods for generating and estimating turbulence in tanks suitable for fish larvae experiments. Sci Mar 65: 151–167.
21. Voulgaris G, Trowbridge JH (1998) Evaluation of the acoustic Doppler velocimeter (ADV) for turbulence measurements. J Atmos Oceanic Tech 15: 272–289. doi: 10.1175/1520-0426(1998)015<0272:EOTADV>2.0.CO;2.
22. Weitzman JS, Aveni-Deforge K, Koseff JR, Thomas FIM (2013) Uptake of DIN by shallow seagrass communities exposed to wave-driven unsteady flow. Mar Ecol Prog Ser 475: 65–83. doi:10.3354/meps09965.
23. Keulegan GH, Carpenter LH (1958) Forces on cylinders and plates in an oscillating fluid. J Res Natl Bur Stand 60: 423–440.
24. Chen CC, Fang FM, Li YC, Huang LM, Chung CY (2009) Fluid forces on a square cylinder in oscillating flows with non-zero-mean velocities. Int J Numer Meth Fluids 60: 79–93. doi: 10.1002/fld.1881.
25. Witman JD, Dayton PK (2001) Rocky subtidal communities. In: Bertness M, Gaines SD, Hay ME, editors. Marine community ecology. Sinauer: pp. 339–366.
26. Palumbi SR (1999) All males are not created equal: Fertility differences depend on gamete recognition polymorphisms in sea urchins. Proc Nat Acad Sci USA 96: 12632–12637. doi: 10.1073/pnas.96.22.12632.
27. Levitan DR, Ferrell DL (2006) Selection on gamete recognition proteins depends on sex, density and genotype frequency. Science 312: 267–269. doi: 10.1126/science.1122183.
28. IPCC (2007) Climate change 2007: The physical science basis. Solomon S, Qin D, Manning M, Chen Z, Marquis M, Averyt KB, Tignor M, Miller HL, editors. Contribution of Working Group I to the Fourth Assessment Report of the Intergovernmental Panel on Climate Change. Cambridge University Press.
29. Shields MA, Woolf DK, Grist EPM, Kerr SA, Jackson AC, et al. (2011) Marine renewable energy: The ecological implications of altering the hydrodynamics of the marine environment. Ocean Coast Manage 54: 2–9. doi:10.1016/j.ocecoaman.2010.10.036.
30. Mead KS, Denny MW (1995) The effects of hydrodynamic shear stress on fertilization and early development of the purple sea urchin Strongylocentrotus purpuratus. Biol Bull 188: 46–56. doi: 10.2307/1542066.
31. Riffell JA, Zimmer RK (2007) Sex and flow: the consequences of fluid shear for sperm-egg interactions. J Exp Biol 210: 3644–3660. doi: 10.1242/jeb.008516.
32. Zimmer RK, Riffel JA (2011) Sperm chemotaxis, fluid shear, and the evolution of sexual reproduction. Proc Natl Acad Sci USA 108: 13200–13205. doi: 10.1073/pnas.1018666108.
33. Levitan DR (1988) Asynchronous spawning and aggregative behaviour in the sea urchin Diadema antillarum (Philippi). In: Burke RD, Mladenov PV, Lambert P, Parsley RL, editors. Echinoderm Biology: Balkema. pp. 181–186.
34. Lamare MD, Stewart BG (1998) Mass spawning by the sea urchin Evechinus chloroticus (Echinodermata: Echinoidea) in a New Zealand fiord. Mar Biol 132: 135–140. doi: 10.1007/s002270050379.
35. Babcock RC, Mundy CN (1992) Reproductive biology, spawning and field fertilization rates of Acanthaster planci. Aust J Mar Fresh Res 43: 525–534.
36. Himmelman JH, Dumont CP, Gaymer CF, Vallières C, Drolet D (2008) Spawning synchrony and aggregative behavior of cold-water echinoderms during multi-species mass spawnings. Mar Ecol Prog Ser 361: 161–168. doi: 10.3354/meps07415.
37. Keesing JK, Graham F, Irvine TR, Crossing R (2011) Synchronous aggregated pseudo-copulation of the sea star Archaster angulatus Müller & Troschel, 1842 (Echinodermata: Asteroidea) and its reproductive cycle in south-western Australia. Mar Biol 158: 1163–1173. doi: 10.1007/s00227-011-1638-2.
38. Gaudette J, Wahle RA, Himmelman JH (2006) Spawning events in small and large populations of the green sea urchin Strongylocentrotus droebachiensis as recorded using fertilization assays. Limnol Oceanogr 51: 1485–1496.
39. Reuter KE, Levitan DR (2010) Influence of sperm and phytoplankton on spawning in the echinoid Lytechinus variegatus. Biol Bull 219: 198–206.
40. Yund PO, McCartney MA (1994) Male reproductive success in sessile invertebrates: competition for fertilizations. Ecology 75: 2151–2167. doi: 10.2307/1940874.
41. Levitan DR (2002) Density-dependent selection on gamete traits in three congeneric sea urchins. Ecology. 83: 464–479. doi: 10.1890/0012 9658(2002)083[0464:DDSOGT]2.0.CO;2.
42. Serrão EA, Pearson G, Kautsky L, Brawley SH (1996) Successful external fertilization in turbulent environments. Proc Natl Acad Sci USA 93: 5286–5290. doi: 10.1073/pnas.93.11.5286.
43. Pearson GA, Brawley SH (1996) Reproductive ecology of Fucus distichus (Phaeophyceae): An intertidal alga with successful external fertilization. Mar Ecol Prog Ser 143: 211–223. doi: 10.3354/meps143211.
44. Marshall DJ (2002) In situ measures of spawning synchrony and fertilization success in an intertidal, free-spawning invertebrate. Mar Ecol Prog Ser 236: 113–119. doi: 10.3354/meps236113.

Permissions

List of Contributors

Isabel Ceballos, Michael Ruiz and Alia Rodríguez
Soil Microbiology, Universidad Nacional de Colombia, Bogotá, Colombia

Cristhian Fernández and Ricardo Peña
Utopía, Universidad de La Salle, Yopal, Colombia

Ian R. Sanders
Department of Ecology and Evolution, University of Lausanne, Lausanne, Switzerland

Sikander Khan Tanveer, Xiaoxia Wen, Xing Li Lu, Junli Zhang and Yuncheng Liao
College of Agronomy, Northwest A&F University Yangling, Shaanxi, P.R. China

Yongsheng Wang, Xusheng Dang and Lei Wang
Key Laboratory of Ecosystem Network Observation and Modeling, Institute of Geographical Sciences and Natural Resources Research, Chinese Academy of Sciences, Beijing, China
University of Chinese Academy of Sciences, Beijing, China

Shulan Cheng and Minjie Xu
University of Chinese Academy of Sciences, Beijing, China

Huajun Fang, Guirui Yu and Linsen Li
Key Laboratory of Ecosystem Network Observation and Modeling, Institute of Geographical Sciences and Natural Resources Research, Chinese Academy of Sciences, Beijing, China

Benjamin D. Duval
Energy Biosciences Institute, University of Illinois at Urbana-Champaign, Urbana, Illinois, United States of America
Global Change Solutions, Urbana, Illinois, United States of America
Dairy Forage Research Center, United States Department of Agriculture, Agricultural Research Service, Madison, Wisconsin, United States of America

Kristina J. Anderson-Teixeira
Energy Biosciences Institute, University of Illinois at Urbana-Champaign, Urbana, Illinois, United States of America
Conservation Ecology Center, Smithsonian Conservation Biology Institute, National Zoological Park, Front Royal, Virginia, United States of America

Sarah C. Davis
Energy Biosciences Institute, University of Illinois at Urbana-Champaign, Urbana, Illinois, United States of America
Voinovich School for Leadership and Public Affairs, Ohio University, Athens, Ohio, United States of America

Cindy Keogh and William J. Parton
Natural Resource Ecology Laboratory, Fort Collins, Colorado, United States of America

Stephen P. Long and Evan H. DeLucia
Energy Biosciences Institute, University of Illinois at Urbana-Champaign, Urbana, Illinois, United States of America
Global Change Solutions, Urbana, Illinois, United States of America
Department of Plant Biology, University of Illinois at Urbana- Champaign, Urbana, Illinois, United States of America

Chao Chai, Hongzhen Cheng, Dong Ma and Yanxi Shi
College of Resources and Environment, Qingdao Agricultural University, Qingdao, China

Wei Ge
College of Life Sciences, Qingdao Agricultural University, Qingdao, China

Guocheng Wang, Tingting Li, Wen Zhang and Yongqiang Yu
State Key Laboratory of Atmospheric Boundary Layer Physics and Atmospheric Chemistry, Institute of Atmospheric Physics, Chinese Academy of Sciences, Beijing, China

Yanfang Xue, Shanchao Yue, Wei Zhang, Dunyi Liu, Zhenling Cui, Xinping Chen and Chunqin Zou
Center for Resources, Environment and Food Security, China Agricultural University, Beijing, China

Youliang Ye
College of Resources and Environmental Sciences, Henan Agricultural University, Zhengzhou, China

Yidong Wang, Changcheng Guo and Qing Chen
Tianjin Key Laboratory of Water Resources and Environment, Tianjin Normal University, Tianjin, China

Zhong-Liang Wang
Tianjin Key Laboratory of Water Resources and Environment, Tianjin Normal University, Tianjin, China
State Key Laboratory of Environmental Geochemistry, Institute of Geochemistry, Chinese Academy of Sciences, Guiyang, China

Xiaoping Feng
Tianjin Key Laboratory of Water Resources and Environment, Tianjin Normal University, Tianjin, China
College of Urban and Environmental Sciences, Tianjin Normal University, Tianjin, China

Karsten Laursen
Institute of Bioscience, Aarhus University, Rønde, Denmark

Anders Pape Møller
Laboratoire d'Ecologie, Syste´matique et Evolution, CNRS UMR 8079, Universite´ Paris-Sud, Orsay, France

Rihuan Cong
Ministry of Agriculture Key Laboratory of Crop Nutrition and Fertilization, Institute of Agricultural Resources and Regional Planning, Chinese Academy of Agricultural Sciences, Beijing, China
College of Resources and Environment, Huazhong Agricultural University, Wuhan, China

Xiujun Wang
State Key Laboratory of Desert and Oasis Ecology, Xinjiang Institute of Ecology and Geography, Chinese Academy of Sciences, Urumqi, China
Earth System Science Interdisciplinary Center, University of Maryland, College Park, Maryland, United States of America

Minggang Xu
Ministry of Agriculture Key Laboratory of Crop Nutrition and Fertilization, Institute of Agricultural Resources and Regional Planning, Chinese Academy of Agricultural Sciences, Beijing, China

Stephen M. Ogle and William J. Parton
Natural Resource Ecology Laboratory, Colorado State University, Fort Collins, Colorado, United States of America

Evgenia Blagodatskaya
Soil Science of Temperate Ecosystems, Büsgen-Institute, University of Göttingen, Göttingen, Germany
Institute of Physicochemical and Biological Problems in Soil Science, Russian Academy of Sciences, Pushchino, Russia

Agricultural Soil Science, Büsgen-Institute, University of Göttingen, Göttingen, Germany

Sergey Blagodatsky
Institute of Physicochemical and Biological Problems in Soil Science, Russian Academy of Sciences, Pushchino, Russia
Institute for Plant Production and Agroecology in the Tropics and Subtropics, University of Hohenheim, Stuttgart, Germany

Traute-Heidi Anderson
Thünen-Institute of Climate-Smart Agriculture (vTI), Braunschweig, Germany

Yakov Kuzyakov
Soil Science of Temperate Ecosystems, Büsgen-Institute, University of Göttingen, Göttingen, Germany
Agricultural Soil Science, Büsgen-Institute, University of Göttingen, Göttingen, Germany

Jin Hua Li, Yu Jie Yang, Bo Wen Li, Wen Jin Li and Gang Wang
State Key Laboratory of Grassland Agro-Ecosystems, School of Life Sciences, Lanzhou University, Lanzhou, P.R. China

Johannes M. H. Knops
School of Biological Sciences, University of Nebraska, Lincoln, Nebraska, United States of America

Joseph E. Knelman
Institute of Arctic and Alpine Research, University of Colorado, Boulder, Colorado, United States of America
Department of Ecology and Evolutionary Biology, University of Colorado, Boulder, Colorado, United States of America

Steven K. Schmidt, Ryan C. Lynch and John L. Darcy
Department of Ecology and Evolutionary Biology, University of Colorado, Boulder, Colorado, United States of America

Sarah C. Castle and Cory C. Cleveland
Department of Ecosystem and Conservation Sciences, University of Montana, Missoula, Montana, United States of America

Diana R. Nemergut
Institute of Arctic and Alpine Research, University of Colorado, Boulder, Colorado, United States of America
Environmental Studies Program, University of Colorado, Boulder, Colorado, United States of America

Michael A. Treberg
Department of Geography and Environmental Studies, Carleton University, Ottawa, ON, Canada

Department of Botany, and Biodiversity Research Center, University of British Columbia, Vancouver, BC, Canada

Roy Turkington
Department of Botany, and Biodiversity Research Center, University of British Columbia, Vancouver, BC, Canada

Dongfeng Ning, Alin Song, Fenliang Fan and Zhaojun Li
Ministry of Agriculture Key Laboratory of Crop Nutrition and Fertilization, Institute of Agricultural Resources and Regional Planning, Chinese Academy of Agricultural Sciences, Beijing, China

Yongchao Liang
Ministry of Agriculture Key Laboratory of Crop Nutrition and Fertilization, Institute of Agricultural Resources and Regional Planning, Chinese Academy of Agricultural Sciences, Beijing, China
Ministry of Education Key Laboratory of Environment Remediation and Ecological Health, College of Environmental and Resource Sciences, Zhejiang University, Hangzhou, China

Jingfeng Huang and Xinxing Li
Institute of Agricultural Remote Sensing & Information Application, Zijingang Campus, Zhejiang University, Hangzhou, China
China Ministry of Education Key Laboratory of Environmental Remediation and Ecological Health, Zhejiang University, Hangzhou, China
Key Laboratory of Agricultural Remote Sensing and Information System, Zhejiang Province, China

Xiuzhen Wang
Institute of Remote Sensing and Earth Sciences, Hangzhou Normal University, Hangzhou, China

Hanqin Tian
International Center for Climate and Global Change Research, Auburn University, Auburn, Alabama, United States of America
Ecosystem Dynamics and Global Ecology (EDGE) Laboratory, School of Forestry and Wildlife Sciences, Auburn University, Auburn, Alabama, United States of America

Zhuokun Pan
Institute of Agricultural Remote Sensing & Information Application, Zijingang Campus, Zhejiang University, Hangzhou, China
Key Laboratory of Agricultural Remote Sensing and Information System, Zhejiang Province, China

Eduardo R. M. Barbosa
Resource Ecology Group, Wageningen University, Wageningen, The Netherlands
Departamento de Botânica, Laborató rio de Termobiologia, Instituto de Ciĕncias Bioló gicas, Universidade de Brasília, Brasília, DF, Brazil

Kyle W. Tomlinson
Resource Ecology Group, Wageningen University, Wageningen, The Netherlands
Community Ecology & Conservation Group, Xishuangbanna Tropical Botanical Garden, Chinese Academy of Sciences, Yunnan, China

Luísa G. Carvalheiro
School of Biology, University of Leeds, Leeds, the United Kingdom
Terrestrial Zoology, Naturalis Biodiversity Center, Leiden, The Netherlands

Kevin Kirkman
School of Life Sciences, University of KwaZulu-Natal, Scottsville, South Africa

Steven de Bie and Frank van Langevelde
Resource Ecology Group, Wageningen University, Wageningen, The Netherlands

Herbert H. T. Prins
Resource Ecology Group, Wageningen University, Wageningen, The Netherlands
School of Life Sciences, University of KwaZulu-Natal, Scottsville, South Africa

Zhiwei Xu, Xinyu Zhang, Juan Xie, Guofu Yuan, Xinzhai Tang, Xiaomin Sun and Guirui Yu
Key Laboratory of Ecosystem Network Observation and Modeling, Institute of Geographic Sciences and Natural Resources Research, Chinese Academy of Sciences, Beijing, China

James J. Elser
School of Life Sciences, Arizona State University, Tempe, Arizona, United States of America

Timothy J. Elser
Flyr, Inc., San Francisco, California, United States of America

Stephen R. Carpenter
Center for Limnology, University of Wisconsin, Madison, Wisconsin, United States of America

William A. Brock
Department of Economics, University of Wisconsin, Madison, Wisconsin, United States of America

Department of Economics, University of Missouri, Columbia, Missouri, United States of America

Huanjun Zhang, Weixin Ding, Hongyan Yu, Jianling Fan and Deyan Liu
State Key Laboratory of Soil and Sustainable Agriculture, Institute of Soil Science, Chinese Academy of Sciences, Nanjing, China

Xinhua He
Forests NSW, NSW Department of Primary Industries, West Pennant Hills, New South Wales, Australia
School of Plant Biology, University of Western Australia, Crawley, Western Australia, Australia

Nafiseh Mahdavi-Arab
Institute of Ecology, Friedrich Schiller University Jena, Jena, Germany
Terrestrial Ecology Research Group, Department of Ecology and Ecosystem Management, School of Life Sciences Weihenstephan, Technische Universität München, Freising, Germany

Sebastian T. Meyer and Wolfgang W. Weisser
Terrestrial Ecology Research Group, Department of Ecology and Ecosystem Management, School of Life Sciences Weihenstephan, Technische Universität München, Freising, Germany

Mohsen Mehrparvar
Department of Biodiversity, Institute of Science and High Technology and Environmental Sciences, Graduate University of Advanced Technology, Kerman, Iran

Zhijian Mu
Chongqing Key Laboratory of Soil Multi-scale Interfacial Processes, College of Resources & Environment, Southwest University, Chongqing, China

Aiying Huang
College of Agronomy & Biotechnology, Southwest University, Chongqing, China

Jiupai Ni and Deti Xie
Chongqing Engineering Research Center for Agricultural Non-point Source Pollution Control in Three -Gorges Region, College of Resources & Environment, Southwest University, Chongqing, China

Louise T. Kregting
Marine Science Center, University of New England, Biddeford, Maine, United States of America
Hawai'i Institute of Marine Biology, University of Hawai'i, Kanéohe, Hawai'i, United States of America
School of Planning, Architecture and Civil Engineering, Queens University of Belfast, Northern Ireland, United Kingdom

Anna L. Bass
Marine Science Center, University of New England, Biddeford, Maine, United States of America

Òscar Guadayol
Hawai'i Institute of Marine Biology, University of Hawai'i, Kanéohe, Hawai'i, United States of America
School of Biological, Biomedical and Environmental Sciences, University of Hull, Kingston-upon-Hull, United Kingdom

Philip O. Yund
Marine Science Center, University of New England, Biddeford, Maine, United States of America
The Downeast Institute, Beals, Maine, United States of America

Florence I. M. Thomas
Hawai'i Institute of Marine Biology, University of Hawai'i, Kanéohe, Hawai'i, United States of America

Index

www.ingramcontent.com/pod-product-compliance
Lightning Source LLC
Chambersburg PA
CBHW080411190526
45161CB00003B/204